Student Solutions Manual, Volume I
for Stewart's

# Calculus

Daniel Anderson
*University of Iowa*

Daniel Drucker
*Wayne State University*

with the assistance of

Lothar Redlin
*Pennsylvania State University*

Terry Tiballi
*North Harris County College*

Brooks/Cole Publishing Company
Monterey, California

Brooks/ Cole Publishing Company
A Division of Wadsworth, Inc.

Printed in the United States of America

10  9  8  7  6  5

QA303.S8825    1986    515'.15    86-6078

ISBN 0-534-06691-7

# Preface

I have edited this solutions manual by taking the solutions
provided by Daniel Anderson, Daniel Drucker, and Terry
Tiballi, comparing them with solutions that were produced
by me and my colleagues at McMaster during the seven years
that we have class-tested the book. A further check was
provided by comparison with solutions produced by several of
my best calculus students whom I hired to work all the
problems: Eric Bosch, Don Callfas, Ron Donaberger,
Sara Lee, Marc Riehm, Martin Sarabura, Jeff Schultheiss,
and Joe Vetrone.

I typed most of the solutions myself, assisted by Lothar
Redlin. As a result of this procedure, I am certain that every
answer in this solutions manual is correct.

*James Stewart*

# Contents

## Chapter 10  Infinite Sequences and Series

# CHAPTER ONE

## Exercises 1.1

1.  $2x+7 > 3 \iff 2x > -4 \iff x > -2 \iff x \in (-2,\infty)$

**1.**    **3.**

3.  $1-x \leq 2 \iff -x \leq 1 \iff x \geq -1 \iff x \in [-1,\infty)$
5.  $2x+1 < 5x-8 \iff 9 < 3x \iff 3 < x \iff x \in (3,\infty)$

**5.**    **7.**

7.  $-1 < 2x-5 < 7 \iff 4 < 2x < 12 \iff 2 < x < 6 \iff x \in (2,6)$
9.  $0 \leq 1-x < 1 \iff -1 \leq -x < 0 \iff 1 \geq x > 0 \iff x \in (0,1]$

**9.**    **11.**

11.  $4x < 2x+1 \leq 3x+2$. So $4x < 2x+1 \iff 2x < 1 \iff x < 1/2$, and
     $2x+1 \leq 3x+2 \iff -1 \leq x$. Thus $x \in [-1,1/2)$.
13.  $1-x \geq 3-2x \geq x-6$. So $1-x \geq 3-2x \iff x \geq 2$, and $3-2x \geq x-6 \iff$
     $9 \geq 3x \iff 3 \geq x$. Thus $x \in [2,3]$.

**13.**    **15.**

15.  $(x-1)(x-2) > 0$. Case (i): $x-1 > 0 \iff x > 1$, and $x-2 > 0 \iff$
     $x > 2$; so $x \in (2,\infty)$. Case (ii): $x-1 < 0 \iff x < 1$, and $x-2 < 0$
     $\iff x < 2$; so $x \in (-\infty,1)$. Solution set: $(-\infty,1) \cup (2,\infty)$
17.  $2x^2+x \leq 1 \iff 2x^2+x-1 \leq 0 \iff (2x-1)(x+1) \leq 0$  Case (i): $2x-1 \geq 0$
     $\iff x \geq 1/2$, and $x \leq -1$; which is impossible.  Case (ii): $2x-1 \leq 0$
     $\iff x \leq 1/2$, and $x \geq -1$; so $x \in [-1,1/2]$.  Solution set:  $[-1,1/2]$

19. $x^2+x+1 > 0 \iff x^2+x+\frac{1}{4}+\frac{3}{4} > 0 \iff (x+\frac{1}{2})^2 + \frac{3}{4} > 0$. But since $(x+\frac{1}{2})^2 \geq 0$ for every real x, the original inequality will be true for all real x as well. Solution set: $(-\infty,\infty)$

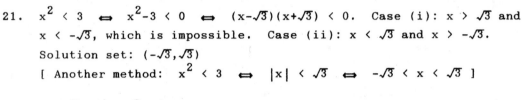

21. $x^2 < 3 \iff x^2-3 < 0 \iff (x-\sqrt{3})(x+\sqrt{3}) < 0$. Case (i): $x > \sqrt{3}$ and $x < -\sqrt{3}$, which is impossible. Case (ii): $x < \sqrt{3}$ and $x > -\sqrt{3}$.
Solution set: $(-\sqrt{3},\sqrt{3})$
[ Another method: $x^2 < 3 \iff |x| < \sqrt{3} \iff -\sqrt{3} < x < \sqrt{3}$ ]

23. $x^3 > x \iff x^3-x > 0 \iff x(x^2-1) > 0 \iff x(x-1)(x+1) > 0$.

| Interval | x | x-1 | x+1 | x(x-1)(x+1) |
|---|---|---|---|---|
| x < -1 | - | - | - | - |
| -1 < x < 0 | - | - | + | + |
| 0 < x < 1 | + | - | + | - |
| x > 1 | + | + | + | + |

Since $x^3 > x$ when the last column is positive, the solution set is: $(-1,0) \cup (1,\infty)$

25. $1/x < 4$. This is clearly true if $x < 0$. So suppose $x > 0$. Then $1/x < 4 \iff 1 < 4x \iff 1/4 < x$. Solution set: $(-\infty,0) \cup (1/4,\infty)$

27. Multiply both sides by x. Case (i): If $x > 0$, then $4/x < x \iff 4 < x^2 \iff 2 < x$ Case (ii): If $x < 0$, then $4/x < x \iff 4 > x^2 \iff -2 < x < 0$ Solution set: $(-2,0) \cup (2,\infty)$

2

**29.** $\frac{2x+1}{x-5} < 3$. Case (i): if x-5 > 0 (i.e. x > 5) then 2x+1 < 3(x-5) $\Longleftrightarrow$

16 < x, so x ∈ (16,∞). Case (ii): if x-5 < 0 (i.e. x < 5) then

2x+1 > 3(x-5) $\Longleftrightarrow$ 16 > x, so in this case x ∈ (-∞,5). Combining

the two cases, the solution set is: (-∞,5) ∪ (16,∞).

**31.** $\frac{x^2-1}{x^2+1} \geq 0$. Since $x^2+1 > 0$ for all real x, this is equivalent to

$x^2-1 \geq 0$ $\Longleftrightarrow$ (x-1)(x+1) $\geq$ 0. Case (i): x $\geq$ 1 and x $\geq$ -1, so

x ∈ [1,∞). Case (ii): x $\leq$ 1 and x $\leq$ -1, so x ∈ (-∞, -1]. Solution

set: (-∞,-1] ∪ [1,∞). [OR: $x^2 \geq 1$ $\Longleftrightarrow$ |x| $\geq$ 1 $\Longleftrightarrow$ x $\geq$ 1 or x $\leq$ -1]

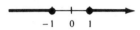

**33.** |2x| = 3 $\Longleftrightarrow$ 2x = 3 or 2x = -3 $\Longleftrightarrow$ x = 3/2 or -3/2

**35.** |x+3| = |2x+1| $\Longleftrightarrow$ x+3 = 2x+1 or x+3 = -(2x+1). In the first

case, x = 2, and in the second 3x = -4 $\Longleftrightarrow$ x = -4/3.

**37.** By (1.6), |x| < 3 $\Longleftrightarrow$ -3 < x < 3, so x ∈ (-3,3).

**39.** |x-4| < 1 $\Longleftrightarrow$ -1 < x-4 < 1 $\Longleftrightarrow$ 3 < x < 5 $\Longleftrightarrow$ x ∈ (3,5)

**41.** |x+5| $\geq$ 2 $\Longleftrightarrow$ x+5 $\geq$ 2 or x+5 $\leq$ -2 $\Longleftrightarrow$ x $\geq$ -3 or x $\leq$ -7 $\Longleftrightarrow$

x ∈ (-∞,-7] ∪ [-3,∞)

**43.** |2x-3| $\leq$ 0.4 $\Longleftrightarrow$ -0.4 $\leq$ 2x-3 $\leq$ 0.4 $\Longleftrightarrow$ 2.6 $\leq$ 2x $\leq$ 3.4 $\Longleftrightarrow$

1.3 $\leq$ x $\leq$ 1.7 $\Longleftrightarrow$ x ∈ [1.3,1.7]

**45.** 1 $\leq$ |x| $\leq$ 4. So either 1 $\leq$ x $\leq$ 4, or 1 $\leq$ -x $\leq$ 4 $\Longleftrightarrow$ -1 $\geq$ x $\geq$ -4.

Solution: x ∈ [-4,-1] ∪ [1,4]

**47.** |x| > |x-1|. Since |x|, |x-1| $\geq$ 0, |x| > |x-1| $\Longleftrightarrow$ $|x|^2 > |x-1|^2$

$\Longleftrightarrow$ $x^2 > (x-1)^2 = x^2-2x+1$ $\Longleftrightarrow$ 0 > -2x+1 $\Longleftrightarrow$ x > $\frac{1}{2}$ $\Longleftrightarrow$ x ∈ ($\frac{1}{2}$,∞)

**49.** $\left|\frac{x}{2+x}\right| < 1$ $\Longleftrightarrow$ $\left[\frac{x}{2+x}\right]^2 < 1$ $\Longleftrightarrow$ $x^2 < (2+x)^2$ $\Longleftrightarrow$ $x^2 < 4+4x+x^2$ $\Longleftrightarrow$

0 < 4+4x $\Longleftrightarrow$ -1 < x $\Longleftrightarrow$ x ∈ (-1,∞)

51. We consider three cases. Case (i): if $x \geq 2$, then $|x| = x$ and $|x-2| = x-2$. So $|x| + |x-2| < 3 \iff x + x-2 < 3 \iff 2x < 5 \iff x < 5/2$. So $x \in [2, 5/2)$. Case (ii): if $0 \leq x < 2$, then $|x| = x$, but $|x-2| = -(x-2)$. So $|x| + |x-2| < 3 \iff x - (x-2) < 3 \iff 2 < 3$. Since this is always true, this case leads to $x \in [0,2)$. Case (iii): if $x < 0$, then $|x| = -x$ and $|x-2| = -(x-2)$. So $|x|+|x-2| < 3 \iff -x - (x-2) < 3 \iff -2x < 1 \iff x > -1/2$. Combining the three cases, we obtain the solution: $x \in (-1/2, 5/2)$.

53. $a(bx-c) \geq bc \iff bx-c \geq \dfrac{bc}{a} \iff bx \geq \dfrac{bc}{a} + c = \dfrac{bc+ac}{a} \iff x \geq \dfrac{bc+ac}{ab}$

55. $ax+b < c \iff ax < c-b \iff x > \dfrac{c-b}{a}$ (since $a < 0$)

57. If $a < b$, then $a+a < a+b$ and $a+b < b+b$. Thus $2a < a+b < 2b$, so dividing by 2 gives $a < \dfrac{a+b}{2} < b$.

59. $|ab| = \sqrt{(ab)^2} = \sqrt{a^2 b^2} = \sqrt{a^2}\sqrt{b^2} = |a||b|$

61. If $0 < a < b$, then $a \cdot a < a \cdot b$ and $a \cdot b < b \cdot b$ (using (1.2(c))). So $a^2 < ab < b^2$ and hence $a^2 < b^2$.

63. By Exercise 62, $-|a| \leq a \leq |a|$ and $-|b| \leq b \leq |b|$. Thus $-(|a|+|b|) \leq a+b \leq |a|+|b|$. If $a = b = 0$, then clearly $|a+b| \leq |a|+|b|$. Otherwise $|a|+|b| > 0$, so by (1.6(b)), $|a+b| \leq |a|+|b|$.

65. $|(x+y)-5| = |(x-2)+(y-3)| \leq |x-2|+|y-3| < 0.01+0.04 = 0.05$

Exercises 1.2

1. From the Distance Formula (1.7) with $x_1 = 1$, $x_2 = 4$, $y_1 = 1$, $y_2 = 5$ we get that the distance is $\sqrt{(4-1)^2+(5-1)^2} = \sqrt{3^2+4^2} = \sqrt{25} = 5$.

3. $\sqrt{(-1-6)^2+(3-(-2))^2} = \sqrt{(-7)^2+5^2} = \sqrt{74}$

5. $\sqrt{(4-2)^2+(-7-5)^2} = \sqrt{2^2+(-12)^2} = \sqrt{148} = 2\sqrt{37}$

7. Since $|AC| = \sqrt{(-4-0)^2+(3-2)^2} = \sqrt{(-4)^2+1^2} = \sqrt{17}$ and

$|BC| = \sqrt{(-4-(-3))^2+(3-(-1))^2} = \sqrt{(-1)^2+4^2} = \sqrt{17}$, the triangle has two sides of equal length and so is isosceles.

9. Label the points A, B, C, D respectively. Then

$|AB| = \sqrt{(4-(-2))^2+(6-9)^2} = \sqrt{6^2+(-3)^2} = 3\sqrt{5},$

$|BC| = \sqrt{(1-4)^2+(0-6)^2} = \sqrt{(-3)^2+(-6)^2} = 3\sqrt{5},$

$|CD| = \sqrt{(-5-1)^2+(3-0)^2} = \sqrt{(-6)^2+3^2} = 3\sqrt{5},$ and

$|DA| = \sqrt{(-2-(-5))^2+(9-3)^2} = \sqrt{3^2+6^2} = 3\sqrt{5}.$ So all sides are of

equal length. Moreover, $|AC| = \sqrt{(1-(-2))^2+(0-9)^2} = \sqrt{3^2+(-9)^2} = 3\sqrt{10}$

and $|BD| = \sqrt{((-5)-4)^2+(3-6)^2} = \sqrt{(-9)^2+(-3)^2} = 3\sqrt{10},$ so the

diagonals are equal also. It must therefore be a square.

11. Let $P(0,y)$ be a point on the y-axis. The distance from P to $(5,-5)$

is $\sqrt{(5-0)^2+(-5-y)^2} = \sqrt{5^2+(y+5)^2}$. The distance from P to $(1,1)$ is

$\sqrt{(1-0)^2+(1-y)^2} = \sqrt{1^2+(y-1)^2}$. We want these distances to be equal:

$\sqrt{5^2+(y+5)^2} = \sqrt{1^2+(y-1)^2} \iff 5^2+(y+5)^2 = 1^2+(y-1)^2 \iff$

$25 + (y^2+10y+25) = 1 + (y^2-2y+1) \iff 12y = -48 \iff y = -4.$ So

the desired point is $(0,-4)$.

13. Using the midpoint formula of Exercise 12, we get:

(a) $\left[\dfrac{1+7}{2}, \dfrac{3+15}{2}\right] = (4,9)$ (b) $\left[\dfrac{-1+8}{2}, \dfrac{6-12}{2}\right] = (\frac{7}{2},-3)$

15. $x = 3$

17. $xy = 0 \iff x = 0$ or $y = 0$

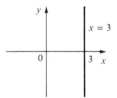

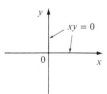

19. $y = x$

21. $xy = 2 \iff y = 2/x$

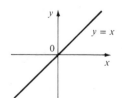

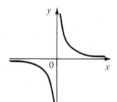

**23.** $x + y^2 = 4 \iff x = 4 - y^2$

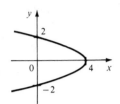

**25.** $\{(x,y) \mid x < 0\}$

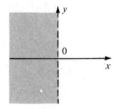

**27.** $\{(x,y) \mid xy < 0\}$

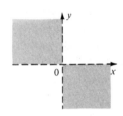

**29.** $\{(x,y) \mid |x| \le 2\}$

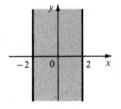

**31.** $\{(x,y) \mid x^2 + y^2 \le 1\}$

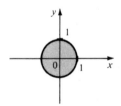

**33.** From (1.8), the equation is $(x-3)^2 + (y+1)^2 = 25$.

**35.** The equation has the form $x^2 + y^2 = r^2$. Since $(4,7)$ lies on the circle, we have $4^2 + 7^2 = r^2 \Rightarrow r^2 = 65$. So the required equation is $x^2 + y^2 = 65$.

**37.** $x^2 - 4x + y^2 + 10y = -13 \iff (x^2 - 4x + 4) + (y^2 + 10y + 25) = -13 + 4 + 25 = 16 \iff$ $(x-2)^2 + (y+5)^2 = 4^2$. The center is $(2,-5)$ and the radius is 4.

**39.** $x^2 + x + y^2 = 0 \iff (x^2 + x + \frac{1}{4}) + y^2 = \frac{1}{4} \iff (x + \frac{1}{2})^2 + y^2 = (1/2)^2$.

Center: $(-1/2, 0)$, radius: $1/2$.

**41.** $2x^2 - x + 2y^2 + y = 1 \iff x^2 - \frac{1}{2}x + y^2 + \frac{1}{2}y = \frac{1}{2} \iff (x^2 - \frac{1}{2}x + \frac{1}{16}) + (y^2 + \frac{1}{2}y + \frac{1}{16}) =$ $\frac{1}{2} + \frac{1}{16} + \frac{1}{16} = \frac{5}{8} \iff (x - \frac{1}{4})^2 + (y + \frac{1}{4})^2 = \frac{5}{8}$

Center: $(1/4, -1/4)$, radius: $\sqrt{5/8} = \sqrt{5}/2\sqrt{2}$

6

## Exercises 1.3

1.  From (1.9), the slope is $\frac{11-5}{4-1} = \frac{6}{3} = 2$    3.   $\frac{-6-3}{-1-(-3)} = \frac{-9}{2}$

5.  From (1.10), the equation of the line is $y-(-3) = 6(x-2)$   or
    $y = 6x-15$.

7.  $y-7 = (2/3)(x-1)$   or   $2x-3y+19 = 0$

9.  The slope is $m = \frac{6-1}{1-2} = -5$, so the equation of the line is
    $y-1 = -5(x-2)$   or   $5x+y = 11$.

11.  From (1.11), the equation is   $y = 3x-2$.

13.  Since the line passes through $(1,0)$ and $(0,-3)$, its slope is
    $m = \frac{-3-0}{0-1} = 3$, so its equation is $y = 3x-3$.

15.  The slope is $m = \tan 30° = \sqrt{3}/3$, so the equation of the line is
    $y-(-4) = \frac{\sqrt{3}}{3}(x-2)$  $\Longleftrightarrow$  $y = \frac{\sqrt{3}}{3}x - \frac{2\sqrt{3}}{3} - 4$  or  $x-\sqrt{3}y = 2+4\sqrt{3}$.

17.  Since $m = 0$, $y-5 = 0(x-4)$   or   $y = 5$.

19.  Putting the line $x+2y = 6$ into slope-intercept form $y = -(1/2)x+3$,
    we see that this line has slope $-1/2$.  So we want the line of slope
    $-1/2$ that passes through the point $(1,-6)$: $y-(-6) = -(1/2)(x-1)$  $\Longleftrightarrow$
    $y = -\frac{1}{2}x - \frac{11}{2}$  or  $x+2y+11 = 0$.

21.  $2x+5y+8 = 0$  $\Longleftrightarrow$  $y = -\frac{2}{5}x - \frac{8}{5}$.  Since this line has slope $-2/5$, a
    line perpendicular to it would have slope $5/2$, so the required line
    is $y-(-2) = \frac{5}{2}(x-(-1))$  $\Longleftrightarrow$  $y = \frac{5}{2}x + \frac{1}{2}$  or  $5x-2y+1 = 0$

23.  The slope of the line segment AB is $\frac{4-1}{7-1} = \frac{1}{2}$, the slope of CD is
    $\frac{7-10}{-1-5} = \frac{1}{2}$,  the slope of BC is $\frac{10-4}{5-7} = -3$, and the slope of DA is
    $\frac{1-7}{1-(-1)} = -3$.  So AB is parallel to CD and BC is parallel to DA.
    Hence ABCD is a parallelogram.

25.  The slopes of the four sides are: $m_{AB} = \frac{3-1}{11-1} = \frac{1}{5}$, $m_{BC} = \frac{8-3}{10-11} = -5$,
    $m_{CD} = \frac{6-8}{0-10} = \frac{1}{5}$, $m_{DA} = \frac{1-6}{1-0} = -5$.  Hence AB ∥ CD, BC ∥ DA, AB ⊥ BC,
    BC ⊥ CD, CD ⊥ DA, and DA ⊥ AB, and so ABCD is a rectangle.

27.  The slope of the segment AB is $\frac{-2-4}{7-1} = -1$, so its perpendicular

bisector has slope 1. The midpoint of AB is $\left[\frac{1+7}{2}, \frac{4-2}{2}\right]$ = (4,1), so the equation of the perpendicular bisector is y-1 = 1(x-4)  or y = x-3.

29.  x+3y = 0  $\iff$  y = $-\frac{1}{3}$x, so the slope is -1/3 and the y-intercept 0.

**29.**  **31.**

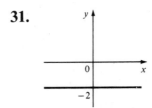

31.  y = -2 is a horizontal line with slope 0 and y-intercept -2.

33.  3x-4y = 12  $\iff$  y = $\frac{3}{4}$x-3, so the slope is $\frac{3}{4}$ and the y-intercept -3.

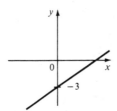

35.  y = x+1 has slope 1 so tan $\theta$ = 1 and $\theta$ = 45°.

37.  2x+3y = 4  $\iff$  y = $-\frac{2}{3}$x+$\frac{4}{3}$, so m = -2/3 = tan $\theta$ and $\theta$ ≈ 146°.

39.  Put y = x into the second equation y = 3-2x to get x = 3-2x  $\iff$  3x = 3  $\iff$  x = 1.  Hence y = x = 1 and the point of intersection is (1,1).  The two lines have slopes $m_1$ = 1 and $m_2$ = -2 respectively, so by (1.16), tan $\theta$ = $\frac{-2 - 1}{1 + (1)(-2)}$ = 3, and so $\theta$ ≈ 72°.  (The supplementary angle 108° is also correct )

41.  In slope-intercept form the lines are y = $\frac{1}{2}$x-$\frac{5}{2}$ and y = $\frac{1}{3}$x-2.  Setting $\frac{1}{2}$x-$\frac{5}{2}$ = $\frac{1}{3}$x-2 gives x = 3 and y = $\frac{1}{3}$(3)-2 = -1, so the lines intersect at (3,-1).  Since $m_1$ = 1/2  and  $m_2$ = 1/3, tan $\theta$ = $\frac{(1/3)-(1/2)}{1 + (1/2)(1/3)}$ = $-\frac{1}{7}$, and so $\theta$ ≈ 172° (or 8°).

43. Since the line passes through the points $(a,0)$ and $(0,b)$, it has slope $\frac{b-0}{0-a} = -b/a$, and hence has equation $y = -\frac{b}{a}x + b$ or $y + \frac{b}{a}x = b$. Dividing through by b gives $\frac{x}{a} + \frac{y}{b} = 1$.

45. $\{(x,y) \mid 1+x \leq y \leq 1-2x\}$

**45.**
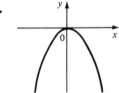

**47.**
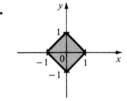

47. Take cases: in quadrants I, II, III, IV, the inequality $|x|+|y| \leq 1$ becomes $y \leq 1-x$, $y \leq 1+x$, $y \geq -x-1$, and $y \geq x-1$.

Section 1.4

Exercises 1.4

1. $y = -x^2$. Parabola.

**1.**
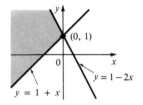

**3.**
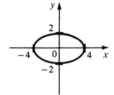

3. $x^2+4y^2 = 16 \iff \frac{x^2}{16} + \frac{y^2}{4} = 1$. Ellipse.

5. $16x^2 - 25y^2 = 400 \iff \dfrac{x^2}{25} - \dfrac{y^2}{16} = 1.$  Hyperbola.

**5.**

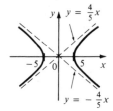

**7.**

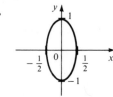

7. $4x^2 + y^2 = 1 \iff \dfrac{x^2}{1/4} + y^2 = 1.$  Ellipse.

9. $x = y^2 - 1.$  Parabola with vertex at $(-1, 0)$.

**9.**

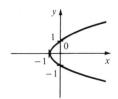

**11.**

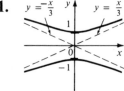

11. $9y^2 - x^2 = 9 \iff y^2 - \dfrac{x^2}{9} = 1.$  Hyperbola.

13. $xy = 4.$  Hyperbola.

**13.**

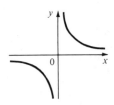

**15.**

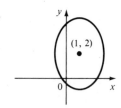

15. $9(x-1)^2 + 4(y-2)^2 = 36 \iff \dfrac{(x-1)^2}{4} + \dfrac{(y-2)^2}{9} = 1.$  Ellipse centered

at $(1,2)$.

17.  $y = x^2-6x+13 = (x^2-6x+9)+4 = (x-3)^2+4.$  Parabola with vertex at (3,4)

**17.**
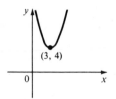

**19.**
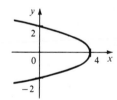

19.  $x = -y^2+4.$  Parabola with vertex at (4,0).

21.  $x^2+4y^2-6x+5 = 0 \iff (x^2-6x+9) + 4y^2 = -5+9 = 4 \iff$
     $\dfrac{(x-3)^2}{4} + y^2 = 1.$  Ellipse centered at (3,0).

**21.**
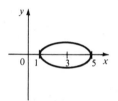

**23.**
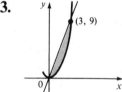

23.  $y = 3x,$  $y = x^2$  intersect where  $3x = x^2 \iff 0 = x^2-3x = x(x-3)$  or at (0,0) and (3,9).

25.  The parabola must have an equation of the form  $y = a(x-1)^2-1.$  Substituting x = 3, y = 3 into the equation gives  $3 = a(3-1)^2-1$  so a = 1, and the equation is  $y = (x-1)^2-1 = x^2 - 2x.$  (Note that using the other point (-1,3) would have given the same value for a and hence the same equation.)

27.  $\{(x,y) \mid y \geq x^2-1\}$

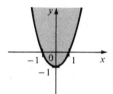

Exercises 1.5

1.  $f(x) = x^2-3x+2$, so $f(1) = 1^2-3(1)+2 = 0$, $f(-2) = (-2)^2-3(-2)+2 = 12$

$f(1/2) = (1/2)^2-3(1/2)+2 = 3/4$, $f(\sqrt{5}) = (\sqrt{5})^2-3(\sqrt{5})+2 = 7-3\sqrt{5}$,

$f(a) = a^2-3a+2$, $f(-a) = (-a)^2-3(-a)+2 = a^2+3a+2$

3.  $g(x) = (1-x)/(1+x)$, so $g(2) = (1-2)/(1+2) = -1/3$,

$g(-2) = (1-(-2))/(1-2) = -3$, $g(\pi) = (1-\pi)/(1+\pi)$,

$g(a) = (1-a)/(1+a)$, $g(a-1) = (1-(a-1))/(1+(a-1)) = (2-a)/a$,

$g(-a) = (1-(-a))/(1-a) = (1+a)/(1-a)$

5.  $f(x) = \sqrt{x}$, $0 \le x \le 4$

Machine diagram              Arrow diagram              Graph

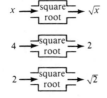     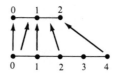     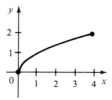

7.  The range of f is the set of values of f, {0,1,2,4}.

Arrow diagram                          Graph

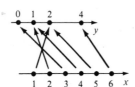                  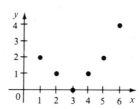

9.  $f(x) = 6-4x$, $-2 \le x \le 3$. The domain is $[-2,3]$. If $-2 \le x \le 3$,

then $14 = 6-4(-2) \ge 6-4x \ge 6-4(3) = -6$, so the range is $[-6,14]$.

11. $g(x) = \dfrac{2}{3x-5}$. This is defined when $3x-5 \ne 0$, so the domain of f is

$\{x \mid x \ne 5/3\}$, and the range is $\{y \mid y \ne 0\} = (-\infty,0) \cup (0,\infty)$.

13. $h(x) = \sqrt{2x-5}$ is defined when $2x-5 \ge 0$ or $x \ge 5/2$, so the domain is

$[5/2,\infty)$ and the range is $[0,\infty)$.

15. $F(x) = \sqrt{1-x^2}$ is defined when $1-x^2 \ge 0 \iff x^2 \le 1 \iff |x| \le 1 \iff$

$-1 \le x \le 1$, so the domain is $[-1,1]$, and the range is $[0,1]$.

17. $f(x) = (x+2)/(x^2-1)$ is defined for all x except when $x^2-1 = 0 \Leftrightarrow$
    x = 1 or -1, so the domain is $\{x \mid x \neq \pm1\}$.

19. $g(x) = \sqrt[4]{x^2-6x}$ is defined when $0 \leq x^2-6x = x(x-6) \Leftrightarrow x \geq 6$ or
    $x \leq 0$, so the domain is $(-\infty,0] \cup [6,\infty)$.

21. $\phi(x) = \sqrt{x/(\pi-x)}$ is defined when $\frac{x}{\pi-x} \geq 0$. So either $x \leq 0$ and
    $\pi-x < 0$ ($\Leftrightarrow x > \pi$), which is impossible, or $x \geq 0$ and $\pi-x > 0$ ($\Leftrightarrow$
    $x < \pi$), and so the domain is $[0,\pi)$.

23. $f(t) = \sqrt[3]{t-1}$ is defined for every t, since every real number has a
    cube root. The domain is the set of all real numbers.

25. $f(x) = 2$. Domain is R.          27. $f(x) = 3-2x$. Domain is R.

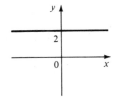          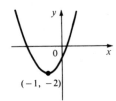

29. $f(x) = -x^2$. Domain is R.          31. $f(x) = x^2+2x-1 = (x^2+2x+1)-2 = (x+1)^2-2$, so the graph is a
                                          parabola with vertex at $(-1,-2)$.
                                          The domain is R.

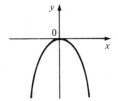

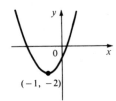

**33.** $g(x) = x^4$.  Domain is R.

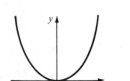

**35.** $g(x) = \sqrt{-x}$.  Domain is $\{x \mid -x \geq 0\} = (-\infty, 0]$.

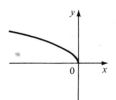

**37.** $h(x) = \sqrt{4-x^2}$.  Now $y = \sqrt{4-x^2}$ $\Rightarrow$ $y^2 = 4-x^2$ $\iff$ $x^2+y^2 = 4$, so the graph is the top half of a circle of radius 2.  The domain is $\{x \mid 4-x^2 \geq 0\} = [-2, 2]$.

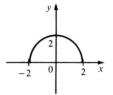

**39.** $F(x) = 1/x$.  Domain is $\{x \mid x \neq 0\}$

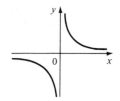

**41.** $G(x) = |x| + x = \begin{cases} 2x, & x \geq 0 \\ 0, & x < 0 \end{cases}$  Domain is R.

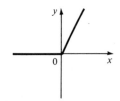

**43.** $H(x) = |2x| = \begin{cases} 2x, & x \geq 0 \\ -2x, & x < 0 \end{cases}$

Domain is R.

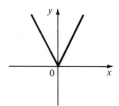

**45.** $f(x) = \dfrac{x}{|x|} = \begin{cases} 1, & x > 0 \\ -1, & x < 0 \end{cases}$

Domain is $\{x \mid x \neq 0\}$

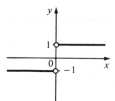

**47.** $f(x) = \dfrac{x^2-1}{x-1} = \dfrac{(x+1)(x-1)}{x-1}$, so for

$x \neq 1$, $f(x) = x+1$. Domain is

$\{x \mid x \neq 1\}$.

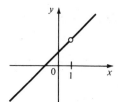

**NOTE:** In each of Problems 49-59, the domain is all of R.

**49.** $f(x) = \begin{cases} x+1, & x \neq 1 \\ 1, & x = 1 \end{cases}$

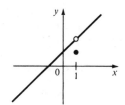

**51.** $f(x) = \begin{cases} 0, & x < 2 \\ 1, & x \geq 2 \end{cases}$

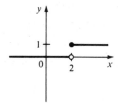

**53.**  $f(x) = \begin{cases} x, & x \le 0 \\ x+1, & x > 0 \end{cases}$

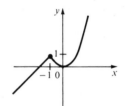

**55.**  $f(x) = \begin{cases} -1, & x < -1 \\ x, & -1 \le x \le 1 \\ 1, & x > 1 \end{cases}$

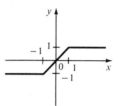

**57.**  $f(x) = \begin{cases} x+2, & x \le -1 \\ x^2, & x > -1 \end{cases}$

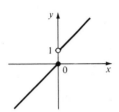

**59.**  $f(x) = \begin{cases} -1, & x \le -1 \\ 3x+2, & -1 < x < 1 \\ 7-2x, & x \ge 1 \end{cases}$

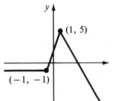

**61.**   Yes, the curve is the graph of a function.  The domain is $[-3,2]$ and the range is $[-2,2]$.

**63.**   No, this is not the graph of a function since for $x = -1$ there are infinitely many points on the curve.

**65.**   The slope of this line segment is $\dfrac{-6-1}{4-(-2)} = -\dfrac{7}{6}$, so its equation is $y-1 = -\dfrac{7}{6}(x+2)$.  The function is $f(x) = -\dfrac{7}{6}x - \dfrac{4}{3}$, $-2 \le x \le 4$.

**67.**   $x+(y-1)^2 = 0 \;\Rightarrow\; y-1 = \pm\sqrt{-x}$.  The bottom half is given by the function $f(x) = 1 - \sqrt{-x}$, $x \le 0$.

**69.**   Let the length and width of the rectangle be $\ell$ and $w$ respectively.  Then the perimeter is $2\ell + 2w = 20$, and the area is $A = \ell w$.  Solving the first equation for $w$ in terms of $\ell$ gives $w = \dfrac{20-2\ell}{2} = 10 - \ell$.  Thus $A(\ell) = \ell(10-\ell) = 10\ell-\ell^2$.  Since lengths are positive, the domain of $A$ is $0 < \ell < 10$.

71. Let the length of a side of the equilateral triangle be x. Then by Pythagoras' Theorem the height y of the triangle satisfies $y^2 + (x/2)^2 = x^2$, so that $y = \frac{\sqrt{3}}{2} x$. Thus the area of the triangle is $A = \frac{1}{2}yx$ and so $A(x) = \frac{1}{2}\left[\frac{\sqrt{3}}{2} x\right]x = \frac{\sqrt{3}x^2}{4}$, with domain $x > 0$.

73. Let each side of the base of the box have length x, and let the height of the box be h. Since the volume is 2, we know that $2 = hx^2$, so that $h = 2/x^2$, and the surface area is $S = x^2 + 4xh$. Thus $S(x) = x^2 + 4x(2/x^2) = x^2 + 8/x$, with domain $x > 0$.

75. $f(-x) = \dfrac{1}{(-x)^2} = \dfrac{1}{x^2} = f(x)$,

so f is an even function.

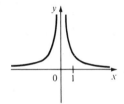

77. $f(-x) = (-x)^2 + (-x) = x^2 - x$. Since this is neither f(x) nor -f(x), the function f is neither even nor odd.

79. $f(-x) = (-x)^3 - (-x) = -x^3 + x$
$= -(x^3 - x) = -f(x)$, so f is odd.

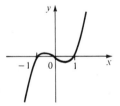

17

Section 1.6

## Exercises 1.6

In this problem set, "D =" stands for "the domain of the function is".

1.  $f(x) = x^2-x$, $g(x) = x+5$. $(f+g)(x) = x^2-x+x+5 = x^2+5$, D = R

   $(f-g)(x) = x^2-x-(x+5) = x^2-2x-5$, D = R

   $(fg)(x) = (x^2-x)(x+5) = x^3+4x^2-5x$, D = R

   $(f/g)(x) = (x^2-x)/(x+5)$, D = $\{x|\ x \neq -5\}$

3.  $f(x) = \sqrt{1+x}$, D = $[-1,\infty)$; $g(x) = \sqrt{1-x}$, D = $(-\infty,1]$

   $(f+g)(x) = \sqrt{1+x} + \sqrt{1-x}$, D = $(-\infty,1] \cap [-1,\infty) = [-1,1]$

   $(f-g)(x) = \sqrt{1+x} - \sqrt{1-x}$, D = $[-1,1]$

   $(fg)(x) = \sqrt{1+x}\cdot\sqrt{1-x} = \sqrt{1-x^2}$, D = $[-1,1]$

   $(f/g)(x) = \sqrt{1+x}/\sqrt{1-x}$, D = $[-1,1)$

5.  $f(x) = \sqrt{x}$, D = $[0,\infty)$; $g(x) = \sqrt[3]{x}$, D = R. $(f+g)(x) = \sqrt{x} + \sqrt[3]{x}$,

   D = $[0,\infty)$; $(f-g)(x) = \sqrt{x} - \sqrt[3]{x}$, D = $[0,\infty)$; $(fg)(x) = \sqrt{x}\cdot\sqrt[3]{x} = x^{5/6}$,

   D = $[0,\infty)$; $(f/g)(x) = \sqrt{x}/\sqrt[3]{x} = x^{1/6}$, D = $(0,\infty)$

7.  $F(x) = (\sqrt{4-x} + \sqrt{3+x})/(x^2-2)$.  Domain of numerator is

   $(-\infty,4] \cap [-3,\infty) = [-3,4]$.  Since denominator is 0 when $x = \pm\sqrt{2}$,

   domain of F is $[-3,-\sqrt{2}) \cup (-\sqrt{2},\sqrt{2}) \cup (\sqrt{2},4]$.

9.  $f(x) = x^3$, $g(x) = 1$          11. $f(x) = x$, $g(x) = 1/x$

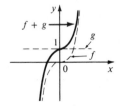     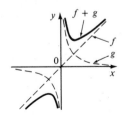

13.  $f(x) = 2x+3$, $g(x) = 4x-1$.  Since both f and g have domain and range

   R, so will all composite functions.

   $(f\circ g)(x) = f(g(x)) = f(4x-1) = 2(4x-1)+3 = 8x+1$.

   $(g\circ f)(x) = g(f(x)) = g(2x+3) = 4(2x+3)-1 = 8x+11$.

   $(f\circ f)(x) = f(f(x)) = f(2x+3) = 2(2x+3)+3 = 4x+9$.

   $(g\circ g)(x) = g(g(x)) = g(4x-1) = 4(4x-1)-1 = 16x-5$.

15. $f(x) = 2x^2-x$, $g(x) = 3x+2$, $D = R$ for both $f$ and $g$, and hence for their composites as well.

$(f \circ g)(x) = f(g(x)) = f(3x+2) = 2(3x+2)^2-(3x+2) = 18x^2+21x+6$

$(g \circ f)(x) = g(f(x)) = g(2x^2-x) = 3(2x^2-x)+2 = 6x^2-3x+2$

$(f \circ f)(x) = f(f(x)) = f(2x^2-x) = 2(2x^2-x)^2-(2x^2-x) = 8x^4-8x^3+x$

$(g \circ g)(x) = g(g(x)) = g(3x+2) = 3(3x+2)+2 = 9x+8$

17. $f(x) = 1/x$, $D = \{x \mid x \neq 0\}$; $g(x) = x^3+2x$, $D = R$.

$(f \circ g)(x) = f(g(x)) = f(x^3+2x) = 1/(x^3+2x)$. $D = \{x \mid x^3+2x \neq 0\} = \{x \mid x \neq 0\}$. $(g \circ f)(x) = g(f(x)) = g(1/x) = 1/x^3 + 2/x$ $D = \{x \mid x \neq 0\}$

$(f \circ f)(x) = f(f(x)) = f(1/x) = 1/(1/x) = x$. $D = \{x \mid x \neq 0\}$.

$(g \circ g)(x) = g(g(x)) = g(x^3+2x) = (x^3+2x)^3+2(x^3+2x)$
$= x^9+6x^7+12x^5+10x^3+4x$. $D = R$.

19. $f(x) = \sqrt[3]{x}$, $D = R$; $g(x) = 1 - \sqrt{x}$, $D = [0,\infty)$.

$(f \circ g)(x) = f(g(x)) = f(1-\sqrt{x}) = \sqrt[3]{1-\sqrt{x}}$ $D = [0,\infty)$

$(g \circ f)(x) = g(f(x)) = g(\sqrt[3]{x}) = 1 - x^{1/6}$ $D = [0,\infty)$

$(f \circ f)(x) = f(f(x)) = f(\sqrt[3]{x}) = x^{1/9}$ $D = R$

$(g \circ g)(x) = g(g(x)) = g(1-\sqrt{x}) = 1-\sqrt{1-\sqrt{x}}$ $D = \{x \geq 0 \mid 1-\sqrt{x} \geq 0\} = [0,1]$

21. $f(x) = \dfrac{x+2}{2x+1}$, $D = \{x \mid x \neq -1/2\}$; $g(x) = \dfrac{x}{x-2}$, $D = \{x \mid x \neq 2\}$.

$(f \circ g)(x) = f(g(x)) = f\left[\dfrac{x}{x-2}\right] = \dfrac{x/(x-2) + 2}{2x/(x-2) + 1} = \dfrac{3x-4}{3x-2}$ $D = \{x \mid x \neq 2, 2/3\}$

$(g \circ f)(x) = g(f(x)) = g\left[\dfrac{x+2}{2x+1}\right] = \dfrac{(x+2)/(2x+1)}{(x+2)/(2x+1) - 2} = \dfrac{-x-2}{3x}$

$D = \{x \mid x \neq 0, -1/2\}$

$(f \circ f)(x) = f(f(x)) = f\left[\dfrac{x+2}{2x+1}\right] = \dfrac{(x+2)/(2x+1) + 2}{2(x+2)/(2x+1) + 1} = \dfrac{5x+4}{4x+5}$

$D = \{x \mid x \neq -1/2, -5/4\}$

$(g \circ g)(x) = g(g(x)) = g\left[\dfrac{x}{x-2}\right] = \dfrac{x/(x-2)}{x/(x-2) - 2} = \dfrac{x}{4-x}$ $D = \{x \mid x \neq 2, 4\}$

23. $(f \circ g \circ h)(x) = f(g(h(x))) = f(g(x-1)) = f(\sqrt{x-1}) = \sqrt{x-1} - 1$

25. $(f \circ g \circ h)(x) = f(g(h(x))) = f(g(\sqrt{x})) = f(\sqrt{x}-5) = (\sqrt{x}-5)^4 + 1$

27. Let $g(x) = x-9$ and $f(x) = x^5$. Then $(f \circ g)(x) = (x-9)^5 = F(x)$

29. Let $g(x) = x^2$ and $f(x) = \dfrac{x}{x+4}$. Then $(f \circ g)(x) = \dfrac{x^2}{x^2+4} = G(x)$

31. Let $h(x) = x^2$, $g(x) = x+1$, $f(x) = \dfrac{1}{x}$. Then $(f \circ g \circ h)(x) = \dfrac{1}{x^2+1} = H(x)$.

33. We need a function $g$ so that $f(g(x)) = 3(g(x))+5 = h(x) = 3x^2+3x+2$
$= 3(x^2+x)+2 = 3(x^2+x-1)+5$. So we see that $g(x) = x^2+x-1$.

## Exercises 1.7

**1.** $y = x^8$

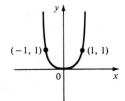

**3.** $y = -1/x$

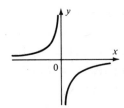

**5.** $y = 2 \sin x$

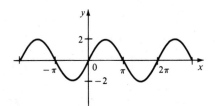

**7.** $y = (x-1)^3 + 2$

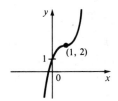

**9.** $y = 1 + \sqrt[6]{-x}$

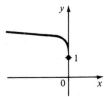

**11.** $y = \cos(x/2)$

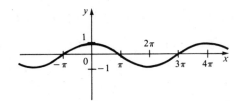

**13.** $y = 1/(x-3)$

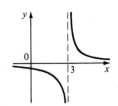

**15.** $y = \frac{1}{3} \sin\left(x - \frac{\pi}{6}\right)$

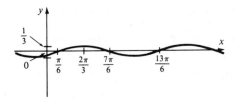

**17.** $y = 1+2x-x^2 = -(x-1)^2+2$

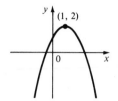

**19.** $y = 2 - \sqrt{x+1}$

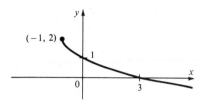

Chapter 1 Review

## Review Exercises for Chapter 1

**1.** $2+5x \leq 9-2x \iff 7x \leq 7 \iff x \leq 1 \iff x \in (-\infty,1]$

**3.** $|x+3| < 7 \iff -7 < x+3 < 7 \iff -10 < x < 4 \iff x \in (-10,4)$

**5.** $x^2-5x+6 > 0 \iff (x-3)(x-2) > 0$. Case (i): $x > 3$ and $x > 2$, so $x \in (3,\infty)$. Case (ii): $x < 3$ and $x < 2$, so $x \in (-\infty,2)$. Solution set: $(-\infty,2) \cup (3,\infty)$.

**7.** $\sqrt{(-4-4)^2 + (-4-2)^2} = \sqrt{8^2+6^2} = \sqrt{100} = 10$

**9.** $(x-2)^2 + (y-1)^2 = 3^2 = 9$

**11.** $x^2+2x+y^2-8y = -8 \iff (x^2+2x+1) + (y^2-8y+16) = -8+1+16 \iff (x+1)^2 + (y-4)^2 = 9 = 3^2$. Center $(-1,4)$, radius 3.

**13.** Slope $m = \dfrac{-1-(-6)}{2-(-1)} = \dfrac{5}{3}$, so $y-(-1) = \dfrac{5}{3}(x-2)$ or $5x-3y = 13$

**15.** $x+2y = 1 \iff y = -\dfrac{1}{2}x+\dfrac{1}{2}$; slope is $-\dfrac{1}{2}$. So $y-3 = -\dfrac{1}{2}(x-2)$ or $x+2y = 8$

**17.** $3x+5y = 10 \iff y = -\dfrac{3}{5}x+2$, so slope is $-\dfrac{3}{5}$ and $y$-intercept is 2.

**19.** $y = 8-2x^2$, parabola

**21.** $x^2-4y^2 = 4 \iff \dfrac{x^2}{4} - y^2 = 1$, hyperbola.

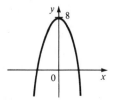

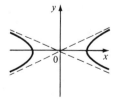

23. $2x^2+y^2-16x+30 = 0 \iff$

    $2(x^2-8x)+y^2 = -30 \iff$

    $2(x^2-8x+16)+y^2 = -30+32 = 2 \iff$

    $2(x-4)^2+y^2 = 2 \iff$

    $(x-4)^2 + \dfrac{y^2}{2} = 1.$  Ellipse.

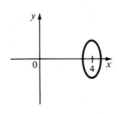

25. $f(x) = 1+\sqrt{x-1}$, so $f(5) = 1+\sqrt{5-1} = 3$, $f(9) = 1+\sqrt{9-1} = 1+2\sqrt{2}$,

    $f(-x) = 1+\sqrt{-x-1}$, $f(x^2) = 1+\sqrt{x^2-1}$, $[f(x)]^2 = [1+\sqrt{x-1}]^2 = x+2\sqrt{x-1}$

27. $g(x)$ is defined unless the denominator is 0.  $x^2+x-1 = 0 \iff$

    $x = \dfrac{-1\pm\sqrt{1^2-4(-1)}}{2} = \dfrac{-1\pm\sqrt{5}}{2}$.  Domain is $\{x\mid x \neq \dfrac{-1\pm\sqrt{5}}{2}\}$.

29. $f(x) = -1$

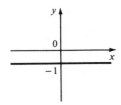

31. $g(x) = x^2+2$

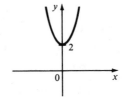

33. $h(x) = \sqrt{x-5}$

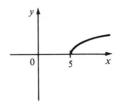

35. $y = -\sin 2x$

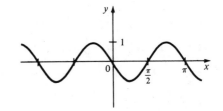

37. $f(x) = x^2$, $g(x) = x+2$; both have domain R.

    (a) $(f+g)(x) = x^2+x+2$.  Domain is R.

    (b) $(f/g)(x) = x^2/(x+2)$.  Domain is $\{x\mid x \neq -2\}$.

    (c) $(f\circ g)(x) = f(g(x)) = f(x+2) = (x+2)^2$.  $D = R$

    (d) $(g\circ f)(x) = g(f(x)) = g(x^2) = x^2+2$.  $D = R$

39. Let $f(x) = \sqrt{x}$ and $g(x) = x^2+x+9$.  Then $(f\circ g)(x) = \sqrt{x^2+x+9} = h(x)$.

# CHAPTER TWO

## Exercises 2.1

1.  $f(x) = 1+x+x^2$, $(1,3)$

    (a) $m_{PQ} = \dfrac{f(x)-f(1)}{x-1} = \dfrac{x^2+x+1-3}{x-1} = \dfrac{x^2+x-2}{x-1} = \dfrac{(x+2)(x-1)}{x-1} = x+2$ if $x \neq 1$

    (b)

    | .x-1. | ..x.. | $m_{PQ}$. |
    |-------|-------|-----------|
    | 0.5   | 1.5   | 3.5       |
    | -0.5  | 0.5   | 2.5       |
    | 0.1   | 1.1   | 3.1       |
    | -0.1  | 0.9   | 2.9       |
    | 0.01  | 1.01  | 3.01      |
    | -0.01 | 0.99  | 2.99      |
    | 0.001 | 1.001 | 3.001     |
    | -0.001| 0.999 | 2.999     |

    (c) 3

    (d) $y-3 = 3(x-1)$   or   $y = 3x$

3.  $y = x^3$, $(2,8)$

    (a) $m_{PQ} = \dfrac{f(x)-f(2)}{x-2} = \dfrac{x^3-8}{x-2} = \dfrac{(x-2)(x^2+2x+4)}{x-2} = x^2+2x+4$   if $x \neq 2$

    (b)

    | .x-2. | ..x.. | $m_{PQ}$.... |
    |-------|-------|--------------|
    | 0.5   | 2.5   | 15.25        |
    | -0.5  | 1.5   | 9.25         |
    | 0.1   | 2.1   | 12.61        |
    | -0.1  | 1.9   | 11.41        |
    | 0.01  | 2.01  | 12.0601      |
    | -0.01 | 1.99  | 11.9401      |
    | 0.001 | 2.001 | 12.006001    |
    | -0.001| 1.999 | 11.994001    |

    (c) 12

    (d) $y-8 = 12(x-2)$
        or $y = 12x-16$

5.  $y = 4x^2-x+1$, $(0,1)$

    (a) $m_{PQ} = \dfrac{f(x)-f(0)}{x-0} = \dfrac{4x^2-x+1-1}{x} = \dfrac{4x^2-x}{x} = 4x-1$ if $x \neq 0$

    (b)

    | .x-0. | ...x... | $m_{PQ}$.. |
    |-------|---------|------------|
    | 0.5   | 0.5     | 1          |
    | -0.5  | -0.5    | -3         |
    | 0.1   | 0.1     | -0.6       |
    | -0.1  | -0.1    | -1.4       |
    | 0.01  | 0.01    | -0.96      |
    | -0.01 | -0.01   | -1.04      |
    | 0.001 | 0.001   | -0.996     |
    | -0.001| -0.001  | -1.004     |

    (c) -1

    (d) $y-1 = -1(x-0)$   or   $x+y = 1$

7.  $y = 1/x$, $(1,1)$

    (a) $m_{PQ} = \dfrac{f(x)-f(1)}{x-1} = \dfrac{(1/x)-1}{x-1} = \dfrac{1-x}{x(x-1)} = -\dfrac{1}{x}$   if $x \neq 1, 0$

    (b)

| .x-1. | ..x.. | ..$m_{PQ}$.... |
|---|---|---|
| 0.5 | 1.5 | -2/3 |
| -0.5 | 0.5 | -2 |
| 0.1 | 1.1 | -0.909091 |
| -0.1 | 0.9 | -1.111111 |
| 0.01 | 1.01 | -0.990099 |
| -0.01 | 0.99 | -1.010101 |
| 0.001 | 1.001 | -0.999001 |
| -0.001 | 0.999 | -1.001001 |

    (c) $-1$

    (d) $y-1 = -1(x-1)$   or $x+y = 2$

9.  $y = \sqrt{x}$, $(4,2)$

    (a) $m_{PQ} = \dfrac{f(x)-f(4)}{x-4} = \dfrac{\sqrt{x}-2}{x-4} = \dfrac{\sqrt{x}-2}{(\sqrt{x}-2)(\sqrt{x}+2)} = \dfrac{1}{\sqrt{x}+2}$   if $x \neq 4$.

    (b)

| .x-4. | ..x.. | ..$m_{PQ}$.... |
|---|---|---|
| 0.5 | 4.5 | 0.242641 |
| -0.5 | 3.5 | 0.258343 |
| 0.1 | 4.1 | 0.248457 |
| -0.1 | 3.9 | 0.251582 |
| 0.01 | 4.01 | 0.249844 |
| -0.01 | 3.99 | 0.250156 |
| 0.001 | 4.001 | 0.249984 |
| -0.001 | 3.999 | 0.250016 |

    (c) $1/4$

    (d) $y-2 = (1/4)(x-4)$

        or $x-4y+4 = 0$

11.  (a) Average velocity is $\dfrac{40(2+h)-16(2+h)^2-16}{h} = \dfrac{-24h-16h^2}{h} = -24-16h$

    (i) $h = 0.5$, $-32$ ft/s         (ii) $h = 0.1$, $-25.6$ ft/s

    (iii) $h = 0.05$, $-24.8$ ft/s     (iv) $h = 0.01$, $-24.16$ ft/s

    (b) $-24$ ft/s

13. Average velocity between times 0 and h is $\dfrac{s(h)-s(0)}{h} = \dfrac{h^2+h-0}{h} = h+1$

    (a) (i) 2+1 = 3 m/s           (ii) 1+1 = 2 m/s

       (iii) 0.5+1 = 1.5 m/s     (iv) 0.1+1 = 1.1 m/s

    (b) As h approaches 0, the velocity approaches 1 m/s.

(c), (d)

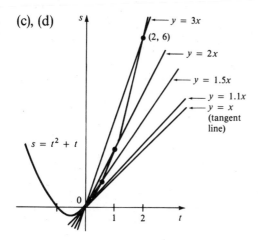

15. (a) (i) $\dfrac{7.9-11.5}{3} = -1.2°/h$    (ii) $\dfrac{9.0-11.5}{2} = -1.25°/h$

       (iii) $\dfrac{10.2-11.5}{1} = -1.3°/h$

    (b) Approximately -1.6 °/h at 8 P.M.

17. (a) (i) $\dfrac{C(105)-C(100)}{5} = \dfrac{6601.25 - 6500}{5} = \$20.25/unit$

       (ii) $\dfrac{C(101)-C(100)}{1} = \dfrac{6520.05 - 6500}{1} = \$20.05/unit$

    (b) $\dfrac{C(100+h)-C(100)}{h} = \dfrac{5000+10(100+h)+0.05(100+h)^2-6500}{h} = 20+0.05h$

so as h approaches 0, the rate of change of C approaches $20/unit.

Exercises 2.2

1. 2                    3. 1                    5. 2

7. 0                    9. 1                    11. 1

13. $\lim\limits_{x \to 2} (x^2 - 2x + 7) = 2^2 - 2(2) + 7 = 7$

15. $\lim\limits_{x \to 12} \sqrt{x-3} = \sqrt{12-3} = \sqrt{9} = 3$

17. $\lim\limits_{x \to 1} \dfrac{x^2+1}{x^2-x+1} = \dfrac{1^2+1}{1^2-1+1} = 2$

19. $\lim\limits_{x \to 0} \dfrac{x+x^2}{x} = \lim\limits_{x \to 0} \dfrac{x(1+x)}{x} = \lim\limits_{x \to 0} (1+x) = 1$

21. $\lim\limits_{x \to 1} \dfrac{x^2-x-2}{x+1} = \dfrac{1^2-1-2}{1+1} = -1$

23. $\lim\limits_{x \to -1} \dfrac{x^2-x-3}{x+1}$ does not exist since as $x \to -1$, numerator $\to -1$ and

 denominator $\to 0$.

25. $\lim\limits_{t \to 1} \dfrac{t^3-t}{t^2-1} = \lim\limits_{t \to 1} \dfrac{t(t^2-1)}{t^2-1} = \lim\limits_{t \to 1} t = 1$

27. $\lim\limits_{h \to 0} \dfrac{(2+h)^3-8}{h} = \lim\limits_{h \to 0} \dfrac{(8+12h+6h^2+h^3)-8}{h} = \lim\limits_{h \to 0} \dfrac{12h+6h^2+h^3}{h}$

 $= \lim\limits_{h \to 0} (12+6h+h^2) = 12$

29. $\lim\limits_{h \to 0} \dfrac{(1+h)^4-1^4}{h} = \lim\limits_{h \to 0} \dfrac{(1+4h+6h^2+4h^3+h^4)-1}{h} = \lim\limits_{h \to 0} \dfrac{4h+6h^2+4h^3+h^4}{h}$

 $= \lim\limits_{h \to 0} (4+6h+4h^2+h^3) = 4$

31. $\lim\limits_{x \to -2} \dfrac{x+2}{x^2-x-6} = \lim\limits_{x \to -2} \dfrac{x+2}{(x-3)(x+2)} = \lim\limits_{x \to -2} \dfrac{1}{x-3} = -\dfrac{1}{5}$

33. $\lim\limits_{x \to -2} \dfrac{1}{x^4} = \dfrac{1}{(-2)^4} = \dfrac{1}{16}$

35. $\lim\limits_{x \to 3} \dfrac{1}{(x-3)^2}$ does not exist since $(x-3)^2 \to 0$ as $x \to 3$.

37. $\lim\limits_{x \to 3^+} \sqrt{x-3} = \sqrt{3-3} = 0$

26

**39.**　(a) $\lim\limits_{x \to 2^-} f(x) = \lim\limits_{x \to 2^-} (-1) = -1$　　(b) $\lim\limits_{x \to 2^+} f(x) = \lim\limits_{x \to 2^+} 1 = 1$

　　　(c) $\lim\limits_{x \to 2} f(x)$ does not exist since $\lim\limits_{x \to 2^-} f(x) \neq \lim\limits_{x \to 2^+} f(x)$

**41.**　(a) $\lim\limits_{x \to 1^-} h(x) = \lim\limits_{x \to 1^-} (x+2) = 3$　　(b) $\lim\limits_{x \to 1^+} h(x) = \lim\limits_{x \to 1^+} (1-x) = 0$

　　　(c) $\lim\limits_{x \to 1} h(x)$ does not exist since $\lim\limits_{x \to 1^-} h(x) \neq \lim\limits_{x \to 1^+} h(x)$

**43.**　$\lim\limits_{x \to 2} \dfrac{\frac{1}{x} - \frac{1}{2}}{x - 2} = \lim\limits_{x \to 2} \dfrac{2 - x}{2x(x-2)} = \lim\limits_{x \to 2} \dfrac{-1}{2x} = -\dfrac{1}{4}$

**45.** $f(x) = \dfrac{1 - \cos x}{x^2}$　　(a)

| ..x.. | ....f(x). |
|-------|-----------|
| 1 | 0.459698 |
| 0.5 | 0.489670 |
| 0.4 | 0.493369 |
| 0.3 | 0.496261 |
| 0.2 | 0.498336 |
| 0.1 | 0.499583 |
| 0.05 | 0.499896 |
| 0.01 | 0.499996 |

　　(b) It appears that $\lim\limits_{x \to 0} f(x) = 0.5$.

**47.** Let $h(x) = (1+x)^{1/x}$

| ........x.. | .h(x)... |
|-------------|----------|
| 1 | 2 |
| 0.1 | 2.59374 |
| 0.01 | 2.70481 |
| 0.001 | 2.71692 |
| 0.0001 | 2.71815 |
| 0.00001 | 2.71827 |
| 0.000001 | 2.71828 |
| 0.0000001 | 2.71828 |
| 0.00000001 | 2.71828 |
| 0.000000001 | 2.71828 |

It appears that $\lim\limits_{x \to 0} (1+x)^{1/x} \approx 2.71828$.

**Exercises 2.3**

1.  (a)  $|(6x + 1) - 19| < 0.1 \iff |6x - 18| < 0.1 \iff 6|x - 3| < 0.1$
    $\iff |x - 3| < (0.1)/6 = 1/60$

    (b)  $|(6x + 1) - 19| < 0.01 \iff |x - 3| < (0.01)/6 = 1/600$

3.  Given $\epsilon > 0$, we need $\delta > 0$ so
    that if $|x-2| < \delta$, then
    $|(3x-2) - 4| < \epsilon \iff |3x-6| < \epsilon$
    $\iff 3|x-2| < \epsilon \iff |x-2| < \epsilon/3$.
    So if we choose $\delta = \epsilon/3$, then
    $|x-2| < \delta \Rightarrow |(3x-2) - 4| < \epsilon$.

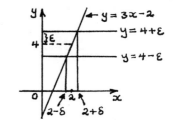

5.  Given $\epsilon > 0$, we need $\delta > 0$ so
    that if $|x-(-1)| < \delta$, then
    $|(5x+8) - 3| < \epsilon \iff |5x+5| < \epsilon$
    $\iff 5|x+1| < \epsilon \iff$
    $|x-(-1)| < \epsilon/5$.  So if we choose
    $\delta = \epsilon/5$, then $|x-(-1)| < \delta \Rightarrow$
    $|(5x+8) - 3| < \epsilon$.

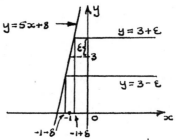

7.  Given $\epsilon > 0$, we need $\delta > 0$ so that if $|x - 2| < \delta$ then $\left|\frac{x}{7} - \frac{2}{7}\right| < \epsilon$
    $\iff \frac{1}{7}|x - 2| < \epsilon \iff |x - 2| < 7\epsilon$.  So take $\delta = 7\epsilon$.  Then
    $|x - 2| < \delta \Rightarrow \left|\frac{x}{7} - \frac{2}{7}\right| < \epsilon$.

9.  Given $\epsilon > 0$, we need $\delta > 0$ so that if $|x - (-5)| < \delta$ then
    $\left|(4 - \frac{3x}{5}) - 7\right| < \epsilon \iff \frac{3}{5}|x + 5| < \epsilon \iff |x - (-5)| < 5\epsilon/3$.  So
    take $\delta = 5\epsilon/3$.  Then $|x - (-5)| < \delta \Rightarrow \left|(4 - \frac{3x}{5}) - 7\right| < \epsilon$.

11. Given $\epsilon > 0$, we need $\delta > 0$ so that if $|x - a| < \delta$ then $|x - a| < \epsilon$.
    So obviously $\delta = \epsilon$ will work.

13. Given $\epsilon > 0$, we need $\delta > 0$ so that if $|x - 2| < \delta$ then $|c - c| < \epsilon$.
    But $|c - c| = 0$, so this will be true no matter what $\delta$ we pick.

15. Given $\epsilon > 0$, we need $\delta > 0$ so that if $|x| < \delta$ then $|x^2 - 0| < \epsilon \iff$
    $x^2 < \epsilon \iff |x| < \sqrt{\epsilon}$.  Take $\delta = \sqrt{\epsilon}$.  Then $|x-0| < \delta \Rightarrow |x^2-0| < \epsilon$.

17. Given $\epsilon > 0$, we need $\delta > 0$ so that if $|x - 0| < \delta$ then

$||x| - 0| < \epsilon$. But $||x|| = |x|$. So this is true if we pick $\delta = \epsilon$.

19. (a) Suppose that $\lim_{x \to 0}(1/x) = L$ for some L. Then, given $\epsilon = 1$, there exists $\delta$ such that $0 < |x| < \delta \Rightarrow |(1/x) - L| < 1$. Thus

$$0 < x < \delta \Rightarrow L - 1 < \frac{1}{x} < L + 1$$

This is impossible since $1/x$ takes on arbitrarily large values in any interval $(-\delta, \delta)$. In fact, $\frac{1}{x} > L + 1$ if $0 < x < \frac{1}{|L|+1}$

(b) Suppose that $\lim_{x \to -4}(x+4)^{-2} = L$ for some L. Then, given $\epsilon = 1$, there exists $\delta$ such that $0 < |x+4| < \delta \Rightarrow |(x+4)^{-2} - L| < 1$. Thus

$$0 < |x+4| < \delta \Rightarrow L - 1 < \frac{1}{(x+4)^2} < L + 1$$

This is impossible since $1/(x+4)^2$ takes on arbitrarily large values in any interval $(-4-\delta, -4+\delta)$. In fact, $(x+4)^{-2} > L + 1$ if $-4 < x < -4 + \frac{1}{\sqrt{L + 1}}$

21. Given $\epsilon > 0$, we need $\delta$ so that if $|x - 3| < \delta$, then $|x^2 - 9| < \epsilon$ $\Longleftrightarrow |x + 3||x - 3| < \epsilon$. To estimate $|x + 3|$ we take $|x - 3| < 1$. Then $|x| < 4$, so $|x + 3| < |x| + 3 < 4 + 3 = 7$. Thus $|x^2 - 9| = |x + 3||x - 3| < 7|x - 3|$. Take $\delta$ to be the smaller of 1 and $\epsilon/7$. If $|x - 3| < \delta$, then $|x - 3| < 1$ and $|x - 3| < \epsilon/7$, so $|x^2 - 9| < 7|x - 3| < 7(\epsilon/7) = \epsilon$.

23. Suppose that $\lim_{x \to 0} f(x) = L$. Given $\epsilon = 1/2$, there exists $\delta > 0$ such that $0 < |x| < \delta \Rightarrow |f(x) - L| < 1/2$. Take any rational number $r$ with $0 < |r| < \delta$. Then $f(r) = 0$, so $|0 - L| < \frac{1}{2}$, so $L \le |L| < \frac{1}{2}$. Now take any irrational number $s$ with $0 < |s| < \delta$. Then $f(s) = 1$, so $|1 - L| < \frac{1}{2}$. Hence $1 - L < \frac{1}{2}$, so $L > \frac{1}{2}$. This contradicts $L < \frac{1}{2}$. Thus $\lim_{x \to 0} f(x)$ does not exist.

## Exercises 2.4

1. $\displaystyle\lim_{x\to 4}(5x^2-2x+3) = \lim_{x\to 4}5x^2 - \lim_{x\to 4}2x + \lim_{x\to 4}3$      (Properties 2 & 1)

$$= 5\lim_{x\to 4}x^2 - 2\lim_{x\to 4}x + 3 \qquad (3\ \&\ 7)$$

$$= 5(4)^2 - 2(4) + 3 = 75 \qquad (9\ \&\ 8)$$

3. $\displaystyle\lim_{x\to 2}(x^2+1)(x^2+4x) = \lim_{x\to 2}(x^2+1)\ \lim_{x\to 2}(x^2+4x)$      (4)

$$= \left[\lim_{x\to 2}x^2 + \lim_{x\to 2}1\right]\left[\lim_{x\to 2}x^2 + 4\lim_{x\to 2}x\right] \quad (1\&3)$$

$$= [(2)^2 + 1]\ [(2)^2 + 4(2)] = 60 \qquad (9,7,\&\ 8)$$

5. $\displaystyle\lim_{x\to -1}\frac{x-2}{x^2+4x-3} = \frac{\displaystyle\lim_{x\to -1}(x-2)}{\displaystyle\lim_{x\to -1}(x^2+4x-3)}$      (5)

$$= \frac{\displaystyle\lim_{x\to -1}x - \lim_{x\to -1}2}{\displaystyle\lim_{x\to -1}x^2 + 4\lim_{x\to -1}x - \lim_{x\to -1}3} \qquad (2,1,\&\ 3)$$

$$= \frac{(-1) - 2}{(-1)^2 + 4(-1) - 3} = \frac{1}{2} \qquad (8,7,\&\ 9)$$

7. $\displaystyle\lim_{x\to -1}\sqrt{x^3+2x+7} = \sqrt{\lim_{x\to -1}(x^3+2x+7)}$      (11)

$$= \sqrt{\lim_{x\to -1}x^3 + 2\lim_{x\to -1}x + \lim_{x\to -1}7} \qquad (1\ \&\ 3)$$

$$= \sqrt{(-1)^3+2(-1)+7} = 2 \qquad (9,8,\&\ 6)$$

9. $\displaystyle\lim_{t\to -2}(t+1)^9(t^2-1) = \lim_{t\to -2}(t+1)^9\ \lim_{t\to -2}(t^2-1)$      (4)

$$= \left[\lim_{t\to -2}(t+1)\right]^9 \lim_{t\to -2}(t^2-1) \qquad (6)$$

$$= \left[\lim_{t\to -2}t + \lim_{t\to -2}1\right]^9\left[\lim_{t\to -2}t^2 - \lim_{t\to -2}1\right] \quad (1\ \&\ 2)$$

$$= [(-2) + 1]^9\ [(-2)^2 - 1] = -3 \qquad (8,7,\&\ 9)$$

11. $\lim\limits_{w \to -2} \sqrt[3]{\dfrac{4w + 3w^3}{3w + 10}} = \sqrt[3]{\lim\limits_{w \to -2} \dfrac{4w + 3w^3}{3w + 10}}$  (11)

$= \sqrt[3]{\dfrac{\lim\limits_{w \to -2} (4w + 3w^3)}{\lim\limits_{w \to -2} (3w + 10)}}$  (5)

$= \sqrt[3]{\dfrac{4 \lim\limits_{w \to -2} w + 3 \lim\limits_{w \to -2} w^3}{3 \lim\limits_{w \to -2} w + \lim\limits_{w \to -2} 10}}$  (1 & 3)

$= \sqrt[3]{\dfrac{4(-2) + 3(-2)^3}{3(-2) + 10}} = -2$  (8, 9, & 7)

13. $\lim\limits_{h \to 1/2} \dfrac{2h}{h + \dfrac{1}{h}} = \dfrac{\lim\limits_{h \to 1/2} 2h}{\lim\limits_{h \to 1/2} \left[ h + \dfrac{1}{h} \right]}$  (5)

$= \dfrac{2 \lim\limits_{h \to 1/2} h}{\lim\limits_{h \to 1/2} h + \dfrac{1}{\lim\limits_{h \to 1/2} h}}$  (3,1,5,&7)

$= \dfrac{2(1/2)}{\dfrac{1}{2} + \dfrac{1}{1/2}} = \dfrac{2}{5}$  (8)

15. (a) $\lim\limits_{x \to a} [f(x) + h(x)] = \lim\limits_{x \to a} f(x) + \lim\limits_{x \to a} h(x) = -3 + 8 = 5$

(b) $\lim\limits_{x \to a} [f(x)]^2 = \left[ \lim\limits_{x \to a} f(x) \right]^2 = (-3)^2 = 9$

(c) $\lim\limits_{x \to a} \sqrt[3]{h(x)} = \sqrt[3]{\lim\limits_{x \to a} h(x)} = \sqrt[3]{8} = 2$

(d) $\lim\limits_{x \to a} \dfrac{1}{f(x)} = \dfrac{1}{\lim\limits_{x \to a} f(x)} = \dfrac{1}{-3} = -\dfrac{1}{3}$

(e) $\lim\limits_{x \to a} \dfrac{f(x)}{h(x)} = \dfrac{\lim\limits_{x \to a} f(x)}{\lim\limits_{x \to a} h(x)} = \dfrac{-3}{8} = -\dfrac{3}{8}$

(f) $\lim\limits_{x \to a} \dfrac{g(x)}{f(x)} = \dfrac{\lim\limits_{x \to a} g(x)}{\lim\limits_{x \to a} f(x)} = \dfrac{0}{-3} = 0$

(g) Does not exist since $\lim\limits_{x \to a} g(x) = 0$ but $\lim\limits_{x \to a} f(x) \neq 0$.

(h) $\lim\limits_{x \to a} \dfrac{2f(x)}{h(x) - f(x)} = \dfrac{2\lim\limits_{x \to a} f(x)}{\lim\limits_{x \to a} h(x) - \lim\limits_{x \to a} f(x)} = \dfrac{2(-3)}{8 - (-3)} = -\dfrac{6}{11}$

17.  $\lim\limits_{x \to -3} \dfrac{x^2 - x + 12}{x + 3}$ does not exist since $x + 3 \to 0$ but $x^2 - x + 12 \to 24$

19.  $\lim\limits_{t \to 9} \dfrac{9 - t}{3 - \sqrt{t}} = \lim\limits_{t \to 9} \dfrac{(3 + \sqrt{t})(3 - \sqrt{t})}{3 - \sqrt{t}} = \lim\limits_{t \to 9} (3 + \sqrt{t}) = 3 + \sqrt{9} = 6$

21.  $\lim\limits_{t \to 1} \dfrac{\sqrt[3]{t}}{t - 1}$ does not exist since $t - 1 \to 0$ but $\sqrt[3]{t} \to 1$.

23.  $\lim\limits_{x \to -2} \left[ \dfrac{x^2}{x+2} + \dfrac{2x}{x+2} \right] = \lim\limits_{x \to -2} \dfrac{x^2 + 2x}{x+2} = \lim\limits_{x \to -2} \dfrac{x(x+2)}{x+2} = \lim\limits_{x \to -2} x = -2$

25.  Factor $x + 8$ as a sum of cubes:

$\lim\limits_{x \to -8} \dfrac{x+8}{\sqrt[3]{x}+2} = \lim\limits_{x \to -8} \dfrac{(\sqrt[3]{x}+2)(x^{2/3} - 2\sqrt[3]{x} + 4)}{\sqrt[3]{x}+2} = \lim\limits_{x \to -8} (x^{2/3} - 2\sqrt[3]{x} + 4)$

$= \lim\limits_{x \to -8} x^{2/3} - 2 \lim\limits_{x \to -8} \sqrt[3]{x} + \lim\limits_{x \to -8} 4 = (-8)^{2/3} - 2\sqrt[3]{-8} + 4 = 12$

27.  $\lim\limits_{x \to 9} \dfrac{x^2 - 81}{\sqrt{x} - 3} = \lim\limits_{x \to 9} \dfrac{(x-9)(x+9)}{\sqrt{x} - 3} = \lim\limits_{x \to 9} \dfrac{(\sqrt{x} - 3)(\sqrt{x} + 3)(x+9)}{\sqrt{x} - 3}$

$= \lim\limits_{x \to 9} (\sqrt{x} + 3)(x+9) = \lim\limits_{x \to 9} (\sqrt{x} + 3) \lim\limits_{x \to 9} (x+9) = (\sqrt{9} + 3)(9+9) = 108$

29.  $\lim\limits_{t \to 0} \left[ \dfrac{1}{t\sqrt{1+t}} - \dfrac{1}{t} \right] = \lim\limits_{t \to 0} \dfrac{1 - \sqrt{1+t}}{t\sqrt{1+t}} = \lim\limits_{t \to 0} \dfrac{(1 - \sqrt{1+t})(1 + \sqrt{1+t})}{t\sqrt{1+t}(1 + \sqrt{1+t})}$

$= \lim\limits_{t \to 0} \dfrac{-t}{t\sqrt{1+t}(1 + \sqrt{1+t})} = \lim\limits_{t \to 0} \dfrac{-1}{\sqrt{1+t}(1 + \sqrt{1+t})} = \dfrac{-1}{\sqrt{1+0}(1 + \sqrt{1+0})} = -\dfrac{1}{2}$

31.  $\lim\limits_{x \to 0} \dfrac{x}{\sqrt{1+3x} - 1} = \lim\limits_{x \to 0} \dfrac{x(\sqrt{1+3x} + 1)}{(\sqrt{1+3x} - 1)(\sqrt{1+3x} + 1)} = \lim\limits_{x \to 0} \dfrac{x(\sqrt{1+3x} + 1)}{3x}$

$= \lim\limits_{x \to 0} \dfrac{\sqrt{1+3x} + 1}{3} = \dfrac{\sqrt{1} + 1}{3} = \dfrac{2}{3}$

33.  $3x \le f(x) \le x^3 + 2$ for $0 \le x \le 2$, and $\lim\limits_{x \to 1} 3x = 3$ & $\lim\limits_{x \to 1} (x^3 + 2)$

$= \lim\limits_{x \to 1} x^3 + \lim\limits_{x \to 1} 2 = 1^3 + 2 = 3$. Therefore, by the Squeeze Theorem,

$\lim\limits_{x \to 1} f(x) = 3$.

35.  $-1 \le \sin \dfrac{1}{\sqrt[3]{x}} \le 1 \;\Rightarrow\; -\sqrt[3]{|x|} \le \sqrt[3]{x} \sin \dfrac{1}{\sqrt[3]{x}} \le \sqrt[3]{|x|}$. But $\lim\limits_{x \to 0} \sqrt[3]{|x|} =$

$\sqrt[3]{\lim_{x\to 0} |x|} = \sqrt[3]{0} = 0 = \lim_{x\to 0} - \sqrt[3]{|x|}$. Thus, by the Squeeze Theorem,

$\lim_{x\to 0} \sqrt[3]{x} \sin \dfrac{1}{\sqrt[3]{x}} = 0$.

37. $\lim_{x\to 5^+}(\sqrt{x-5} + \sqrt{5x}) = \sqrt{\lim_{x\to 5^+}(x-5)} + \sqrt{\lim_{x\to 5^+}5x} = \sqrt{5-5} + \sqrt{5 \cdot 5} = 5$

39. $\lim_{x\to -1.5^+}(\sqrt{3+2x} + x) = \sqrt{\lim_{x\to -1.5+}3 + 2\lim_{x\to -1.5^+}x} + \lim_{x\to -1.5^+}x$

$= \sqrt{3+2(-1.5)} - 1.5 = -1.5$

41. If $x < -4$, then $|x+4| = -(x+4)$, so

$\lim_{x\to -4^-} \dfrac{|x+4|}{x+4} = \lim_{x\to -4^-} \dfrac{-(x+4)}{x+4} = \lim_{x\to -4^-}(-1) = -1$

43. As $x \to 2^+$, $x-2 \to 0$, so $\dfrac{1}{x-2}$ becomes very large and $\lim_{x\to 2} \dfrac{1}{x-2}$ doesn't exist.

45. $[\![x]\!] = 9$ for $9 \le x < 10$, so $\lim_{x\to 9^+}[\![x]\!] = \lim_{x\to 9^+}9 = 9$.

47. $[\![x]\!] = -2$ for $-2 \le x < -1$, so $\lim_{x\to -2^+}[\![x]\!] = \lim_{x\to -2^+}(-2) = -2$.

$[\![x]\!] = -3$ for $-3 \le x < -2$, so $\lim_{x\to -2^-}[\![x]\!] = \lim_{x\to -2^-}(-3) = -3$.

The right and left limits are different, so $\lim_{x\to -2}[\![x]\!]$ doesn't exist.

49. $\lim_{x\to 8^+}(\sqrt{x-8} + [\![x+1]\!]) = \lim_{x\to 8^+}\sqrt{x-8} + \lim_{x\to 8^+}[\![x+1]\!] = 0 + 9 = 9$ because

$[\![x+1]\!] = 9$ for $8 \le x < 9$.

51. $\lim_{x\to -2^-}\sqrt{x^2+x-2} = \sqrt{\lim_{x\to -2^-}x^2 + \lim_{x\to -2^-}x - \lim_{x\to -2^-}2} = \sqrt{(-2)^2-2-2} = 0$.

[Notice that the domain of $\sqrt{x^2+x-2}$ is $(-\infty,-2] \cup [1,\infty)$.]

53. $\lim_{x\to 5^+} x \sqrt[6]{x^2-25} = \lim_{x\to 5^+}x \sqrt[6]{\lim_{x\to 5^+}x^2 - \lim_{x\to 5^+}25} = 5 \sqrt[6]{5^2-25} = 0$

55. Since $|x| = x$ for $x > 0$, we have

$\lim_{x\to 0^+}\left[\dfrac{1}{x} - \dfrac{1}{|x|}\right] = \lim_{x\to 0^+}\left[\dfrac{1}{x} - \dfrac{1}{x}\right] = \lim_{x\to 0^+}0 = 0$.

57. $H(t) = 1$ for $t \ge 0$, so $\lim_{t\to 0^+}H(t) = \lim_{t\to 0^+}1 = 1$. $H(t) = 0$ for $t < 0$,

so $\lim_{t\to 0^-}H(t) = \lim_{t\to 0^-}0 = 0$. By Theorem 2.18, $\lim_{t\to 0}H(t)$ doesn't exist.

59. (a) $\lim\limits_{x \to -1^-} g(x) = \lim\limits_{x \to -1^-} (-x^3) = -(-1)^3 = 1$    (c)

$\lim\limits_{x \to -1^+} g(x) = \lim\limits_{x \to -1^+} (x+2)^2 = (-1+2)^2 = 1$

(b) Therefore $\lim\limits_{x \to -1} g(x) = 1$.

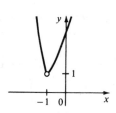

61. (a)(i) $[\![x]\!] = n-1$ for $n-1 \leq x < n$, so $\lim\limits_{x \to n^-} [\![x]\!] = \lim\limits_{x \to n^-} (n-1) = n-1$.

    (ii) $[\![x]\!] = n$ for $n \leq x < n+1$, so $\lim\limits_{x \to n^+} [\![x]\!] = \lim\limits_{x \to n^+} n = n$.

(b) $\lim\limits_{x \to a} [\![x]\!]$ exists $\iff$ $a$ is not an integer.

63. (a)(i) $\lim\limits_{x \to 1^+} \dfrac{x^2-1}{|x-1|} = \lim\limits_{x \to 1^+} \dfrac{x^2-1}{x-1} = \lim\limits_{x \to 1^+} (x+1) = 2$

    (ii) $\lim\limits_{x \to 1^-} \dfrac{x^2-1}{|x-1|} = \lim\limits_{x \to 1^-} \dfrac{x^2-1}{-(x-1)} = \lim\limits_{x \to 1^-} -(x+1) = -2$

(b) No, since $\lim\limits_{x \to 1^+} F(x) \neq \lim\limits_{x \to 1^-} F(x)$.

(c)

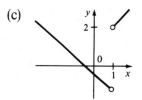

65. Observe that $0 \leq f(x) \leq x^2$ for all $x$, and $\lim\limits_{x \to 0} 0 = 0 = \lim\limits_{x \to 0} x^2$. So, by the Squeeze Theorem, $\lim\limits_{x \to 0} f(x) = 0$.

67. Let $f(x) = H(x)$ and $g(x) = 1 - H(x)$, where $H$ is the Heaviside function defined in Example 9 in Section 2.2. Then $\lim\limits_{x \to 0} f(x)$ and $\lim\limits_{x \to 0} g(x)$ do not exist but $\lim\limits_{x \to 0} [f(x)g(x)] = \lim\limits_{x \to 0} 0 = 0$.

69. Let $t = \sqrt[6]{x}$. Then $t \to 1$ as $x \to 1$, so $\lim\limits_{x \to 1} \dfrac{\sqrt[3]{x}-1}{\sqrt{x}-1} = \lim\limits_{t \to 1} \dfrac{t^2-1}{t^3-1}$

$= \lim\limits_{t \to 1} \dfrac{(t-1)(t+1)}{(t-1)(t^2+t+1)} = \lim\limits_{t \to 1} \dfrac{t+1}{t^2+t+1} = \dfrac{1+1}{1^2+1+1} = \dfrac{2}{3}$

[Another method: multiply numerator and denominator by $\sqrt{x} + 1$.]

**Exercises 2.5**

1.  (a) $-5$, $-3$, $-1$, $3$, $5$, $8$, $10$

    (b) $-5$: from the left; $-3$: from the left; $-1$: neither; $3$: neither; $5$: neither; $8$: from the right; $10$: neither.

3.  $\displaystyle\lim_{x\to5} f(x) = \lim_{x\to5}\left[1 + \sqrt{x^2-9}\,\right] = \lim_{x\to5} 1 + \sqrt{\lim_{x\to5} x^2 - \lim_{x\to5} 9} = 1 + \sqrt{5^2-9}$

    $= 5 = f(5)$   Thus $f$ is continuous at $5$.

5.  $\displaystyle\lim_{t\to-8} g(t) = \lim_{t\to-8}\frac{\sqrt[3]{t}}{(t+1)^4} = \frac{\sqrt[3]{\lim\limits_{t\to-8} t}}{\left[\lim\limits_{t\to-8} t + 1\right]^4} = \frac{\sqrt[3]{-8}}{(-8+1)^4} = -\frac{2}{2401} = g(-8)$

    Thus $g$ is continuous at $-8$.

7.  For $-4 < a < 4$ we have $\displaystyle\lim_{x\to a} f(x) = \lim_{x\to a} x\sqrt{16-x^2}$

    $= \displaystyle\lim_{x\to a} x \cdot \sqrt{\lim_{x\to a} 16 - \lim_{x\to a} x^2} = a\sqrt{16-a^2} = f(a)$, so $f$ is continuous on

    $(-4,4)$.  Similarly, we get $\displaystyle\lim_{x\to4^-} f(x) = 0 = f(4)$ and $\displaystyle\lim_{x\to-4^+} f(x) = 0$

    $= f(-4)$, so $f$ is continuous from the left at $4$ and from the right at $-4$.  Thus $f$ is continuous on $[-4,4]$.

9.  For any $a \in \mathbb{R}$ we have $\displaystyle\lim_{x\to a} f(x) = \lim_{x\to a} (x^2-1)^8 = \left[\lim_{x\to a} x^2 - \lim_{x\to a} 1\right]^8$

    $= (a^2-1)^8 = f(a)$.  Thus $f$ is continuous on $(-\infty,\infty)$.

11. $f(x) = \dfrac{3x^2-5x-2}{x-2}$ is discontinuous at $2$

    since $f(2)$ is not defined.

    For the graph, note that
    $f(x) = \dfrac{(3x+1)(x-2)}{x-2} = 3x+1,\ x \neq 2$

    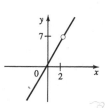

13. $f(x) = -\dfrac{1}{(x-1)^2}$ is discontinuous at $1$

    since $f(1)$ is not defined.

    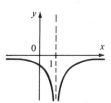

15. $\lim\limits_{x \to 1} f(x) = \lim\limits_{x \to 1} - \dfrac{1}{(x-1)^2}$ does not exist. Therefore f is discontinuous at 1.

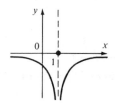

17. Since $f(x) = x^2 - 2$ for $x \neq -3$,
$\lim\limits_{x \to -3} f(x) = \lim\limits_{x \to -3} (x^2 - 2) = (-3)^2 - 2 = 7$
But $f(-3) = 5$, so $\lim\limits_{x \to -3} f(x) \neq f(-3)$.
Therefore f is discontinuous at -3.

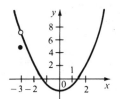

19. $f(x) = (x+1)(x^3 + 8x + 9)$ is a polynomial, so by Theorem 2.23 it is continuous on R.

21. $h(x) = \dfrac{x^2 + 2x - 1}{x+1}$ is a rational function, so by Theorem 2.23 it is continuous on its domain, which is $\{x \mid x \neq -1\}$.

23. $g(x) = x+1$ is continuous (since it is a polynomial) and $f(x) = \sqrt{x}$ is continuous on $[0, \infty)$ by Theorem 2.25, so $f(g(x)) = \sqrt{x+1}$ is continuous on $[-1, \infty)$ by Theorem 2.27. By Theorem 2.22(e), $H(x) = 1/\sqrt{x+1}$ is continuous on $(-1, \infty)$.

25. $g(x) = x-1$ and $G(x) = x^2 - 2$ are polynomials, so by Theorem 2.23 they are continuous. Also $f(x) = \sqrt[5]{x}$ is continuous by Theorem 2.25, so $f(g(x)) = \sqrt[5]{x-1}$ is continuous on R by Theorem 2.27. Thus the product $h(x) = \sqrt[5]{x-1}(x^2 - 2)$ is continuous on R by Theorem 2.22(d).

27. Since the discriminant of $t^2 + t + 1$ is negative, $t^2 + t + 1$ is always positive. So the domain of $F(t)$ is R. By Theorem 2.23 the polynomial $(t^2 + t + 1)^3$ is continuous. By Theorems 2.25 and 2.27 the composition $F(t) = \sqrt{(t^2 + t + 1)^3}$ is continuous.

29. $H(x) = \sqrt{(x-2)/(5+x)}$. The domain is $\{x \mid (x-2)/(5+x) > 0\} = (-\infty, -5) \cup [2, \infty)$ by the methods of Section 1.1. By Theorem 2.23 the rational function $(x-2)/(5+x)$ is continuous. Since the square root function is continuous (Theorem 2.25), the composition $H(x) = \sqrt{(x-2)/(5+x)}$ is continuous by Theorem 2.27.

31.  f is continuous on $(-\infty,3)$ and $(3,\infty)$ since on each of these intervals it is a polynomial. Also $\lim_{x\to3^+}f(x) = \lim_{x\to3^+}(5-x) = 2$ and $\lim_{x\to3^-}f(x) = \lim_{x\to3^-}(x-1) = 2$, so $\lim_{x\to3}f(x) = 2$. Since $f(3) = 5-3 = 2$, f is also continuous at 3. Thus f is continuous on $(-\infty,\infty)$.

33.  f is continuous on $(-\infty,0)$ and $(0,\infty)$ since on each of these intervals it is a polynomial. Now $\lim_{x\to0^-}f(x) = \lim_{x\to0^-}(x-1)^3 = -1$ and $\lim_{x\to0^+}f(x) = \lim_{x\to0^+}(x+1)^3 = 1$. Thus $\lim_{x\to0}f(x)$ does not exist, so f is discontinuous at 0. Since $f(0) = 1$, f is continuous from the right at 0.

**33.**

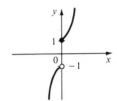

**35.**

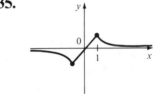

35.  f is continuous on $(-\infty,-1)$, $(-1,1)$, and $(1,\infty)$. Now $\lim_{x\to-1^-}f(x) = \lim_{x\to-1^-}\frac{1}{x} = -1$ and $\lim_{x\to-1^+}f(x) = \lim_{x\to-1^+}x = -1$, so $\lim_{x\to-1}f(x) = -1 = f(-1)$ and f is continuous at -1. Also $\lim_{x\to1^-}f(x) = \lim_{x\to1^-}x = 1$ and $\lim_{x\to1^+}f(x) = \lim_{x\to1^+}\frac{1}{x^2} = 1$, so $\lim_{x\to1}f(x) = 1 = f(1)$ and f is continuous at 1. Thus f has no discontinuities.

37. $f(x) = H(x) - H(x-1) = \begin{cases} 0 & \text{if } x < 0 \\ 1 & \text{if } 0 \le x < 1 \\ 0 & \text{if } x \ge 1 \end{cases}$

f is continuous on $(-\infty,0)$, $(0,1)$, and $(1,\infty)$.

$\lim\limits_{x\to 0^-} f(x) = \lim\limits_{x\to 0^-} 0 = 0$ and $\lim\limits_{x\to 0^+} f(x) = \lim\limits_{x\to 0^+} 1 = 1$ Since $f(0) = 1$, f

is continuous only from the right at 0. Also $\lim\limits_{x\to 1^-} f(x) = 1$ and

$\lim\limits_{x\to 1^+} f(x) = 0$, so f is continuous only from the right at 1.

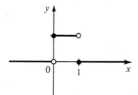

39. $f(x) = [2x]$ is continuous except when $2x = n \iff x = n/2$, n an

integer. In fact, $\lim\limits_{x\to n/2^-} [2x] = n-1$ and $\lim\limits_{x\to n/2^+} [2x] = n = f(n)$, so f

is continuous only from the right at n/2.

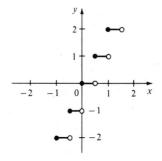

41. f is continuous on $(-\infty,3)$ and $(3,\infty)$. Now $\lim\limits_{x\to 3^-} f(x) = \lim\limits_{x\to 3^-} (cx+1)$

$= 3c+1$ and $\lim\limits_{x\to 3^+} f(x) = \lim\limits_{x\to 3^+} (cx^2-1) = 9c-1$. So f is continuous $\iff$

$3c+1 = 9c-1 \iff 6c = 2 \iff c = 1/3$.

**43.** (a) $\lim\limits_{x \to 1^-} f(x) = \lim\limits_{x \to 1^-} (1-x^2) = 0$ and $\lim\limits_{x \to 1^+} f(x) = \lim\limits_{x \to 1^+} (1+x/2) = 3/2$.

Thus $\lim\limits_{x \to 1} f(x)$ does not exist, so f is not continuous at 1.

(b) $f(0) = 1$ and $f(2) = 2$. For $0 \le x \le 1$, f takes the values in $[0,1]$. For $1 < x \le 2$, f takes the values in $(1.5,2]$. Thus f does not take on the value 1.5 (or any other value in $(1,1.5]$).

**45.** $f(x) = x^3 - x^2 + x$ is continuous on $[2,3]$ and $f(2) = 6$, $f(3) = 21$. Since $6 < 10 < 21$, there is a number c in $(2,3)$ such that $f(c) = 10$ by the Intermediate Value Theorem.

**47.** $f(x) = x^3 - 3x + 1$ is continuous on $[0,1]$ and $f(0) = 1$, $f(1) = -1$. Since $-1 < 0 < 1$, there is a number c in $(0,1)$ such that $f(c) = 0$ by the Intermediate Value Theorem. Thus there is a root of the equation $x^3 - 3x + 1 = 0$ in the interval $(0,1)$.

**49.** $f(x) = x^4 - 3x^3 - 2x^2 - 1$ is continuous on $[3,4]$ and $f(3) = -19$, $f(4) = 31$. Since $-19 < 0 < 31$, there is a number c in $(3,4)$ such that $f(c) = 0$ by the Intermediate Value Theorem. Thus there is a root of the equation $x^4 - 3x^3 - 2x^2 - 1 = 0$ in the interval $(3,4)$.

**51.** $f(x) = x^3 + 2x - (x^2 + 1) = x^3 + 2x - x^2 - 1$ is continuous on $[0,1]$ and $f(0) = -1$, $f(1) = 1$. Since $-1 < 0 < 1$, there is a number c in $(0,1)$ such that $f(c) = 0$ by the Intermediate Value Theorem. Thus there is a root of the equation $x^3 + 2x - x^2 - 1 = 0$, or equivalently, $x^3 + 2x = x^2 + 1$ in the interval $(0,1)$.

**53.** (a) $f(x) = x^3 - x + 1$ is continuous on $[-2,-1]$ and $f(-2) = -5$, $f(-1) = 1$. Since $-5 < 0 < 1$, there is a number c in $(-2,-1)$ such that $f(c) = 0$ by the Intermediate Value Theorem. Thus there is a root of the equation $x^3 - x + 1 = 0$ in the interval $(-2,-1)$.

(b) $f(-1.33) \approx -0.0226$ and $f(-1.32) \approx 0.0200$, so there is a root between $-1.33$ and $-1.32$.

**55.** $f(x) = \begin{cases} 0 \text{ if } x \text{ is rational} \\ 1 \text{ if } x \text{ is irrational} \end{cases}$ is continuous nowhere. For, given any number a and any $\delta > 0$, the interval $(a-\delta, a+\delta)$ contains both infinitely many rational and infinitely many irrational numbers. Since $f(a) = 0$ or 1, there are infinitely many numbers x with $|x-a| < \delta$ and $|f(x) - f(a)| = 1$. Thus $\lim\limits_{x \to a} f(x) \ne f(a)$. (In fact,

$\lim\limits_{x \to a} f(x)$ does not even exist.)

57.  ($\Rightarrow$) If $f$ is continuous at $a$, then by Theorem 2.26 with $g(h) = a+h$, we have $\lim\limits_{h \to 0} f(a+h) = f\left[\lim\limits_{h \to 0} (a+h)\right] = f(a)$.

($\Leftarrow$) Let $\epsilon > 0$. Since $\lim\limits_{h \to 0} f(a+h) = f(a)$, there exists $\delta > 0$ such that $|h| < \delta \Rightarrow |f(a+h) - f(a)| < \epsilon$. So if $|x-a| < \delta$, then $|f(x) - f(a)| = |f(a+(x-a)) - f(a)| < \epsilon$. Thus $\lim\limits_{x \to a} f(x) = f(a)$ and so $f$ is continuous at $a$.

# Chapter 2 Review

## Review Exercises for Chapter 2

1.  (a) The slope of the tangent line at $(2,1)$ is

$$\lim_{x \to 2} \frac{f(x)-f(2)}{x-2} = \lim_{x \to 2} \frac{9-2x^2-1}{x-2} = \lim_{x \to 2} \frac{8-2x^2}{x-2} = \lim_{x \to 2} \frac{-2(x^2-4)}{x-2} =$$

$$\lim_{x \to 2} \frac{-2(x-2)(x+2)}{x-2} = \lim_{x \to 2} -2(x+2) = -8$$

(b) The equation of this tangent line is $y-1 = -8(x-2)$ or $8x+y = 17$

3.  $\lim\limits_{x \to 4} \sqrt{x+\sqrt{x}} = \sqrt{4+\sqrt{4}} = \sqrt{6}$ since the function is continuous.

5.  $\lim\limits_{t \to -1} \frac{t+1}{t^3-t} = \lim\limits_{t \to -1} \frac{t+1}{t(t+1)(t-1)} = \lim\limits_{t \to -1} \frac{1}{t(t-1)} = \frac{1}{(-1)(-2)} = \frac{1}{2}$

7.  $\lim\limits_{h \to 0} \frac{(1+h)^2 - 1}{h} = \lim\limits_{h \to 0} \frac{1+2h+h^2-1}{h} = \lim\limits_{h \to 0} \frac{2h+h^2}{h} = \lim\limits_{h \to 0} (2+h) = 2$

9.  $\lim\limits_{x \to -1} \frac{x^2-x-2}{x^2+3x-2} = \frac{(-1)^2-(-1)-2}{(-1)^2+3(-1)-2} = \frac{0}{-4} = 0$

11.  $\lim\limits_{s \to 16} \frac{4-\sqrt{s}}{s-16} = \lim\limits_{s \to 16} \frac{4-\sqrt{s}}{(\sqrt{s}+4)(\sqrt{s}-4)} = \lim\limits_{s \to 16} \frac{-1}{\sqrt{s}+4} = \frac{-1}{\sqrt{16}+4} = -\frac{1}{8}$

13.  $\lim\limits_{x \to 8^-} \frac{|x-8|}{x-8} = \lim\limits_{x \to 8^-} \frac{-(x-8)}{x-8} = \lim\limits_{x \to 8^-} (-1) = -1$

15.  Given $\epsilon > 0$, we need $\delta > 0$ so that if $|x-5| < \delta$ then $|(7x-27) - 8| < \epsilon \Longleftrightarrow |7x-35| < \epsilon \Longleftrightarrow |x-5| < \epsilon/7$. So take $\delta = \epsilon/7$. Then $|x-5| < \delta \Rightarrow |(7x-27) - 8| < \epsilon$.

17.   (a) $f(x) = \sqrt{-x}$ if $x < 0$, $f(x) = 3-x$ if $0 \le x < 3$, $f(x) = (x-3)^2$ if $x > 3$.   Therefore

(i) $\lim\limits_{x \to 0^+} f(x) = \lim\limits_{x \to 0^+} (3-x) = 3$     (ii) $\lim\limits_{x \to 0^-} f(x) = \lim\limits_{x \to 0^-} \sqrt{-x} = 0$

(iii) Because of (i) and (ii), $\lim\limits_{x \to 0} f(x)$ does not exist.

(iv) $\lim\limits_{x \to 3^-} f(x) = \lim\limits_{x \to 3^-} (3-x) = 0$     (v) $\lim\limits_{x \to 3^+} f(x) = \lim\limits_{x \to 3^+} (3-x)^2 = 0$

(vi) Because of (iv) and (v), $\lim\limits_{x \to 3} f(x) = 0$.

(b) f is discontinuous at 0          (c)
since $\lim\limits_{x \to 0} f(x)$ does not exist.

f is discontinuous at 3 since
f(3) does not exist.

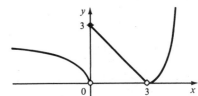

19.   $f(x) = \dfrac{x+1}{x^2+x+1}$ is rational so it is continuous on its domain which

is R.   (Note that $x^2+x+1 = 0$ has no real roots.)

21.   $f(x) = 2x^3+x^2+2$ is a polynomial, so it is continuous on $[-2,-1]$ and
$f(-2) = -10 < 0 < 1 = f(-1)$.   So by the Intermediate Value Theorem
there is a number c in $(-2,-1)$ such that $f(c) = 0$, i.e., the
equation $2x^3+x^2+2 = 0$ has a root in $(-2,-1)$.

## CHAPTER THREE

## Exercises 3.1

1.  slope = $f'(-3) = \lim\limits_{h\to 0} \dfrac{f(-3+h)-f(-3)}{h} = \lim\limits_{h\to 0} \dfrac{(-3+h)^2+2(-3+h) - 3}{h}$

    $= \lim\limits_{h\to 0} \dfrac{9-6h+h^2-6+2h-3}{h} = \lim\limits_{h\to 0} \dfrac{h^2-4h}{h} = \lim\limits_{h\to 0} (h-4) = -4.$  So the equation

    of the tangent line at $(-3,3)$ is $y-3 = (-4)(x+3)$ or $4x+y+9 = 0$.

**1.**

**3.**

3.  slope = $f'(2) = \lim\limits_{h\to 0} \dfrac{f(2+h)-f(2)}{h} = \lim\limits_{h\to 0} \dfrac{\sqrt{2+h+7}-3}{h}$

    $= \lim\limits_{h\to 0} \dfrac{(\sqrt{9+h}-3)(\sqrt{9+h}+3)}{h(\sqrt{9+h}+3)} = \lim\limits_{h\to 0} \dfrac{9+h-9}{h(\sqrt{9+h}+3)} = \lim\limits_{h\to 0} \dfrac{1}{\sqrt{9+h}+3} = \dfrac{1}{\sqrt{9}+3} = \dfrac{1}{6}$

    So the equation of the tangent line at $(2,3)$ is $y-3 = \dfrac{1}{6}(x-2)$ or

    $x-6y+16 = 0$.

5.  slope = $f'(-2) = \lim\limits_{h\to 0} \dfrac{f(-2+h)-f(-2)}{h} = \lim\limits_{h\to 0} \dfrac{1-2(-2+h)-3(-2+h)^2-(-7)}{h}$

    $= \lim\limits_{h\to 0} \dfrac{10h-3h^2}{h} = \lim\limits_{h\to 0} (10-3h) = 10.$  So the equation of the tangent

    line at $(-2,-7)$ is $y+7 = 10(x+2)$ or $y = 10x+13$.

7.  slope = $g'(1) = \lim\limits_{h\to 0} \dfrac{g(1+h)-g(1)}{h} = \lim\limits_{h\to 0} \dfrac{\frac{2}{1+h+3} - \frac{1}{2}}{h} = \lim\limits_{h\to 0} \dfrac{4-(4+h)}{2h(4+h)}$

    $= \lim\limits_{h\to 0} \dfrac{-h}{2h(4+h)} = \lim\limits_{h\to 0} \dfrac{-1}{2(4+h)} = -\dfrac{1}{8}$  So the equation of the tangent

    line at $(1,\frac{1}{2})$ is $y-\dfrac{1}{2} = -\dfrac{1}{8}(x-1)$ or $x+8y-5 = 0$.

9. Average velocity over $[5, 5+h]$ is $\dfrac{f(5+h)-f(5)}{h} = \dfrac{(5+h)^2-4(5+h)-5}{h}$

$= \dfrac{h^2+6h}{h} = h+6$. So the average velocity over $[5,6]$ is $1+6 = 7$ m/s, over $[5,5.1]$ is $0.1+6 = 6.1$ m/s, over $[5,5.01]$ is $0.01+6 = 6.01$ m/s Instantaneous velocity when $t = 5$ is $f'(5) = \lim\limits_{h \to 0} \dfrac{f(5+h)-f(5)}{h}$

$= \lim\limits_{h \to 0} (h+6) = 6$ m/s.

11. Average velocity over $[1, 1+h]$ is $\dfrac{f(1+h)-f(1)}{h} = \dfrac{(1+h)^3-1}{h} = \dfrac{3h+3h^2+h^3}{h}$

$= 3+3h+h^2$. So the average velocity over $[1,1.1]$ is $3+3(.1)+(.1)^2$ $= 3.31$ m/s, over $[1,1.01]$ is $3+3(.01)+(.01)^2 = 3.0301$ m/s, over $[1,1.001]$ is $3+3(.001)+(.001)^2 = 3.003001$ m/s. Instantaneous velocity when $t = 1$ is $f'(1) = \lim\limits_{h \to 0} \dfrac{f(1+h)-f(1)}{h} = \lim\limits_{h \to 0} (3+3h+h^2)$

$= 3$ m/s

13. The velocity at time $t$ is $h'(t) = \lim\limits_{x \to t} \dfrac{h(x)-h(t)}{x-t} =$

$\lim\limits_{x \to t} \dfrac{550-4.9x^2-(550-4.9t^2)}{x-t} = \lim\limits_{x \to t} \dfrac{4.9(t^2-x^2)}{x-t} = \lim\limits_{x \to t} \dfrac{-4.9(x-t)(x+t)}{x-t}$

$= \lim\limits_{x \to t} -4.9(x+t) = -9.8t$. So the speed is $9.8t$ after $t$ seconds.

(a) $9.8(1) = 9.8$ m/s after 1 s. (b) $9.8(2) = 19.6$ m/s after 2s.
(c) $9.8(5) = 49$ m/s after 5 s. (d) The ball hits the ground when $550-4.9t^2 = 0 \iff t = \sqrt{550/4.9}$. The speed is then $9.8\sqrt{550/4.9}$ $= 2\sqrt{2695} \approx 104$ m/s.

**15.** (a) $v(a) = \lim\limits_{h \to 0} \dfrac{f(a+h)-f(a)}{h} = \lim\limits_{h \to 0} \dfrac{(a+h)^2-6(a+h)+9-(a^2-6a+9)}{h}$

$= \lim\limits_{h \to 0} \dfrac{2ah+h^2-6h}{h} = \lim\limits_{h \to 0} (2a+h-6) = 2a-6$

So $v(1) = 2(1)-6 = -4$ ft/s, $\quad v(2) = 2(2)-6 = -2$ ft/s

(b) It is at rest when $v(t) = 2t-6 = 0 \iff t = 3$.

(c) It moves in the positive direction when $2t-6 > 0 \iff t > 3$.

(d) Distance in positive direction $= |f(4)-f(3)| = |1-0| = 1$ ft.

Distance in negative direction $= |f(3)-f(0)| = |0-9| = 9$ ft

Total distance traveled $= 1+9 = 10$ ft.

(e)

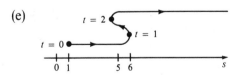

**17.** (a) $v(a) = \lim\limits_{h \to 0} \dfrac{f(a+h)-f(a)}{h}$

$= \lim\limits_{h \to 0} \dfrac{2(a+h)^3-9(a+h)^2+12(a+h)+1-(2a^3-9a^2+12a+1)}{h}$

$= \lim\limits_{h \to 0} \dfrac{6a^2h+6ah^2+2h^3-18ah-9h^2+12h}{h} = \lim\limits_{h \to 0} (6a^2+6ah+2h^2-18a-9h+12)$

$= 6a^2-18a+12 \quad$ So $v(1) = 6(1)^2-18(1)+12 = 0$ ft/s and

$v(2) = 6(2)^2-18(2)+12 = 0$ ft/s.

(b) It is at rest when $v(t) = 6t^2-18t+12 = 6(t-1)(t-2) = 0 \iff$

$t = 1$ or $2$

(c) It moves in the positive direction when $6(t-1)(t-2) > 0 \iff$

$0 \le t < 1$ or $t > 2$.

(d) Distance in positive direction $= |f(4)-f(2)|+|f(1)-f(0)|$

$= |33-5|+|6-1| = 33$ ft. Distance in negative direction $=$

$|f(2)-f(1)| = |5-6| = 1$ ft. Total distance traveled $= 33+1= 34$ ft.

(e)

19. $f'(a) = \lim\limits_{h \to 0} \dfrac{f(a+h)-f(a)}{h} = \lim\limits_{h \to 0} \dfrac{\dfrac{a+h}{2(a+h)-1} - \dfrac{a}{2a-1}}{h}$

$= \lim\limits_{h \to 0} \dfrac{(a+h)(2a-1)-a(2a+2h-1)}{h(2a+2h-1)(2a-1)} = \lim\limits_{h \to 0} \dfrac{-h}{h(2a+2h-1)(2a-1)}$

$= \lim\limits_{h \to 0} \dfrac{-1}{(2a+2h-1)(2a-1)} = -\dfrac{1}{(2a-1)^2}$

21. $f'(a) = \lim\limits_{h \to 0} \dfrac{f(a+h)-f(a)}{h} = \lim\limits_{h \to 0} \dfrac{\dfrac{2}{\sqrt{3-(a+h)}} - \dfrac{2}{\sqrt{3-a}}}{h} = \lim\limits_{h \to 0} \dfrac{2(\sqrt{3-a}-\sqrt{3-a-h})}{h\sqrt{3-a-h}\,\sqrt{3-a}}$

$= \lim\limits_{h \to 0} \dfrac{2(\sqrt{3-a}-\sqrt{3-a-h})}{h\sqrt{3-a-h}\,\sqrt{3-a}} \cdot \dfrac{(\sqrt{3-a}+\sqrt{3-a-h})}{(\sqrt{3-a}+\sqrt{3-a-h})} = \lim\limits_{h \to 0} \dfrac{2[3-a-(3-a-h)]}{h\sqrt{3-a-h}\,\sqrt{3-a}(\sqrt{3-a}+\sqrt{3-a-h})}$

$= \lim\limits_{h \to 0} \dfrac{2}{\sqrt{3-a-h}\,\sqrt{3-a}\,(\sqrt{3-a}+\sqrt{3-a-h})} = \dfrac{2}{\sqrt{3-a}\,\sqrt{3-a}\,(2\sqrt{3-a})} = \dfrac{1}{(3-a)^{3/2}}$

23. (a) $f'(a) = \lim\limits_{x \to a} \dfrac{f(x)-f(a)}{x-a} = \lim\limits_{x \to a} \dfrac{\sqrt[3]{x}-\sqrt[3]{a}}{x-a} = \lim\limits_{x \to a} \dfrac{\sqrt[3]{x}-\sqrt[3]{a}}{(\sqrt[3]{x})^3-(\sqrt[3]{a})^3}$

$= \lim\limits_{x \to a} \dfrac{\sqrt[3]{x}-\sqrt[3]{a}}{(\sqrt[3]{x}-\sqrt[3]{a})((\sqrt[3]{x})^2+\sqrt[3]{x}\cdot\sqrt[3]{a}+(\sqrt[3]{a})^2)} = \lim\limits_{x \to a} \dfrac{1}{x^{2/3}+x^{1/3}a^{1/3}+a^{2/3}}$

$= \dfrac{1}{a^{2/3}+a^{1/3}a^{1/3}+a^{2/3}} = \dfrac{1}{3a^{2/3}}$

(b) $f'(a) = \lim\limits_{h \to 0} \dfrac{f(a+h)-f(a)}{h} = \lim\limits_{h \to 0} \dfrac{\sqrt[3]{a+h}-\sqrt[3]{a}}{h} = \lim\limits_{h \to 0} \dfrac{\sqrt[3]{a+h}-\sqrt[3]{a}}{(a+h)-a}$

$= \lim\limits_{h \to 0} \dfrac{\sqrt[3]{a+h}-\sqrt[3]{a}}{(\sqrt[3]{a+h}-\sqrt[3]{a})((\sqrt[3]{a+h})^2+\sqrt[3]{a+h}\sqrt[3]{a}+(\sqrt[3]{a})^2)}$

$= \lim\limits_{h \to 0} \dfrac{1}{(a+h)^{2/3}+(a+h)^{1/3}a^{1/3}+a^{1/3}} = \dfrac{1}{a^{2/3}+a^{1/3}a^{1/3}+a^{2/3}} = \dfrac{1}{3a^{2/3}}$

25. $\lim\limits_{h \to 0} \dfrac{\sqrt{1+h}-1}{h} = f'(1)$ where $f(x) = \sqrt{x}$ (Or $f'(0)$ where $f(x) = \sqrt{1+x}$.

The answers to Exercises 25-30 are not unique.)

27. $\lim\limits_{x \to 1} \dfrac{x^9-1}{x-1} = f'(1)$ where $f(x) = x^9$

29. $\lim\limits_{t \to 0} \dfrac{\sin\left[\dfrac{\pi}{2}+t\right]-1}{t} = f'(\pi/2)$ where $f(x) = \sin x$

31. $f'(x) = \lim\limits_{h \to 0} \dfrac{f(x+h)-f(x)}{h} = \lim\limits_{h \to 0} \dfrac{5(x+h)+3-(5x+3)}{h} = \lim\limits_{h \to 0} \dfrac{5h}{h} = \lim\limits_{h \to 0} 5 = 5$

Domain of $f$ = domain of $f'$ = R

33. $f'(x) = \lim_{h \to 0} \dfrac{f(x+h)-f(x)}{h} = \lim_{h \to 0} \dfrac{(x+h)^3-(x+h)^2+2(x+h)-(x^3-x^2+2x)}{h}$

$= \lim_{h \to 0} \dfrac{3x^2h+3xh^2+h^3-2xh-h^2+2h}{h} = \lim_{h \to 0} (3x^2+3xh+h^2-2x-h+2) = 3x^2-2x+2$

Domain of f = domain of f' = R

35. $f'(x) = \lim_{h \to 0} \dfrac{f(x+h)-f(x)}{h} = \lim_{h \to 0} \dfrac{x+h-\frac{2}{x+h} - \left[x-\frac{2}{x}\right]}{h} = \lim_{h \to 0} \dfrac{h+\frac{2(x+h)-2x}{x(x+h)}}{h}$

$= \lim_{h \to 0} \left[1 + \dfrac{2}{x(x+h)}\right] = 1 + \dfrac{2}{x^2}$   Domain of f = domain of f' = $\{x \mid x \neq 0\}$

37. $g'(x) = \lim_{h \to 0} \dfrac{g(x+h)-g(x)}{h} = \lim_{h \to 0} \dfrac{\sqrt{1+2(x+h)}-\sqrt{1+2x}}{h}\left[\dfrac{\sqrt{1+2(x+h)}+\sqrt{1+2x}}{\sqrt{1+2(x+h)}+\sqrt{1+2x}}\right]$

$= \lim_{h \to 0} \dfrac{1+2x+2h-(1+2x)}{h(\sqrt{1+2(x+h)}+\sqrt{1+2x})} = \lim_{h \to 0} \dfrac{2}{\sqrt{1+2(x+h)}+\sqrt{1+2x}} = \dfrac{1}{\sqrt{1+2x}}$

Dom(g) = $[-1/2, \infty)$,   dom(g') = $(-1/2, \infty)$

39. $G'(x) = \lim_{h \to 0} \dfrac{G(x+h)-G(x)}{h} = \lim_{h \to 0} \dfrac{\frac{4-3(x+h)}{2+(x+h)} - \frac{4-3x}{2+x}}{h}$

$= \lim_{h \to 0} \dfrac{(4-3x-3h)(2+x)-(4-3x)(2+x+h)}{h(2+x+h)(2+x)} = \lim_{h \to 0} \dfrac{-10h}{h(2+x+h)(2+x)}$

$= \lim_{h \to 0} \dfrac{-10}{(2+x+h)(2+x)} = \dfrac{-10}{(2+x)^2}$   Dom(G) = dom(G') = $\{x \mid x \neq -2\}$

41. $f'(t) = \lim_{h \to 0} \dfrac{f(t+h)-f(t)}{h} = \lim_{h \to 0} \dfrac{(t+h+1)^3-(t+1)^3}{h}$

$= \lim_{h \to 0} \dfrac{h[(t+h+1)^2+(t+h+1)(t+1)+(t+1)^2]}{h} = (t+1)^2+(t+1)(t+1)+(t+1)^2$

$= 3(t+1)^2$   Domain of f = domain of f' = R

43. $f'(x) = \lim_{h \to 0} \dfrac{f(x+h)-f(x)}{h} = \lim_{h \to 0} \dfrac{a(x+h)^2+b(x+h)+c-(ax^2+bx+c)}{h}$

$= \lim_{h \to 0} \dfrac{2axh+ah^2+bh}{h} = \lim_{h \to 0} (2ax+ah+b) = 2ax+b$

Domain of f = domain of f' = R

**45.** $f(x) = x$: $f'(x) = \lim\limits_{h \to 0} \dfrac{x+h-x}{h} = \lim\limits_{h \to 0} 1 = 1$

$f(x) = x^2$: $f'(x) = \lim\limits_{h \to 0} \dfrac{(x+h)^2 - x^2}{h} = \lim\limits_{h \to 0} \dfrac{2xh + h^2}{h} = \lim\limits_{h \to 0} (2x+h) = 2x$

$f(x) = x^3$: $f'(x) = \lim\limits_{h \to 0} \dfrac{(x+h)^3 - x^3}{h} = \lim\limits_{h \to 0} \dfrac{3x^2 h + 3xh^2 + h^3}{h}$

$= \lim\limits_{h \to 0} (3x^2 + 3xh + h^2) = 3x^2 \qquad f(x) = x^4$: $f'(x) = 4x^3$ from Exercise 34

Guess: The derivative of $f(x) = x^n$ is $f'(x) = nx^{n-1}$. Test for $n=5$:

$f(x) = x^5$: $f'(x) = \lim\limits_{h \to 0} \dfrac{(x+h)^5 - x^5}{h} = \lim\limits_{h \to 0} \dfrac{5x^4 h + 10x^3 h^2 + 10x^2 h^3 + 5xh^4 + h^5}{h}$

$\lim\limits_{h \to 0} (5x^4 + 10x^3 h + 10x^2 h^2 + 5xh^3 + h^4) = 5x^4$

**47.**

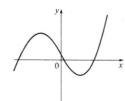

**49.**

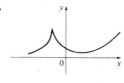

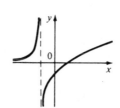

**51.**

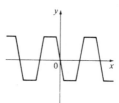

**53.** $f(x) = |x-6| = \begin{cases} x-6 & \text{if } x \geq 6 \\ 6-x & \text{if } x < 6 \end{cases}$

$\lim\limits_{x \to 6^+} \dfrac{f(x) - f(6)}{x-6} = \lim\limits_{x \to 6^+} \dfrac{|x-6| - 0}{x-6} = \lim\limits_{x \to 6^+} \dfrac{x-6}{x-6} = \lim\limits_{x \to 6^+} 1 = 1 \qquad$ But

$\lim\limits_{x \to 6^-} \dfrac{f(x) - f(6)}{x-6} = \lim\limits_{x \to 6^-} \dfrac{|x-6| - 0}{x-6} = \lim\limits_{x \to 6^-} \dfrac{6-x}{x-6} = \lim\limits_{x \to 6^-} (-1) = -1$

So $f'(6) = \lim\limits_{x \to 6} \dfrac{f(x) - f(6)}{x-6}$ does not exist.

However $f'(x) = \begin{cases} 1 & \text{if } x > 6 \\ -1 & \text{if } x < 6 \end{cases}$

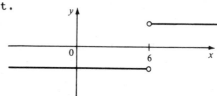

**55.** (a) $f(x) = x|x| = \begin{cases} x^2 & \text{if } x \geq 0 \\ -x^2 & \text{if } x < 0 \end{cases}$

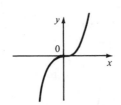

(b) Since $f(x) = x^2$ for $x \geq 0$, we have $f'(x) = 2x$ for $x > 0$. Since $f(x) = -x^2$ for $x < 0$, we have $f'(x) = -2x$ for $x < 0$. At $x = 0$, we have $f'(0) = \lim\limits_{x \to 0} \dfrac{f(x)-f(0)}{x-0} = \lim\limits_{x \to 0} \dfrac{x|x|}{x} = \lim\limits_{x \to 0} |x| = 0$ (Example 2.4.5)

So f is differentiable at 0. Thus f is differentiable for all x.

(c) From part (b) we have $f'(x) = \begin{cases} 2x & \text{if } x \geq 0 \\ -2x & \text{if } x < 0 \end{cases} = 2|x|$

**57.** Since $f(x) = x \sin(1/x)$ when $x \neq 0$ and $f(0) = 0$, we have $f'(0) = \lim\limits_{h \to 0} \dfrac{f(0+h)-f(0)}{h} = \lim\limits_{h \to 0} \dfrac{h \sin(1/h)-0}{h} = \lim\limits_{h \to 0} \sin\dfrac{1}{h}$ This limit does not exist since $\sin(1/h)$ takes the values $-1$ and $1$ on any interval containing 0. (Compare with Example 2.2.7.)

**59.** (a) If f is even, then $f'(-x) = \lim\limits_{h \to 0} \dfrac{f(-x+h)-f(-x)}{h} = \lim\limits_{h \to 0} \dfrac{f(x-h)-f(x)}{h}$

$= -\lim\limits_{h \to 0} \dfrac{f(x-h)-f(x)}{-h}$  [Let $\Delta x = -h$.]  $= -\lim\limits_{\Delta x \to 0} \dfrac{f(x+\Delta x)-f(x)}{\Delta x} = -f'(x)$

Therefore f' is odd.

(b) If f is odd, then $f'(-x) = \lim\limits_{h \to 0} \dfrac{f(-x+h)-f(-x)}{h} = \lim\limits_{h \to 0} \dfrac{-f(x-h)+f(x)}{h}$

$= \lim\limits_{h \to 0} \dfrac{f(x-h)-f(x)}{-h}$  [Let $\Delta x = -h$.]  $= \lim\limits_{\Delta x \to 0} \dfrac{f(x+\Delta x)-f(x)}{\Delta x} = f'(x)$

Therefore f' is even.

Section 3.2

## Exercises 3.2

**1.** $f(x) = x^2 - 10x + 100 \Rightarrow f'(x) = 2x - 10$

**3.** $V(r) = \dfrac{4}{3}\pi r^3 \Rightarrow V'(r) = \dfrac{4}{3}\pi(3r^2) = 4\pi r^2$

5. $F(x) = (16x)^3 = 4096x^3 \Rightarrow F'(x) = 4096(3x^2) = 12,288x^2$

7. $Y(t) = 6t^{-9} \Rightarrow Y'(t) = 6(-9)t^{-10} = -54t^{-10}$

9. $g(x) = x^2 + 1/x^2 = x^2 + x^{-2} \Rightarrow g'(x) = 2x + (-2)x^{-3} = 2x - 2/x^3$

11. $h(x) = \dfrac{x+2}{x-1} \Rightarrow h'(x) = \dfrac{(x-1)D(x+2)-(x+2)D(x-1)}{(x-1)^2} = \dfrac{x-1-(x+2)}{(x-1)^2} = \dfrac{-3}{(x-1)^2}$

13. $G(s) = (s^2+s+1)(s^2+2) \Rightarrow G'(s) = (2s+1)(s^2+2)+(s^2+s+1)(2s)$

   $= 4s^3+3s^2+6s+2$

15. $y = (x^2+4x+3)/\sqrt{x} = x^{3/2}+4x^{1/2}+3x^{-1/2} \Rightarrow$

   $y' = (3/2)x^{1/2}+4(1/2)x^{-1/2}+3(-1/2)x^{-3/2} = (3/2)\sqrt{x} + 2/\sqrt{x} - 3/2x\sqrt{x}$

   [Another method: Use the Quotient Rule.]

17. $y = \sqrt{5x} = \sqrt{5}x^{1/2} \Rightarrow y' = \sqrt{5}(1/2)x^{-1/2} = \sqrt{5}/2\sqrt{x}$

19. $y = \dfrac{1}{x^4+x^2+1} \Rightarrow y' = \dfrac{(x^4+x^2+1)(0)-1(4x^3+2x+1)}{(x^4+x^2+1)^2} = -\dfrac{4x^3+2x}{(x^4+x^2+1)^2}$

21. $y = ax^2+bx+c \Rightarrow y' = 2ax+b$

23. $y = \dfrac{3t-7}{t^2+5t-4} \Rightarrow y' = \dfrac{(t^2+5t-4)(3)-(3t-7)(2t+5)}{(t^2+5t-4)^2} = \dfrac{-3t^2+14t+23}{(t^2+5t-4)^2}$

25. $y = x+\sqrt[5]{x^2} = x+x^{2/5} \Rightarrow y' = 1+(2/5)x^{-3/5} = 1+2/5\sqrt[5]{x^3}$

27. $u = x^{\sqrt{2}} \Rightarrow u' = \sqrt{2}x^{\sqrt{2}-1}$

29. $v = x\sqrt{x}+1/x^2\sqrt{x} = x^{3/2}+x^{-5/2} \Rightarrow v' = (3/2)x^{1/2}-(5/2)x^{-7/2}$

   $= (3/2)\sqrt{x} - 5/2x^3\sqrt{x}$

31. $f(x) = \dfrac{x}{x+c/x} \Rightarrow f'(x) = \dfrac{(x+c/x)(1)-x(1-c/x^2)}{(x+c/x)^2} = \dfrac{2cx}{(x^2+c)^2}$

33. $f(x) = \dfrac{x^5}{x^3-2} \Rightarrow f'(x) = \dfrac{(x^3-2)5x^4-x^5(3x^2)}{(x^3-2)^2} = \dfrac{2x^4(x^3-5)}{(x^3-2)^2}$

35. $s = \dfrac{2-1/t}{t+1} \Rightarrow s' = \dfrac{(t+1)(1/t^2)-(2-1/t)(1)}{(t+1)^2} = \dfrac{1+2t-2t^2}{t^2(t+1)^2}$

37. $P(x) = a_n x^n + a_{n-1}x^{n-1} + \ldots + a_2x^2 + a_1x + a_0 \Rightarrow$

   $P'(x) = na_n x^{n-1} + (n-1)a_{n-1}x^{n-2} + \ldots + 2a_2x + a_1$

39. $y = f(x) = x+4/x \Rightarrow f'(x) = 1-4/x^2$ So the slope of the tangent

   line at $(2,4)$ is $f'(2) = 0$ and the equation is $y-4 = 0$ or $y = 4$.

41. $y = f(x) = \dfrac{1}{x^2+1} \Rightarrow f'(x) = \dfrac{(x^2+1)(0)-1(2x)}{(x^2+1)^2} = \dfrac{-2x}{(x^2+1)^2}$ So the slope

   of the tangent line at $(-1,1/2)$ is $f'(-1) = 1/2$ and the equation is

   $y-1/2 = (1/2)(x+1)$ or $x-2y+2 = 0$.

43. If $y = f(x) = \frac{x}{x+1}$ then $f'(x) = \frac{(x+1)(1)-x(1)}{(x+1)^2} = \frac{1}{(x+1)^2}$. When $x = a$, the equation of the tangent line is $y - \frac{a}{a+1} = \frac{1}{(a+1)^2}(x-a)$. This line passes through $(1,2)$ when $2 - \frac{a}{a+1} = \frac{1}{(a+1)^2}(1-a)$ ⟺ $2(a+1)^2 = a(a+1)+(1-a) = a^2+1$ ⟺ $a^2+4a+1 = 0$. The quadratic formula gives the roots of this equation as $-2\pm\sqrt{3}$, so there are 2 such tangent lines, which touch the curve at $(-2+\sqrt{3},(1-\sqrt{3})/2)$ and $(-2-\sqrt{3},(1+\sqrt{3})/2)$.

45. $y = x^3-x^2-x+1$ has a horizontal tangent when $y' = 3x^2-2x-1 = 0$ ⟺ $(3x+1)(x-1) = 0$ ⟺ $x = 1$ or $-1/3$. Therefore the points are $(1,0)$ and $(-1/3,32/27)$.

47. $y = x\sqrt{x} = x^{3/2}$ ⟹ $y' = (3/2)\sqrt{x}$ so the tangent is parallel to $3x-y+6 = 0$ when $(3/2)\sqrt{x} = 3$ ⟺ $\sqrt{x} = 2$ ⟺ $x = 4$. So the point is $(4,8)$.

49. $y = f(x) = 1-x^2$ ⟹ $f'(x) = -2x$, so the tangent line at $(2,-3)$ has slope $f'(2) = -4$. The normal line has slope $-1/(-4) = 1/4$ and equation $y+3 = (1/4)(x-2)$ or $x-4y = 14$.

**49.**

**51.**

51. $y = f(x) = \sqrt[3]{x} = x^{1/3}$ ⟹ $f'(x) = (1/3)x^{-2/3}$, so the tangent line at $(-8,-2)$ has slope $f'(-8) = 1/12$. The normal line has slope $-1/(1/12) = -12$ and equation $y+2 = -12(x+8)$ or $12x+y+98 = 0$.

53. If the normal line has slope 16, then the tangent has slope $-1/16$, so $y' = 4x^3 = -1/16$ ⟹ $x^3 = -1/64$ ⟹ $x = -1/4$. The point is $(-1/4,1/256)$.

55. (a) $s = f(t) = 3t^4-16t^3+18t^2$ ⟹ $v(t) = f'(t) = 12t^3-48t^2+36t$

(b) It is at rest when $v(t) = 12t(t-1)(t-3) = 0$ ⟺ $t = 0,1$, or 3

(c) It moves in the positive direction when $v(t) = 12t(t-1)(t-3)>0$ ⟺ $0 < t < 1$ or $t > 3$.

57. (a) $s = f(t) = \dfrac{t}{t^2+1} \Rightarrow v = f'(t) = \dfrac{(t^2+1)(1)-t(2t)}{(t^2+1)^2} = \dfrac{1-t^2}{(t^2+1)^2}$

(b) It is at rest when $v = 0 \Leftrightarrow 1-t^2 = 0 \Leftrightarrow t = 1$.

(c) It moves in the positive direction when $v > 0 \Leftrightarrow 1-t^2 > 0$
$\Leftrightarrow t^2 < 1 \Leftrightarrow 0 \leq t < 1$.

59. $f(x) = 2-x$ if $x \leq 1$ and $f(x) = x^2-2x+2$ if $x > 1$. Therefore
$$\lim_{h \to 0^-} \frac{f(1+h)-f(1)}{h} = \lim_{h \to 0^-} \frac{2-(1+h)-1}{h} = \lim_{h \to 0^-} \frac{-h}{h} = \lim_{h \to 0^-} (-1) = -1 \text{ and}$$

$$\lim_{h \to 0^+} \frac{f(1+h)-f(1)}{h} = \lim_{h \to 0^+} \frac{(1+h)^2-2(1+h)+2-1}{h} = \lim_{h \to 0^+} \frac{h^2}{h} = \lim_{h \to 0^+} h = 0$$

Thus $f'(1)$ does not exist, so $f$ is not differentiable at 1.
But $f'(x) = -1$ for $x < 1$ and $f'(x) = 2x-2$ if $x > 1$.

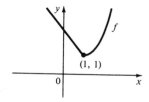 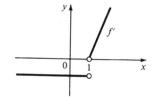

61. (a) $(fg)'(5) = f'(5)g(5)+f(5)g'(5) = 6(-3)+1(2) = -16$

(b) $\left[\dfrac{f}{g}\right]'(5) = \dfrac{f'(5)g(5)-f(5)g'(5)}{[g(5)]^2} = \dfrac{6(-3)-1(2)}{(-3)^2} = -\dfrac{20}{9}$

(c) $\left[\dfrac{g}{f}\right]'(5) = \dfrac{g'(5)f(5)-g(5)f'(5)}{[f(5)]^2} = \dfrac{2(1)-(-3)(6)}{1^2} = 20$

63. (a) $(fgh)' = ((fg)h)' = (fg)'h + (fg)h' = (f'g+fg')h + (fg)h'$
$= f'gh + fg'h + fgh'$

(b) Putting $f = g = h$ in part (a), we have $\dfrac{d}{dx}[f(x)]^3 = (fff)'$
$= f'ff + ff'f + fff' = 3fff' = 3[f(x)]^2f'(x)$

65. $y = \sqrt{x}(x^4+x+1)(2x-3)$ Using Exercise 63(a), we have
$y' = (1/2\sqrt{x})(x^4+x+1)(2x-3) + \sqrt{x}(4x^3+1)(2x-3) + \sqrt{x}(x^4+x+1)(2)$
$= (x^4+x+1)(2x-3)/2\sqrt{x} + \sqrt{x}[(4x^3+1)(2x-3)+2(x^4+x+1)]$

67. $F = f/g \Rightarrow f = Fg \Rightarrow f' = F'g+Fg' \Rightarrow$
$F' = \dfrac{f'-Fg'}{g} = \dfrac{f'-(f/g)g'}{g} = \dfrac{f'g-fg'}{g^2}$

**Exercises 3.3**

1.  $\lim\limits_{x\to 0} (x^2 + \cos x) = \lim\limits_{x\to 0} x^2 + \lim\limits_{x\to 0} \cos x = 0^2 + \cos 0 = 0 + 1 = 1$

3.  $\lim\limits_{x\to \pi/3} (\sin x - \cos x) = \sin \frac{\pi}{3} - \cos \frac{\pi}{3} = \frac{\sqrt{3}}{2} - \frac{1}{2}$

5.  $\lim\limits_{x\to \pi/4} \frac{\sin x}{3x} = \frac{\sin(\pi/4)}{3\pi/4} = \frac{1/\sqrt{2}}{3\pi/4} = \frac{2\sqrt{2}}{3\pi}$

7.  $\lim\limits_{t\to -3\pi} t^3 \sin^4 t = \left[\lim\limits_{t\to -3\pi} t\right]^3 \left[\lim\limits_{t\to -3\pi} \sin t\right]^4 = (-3\pi)^3 (0)^4 = 0$

9.  $\lim\limits_{t\to 0} \frac{\sin 5t}{t} = \lim\limits_{t\to 0} \frac{5 \sin 5t}{5t} = 5 \lim\limits_{t\to 0} \frac{\sin 5t}{5t} = 5 \cdot 1 = 5$

11. $\lim\limits_{\theta \to 0} \frac{\sin(\cos \theta)}{\sec \theta} = \frac{\sin\left[\lim\limits_{\theta\to 0} \cos \theta\right]}{\lim\limits_{\theta\to 0} \sec \theta} = \frac{\sin 1}{1} = \sin 1$

13. $\lim\limits_{x\to \pi/4} \frac{\tan x}{4x} = \frac{\tan(\pi/4)}{4(\pi/4)} = \frac{1}{\pi}$

15. $\lim\limits_{x\to 0} \frac{\tan 3x}{3 \tan 2x} = \lim\limits_{x\to 0} \frac{\frac{\tan 3x}{3x}}{2\frac{\tan 2x}{2x}} = \frac{1}{2} \frac{\lim\limits_{x\to 0} \frac{\sin 3x}{3x} \cdot \frac{1}{\cos 3x}}{\lim\limits_{x\to 0} \frac{\sin 2x}{2x} \cdot \frac{1}{\cos 2x}} = \frac{1}{2} \cdot \frac{1 \cdot 1}{1 \cdot 1} = \frac{1}{2}$

17. $\lim\limits_{t\to 0} \frac{\sin^2 3t}{t^2} = \lim\limits_{t\to 0} 9 \left[\frac{\sin 3t}{3t}\right]^2 = 9 \left[\lim\limits_{t\to 0} \frac{\sin 3t}{3t}\right]^2 = 9(1)^2 = 9$

19. $\frac{d}{dx}(\csc x) = \frac{d}{dx}\left[\frac{1}{\sin x}\right] = \frac{(\sin x)(0) - 1(\cos x)}{\sin^2 x} = \frac{-\cos x}{\sin^2 x}$

    $= -\frac{1}{\sin x} \cdot \frac{\cos x}{\sin x} = -\csc x \cot x$

21. $\frac{d}{dx}(\cot x) = \frac{d}{dx}\left[\frac{\cos x}{\sin x}\right] = \frac{(\sin x)(-\sin x) - (\cos x)(\cos x)}{\sin^2 x}$

    $= -\frac{\sin^2 x + \cos^2 x}{\sin^2 x} = -\frac{1}{\sin^2 x} = -\csc^2 x$

23. $y = \sin x + \cos x \Rightarrow dy/dx = \cos x - \sin x$

25. $y = \csc x \cot x \Rightarrow dy/dx = (-\csc x \cot x) \cot x + \csc x (-\csc^2 x)$

    $= -\csc x (\cot^2 x + \csc^2 x)$

27. $y = \frac{\tan x}{x} \Rightarrow \frac{dy}{dx} = \frac{x \sec^2 x - \tan x}{x^2}$

29. $y = \frac{x}{\sin x + \cos x} \Rightarrow \frac{dy}{dx} = \frac{(\sin x + \cos x) - x(\cos x - \sin x)}{(\sin x + \cos x)^2}$

$$= \frac{(1 + x) \sin x + (1 - x) \cos x}{\sin^2 x + \cos^2 x + 2 \sin x \cos x} = \frac{(1 + x) \sin x + (1 - x) \cos x}{1 + \sin 2x}$$

31. $y = x^{-3} \sin x \tan x \Rightarrow dy/dx = -3x^{-4} \sin x \tan x + x^{-3} \cos x \tan x$

   $+ x^{-3} \sin x \sec^2 x = x^{-4} \sin x (-3 \tan x + x + x \sec^2 x)$

33. $y = \dfrac{x^2 \tan x}{\sec x} \Rightarrow \dfrac{dy}{dx} = \dfrac{\sec x (2x \tan x + x^2 \sec^2 x) - x^2 \tan x \sec x \tan x}{\sec^2 x}$

   $= \dfrac{2x \tan x + x^2 (\sec^2 x - \tan^2 x)}{\sec x} = \dfrac{2x \tan x + x^2}{\sec x}$

   [Another method: Write $y = x^2 \sin x$. Then $y' = 2x \sin x + x^2 \cos x$.]

35. $y = \tan x \Rightarrow y' = \sec^2 x \Rightarrow$ The slope of the tangent line at
   $(\pi/4, 1)$ is $\sec^2(\pi/4) = 2$ and the equation is $y - 1 = 2(x - \pi/4)$ or
   $4x - 2y = \pi - 2$.

37. $y = x + 2 \sin x$ has a horizontal tangent when $y' = 1 + 2 \cos x = 0$
   $\Longleftrightarrow \cos x = -1/2 \Longleftrightarrow x = (2n+1)\pi \pm \pi/3$, n an integer.

39. $\lim\limits_{x \to 0} \dfrac{\cot 2x}{\csc x} = \lim\limits_{x \to 0} \dfrac{\cos 2x \sin x}{\sin 2x} = \lim\limits_{x \to 0} \cos 2x \left[ \dfrac{\frac{\sin x}{x}}{\frac{\sin 2x}{x}} \right]$

   $= \lim\limits_{x \to 0} \cos 2x \dfrac{\lim\limits_{x \to 0} \frac{\sin x}{x}}{2 \lim\limits_{x \to 0} \frac{\sin 2x}{2x}} = 1 \cdot \dfrac{1}{2 \cdot 1} = \dfrac{1}{2}$

41. $\lim\limits_{x \to \pi} \dfrac{\tan x}{\sin 2x} = \lim\limits_{x \to \pi} \dfrac{\sin x}{\cos x (2 \sin x \cos x)} = \lim\limits_{x \to \pi} \dfrac{1}{2 \cos^2 x} = \dfrac{1}{2(-1)^2} = \dfrac{1}{2}$

43. Divide numerator and denominator by $\theta$. ($\sin \theta$ also works.)

   $\lim\limits_{\theta \to 0} \dfrac{\sin \theta}{\theta + \tan \theta} = \lim\limits_{\theta \to 0} \dfrac{\frac{\sin \theta}{\theta}}{1 + \frac{\sin \theta}{\theta} \cdot \frac{1}{\cos \theta}} = \dfrac{\lim\limits_{\theta \to 0} \frac{\sin \theta}{\theta}}{1 + \lim\limits_{\theta \to 0} \frac{\sin \theta}{\theta} \lim\limits_{\theta \to 0} \frac{1}{\cos \theta}}$

   $= \dfrac{1}{1 + 1 \cdot 1} = \dfrac{1}{2}$

45. $\lim\limits_{x \to 0^+} \sqrt{x} \csc \sqrt{x} = \lim\limits_{x \to 0^+} \dfrac{\sqrt{x}}{\sin \sqrt{x}} = \left[ \lim\limits_{x \to 0^+} \dfrac{\sin \sqrt{x}}{\sqrt{x}} \right]^{-1} = 1^{-1} = 1$

47. (a) $\frac{d}{dx} \tan x = \frac{d}{dx} \frac{\sin x}{\cos x}$ ➤ $\sec^2 x = \frac{\cos x \cos x - \sin x (-\sin x)}{\cos^2 x}$

$= \frac{\cos^2 x + \sin^2 x}{\cos^2 x}$   So $\sec^2 x = \frac{1}{\cos^2 x}$

(b) $\frac{d}{dx} \sec x = \frac{d}{dx} \frac{1}{\cos x}$ ➤ $\sec x \tan x = \frac{(\cos x)(0) - 1(-\sin x)}{\cos^2 x}$

So $\sec x \tan x = \frac{\sin x}{\cos^2 x}$

(c) $\frac{d}{dx}(\sin x + \cos x) = \frac{d}{dx} \frac{1 + \cot x}{\csc x}$ ➤

$\cos x - \sin x = \frac{\csc x (-\csc^2 x) - (1 + \cot x)(-\csc x \cot x)}{\csc^2 x}$

$= \frac{-\csc^2 x + \cot^2 x + \cot x}{\csc x}$   So   $\cos x - \sin x = \frac{\cot x - 1}{\csc x}$

Section 3.4

Exercises 3.4

1.   $y = u^2$, $u = x^2 + 2x + 3$   (a) $\frac{dy}{dx} = \frac{dy}{du} \frac{dx}{dx} = 2u(2x+2) = 4u(x+1)$

When $x = 1$, $u = 1^2 + 2(1) + 3 = 6$, so $\frac{dy}{dx}\Big]_{x=1} = 4(6)(1+1) = 48$

(b) $y = u^2 = (x^2 + 2x + 3)^2 = x^4 + 4x^2 + 9 + 4x^3 + 6x^2 + 12x = x^4 + 4x^3 + 10x^2 + 12x + 9$

so $\frac{dy}{dx} = 4x^3 + 12x^2 + 20x + 12$ and $\frac{dy}{dx}\Big]_{x=1} = 4(1)^3 + 12(1)^2 + 20(1) + 12 = 48$

3.   $y = u^3$, $u = x + \frac{1}{x}$   (a) $\frac{dy}{dx} = \frac{dy}{du} \frac{du}{dx} = 3u^2(1 - 1/x^2)$

When $x = 1$, $u = 1 + \frac{1}{1} = 2$, so $\frac{dy}{dx}\Big]_{x=1} = 3(2)^2(1 - 1/1^2) = 0$

(b) $y = u^3 = \left[x + \frac{1}{x}\right]^3 = x^3 + 3x^2\left[\frac{1}{x}\right] + 3x\left[\frac{1}{x}\right]^2 + \left[\frac{1}{x}\right]^3 = x^3 + 3x + 3x^{-1} + x^{-3}$, so

$\frac{dy}{dx} = 3x^2 + 3 - 3x^{-2} - 3x^{-4}$ and $\frac{dy}{dx}\Big]_{x=1} = 3(1)^2 + 3 - 3(1)^{-2} - 3(1)^{-4} = 0$

5.   $F(x) = (x^2 + 4x + 6)^5$ ➤ $F'(x) = 5(x^2 + 4x + 6)^4 \frac{d}{dx}(x^2 + 4x + 6)$

$= 5(x^2 + 4x + 6)^4(2x+4) = 10(x^2 + 4x + 6)^4(x+2)$

7.   $G(x) = (3x-2)^{10}(5x^2 - x + 1)^{12}$

$G'(x) = 10(3x-2)^9(3)(5x^2 - x + 1)^{12} + (3x-2)^{10}12(5x^2 - x + 1)^{11}(10x-1)$

$= 30(3x-2)^9(5x^2 - x + 1)^{12} + 12(3x-2)^{10}(5x^2 - x + 1)^{11}(10x-1)$

[This can be simplified to $6(3x-2)^9(5x^2 - x + 1)^{11}(85x^2 - 51x + 9)$.]

9. $f(t) = (2t^2 - 6t + 1)^{-8}$

$f'(t) = -8(2t^2 - 6t + 1)^{-9}(4t - 6) = -16(2t^2 - 6t + 1)^{-9}(2t - 3)$

11. $g(x) = \sqrt{x^2 - 7x} = (x^2 - 7x)^{1/2}$

$g'(x) = (1/2)(x^2 - 7x)^{-1/2}(2x - 7) = (2x - 7)/2\sqrt{x^2 - 7x}$

13. $h(t) = (t - 1/t)^{3/2} \Rightarrow h'(t) = (3/2)(t - 1/t)^{1/2}(1 + 1/t^2)$

15. $F(y) = \left[\dfrac{y-6}{y+7}\right]^3$

$F'(y) = 3\left[\dfrac{y-6}{y+7}\right]^2 \dfrac{(y+7)(1)-(y-6)(1)}{(y+7)^2} = 3\left[\dfrac{y-6}{y+7}\right]^2 \dfrac{13}{(y+7)^2} = \dfrac{39(y-6)^2}{(y+7)^4}$

17. $f(z) = (2z - 1)^{-1/5} \Rightarrow f'(z) = -\dfrac{1}{5}(2z - 1)^{-6/5}(2) = -\dfrac{2}{5}(2z - 1)^{-6/5}$

19. $y = (2x - 5)^4(8x^2 - 5)^{-3}$

$y' = 4(2x - 5)^3(2)(8x^2 - 5)^{-3} + (2x - 5)^4(-3)(8x^2 - 5)^{-4}(16x)$

$= 8(2x - 5)^3(8x^2 - 5)^{-3} - 48x(2x - 5)^4(8x^2 - 5)^{-4}$

[This simplifies to $8(2x - 5)^3(8x^2 - 5)^{-4}(-4x^2 + 30x - 5)$.]

21. $y = \tan 3x \Rightarrow y' = \sec^2 3x \dfrac{d}{dx}(3x) = 3 \sec^2 3x$

23. $y = \cos(x^3) \Rightarrow y' = -\sin(x^3)(3x^2) = -3x^2 \sin(x^3)$

25. $y = (1 + \cos^2 x)^6$

$y' = 6(1 + \cos^2 x)^5 \, 2 \cos x \, (-\sin x) = -12 \cos x \sin x \, (1 + \cos^2 x)^5$

27. $y = \cos(\tan x) \Rightarrow y' = -\sin(\tan x) \sec^2 x$

29. $y = \sec^2 2x - \tan^2 2x$

$y' = 2 \sec 2x (\sec 2x \tan 2x)(2) - 2 \tan 2x \sec^2(2x)(2) = 0$

[Easier method: $y = \sec^2 2x - \tan^2 2x = 1 \Rightarrow y' = 0$.]

31. $y = \csc(x/3) \Rightarrow y' = -(1/3) \csc(x/3) \cot(x/3)$

33. $y = \sin^3 x + \cos^3 x \Rightarrow y' = 3 \sin^2 x \cos x + 3 \cos^2 x \, (-\sin x)$

$= 3 \sin x \cos x (\sin x - \cos x)$

35. $y = \sin(1/x) \Rightarrow y' = \cos(1/x) \, (-1/x^2) = -\cos(1/x)/x^2$

37. $y = \dfrac{1 + \sin 2x}{1 - \sin 2x}$

$y' = \dfrac{(1 - \sin 2x)(2 \cos 2x) - (1 + \sin 2x)(-2 \cos 2x)}{(1 - \sin 2x)^2}$

$= \dfrac{4 \cos 2x}{(1 - \sin 2x)^2}$

39. $y = \tan^2(x^3)$

$y' = 2 \tan(x^3) \sec^2(x^3) (3x^2) = 6x^2 \tan(x^3) \sec^2(x^3)$

41. $y = \cos^2\left[\dfrac{1-\sqrt{x}}{1+\sqrt{x}}\right]$

$y' = 2 \cos\left[\dfrac{1-\sqrt{x}}{1+\sqrt{x}}\right] (-1) \sin\left[\dfrac{1-\sqrt{x}}{1+\sqrt{x}}\right] \dfrac{(1+\sqrt{x})(-1/2\sqrt{x}) - (1-\sqrt{x})(1/2\sqrt{x})}{(1+\sqrt{x})^2}$

$= 2 \cos\left[\dfrac{1-\sqrt{x}}{1+\sqrt{x}}\right] \sin\left[\dfrac{1-\sqrt{x}}{1+\sqrt{x}}\right] / \sqrt{x}(1+\sqrt{x})^2$

43. $y = \cos^2(\cos x) + \sin^2(\cos x) = 1 \Rightarrow y' = 0$

45. $y = \sqrt{x+\sqrt{x}} \Rightarrow y' = \frac{1}{2}(x+\sqrt{x})^{-1/2}(1+\frac{1}{2}x^{-1/2}) = \dfrac{1}{2\sqrt{x+\sqrt{x}}}\left[1 + \dfrac{1}{2\sqrt{x}}\right]$

47. $f(x) = [x^3+(2x-1)^3]^3$

$f'(x) = 3[x^3+(2x-1)^3]^2[3x^2+3(2x-1)^2(2)] = 9[x^3+(2x-1)^3]^2(9x^2-8x+2)$

49. $p(t) = [(1+2/t)^{-1}+3t]^{-2}$

$p'(t) = -2[(1+2/t)^{-1}+3t]^{-3}[-(1+2/t)^{-2}(-2/t^2)+3]$

$= -2[(1+2/t)^{-1}+3t]^{-3}[2(t+2)^{-2}+3]$

51. $y = \sin(\tan\sqrt{\sin x})$

$y' = \cos(\tan\sqrt{\sin x})(\sec^2\sqrt{\sin x})(1/2\sqrt{\sin x})(\cos x)$

53. $y = f(x) = (x^3-x^2+x-1)^{10} \Rightarrow f'(x) = 10(x^3-x^2+x-1)^9(3x^2-2x+1)$

The slope of the tangent at $(1,0)$ is $f'(1) = 0$ and its equation is

$y-0 = 0(x-1)$ or $y = 0$.

55. $y = f(x) = 8/\sqrt{4+3x} \Rightarrow f'(x) = 8(-1/2)(4+3x)^{-3/2}(3) = -12(4+3x)^{-3/2}$

The slope of the tangent at $(4,2)$ is $f'(4) = -3/16$ and its equation

is $y-2 = (-3/16)(x-4)$ or $3x+16y = 44$.

57. $y = f(x) = \cot^2 x \Rightarrow y' = 2 \cot x (-\csc^2 x) = -2 \cot x \csc^2 x$

The slope of the tangent at $(\pi/4,1)$ is $f'(\pi/4) = -2(1)(\sqrt{2})^2 = -4$

and its equation is $y-1 = -4(x-\pi/4)$ or $4x+y = \pi+1$.

59. $F(x) = f(g(x)) \Rightarrow F'(x) = f'(g(x))g'(x)$, so $F'(3) = f'(g(3))g'(3)$

$= f'(6)g'(3) = 7\cdot4 = 28$.

61. $f(x) = x^2\sec^2 3x \Rightarrow f'(x) = 2x \sec^2 3x + x^2(2 \sec 3x)(\sec 3x \tan 3x)3$

$= 2x \sec^2 3x (1 + 3x \tan 3x)$ $\qquad$ Domain of $f$ = domain of $f'$

$= \{x | \cos 3x \neq 0\} = \{x | x \neq (2n-1)\pi/6, \text{ n an integer}\}$

63. $f(x) = \sqrt{\cos\sqrt{x}}$ ➔ $f'(x) = (1/2)(\cos\sqrt{x})^{-1/2}(-\sin\sqrt{x})(1/2)x^{-1/2}$

   $= -\sin\sqrt{x}/4\sqrt{x}\sqrt{\cos\sqrt{x}}$     Domain of f = $\{x\,|\,x\geq 0$ and $\cos\sqrt{x}\geq 0\}$

   $= \{x\,|\,0\leq x\leq\pi^2/4$ or $[(4n-1)\pi/2]^2\leq x\leq[(4n+1)\pi/2]^2$ for some $n = 1,2,\cdots\}$

   Domain of f' = $\{x\,|\,x>0$ and $\cos\sqrt{x}>0\}$

   $= \{x\,|\,0<x<\pi^2/4$ or $[(4n-1)\pi/2]^2<x<[(4n+1)\pi/2]^2$ for some $n = 1,2,\cdots\}$

65. $s(t) = 10 + (1/4)\sin(10\pi t)$ ➔ the velocity after t seconds is

   $v(t) = s'(t) = (1/4)\cos(10\pi t)(10\pi) = (5\pi/2)\cos(10\pi t)$ cm/s.

67. (a) Since h is differentiable on $[0,\infty)$ and $\sqrt{x}$ is differentiable on

   $(0,\infty)$, it follows that $G(x) = h(\sqrt{x})$ is differentiable on $(0,\infty)$.

   (b) By the Chain Rule, $G'(x) = h'(\sqrt{x})(d/dx)\sqrt{x} = h'(\sqrt{x})/2\sqrt{x}$

69. (a) $F(x) = f(\cos x)$ ➔ $F'(x) = f'(\cos x)(d/dx)(\cos x)$

   $= -\sin x \, f'(\cos x)$

   (b) $G(x) = \cos(f(x))$ ➔ $G'(x) = -\sin(f(x))f'(x)$

71. (a) If f is even, then $f(x) = f(-x)$. Using the Chain Rule to

   differentiate this equation, we get $f'(x) = f'(-x)(d/dx)(-x)$

   $= -f'(-x)$. Thus $f'(-x) = -f'(x)$, so f' is odd.

   (b) If f is odd, then $f(x) = -f(-x)$. Differentiating this

   equation, we get $f'(x) = -f'(-x)(-1) = f'(-x)$, so f' is even.

73. $f(x) = |x| = \sqrt{x^2}$ ➔ $f'(x) = \frac{1}{2}(x^2)^{-1/2}(2x) = x/\sqrt{x^2} = x/|x|$

75. Using Exercise 73, we have $f(x) = x|2x-1|$ ➔

   $f'(x) = |2x-1| + x\frac{2x-1}{|2x-1|}(2) = |2x-1| + \frac{2x(2x-1)}{|2x-1|}$

77. Since $\theta° = (\pi/180)\theta$ rad, we have

   $\frac{d}{d\theta}(\sin\theta°) = \frac{d}{d\theta}(\sin\frac{\pi}{180}\theta) = \frac{\pi}{180}\cos\frac{\pi}{180}\theta = \frac{\pi}{180}\cos\theta°$

Section 3.5

Exercises 3.5

1. (a) $x^2+3x+xy = 5$ ➔ $2x+3+y+xy' = 0$ ➔ $y' = -(2x+y+3)/x$

   (b) $x^2+3x+xy = 5$ ➔ $y = \frac{5-x^2-3x}{x} = \frac{5}{x} - x - 3$ ➔ $y' = -5/x^2 - 1$

   (c) $y' = -\frac{2x+y+3}{x} = \frac{-2x-3-(-3-x+5/x)}{x} = -1-5/x^2$

3. (a) $\frac{1}{x} + \frac{1}{y} = 3$ ⟹ $-\frac{1}{x^2} - \frac{1}{y^2}y' = 0$ ⟹ $y' = -\frac{y^2}{x^2}$

(b) $\frac{1}{y} = 3 - \frac{1}{x} = \frac{3x-1}{x}$ ⟹ $y = \frac{x}{3x-1}$ ⟹ $y' = \frac{(3x-1)-x(3)}{(3x-1)^2} = -\frac{1}{(3x-1)^2}$

(c) $y' = -\frac{y^2}{x^2} = -\frac{x^2/(3x-1)^2}{x^2} = -\frac{1}{(3x-1)^2}$

5. (a) $2y^2+xy = x^2+3$ ⟹ $4yy'+y+xy' = 2x$ ⟹ $y' = \frac{2x-y}{x+4y}$

(b) Use the quadratic formula: $2y^2+xy-(x^2+3) = 0$ ⟹

$y = \frac{1}{4}\left[-x\pm\sqrt{x^2+8(x^2+3)}\right] = \frac{1}{4}\left[-x\pm\sqrt{9x^2+24}\right]$ ⟹ $y' = \frac{1}{4}\left[-1 \pm 9x/\sqrt{9x^2+24}\right]$

(c) $y' = \frac{2x-y}{x+4y} = \frac{2x-(-x\pm\sqrt{9x^2+24})/4}{x + (-x\pm\sqrt{9x^2+24})} = \frac{1}{4}\left[-1 \pm \frac{9x}{\sqrt{9x^2+24}}\right]$

7. $x^2-xy+y^3 = 8$ ⟹ $2x-y-xy'+3y^2y' = 0$ ⟹ $y' = (y-2x)/(3y^2-x)$

9. $2y^2+\sqrt[3]{xy} = 3x^2+17$ ⟹ $4yy' + (1/3)x^{-2/3}y^{1/3} + (1/3)x^{1/3}y^{-2/3}y' = 6x$

⟹ $y' = \frac{6x - (1/3)x^{-2/3}y^{1/3}}{4y + (1/3)x^{1/3}y^{-2/3}} = \frac{18x - x^{-2/3}y^{1/3}}{12y + x^{1/3}y^{-2/3}}$

11. $x^4+y^4 = 16$ ⟹ $4x^3+4y^3y' = 0$ ⟹ $y' = -x^3/y^3$

13. $2xy = (x^2+y^2)^{3/2}$ ⟹ $2y+2xy' = (3/2)(x^2+y^2)^{1/2}(2x+2yy')$ ⟹

$y' = \frac{3x(x^2+y^2)^{1/2} - 2y}{2x - 3y(x^2+y^2)^{1/2}}$

15. $\frac{y}{x-y} = x^2+1$ ⟹ $2x = \frac{(x-y)y'-y(1-y')}{(x-y)^2} = \frac{xy'-y}{(x-y)^2}$ ⟹ $y' = \frac{y}{x} + 2(x-y)^2$

[Another method: Write the equation as $y = (x-y)(x^2+1) = x^3+x-yx^2-y$
This gives $y' = (3x^2+1-2xy)/(x^2+2)$.]

17. $\cos(x-y) = y \sin x$ ⟹ $-\sin(x-y)\cdot(1-y') = y' \sin x + y \cos x$

⟹ $y' = \frac{\sin(x-y) + y \cos x}{\sin(x-y) - \sin x}$

19. $xy = \cot(xy)$ ⟹ $y+xy' = -\csc^2(xy)\cdot(y+xy')$ ⟹

$(y+xy')(1 + \csc^2(xy)) = 0$ ⟹ $y+xy' = 0$ ⟹ $y' = -y/x$

21. $y^4+x^2y^2+yx^4 = y+1$ ⟹ $4y^3 + 2x\frac{dx}{dy}y^2 + 2x^2y + x^4 + 4yx^3\frac{dx}{dy} = 1$

⟹ $\frac{dx}{dy} = \frac{1-4y^3-2x^2y-x^4}{2xy^2+4yx^3}$

23. $x[f(x)]^3+xf(x) = 6$ ⟹ $[f(x)]^3+3x[f(x)]^2f'(x)+f(x)+xf'(x) = 0$ ⟹

$f'(x) = -\frac{[f(x)]^3+f(x)}{3x[f(x)]^2+x}$ ⟹ $f'(3) = -\frac{(1)^3+1}{3(3)(1)^2+3} = -\frac{1}{6}$

25. $\frac{x^2}{16} - \frac{y^2}{9} = 1$ ➤ $\frac{x}{8} - \frac{2yy'}{9} = 0$ ➤ $y' = \frac{9x}{16y}$ When $x = -5$ and $y = \frac{9}{4}$

we have $y' = \frac{9(-5)}{16(9/4)} = -\frac{5}{4}$ so the equation of the tangent is

$y-(9/4) = (-5/4)(x+5)$ or $5x+4y+16 = 0$.

**25.**

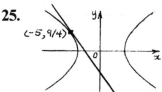

(-5, 9/4)

**27.**

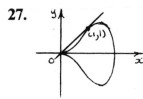

(1,1)

27. $y^2 = x^3(2-x) = 2x^3 - x^4$ ➤ $2yy' = 6x^2 - 4x^3$ ➤ $y' = (3x^2 - 2x^3)/y$

When $x = y = 1$, $y' = [3(1)^2 - 2(1)^3]/1 = 1$, so the equation of the

tangent line is $y-1 = 1(x-1)$ or $y = x$.

29. $2(x^2+y^2)^2 = 25(x^2-y^2)$ ➤ $4(x^2+y^2)(2x+2yy') = 25(2x-2yy')$ ➤

$y' = \frac{25x-4x(x^2+y^2)}{25y+4y(x^2+y^2)}$ When $x = 3$ and $y = 1$, $y' = \frac{75-120}{25+40} = -\frac{9}{13}$ so

the equation of the tangent is $y-1 = (-9/13)(x-3)$ or $9x+13y = 40$.

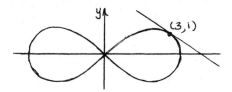

(3,1)

31. From question 29, a tangent to the lemniscate will be horizontal

$\Longleftrightarrow$ $y' = 0$ ➤ $25x-4x(x^2+y^2) = 0$ ➤ $x^2+y^2 = 25/4$ (Note that $x = 0$

➤ $y = 0$ and there is no horizontal tangent at the origin.) Putting

this in the equation of the lemniscate, we get $x^2-y^2 = 25/8$.

Solving these two equations, we have $x^2 = 75/16$ and $y^2 = 25/16$, so

the points are $(\pm 5\sqrt{3}/4, \pm 5/4)$.

33. $\frac{x^2}{a^2} - \frac{y^2}{b^2} = 1$ ➤ $\frac{2x}{a^2} - \frac{2yy'}{b^2} = 0$ ➤ $y' = \frac{b^2x}{a^2y}$ ➤ the equation of the

tangent at $(x_0,y_0)$ is $y-y_0 = (b^2x_0/a^2y_0)(x-x_0)$ Multiplying both

sides by $\frac{y_0}{b^2}$ gives $\frac{y_0y}{b^2} - \frac{y_0^2}{b^2} = \frac{x_0x}{a^2} - \frac{x_0^2}{a^2}$ Since $(x_0,y_0)$ lies on the

hyperbola, we have $\frac{x_0x}{a^2} - \frac{y_0y}{b^2} = \frac{x_0^2}{a^2} - \frac{y_0^2}{b^2} = 1$

35. If the circle has radius r, its equation is $x^2+y^2 = r^2$ ➙ $2x+2yy' = 0$ ➙ $y' = -x/y$, so the slope of the tangent line at $P(x_0,y_0)$ is $-x_0/y_0$. The slope of OP is $y_0/x_0 = -1/(-x_0/y_0)$, so the tangent line is perpendicular to OP.

37. $y = 2x$ has slope $m_1 = 2$ and $y = -3x+5$ has slope $m_2 = -3$, so Formula 3.40 gives $\tan \alpha = \frac{m_2-m_1}{1+m_1 m_2} = \frac{-3-2}{1+2(-3)} = 1$ ➙ $\alpha = 45°$ (or $135°$).

39. $y = x^2$ and $y = x^3$ intersect at $(0,0)$ and $(1,1)$. $y = x^2$ ➙ $y' = 2x$ and $y = x^3$ ➙ $y' = 3x^2$. At $(0,0)$ the slopes of the tangents are $m_1 = 0 = m_2$ so the angle is $0°$. At $(1,1)$ the slopes are $m_1 = 2$, $m_2 = 3$, so $\tan \alpha = (3-2)/(1+2\cdot 3) = 1/7$ ➙ $\alpha \approx 8°$ (or $172°$).

41. $y = x^2$ and $y = 1/x$ intersect at $(1,1)$. $y = x^2$ ➙ $y' = 2x$ and $y = 1/x$ ➙ $y' = -1/x^2$. At $(1,1)$ the slopes of the tangents are $m_1 = 2$, $m_2 = -1$, so Formula 3.40 gives $\tan \alpha = (-1-2)/(1+2(-1)) = 3$ ➙ $\alpha \approx 72°$ (or $108°$).

43. $x-2y+5 = 0$ and $x^2+y^2 = 25$ intersect when $(2y-5)^2+y^2 = 25$ ⟺ $5y(y-4) = 0$ ⟺ $y = 0$ or $4$. The slope of $x-2y+5 = 0$ is $1/2$ and $x^2+y^2 = 25$ ➙ $2x+2yy' = 0$ ➙ $y' = -x/y$. At $(-5,0)$ the tangent line to the circle is vertical and the angle of inclination of the line is $\phi$, where $\tan \phi = 1/2$ ➙ $\phi \approx 27°$. So the required angle is $\alpha \approx 90°-27° = 63°$. At $(3,4)$ the slopes are $m_1 = 1/2$ and $m_2 = -3/4$, so $\tan \alpha = (-3/4-1/2)/[1+(1/2)(-3/4)] = -2$ ➙ $\alpha \approx 117°$ (or $63°$).

45. $x^2-y^2 = 3$ and $x^2-4x+y^2+3 = 0$ intersect when $x^2-4x+x^2 = 0$ ⟺ $2x(x-2) = 0$ ⟺ $x = 0$ or $2$, but $0$ is extraneous. $x^2-y^2 = 3$ ➙ $2x-2yy' = 0$ ➙ $y' = x/y$ and $x^2-4x+y^2+3 = 0$ ➙ $2x-4+2yy' = 0$ ➙ $y' = (2-x)/y$. At $(2,1)$ the slopes are $m_1 = 2$ and $m_2 = 0$ so $\tan \alpha = (0-2)/(1+2\cdot 0) = -2$ ➙ $\alpha \approx 117°$ (or $63°$). By symmetry, the angle at $(2,-1)$ is the same.

47. $xy = 1$ and $x^2y = 1$ intersect at $(1,1)$. $y = 1/x$ ➙ $y' = -1/x^2$ and $y = 1/x^2$ ➙ $y' = -2/x^3$. At $(1,1)$ the slopes are $m_1 = -1$ and $m_2 = -2$, so $\tan \alpha = (-2+1)/(1+(-1)(-2)) = -1/3$ ➙ $\alpha \approx 162°$ (or $18°$)

49. $y = \sin x$ and $y = \cos x$ intersect when $\sin x = \cos x$ ⟺ $x = n\pi+\pi/4$, n an integer. $y = \sin x$ ➙ $y' = \cos x$ and $y = \cos x$ ➙ $y' = -\sin x$. At $(2n\pi+\pi/4, 1/\sqrt{2})$ the slopes are $m_1 = 1/\sqrt{2}$ and $m_2 = -1/\sqrt{2}$, so $\tan \alpha = \dfrac{-1/\sqrt{2} - 1/\sqrt{2}}{1+(1/\sqrt{2})(-1/\sqrt{2})} = -2\sqrt{2}$ ➙ $\alpha \approx 109°$ (or $71°$)

By symmetry, at $((2n+1)\pi+\pi/4, -1/\sqrt{2})$ the angle is the same.

51.  The curves intersect when $x^3+x = 2x^2+x \iff x^3-2x^2 = x^2(x-2) = 0$,
     so $x = 0$ or 2.  $y = x^3+x \Rightarrow y' = 3x^2+1$ and $y = 2x^2+x \Rightarrow y' = 4x+1$
     At $(0,0)$ the slopes are $m_1 = 1 = m_2$ so the curves are tangent
     there.  At $(2,10)$ the slopes are $m_1 = 13$ and $m_2 = 9$, so the curves
     are not tangent there.

53.  $2x^2+y^2 = 3$ and $x = y^2$ intersect when $2x^2+x-3 = (2x+3)(x-1) = 0 \iff$
     $x = -3/2$ or 1, but $-3/2$ is extraneous.  $2x^2+y^2 = 3 \Rightarrow 4x+2yy' = 0$
     $\Rightarrow y' = -2x/y$ and $x = y^2 \Rightarrow 1 = 2yy' \Rightarrow y' = 1/2y$.  At $(1,1)$ the
     slopes are $m_1 = -2$ and $m_2 = 1/2$, so the curves are orthogonal
     there.  By symmetry they are also orthogonal at $(1,-1)$.

55.  $x^2+y^2 = r^2$ is a circle with center O and $ax+by = 0$ is a line
     through O.  By Exercise 35, the curves are orthogonal.

**55.** 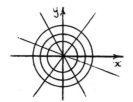   **57.**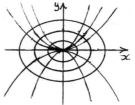

57.  $y = cx^2 \Rightarrow y' = 2cx$ and $x^2+2y^2 = k \Rightarrow 2x+4yy' = 0 \Rightarrow y' = -\dfrac{x}{2y}$
     $= -\dfrac{x}{2cx^2} = -\dfrac{1}{2cx}$ so the curves are orthogonal.

59.  $y = cx^2 \Rightarrow y' = 2cx$ and $y = 1-x$ has slope $-1$, so they will be
     orthogonal when $2cx = 1 \Rightarrow x = 1/2c$.  But the curves intersect when
     $cx^2 = 1-x$.  Putting $x = 1/2c$ in this equation we get $1/4c = 1-1/2c$
     $\Rightarrow c = 3/4$.

<br>

Section 3.6

<br>

Exercises 3.6

1.  $f(x) = x^4-3x^3+16x \Rightarrow f'(x) = 4x^3-9x^2+16 \Rightarrow f''(x) = 12x^2-18x$

3.  $h(x) = \sqrt{x^2+1} \Rightarrow h'(x) = (1/2)(x^2+1)^{-1/2}(2x) = x/\sqrt{x^2+1} \Rightarrow$

   $h''(x) = \dfrac{\sqrt{x^2+1} - x(x/\sqrt{x^2+1})}{x^2+1} = \dfrac{x^2+1-x^2}{(x^2+1)^{3/2}} = \dfrac{1}{(x^2+1)^{3/2}}$

5. $F(s) = (3s+5)^8 \Rightarrow F'(s) = 8(3s+5)^7(3) = 24(3s+5)^7 \Rightarrow$
$F''(s) = 168(3s+5)^6(3) = 504(3s+5)^6$

7. $y = \frac{x}{1-x} \Rightarrow y' = \frac{1(1-x)-x(-1)}{(1-x)^2} = \frac{1}{(1-x)^2} \Rightarrow y'' = -2(1-x)^{-3}(-1) = \frac{2}{(1-x)^3}$

9. $y = (1-x^2)^{3/4} \Rightarrow y' = (3/4)(1-x^2)^{-1/4}(-2x) = -(3/2)x(1-x^2)^{-1/4} \Rightarrow$
$y'' = -(3/2)(1-x^2)^{-1/4}-(3/2)x(-1/4)(1-x^2)^{-5/4}(-2x)$
$\quad = -(3/2)(1-x^2)^{-1/4}-(3/4)x^2(1-x^2)^{-5/4} = (3/4)(1-x^2)^{-5/4}(x^2-2)$

11. $H(t) = \tan^3(2t-1) \Rightarrow$
$H'(t) = 3\tan^2(2t-1)\sec^2(2t-1)(2) = 6\tan^2(2t-1)\sec^2(2t-1)$
$H''(t) = 12\tan(2t-1)\sec^2(2t-1)(2)\sec^2(2t-1)$
$\qquad\quad + 6\tan^2(2t-1)\,2\sec(2t-1)\sec(2t-1)\tan(2t-1)(2)$
$\qquad = 24\tan(2t-1)\sec^4(2t-1) + 24\tan^3(2t-1)\sec^2(2t-1)$

13. $F(r) = \sec\sqrt{r} \Rightarrow F'(r) = (\sec\sqrt{r}\,\tan\sqrt{r})/2\sqrt{r} \Rightarrow F''(r) =$
$$\frac{2\sqrt{r}[\{(\sec\sqrt{r}\,\tan\sqrt{r})/2\sqrt{r}\}\tan\sqrt{r}+\ \sec\sqrt{r}(\sec^2\sqrt{r})/2\sqrt{r}] - (\sec\sqrt{r}\,\tan\sqrt{r})/\sqrt{r}}{4r}$$
$\quad = (\sec\sqrt{r}\,\tan^2\sqrt{r} + \sec^3\sqrt{r} - r^{-1/2}\sec\sqrt{r}\,\tan\sqrt{r})/4r$

15. $y = ax^2+bx+c \Rightarrow y' = 2ax+b \Rightarrow y'' = 2a \Rightarrow y''' = 0$

17. $y = \sqrt{5t-1} \Rightarrow y' = (1/2)(5t-1)^{-1/2}(5) = (5/2)(5t-1)^{-1/2} \Rightarrow$
$y'' = -(5/4)(5t-1)^{-3/2}(5) = -(25/4)(5t-1)^{-3/2} \Rightarrow$
$y''' = (75/8)(5t-1)^{-5/2}(5) = (375/8)(5t-1)^{-5/2}$

19. $f(x) = (2-3x)^{-1/2} \Rightarrow f(0) = 2^{-1/2} = 1/\sqrt{2}$
$f'(x) = -\frac{1}{2}(2-3x)^{-3/2}(-3) = \frac{3}{2}(2-3x)^{-3/2} \Rightarrow f'(0) = \frac{3}{2}(2)^{-3/2} = 3/4\sqrt{2}$
$f''(x) = -\frac{9}{4}(2-3x)^{-5/2}(-3) = \frac{27}{4}(2-3x)^{-5/2} \Rightarrow f''(0) = \frac{27}{4}2^{-5/2} = 27/16\sqrt{2}$
$f'''(x) = \frac{405}{8}(2-3x)^{-7/2} \Rightarrow f'''(0) = \frac{405}{8}(2)^{-7/2} = 405/64\sqrt{2}$

21. $f(\theta) = \cot\theta \Rightarrow f'(\theta) = -\csc^2\theta \Rightarrow f''(\theta) = -2\csc\theta(-\csc\theta\cot\theta)$
$= 2\csc^2\theta\cot\theta \Rightarrow f'''(\theta) = 2(-2\csc^2\theta\cot\theta)\cot\theta$
$+ 2\csc^2\theta(-\csc^2\theta) = -2\csc^2\theta(2\cot^2\theta + \csc^2\theta) \Rightarrow$
$f'''(\pi/6) = -2(2)^2[2(\sqrt{3})^2+(2)^2] = -80$

23. $x^3+y^3 = 1 \Rightarrow 3x^2+3y^2y' = 0 \Rightarrow y' = -x^2/y^2 \Rightarrow$
$y'' = -\dfrac{2xy^2-2x^2yy'}{y^4} = -\dfrac{2xy^2-2x^2y(-x^2/y^2)}{y^4} = -\dfrac{2xy^3+2x^4}{y^5} = -\dfrac{2x(y^3+x^3)}{y^5}$
$= -2x/y^5$ since x, y must satisfy the original equation $x^3+y^3 = 1$.

25. $x^2+6xy+y^2 = 8 \Rightarrow 2x+6y+6xy'+2yy' = 0 \Rightarrow y' = -(x+3y)/(3x+y) \Rightarrow$

$y'' = -\dfrac{(1+3y')(3x+y)-(x+3y)(3+y')}{(3x+y)^2} = \dfrac{8(y-xy')}{(3x+y)^2} = \dfrac{8[y-x(-x-3y)/(3x+y)]}{(3x+y)^2}$

$= \dfrac{8[y(3x+y)+x(x+3y)]}{(3x+y)^3} = \dfrac{8(x^2+6xy+y^2)}{(3x+y)^3} = \dfrac{64}{(3x+y)^3}$

since $x$, $y$ must satisfy the original equation $x^2+6xy+y^2 = 8$.

27. $f(x) = x-x^2+x^3-x^4+x^5-x^6 \Rightarrow f'(x) = 1-2x+3x^2-4x^3+5x^4-6x^5 \Rightarrow$

$f''(x) = -2+6x-12x^2+20x^3-30x^4 \Rightarrow f'''(x) = 6-24x+60x^2-120x^3 \Rightarrow$

$f^{(4)}(x) = -24+120x-360x^2 \Rightarrow f^{(5)}(x) = 120-720x \Rightarrow f^{(6)}(x) = -720$

$\Rightarrow f^{(n)}(x) = 0$ for $7 \le n \le 73$

29. $f(x) = x^n \Rightarrow f'(x) = nx^{n-1} \Rightarrow f''(x) = n(n-1)x^{n-2} \Rightarrow \cdots \Rightarrow$

$f^{(n)}(x) = n(n-1)(n-2)\cdots 2\cdot 1\ x^{n-n} = n!$

31. $f(x) = 1/3x^3 = (1/3)x^{-3} \Rightarrow f'(x) = (1/3)(-3)x^{-4} \Rightarrow$

$f''(x) = (1/3)(-3)(-4)x^{-5} \Rightarrow f'''(x) = (1/3)(-3)(-4)(-5)x^{-6} \Rightarrow \cdots$

$\Rightarrow f^{(n)}(x) = (1/3)(-3)(-4)\cdots[-(n+2)]x^{-(n+3)}$

$= (-1)^n 3\cdot 4\cdot 5 \cdots (n+2)/3x^{n+3} = (-1)^n(n+2)!/6x^{n+3}$

33. In general, $Df(2x) = 2f'(2x)$, $D^2f(2x) = 4f''(2x)$, $\cdots$

$D^nf(2x) = 2^nf^{(n)}(2x)$. Since $f(x) = \cos x$ and $50 = 4(12)+2$, we have

$f^{(50)}(x) = f^{(2)}(x) = -\cos x$, so $D^{50}\cos 2x = -2^{50}\cos 2x$.

35. (a) $s = t^3-3t \Rightarrow v(t) = s'(t) = 3t^2-3 \Rightarrow a(t) = v'(t) = 6t$

(b) $a(1) = 6(1) = 6$ m/s$^2$

(c) $v(t) = 3t^2-3 = 0$ when $t^2 = 1$, i.e., $t = 1$ and $a(1) = 6$ m/s$^2$

37. (a) $s = At^2+Bt+C \Rightarrow v(t) = s'(t) = 2At+B \Rightarrow a(t) = 2A$

(b) $a(1) = 2A$ m/s$^2$

(c) $2A$ m/s$^2$ (since $a(t)$ is constant)

39. (a) $s(t) = t^4-4t^3+2 \Rightarrow v(t) = s'(t) = 4t^3-12t^2 \Rightarrow$

$a(t) = v'(t) = 12t^2-24t = 12t(t-2) = 0$ when $t = 0$, 2

(b) $s(0) = 2$ m, $v(0) = 0$ m/s, $s(2) = -14$ m, $v(2) = -16$ m/s

41. (a) $y(t) = A \sin \omega t \Rightarrow v(t) = y'(t) = A\omega \cos \omega t \Rightarrow$

$a(t) = v'(t) = -A\omega^2 \sin \omega t$

(b) $a(t) = -A\omega^2 \sin \omega t = -\omega^2 y(t)$

(c) $|v(t)| = A\omega|\cos \omega t|$ is a maximum when $\cos \omega t = \pm 1 \iff$

$\sin \omega t = 0 \iff a(t) = -A\omega^2 \sin \omega t = 0$

43. Let $P(x) = ax^2+bx+c$. Then $P'(x) = 2ax+b$ and $P''(x) = 2a$.

$P''(2) = 2 \Rightarrow 2a = 2 \Rightarrow a = 1$. $P'(2) = 3 \Rightarrow 4a+b = 4+b = 3 \Rightarrow$

$b = -1$. $P(2) = 5 \Rightarrow 2^2-2+c = 5 \Rightarrow c = 3$. So $P(x) = x^2-x+3$.

45.  $P(x) = c_n x^n + c_{n-1} x^{n-1} + \cdots + c_1 x + c_0 \Rightarrow P'(x) = n c_n x^{n-1} + (n-1) c_{n-1} x^{n-2} + \cdots$

$\Rightarrow P''(x) = n(n-1) c_n x^{n-2} + \cdots \Rightarrow P^{(n)}(x) = n(n-1)(n-2)\cdots(1) c_n x^{n-n}$

$= n! c_n$ which is a constant. Therefore $P^{(m)}(x) = 0$ for $m > n$.

47.  The Chain Rule says $\dfrac{dy}{dx} = \dfrac{dy}{du}\dfrac{du}{dx}$ and so

$$\frac{d^2 y}{dx^2} = \frac{d}{dx}\left[\frac{dy}{dx}\right] = \frac{d}{dx}\left[\frac{dy}{du}\frac{du}{dx}\right] = \left[\frac{d}{dx}\left[\frac{dy}{du}\right]\right]\frac{du}{dx} + \frac{dy}{du}\frac{d}{dx}\left[\frac{du}{dx}\right] \quad \text{(Product Rule)}$$

$$= \left[\frac{d}{du}\left[\frac{dy}{du}\right]\frac{du}{dx}\right]\frac{du}{dx} + \frac{dy}{du}\frac{d^2 u}{dx^2} = \frac{d^2 y}{du^2}\left[\frac{du}{dx}\right]^2 + \frac{dy}{du}\frac{d^2 u}{dx^2}$$

Section 3.7

Exercises 3.7

1.   (a) $V(x) = x^3 \Rightarrow$ the average rate of change is

(i) $\dfrac{V(6)-V(5)}{6-5} = 6^3 - 5^3 = 216 - 125 = 91$

(ii) $\dfrac{V(5.1)-V(5)}{5.1-5} = \dfrac{(5.1)^3 - 5^3}{0.1} = 76.51$

(iii) $\dfrac{V(5.01)-V(5)}{5.01-5} = \dfrac{(5.01)^3 - 5^3}{0.01} = 75.1501$

(b) $V'(x) = 3x^2$, so $V'(5) = 75$.

(c) The surface area is $S(x) = 6x^2$, so $V'(x) = 3x^2 = \frac{1}{2}(6x^2) = \frac{1}{2}S(x)$

3.   After $t$ seconds the radius is $r = 60t$, so the area is

$A(t) = \pi(60t)^2 = 3600\pi t^2 \Rightarrow A'(t) = 7200\pi t \Rightarrow$

(a) $A'(1) = 7200\pi$ cm$^2$/s    (b) $A'(3) = 21600\pi$ cm$^2$/s

(c) $A'(5) = 36000\pi$ cm$^2$/s

5.   $S(r) = 4\pi r^2 \Rightarrow S'(r) = 8\pi r$.    (a) $S'(1) = 8\pi$ ft$^2$/ft

(b) $S'(2) = 16\pi$ ft$^2$/ft    (c) $S'(3) = 24\pi$ ft$^2$/ft

7.   $f(x) = 3x^2$, so the linear density at $x$ is $\rho(x) = f'(x) = 6x$.

(a) $\rho(1) = 6$ kg/m    (b) $\rho(2) = 12$ kg/m    (c) $\rho(3) = 18$ kg/m

9.   $Q(t) = t^3 - 2t^2 + 6t + 2$, so the current is $Q'(t) = 3t^2 - 4t + 6$

(a) $Q'(0.5) = 3(.5)^2 - 4(.5) + 6 = 4.75$ A

(b) $Q'(1) = 3(1)^2 - 4(1) + 6 = 5$ A

11.　(a) $PV = C \Rightarrow V = \dfrac{C}{P} \Rightarrow \dfrac{dV}{dP} = - C/P^2$

　　(b) $\beta = - \dfrac{1}{V} \dfrac{dV}{dP} = - \dfrac{1}{V}\left[- \dfrac{C}{P^2}\right] = \dfrac{C}{(PV)P} = \dfrac{C}{CP} = \dfrac{1}{P}$

13.　$m(t) = 5 - 0.02t^2 \Rightarrow m'(t) = -0.04t \Rightarrow m'(1) = -0.04$

15.　$v(r) = \dfrac{P}{4\eta\ell}(R^2 - r^2) \Rightarrow v'(r) = \dfrac{P}{4\eta\ell}(-2r) = - \dfrac{Pr}{2\eta\ell}$ When $\ell = 3$,

　　$P = 3000,\ \eta = .027$, we have $v'(.005) = - \dfrac{3000(.005)}{2(.027)(3)} \approx -92.6$ cm/s/cm

17.　$C(x) = 420 + 1.5x + 0.002x^2 \Rightarrow C'(x) = 1.5 + 0.004x \Rightarrow$

　　$C'(100) = 1.5 + (.004)(100) = \$1.90/\text{item}$

　　$C(101) - C(100) = (420 + 151.5 + 20.402) - (420 + 150 + 20) = \$1.902/\text{item}$

19.　$C(x) = 2000 + 3x + 0.01x^2 + 0.0002x^3 \Rightarrow C'(x) = 3 + 0.02x + 0.0006x^2 \Rightarrow$

　　$C'(100) = 3 + 0.02(100) + 0.0006(10000) = 3 + 2 + 6 = \$11/\text{item}$

　　$C(101) - C(100) = (2000 + 303 + 102.1 + 206.0602) - (2000 + 300 + 100 + 200)$

　　$= 11.0702 \approx \$11.07/\text{item}$

Section 3.8

Exercises 3.8

1.　$V = x^3 \Rightarrow \dfrac{dV}{dt} = 3x^2 \dfrac{dx}{dt}$

3.　$xy = 1 \Rightarrow x\dfrac{dy}{dt} + y\dfrac{dx}{dt} = 0$ If $\dfrac{dx}{dt} = 4$ and $x = 2$, then $y = \dfrac{1}{2}$, so

　　$\dfrac{dy}{dt} = - \dfrac{y}{x} \dfrac{dx}{dt} = - \dfrac{1/2}{2}(4) = -1$

5.　If the radius is r and the diameter x, then $V = \dfrac{4}{3}\pi r^3 = \dfrac{\pi}{6}x^3 \Rightarrow$

　　$-1 = \dfrac{dV}{dt} = \dfrac{\pi}{2}x^2 \dfrac{dx}{dt} \Rightarrow \dfrac{dx}{dt} = - \dfrac{2}{\pi x^2}$ When $x = 10$, $\dfrac{dx}{dt} = - \dfrac{2}{\pi(100)} = - \dfrac{1}{50\pi}$

　　So the rate of decrease is $1/50\pi$ cm/min.

7.

We are given that $dx/dt = 5$ ft/s.

By similar triangles, $\dfrac{15}{6} = \dfrac{x+y}{y} \Rightarrow y = \dfrac{2}{3}x$

(a) The shadow moves at a rate of $\dfrac{d}{dt}(x+y)$

$= \dfrac{d}{dt}(x + \dfrac{2}{3}x) = \dfrac{5}{3}\dfrac{dx}{dt} = \dfrac{5}{3}(5) = \dfrac{25}{3}$ ft/s

(b) The shadow lengthens at a rate of

$\dfrac{dy}{dt} = \dfrac{d}{dt}\left[\dfrac{2}{3}x\right] = \dfrac{2}{3}\dfrac{dx}{dt} = \dfrac{2}{3}(5) = \dfrac{10}{3}$ ft/s

9.

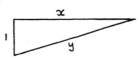

We are given that $dx/dt = 500$ mi/h.

By the Pythagorean theorem, $y^2 = x^2 + 1$, so

$2y\frac{dy}{dt} = 2x\frac{dx}{dt} \Rightarrow \frac{dy}{dt} = \frac{x}{y}\frac{dx}{dt} = 500\frac{x}{y}$ When $y = 2$

$x = \sqrt{3}$, so $\frac{dy}{dt} = 500(\sqrt{3}/2) = 250\sqrt{3}$ mi/h.

11.

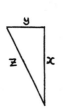

We are given that $\frac{dx}{dt} = 60$ mi/h, $\frac{dy}{dt} = 25$ mi/h.

$z^2 = x^2 + y^2 \Rightarrow 2z\frac{dz}{dt} = 2x\frac{dx}{dt} + 2y\frac{dy}{dt}$ After 2

hours, $x = 120$, $y = 50 \Rightarrow z = 130$, so

$\frac{dz}{dt} = \frac{1}{z}\left[x\frac{dx}{dt} + y\frac{dy}{dt}\right] = \frac{120(60)+50(25)}{130} = 65$ mi/h.

13.

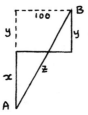

We are given that $\frac{dx}{dt} = 35$ km/h, $\frac{dy}{dt} = 25$ km/h.

$z^2 = (x+y)^2 + 100^2 \Rightarrow 2z\frac{dz}{dt} = 2(x+y)\left[\frac{dx}{dt} + \frac{dy}{dt}\right]$

At 4:00 P.M., $x = 140$, $y = 100 \Rightarrow z = 260$,

so $\frac{dz}{dt} = \frac{x+y}{z}\left[\frac{dx}{dt} + \frac{dy}{dt}\right] = \frac{140+100}{260}(35+25) = \frac{720}{13}$

$\approx 55.4$ km/h

15. $A = bh/2$, where $b$ is the base and $h$ is the altitude. We are given

that $\frac{dh}{dt} = 1$, $\frac{dA}{dt} = 2$. So $2 = \frac{dA}{dt} = \frac{1}{2}b\frac{dh}{dt} + \frac{1}{2}h\frac{db}{dt} = \frac{1}{2}b + \frac{1}{2}h\frac{db}{dt} \Rightarrow \frac{db}{dt} = \frac{4-b}{h}$

When $h = 10$ and $A = 100$, we have $b = 20$, so $\frac{db}{dt} = \frac{4-20}{10} = -1.6$ cm/min

17.

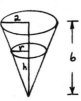

If C = rate at which water is pumped in, then

$dV/dt = C - 10{,}000$, where $V = (1/3)\pi r^2 h$ is the

volume at time t. By similar triangles,

$\frac{r}{2} = \frac{h}{6} \Rightarrow r = \frac{h}{3} \Rightarrow V = \frac{1}{3}\pi\left[\frac{h}{3}\right]^2 h = \frac{\pi}{27}h^3 \Rightarrow$

$\frac{dV}{dt} = \frac{\pi}{9}h^2\frac{dh}{dt}$ When $h = 200$, $\frac{dh}{dt} = 20$, so

$C - 10{,}000 = \frac{\pi}{9}(200)^2(20) \Rightarrow C = 10{,}000 + \frac{800000}{9}\pi \approx 2.89 \times 10^5$ cm$^3$/min

19.

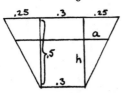

$V = (1/2)[.3 + (.3 + 2a)]h(10)$, where $\frac{a}{h} = \frac{.25}{.5} = \frac{1}{2}$

so $2a = h \Rightarrow V = 5(.6 + h)h = 3h + 5h^2 \Rightarrow$

$.2 = \frac{dV}{dt} = (3 + 10h)\frac{dh}{dt} \Rightarrow \frac{dh}{dt} = \frac{.2}{3+10h}$ When

$h = .3$, $\frac{dh}{dt} = \frac{.2}{3+10(.3)} = \frac{.2}{6}$ m/min $= \frac{10}{3}$ cm/min.

21. $PV = C \Rightarrow P\frac{dV}{dt} + V\frac{dP}{dt} = 0 \Rightarrow \frac{dV}{dt} = -\frac{V}{P}\frac{dP}{dt}$ When $V = 600$, $P = 150$,

and $\frac{dP}{dt} = 20$, we have $\frac{dV}{dt} = -\frac{600}{150}(20) = -80$, so the volume is

decreasing at a rate of 80 cm$^3$/min.

23. 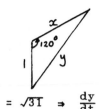 We are given that dx/dt = 300 km/h. By the
Law of Cosines, $y^2 = x^2 + 1 - 2x \cos 120°$
$= x^2 + 1 - 2x(-1/2) = x^2 + x + 1$, so $2y\frac{dy}{dt} = 2x\frac{dx}{dt} + \frac{dx}{dt}$

$\Rightarrow \frac{dy}{dt} = \frac{2x+1}{2y}\frac{dx}{dt}$  After 1 min, $x = \frac{300}{60} = 5 \Rightarrow$

$y = \sqrt{31} \Rightarrow \frac{dy}{dt} = \frac{2(5)+1}{2\sqrt{31}}(300) = \frac{1650}{\sqrt{31}} \approx 296$ km/h.

25.  We are given that dx/dt = 2 ft/s.

$x = 10 \sin \theta \Rightarrow \frac{dx}{dt} = 10 \cos \theta \frac{d\theta}{dt}$

When $\theta = \pi/4$, $\frac{d\theta}{dt} = \frac{2}{10(1/\sqrt{2})} = \frac{\sqrt{2}}{5}$ rad/s.

Section 3.9

Exercises 3.9

1. $y = x^5 \Rightarrow dy = 5x^4 dx$

3. $y = \sqrt{x^4 + x^2 + 1} \Rightarrow dy = \frac{1}{2}(x^4 + x^2 + 1)^{-1/2}(4x^3 + 2x)dx = \frac{2x^3 + x}{\sqrt{x^4 + x^2 + 1}} dx$

5. $y = \frac{x-2}{2x+3} \Rightarrow dy = \frac{(2x+3) - (x-2)(2)}{(2x+3)^2} dx = \frac{7}{(2x+3)^2} dx$

7. $y = \sin 2x \Rightarrow dy = 2 \cos 2x\ dx$

9. (a) $y = 1 - x^2 \Rightarrow dy = -2x\ dx$

(b) When $x = 5$ and $dx = 1/2$, $dy = -2(5)(1/2) = -5$

11. (a) $y = (x^2 + 5)^3 \Rightarrow dy = 3(x^2 + 5)^2 2x\ dx = 6x(x^2 + 5)^2 dx$

(b) When $x = 1$ and $dx = 0.05$, $dy = 6(1)(1^2 + 5)^2(0.05) = 10.8$

13. (a) $y = (3x+2)^{-1/3} \Rightarrow dy = (-1/3)(3x+2)^{-4/3}(3) = -(3x+2)^{-4/3}dx$

(b) When $x = 2$ and $dx = -0.04$, $dy = -8^{-4/3}(-0.04) = 0.0025$

15. (a) $y = \cos x \Rightarrow dy = -\sin x\ dx$

(b) When $x = \pi/6$ and $dx = 0.05$, $dy = -(1/2)(0.05) = -0.025$

17. $y = x^2$, $x = 1$, $\Delta x = 0.5$ $\Rightarrow$ $\Delta y = (1.5)^2 - 1^2 = 1.25$

$dy = 2x\ dx = 2(1)(0.5) = 1$

**17.**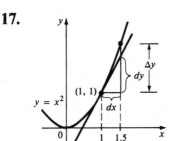

**19.**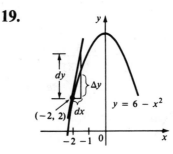

19. $y = 6 - x^2$, $x = -2$, $\Delta x = 0.4$ $\Rightarrow$ $\Delta y = [6-(-1.6)^2]-[6-(-2)^2] = 1.44$

$dy = -2x\ dx = -2(-2)(0.4) = 1.6$

21. $y = f(x) = 2x^3 + 3x - 4$, $x = 3$ $\Rightarrow$ $dy = (6x^2 + 3)dx = 57\ dx$

$\Delta x = 1$ $\Rightarrow$ $\Delta y = f(4)-f(3) = 136-59 = 77$, $dy = 57(1) = 57$

$\Delta y - dy = 77-57 = 20$

$\Delta x = .5$ $\Rightarrow$ $\Delta y = f(3.5)-f(3) = 92.25-59 = 33.25$, $dy = 57(.5) = 28.5$

$\Delta y - dy = 33.25-28.5 = 4.75$

$\Delta x = .1$ $\Rightarrow$ $\Delta y = f(3.1)-f(3) = 64.882-59 = 5.882$

$\Delta x = .01$ $\Rightarrow$ $\Delta y = f(3.01)-f(3) = 59.571802-59 = 0.571802$

$dy = 57(.01) = 0.57$, $\Delta y - dy = 0.571802-0.57 = .001802$

23. $y = f(x) = \sqrt{x}$ $\Rightarrow$ $dy = (1/2\sqrt{x})dx$. When $x = 36$ and $dx = 0.1$,

$dy = (1/2\sqrt{36})(.1) = 1/120$, so $\sqrt{36.1} = f(36.1) \approx f(36)+dy$

$= \sqrt{36} + 1/120 \approx 6.0083$

25. $y = \sqrt[3]{x}$ $\Rightarrow$ $dy = (1/3)x^{-2/3}dx$. When $x = 216$ and $dx = 2$,

$dy = (1/3)(216)^{-2/3}(2) = 1/54$, so $\sqrt[3]{218} = f(218) \approx f(216)+dy$

$= 6 + 1/54 \approx 6.0185$

27. $y = f(x) = 1/x$ $\Rightarrow$ $dy = (-1/x^2)dx$. When $x = 10$ and $dx = 0.1$,

$dy = (-1/100)(0.1) = -.001$, so $1/10.1 = f(10.1) \approx f(10)+dy$

$= 0.1-0.001 = 0.099$

29. $y = f(x) = \sin x$ $\Rightarrow$ $dy = \cos x\ dx$. When $x = \pi/3$ and $dx = -\pi/180$,

$dy = \cos(\pi/3)(-\pi/180) = -\pi/360$, so $\sin 59^\circ = f(59\pi/180) \approx f(\pi/3)+dy$

$= \sqrt{3}/2 - \pi/360 \approx 0.857$

31. $y = f(x) = \tan x$ $\Rightarrow$ $dy = \sec^2 x\ dx$. When $x = \pi/4$ and $dx = \pi/90$,

$dy = \sec^2(\pi/4)(\pi/90) = 2\pi/90 = \pi/45$, so $\tan 47^\circ = f(47\pi/180)$

$\approx f(\pi/4)+dy = 1 + \pi/45 \approx 1.07$

33. (a) If x is the edge length, then $V = x^3 \Rightarrow dV = 3x^2 dx$. When
x = 30 and dx = 0.1, $dV = 3(30)^2(.1) = 270$, so the maximum error
$\approx 270$ cm$^3$       (b) $S = 6x^2 \Rightarrow dS = 12x\ dx$. When x = 30 and
dx = 0.1, $dS = 12(30)(.1) = 36$, so the maximum error $\approx 36$ cm$^2$

35. (a) For a sphere of radius r, the circumference is $C = 2\pi r$ and
surface area is $S = 4\pi r^2$, so $r = C/2\pi \Rightarrow S = 4\pi(C/2\pi)^2 = C^2/\pi \Rightarrow$
$dS = (2/\pi)C\ dC$. When C = 84 and dC = 0.5, $dS = (2/\pi)(84)(.5) = 84/\pi$
so the maximum error $\approx 84/\pi \approx 27$ cm$^2$
(b) Relative error $\approx \dfrac{dS}{S} = \dfrac{84/\pi}{84^2/\pi} = 1/84 \approx 0.012$

37. (a) $V = \pi r^2 h \Rightarrow \varDelta V \approx dV = 2\pi rh\ dr = 2\pi rh\ \varDelta r$
(b) $\varDelta V = \pi(r+\varDelta r)^2 h - \pi r^2 h$, so the error is
$\varDelta V - dV = \pi(r+\varDelta r)^2 h - \pi r^2 h - 2\pi rh\ \varDelta r = \pi(\varDelta r)^2 h$

39. (a) $dc = \dfrac{dc}{dx}\ dx = 0\ dx = 0$       (b) $d(cu) = \dfrac{d}{dx}(cu)dx = c\ \dfrac{du}{dx}\ dx = c\ du$

(c) $d(u+v) = \dfrac{d}{dx}(u+v)dx = \left[\dfrac{du}{dx} + \dfrac{dv}{dx}\right]dx = \dfrac{du}{dx}\ dx + \dfrac{dv}{dx}\ dx = du + dv$

(d) $d(uv) = \dfrac{d}{dx}(uv)dx = \left[u\dfrac{dv}{dx} + v\dfrac{du}{dx}\right]dx = u\ \dfrac{dv}{dx}\ dx + v\ \dfrac{du}{dx}\ dx$

$\qquad\qquad = u\ dv + v\ du$

(e) $d\left[\dfrac{u}{v}\right] = \dfrac{d}{dx}\left[\dfrac{u}{v}\right]dx = \dfrac{v\dfrac{du}{dx} - u\dfrac{dv}{dx}}{v^2}\ dx = \dfrac{v\ \dfrac{du}{dx}\ dx - u\ \dfrac{dv}{dx}\ dx}{v^2} = \dfrac{v\ du - u\ dv}{v^2}$

(f) $d(x^n) = \dfrac{d}{dx}(x^n)dx = nx^{n-1}dx$

Exercises 3.10

1. $f(x) = x^3+x+1 \Rightarrow f'(x) = 3x^2+1$, so $x_{n+1} = x_n - \dfrac{x_n^3+x_n+1}{3x_n^2+1}$       $x_1 = -1$

$\Rightarrow x_2 = -1 - \dfrac{-1-1+1}{3\cdot1+1} = -0.75 \Rightarrow x_3 = -.75 - \dfrac{(-.75)^3-.75+1}{3(-.75)^2+1} \approx -0.6860$

3. $f(x) = x^5 - 10 \Rightarrow f'(x) = 5x^4$, so $x_{n+1} = x_n - \dfrac{x_n^5 - 10}{5x_n^4}$ $\qquad x_1 = 1.5 \Rightarrow$

$x_2 = 1.5 - \dfrac{(1.5)^5 - 10}{5(1.5)^4} \approx 1.5951 \Rightarrow x_3 = 1.5951 - \dfrac{f(1.5951)}{f'(1.5951)} \approx 1.5850$

5. Finding $\sqrt[4]{22}$ is equivalent to finding the positive root of $x^4 - 22 = 0$

so we take $f(x) = x^4 - 22 \Rightarrow f'(x) = 4x^3$ and $x_{n+1} = x_n - \dfrac{x_n^4 - 22}{4x_n^3}$

Taking $x_1 = 2$, we get $x_2 = 2.1875$, $x_3 \approx 2.166059$, $x_4 \approx 2.165737$

$x_5 \approx 2.165737$. Thus $\sqrt[4]{22} \approx 2.165737$ to 6 decimal places.

7. $f(x) = x^3 - 2x - 1 \Rightarrow f'(x) = 3x^2 - 2$, so $x_{n+1} = x_n - \dfrac{x_n^3 - 2x_n - 1}{3x_n^2 - 2}$

Taking $x_1 = 1.5$, we get $x_2 \approx 1.631579$, $x_3 \approx 1.618184$,

$x_4 \approx 1.618034$, $x_5 \approx 1.618034$. So the root is $1.618034$ to 6

decimal places.

9. $f(x) = x^4 + x^3 - 22x^2 - 2x + 41 \Rightarrow f'(x) = 4x^3 + 3x^2 - 44x - 2$, so

$x_{n+1} = x_n - \dfrac{x_n^4 + x_n^3 - 22x_n^2 - 2x_n + 41}{4x_n^3 + 3x_n^2 - 44x_n - 2}$ $\qquad$ Taking $x_1 = 4$, we get $x_2 \approx 3.992063$

$x_3 \approx 3.992020$, $x_4 \approx 3.992020$. So the root in the interval $[3,4]$

is $3.992020$ to 6 decimal places.

11.

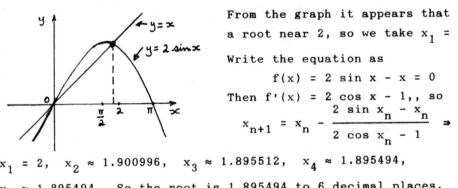

From the graph it appears that there a root near 2, so we take $x_1 = 2$.

Write the equation as
$$f(x) = 2 \sin x - x = 0$$
Then $f'(x) = 2 \cos x - 1$, so
$$x_{n+1} = x_n - \frac{2 \sin x_n - x_n}{2 \cos x_n - 1} \Rightarrow$$

$x_1 = 2$, $x_2 \approx 1.900996$, $x_3 \approx 1.895512$, $x_4 \approx 1.895494$,

$x_5 \approx 1.895494$. So the root is $1.895494$ to 6 decimal places.

13. $f(x) = x^3-4x+1 \Rightarrow f'(x) = 3x^2-4$, so $x_{n+1} = x_n - \dfrac{x_n^3-4x_n+1}{3x_n^2-4}$

Observe that $f(-3) = -14$, $f(-2) = 1$, $f(0) = 1$, $f(1) = -2$, $f(2) = 1$, so there are roots in $[-3,-2]$, $[0,1]$, and $[1,2]$.

| For $[-3,-2]$: | For $[0,1]$: | For $[1,2]$: |
|---|---|---|
| $x_1 = -2$ | $x_1 = 0$ | $x_1 = 2$ |
| $x_2 = -2.125$ | $x_2 = 0.25$ | $x_2 = 1.875$ |
| $x_3 \approx -2.114975$ | $x_3 \approx 0.254098$ | $x_3 \approx 1.860979$ |
| $x_4 \approx -2.114908$ | $x_4 \approx 0.254102$ | $x_4 \approx 1.860806$ |
| $x_5 \approx -2.114908$ | $x_5 \approx 0.254102$ | $x_5 \approx 1.860806$ |

To 6 decimal places, the roots are $-2.114908$, $0.254102$, $1.860806$.

15. $f(x) = x^4+x^2-x-1 \Rightarrow f'(x) = 4x^3+2x-1$, so $x_{n+1} = x_n - \dfrac{x_n^4+x_n^2-x_n-1}{4x_n^3+2x_n-1}$

Note that $f(1) = 0$, so $x = 1$ is a root. Also $f(-1) = 2$ and $f(0) = -1$, so there is a root in $[-1,0]$. A sketch shows that these are the only roots. Taking $x_1 = -0.5$, we have $x_2 = -0.575$, $x_3 \approx -0.569867$, $x_4 \approx -0.569840$, $x_5 \approx -0.569840$. The roots are 1 and $-0.569840$, to 6 decimal places.

17.

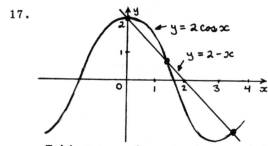

Obviously $x = 0$ is a root. From the sketch, there appear to be roots near 1 and 3.5. Write the equation as $f(x) = 2 \cos x + x - 2 = 0$. Then $f'(x) = -2 \sin x + 1$, so

$$x_{n+1} = x_n - \frac{2 \cos x_n + x_n - 2}{1 - 2 \sin x_n}$$

Taking $x_1 = 1$, we get $x_2 \approx 1.118026$, $x_3 \approx 1.109188$, $x_4 \approx 1.109144$ $x_5 \approx 1.109144$. Taking $x_1 = 3.5$, we get $x_2 \approx 3.719159$, $x_3 \approx 3.698331$, $x_4 \approx 3.698154$, $x_5 \approx 3.698154$. To 6 decimal places the roots are 0, 1.109144, and 3.698154.

19.  (a) $f(x) = x^2-a \Rightarrow f'(x) = 2x$, so Newton's Method gives

$$x_{n+1} = x_n - \frac{x_n^2-a}{2x_n} = x_n - \frac{1}{2}x_n + \frac{a}{2x_n} = \frac{1}{2}\left[x_n + \frac{a}{x_n}\right]$$

(b) Using (a) with $x_1 = 30$, we get $x_2 \approx 31.666667$, $x_3 \approx 31.622807$,

$x_4 \approx 31.622777$, $x_5 \approx 31.622777$. So $\sqrt{1000} \approx 31.622777$.

## Chapter 3 Review

## Review Exercises for Chapter 3

1.  $f(x) = x^3+5x+4 \Rightarrow f'(x) = \lim\limits_{h\to 0} \dfrac{f(x+h)-f(x)}{h}$

$= \lim\limits_{h\to 0} \dfrac{(x+h)^3+5(x+h)+4-(x^3+5x+4)}{h} = \lim\limits_{h\to 0} \dfrac{3x^2h+3xh^2+h^3+5h}{h}$

$= \lim\limits_{h\to 0} (3x^2+3xh+h^2+5) = 3x^2+5$

3.  $f(x) = \sqrt{3-5x} \Rightarrow f'(x) = \lim\limits_{h\to 0} \dfrac{f(x+h)-f(x)}{h} = \lim\limits_{h\to 0} \dfrac{\sqrt{3-5(x+h)} - \sqrt{3-5x}}{h}$

$= \lim\limits_{h\to 0} \dfrac{\sqrt{3-5x-5h} - \sqrt{3-5x}}{h} \left[\dfrac{\sqrt{3-5x-5h} + \sqrt{3-5x}}{\sqrt{3-5x-5h} + \sqrt{3-5x}}\right] = \lim\limits_{h\to 0} \dfrac{-5h}{h(\sqrt{3-5x-5h} + \sqrt{3-5x})}$

$= \lim\limits_{h\to 0} \dfrac{-5}{\sqrt{3-5x-5h} + \sqrt{3-5x}} = \dfrac{-5}{2\sqrt{3-5x}}$

5.  $y = (x+2)^8(x+3)^6 \Rightarrow y' = 6(x+3)^5(x+2)^8 + 8(x+2)^7(x+3)^6$

$= 2(7x+18)(x+2)^7(x+3)^5$

7.  $y = \dfrac{x}{\sqrt{9-4x}} \Rightarrow y' = \dfrac{\sqrt{9-4x} - x(-4/2\sqrt{9-4x})}{9-4x} = \dfrac{9-4x+2x}{(9-4x)^{3/2}} = \dfrac{9-2x}{(9-4x)^{3/2}}$

9.  $x^2y^3+3y^2 = x-4y \Rightarrow 2xy^3+3x^2y^2y'+6yy' = 1-4y' \Rightarrow y' = \dfrac{1-2xy^3}{3x^2y^2+6y+4}$

11.  $y = \sqrt{x\sqrt{x\sqrt{x}}} = [x(x^{3/2})^{1/2}]^{1/2} = [x(x^{3/4})]^{1/2} = x^{7/8} \Rightarrow y' = \dfrac{7}{8} x^{-1/8}$

13.  $y = \dfrac{x}{8-3x} \Rightarrow y' = \dfrac{(8-3x)-x(-3)}{(8-3x)^2} = \dfrac{8}{(8-3x)^2}$

15.  $y = (x \tan x)^{1/5} \Rightarrow y' = (1/5)(x \tan x)^{-4/5}(\tan x + x \sec^2 x)$

17. $x^2 = y(y+1) = y^2+y \Rightarrow 2x = 2yy'+y' \Rightarrow y' = 2x/(2y+1)$

19. $y = \dfrac{(x-1)(x-4)}{(x-2)(x-3)} = \dfrac{x^2-5x+4}{x^2-5x+6} \Rightarrow y' = \dfrac{(x^2-5x+6)(2x-5)-(x^2-5x+4)(2x-5)}{(x^2-5x+6)^2}$

$= \dfrac{2(2x-5)}{(x-2)^2(x-3)^2}$

21. $y = \tan\sqrt{1-x} \Rightarrow y' = (\sec^2\sqrt{1-x})(1/2\sqrt{1-x})(-1) = -(\sec^2\sqrt{1-x})/2\sqrt{1-x}$

23. $y = \sin(\tan\sqrt{1+x^3}) \Rightarrow y' = \cos(\tan\sqrt{1+x^3})(\sec^2\sqrt{1+x^3})(3x^2/2\sqrt{1+x^3})$

25. $y = \cot(3x^2+5) \Rightarrow y' = -\csc^2(3x^2+5)(6x) = -6x\csc^2(3x^2+5)$

27. $y = (\sin mx)/x \Rightarrow y' = (mx\cos mx - \sin mx)/x^2$

29. $y = \cos^2(\tan x) \Rightarrow y' = 2\cos(\tan x)(-\sin(\tan x))\sec^2 x$

$= -\sin(2\tan x)\sec^2 x$

31. $y = \sqrt{7-x^2}(x^3+7)^5 \Rightarrow y' = (1/2)(7-x^2)^{-1/2}(-2x)(x^3+7)^5$

$+ \sqrt{7-x^2}[5(x^3+7)^4(3x^2)] = -x(x^3+7)^5/\sqrt{7-x^2} + 15x^2(x^3+7)^4\sqrt{7-x^2}$

33. $f(x) = (2x-1)^{-5} \Rightarrow f'(x) = -5(2x-1)^{-6}(2) = -10(2x-1)^{-6} \Rightarrow$

$f''(x) = 60(2x-1)^{-7}(2) = 120(2x-1)^{-7} \Rightarrow f''(0) = 120(-1)^{-7} = -120$

35. $x^6+y^6 = 1 \Rightarrow 6x^5+6y^5y' = 0 \Rightarrow y' = -x^5/y^5 \Rightarrow$

$y'' = -\dfrac{5x^4y^5-x^5(5y^4y')}{y^{10}} = -\dfrac{5x^4y^5-5x^5y^4(-x^5/y^5)}{y^{10}} = -\dfrac{5x^4y^6+5x^{10}}{y^{11}}$

$= -\dfrac{5x^4(y^6+x^6)}{y^{11}} = -\dfrac{5x^4}{y^{11}}$

37. $y = \dfrac{x}{x^2-2} \Rightarrow y' = \dfrac{(x^2-2)-x(2x)}{(x^2-2)^2} = \dfrac{-x^2-2}{(x^2-2)^2}$   When $x = 2$, $y' = \dfrac{-2^2-2}{(2^2-2)^2}$

$= -3/2$, so the equation of the tangent at $(2,1)$ is

$y-1 = (-3/2)(x-2)$   or   $3x+2y-8 = 0$.

39. $y = \tan x \Rightarrow y' = \sec^2 x$   When $x = \pi/3$, $y' = 2^2 = 4$, so the tangent

at $(\pi/3,\sqrt{3})$ is $y-\sqrt{3} = 4(x-\pi/3)$   or   $y = 4x+\sqrt{3}-4\pi/3$.

41. $y = \sin x + \cos x \Rightarrow y' = \cos x - \sin x = 0 \Leftrightarrow \cos x = \sin x$ and

$0 \leq x \leq 2\pi \Leftrightarrow x = \pi/4, 5\pi/4$, so the points are $(\pi/4,\sqrt{2})$, $(5\pi/4,-\sqrt{2})$.

43. (a) $y = t^3-12t+3 \Rightarrow v(t) = y' = 3t^2-12$, $a(t) = v'(t) = 6t$

(b) $v(t) = 3(t^2-4)>0$ when $t>2$, so it moves upward when $t>2$ and

downward when $0 \leq t<2$.

(c) Distance upward $= y(3)-y(2) = -6-(-13) = 7$.   Distance downward

$= y(0)-y(2) = 3-(-13) = 16$.   Total distance $= 7+16 = 23$

45. $f(x) = x^2g(x) \Rightarrow f'(x) = 2xg(x) + x^2g'(x)$

47. $f(x) = [g(x)]^2 \Rightarrow f'(x) = 2g(x)g'(x)$

49. $f(x) = g(g(x)) \Rightarrow f'(x) = g'(g(x))g'(x)$

51. $\rho = x(1+\sqrt{x}) = x+x^{3/2} \Rightarrow d\rho/dx = 1+(3/2)\sqrt{x}$, so the density when $x = 4$ is $1+(3/2)\sqrt{4} = 4$ kg/m.

53. If $x$ = edge length, then $V = x^3 \Rightarrow dV/dt = 3x^2 dx/dt = 10 \Rightarrow dx/dt = 10/3x^2$ and $S = 6x^2 \Rightarrow dS/dt = (12x)dx/dt = 12x(10/3x^2) = 40/x$. When $x = 30$, $dS/dt = 40/30 = 4/3$ cm$^2$/min.

55.

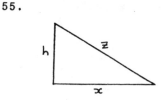

Given $dh/dt = 5$, $dx/dt = 15$, find $dz/dt$.
$z^2 = x^2+h^2 \Rightarrow 2z\dfrac{dz}{dt} = 2x\dfrac{dx}{dt} + 2h\dfrac{dh}{dt} \Rightarrow$

$\dfrac{dz}{dt} = \dfrac{1}{z}(15x+5h)$. When $t = 3$, $h = 45+3(5) = 60$

and $x = 15(3) = 45 \Rightarrow z = 75$, so

$\dfrac{dz}{dt} = \dfrac{1}{75}[15(45)+5(60)] = 13$ ft/s

57. $y = (4-x^2)^{3/2} \Rightarrow dy = (3/2)(4-x^2)^{1/2}(-2x)dx = -3x(4-x^2)^{1/2}dx$

59. Let $y = f(x) = 8+\sqrt{x}$. Then $dy = (1/2\sqrt{x})dx$. When $x = 144$ and $dx = \Delta x = -0.4$, $dy = (1/24)(-0.4) = -1/60$. Thus $8+\sqrt{143.6} = f(143.6) \approx f(144)+dy = 20-1/60 \approx 19.983$.

61. $f(x) = x^4+x-1 \Rightarrow f'(x) = 4x^3+1 \Rightarrow x_{n+1} = x_n - \dfrac{x_n^4+x_n-1}{4x_n^3+1}$ If $x_1 = .5$

then $x_2 \approx .791667$, $x_3 \approx .729862$, $x_4 \approx .724528$, $x_5 \approx .724492$,

$x_6 \approx .724492$, so, to 6 decimal places, the root is $.724492$.

63. $y = x^6+2x^2-8x+3$ has a horizontal tangent when $y' = 6x^5+4x-8 = 0$.

Let $f(x) = 6x^5+4x-8$. Then $f'(x) = 30x^4+4$, so $x_{n+1} = x_n - \dfrac{6x_n^5+4x_n-8}{30x_n^4+4}$

A sketch shows the root is near 1, so we take $x_1 = 1$. Then

$x_2 \approx 0.9412$, $x_3 \approx 0.9341$, $x_4 \approx 0.9340$, $x_5 \approx 0.9340$.

Thus, to 4 decimal places, the point is $(0.9340,-2.0634)$.

Exercises 4.1

1. Absolute maximum at e; absolute minimum at d; local maximum at b,e; local minimum at d, s.

3. $f(x) = 1+2x$, $x \geq -1$. Absolute minimum $f(-1) = -1$; no local minimum. No local or absolute maximum.

**3.**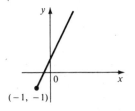

**5.**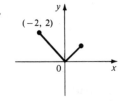

5. $f(x) = |x|$, $-2 \leq x \leq 1$. Absolute maximum $f(-2) = 2$; no local maximum. Absolute and local minimum $f(0) = 0$.

7. $f(x) = 1-x^2$, $0 < x < 1$. No extrema.

**7.**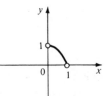

**9.**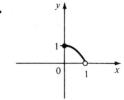

9. $f(x) = 1-x^2$, $0 \leq x < 1$. Absolute maximum $f(0) = 1$; no local maximum. No absolute or local minimum.

11. $f(x) = 1-x^2$, $-2 \leq x \leq 1$. Absolute and local maximum $f(0) = 1$. Absolute minimum $f(-2) = -3$; no local minimum.

**11.**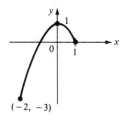

**13.**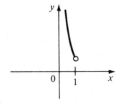

13. $f(t) = 1/t$, $0 < t < 1$. No extrema.

15.  $f(\theta) = \sin \theta$, $-2\pi \le \theta \le 2\pi$.  Absolute and local maxima
$f(-3\pi/2) = f(\pi/2) = 1$.  Absolute and local minima
$f(-\pi/2) = f(3\pi/2) = -1$.

**15.**

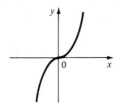

**17.**

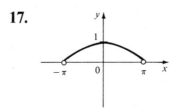

17.  $f(\theta) = \cos(\theta/2)$, $-\pi < \theta < \pi$.  Local and absolute maximum $f(0) = 1$.
No local or absolute minimum.

19.  $f(x) = x^5$.  No extrema.

**19.**

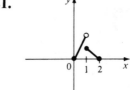

**21.**

21.  $f(x) = 2x$ if $0 \le x < 1$, $f(x) = 2 - x$ if $1 \le x \le 2$.  Absolute minima
$f(0) = f(2) = 0$; no local minima.  No absolute or local maximum.

23.  $f(x) = 2x - 3x^2$ ➔ $f'(x) = 2 - 6x = 0$ ⟺ $x = 1/3$.  So the critical
number is 1/3.

25.  $f(x) = x^3 - 3x + 1$ ➔ $f'(x) = 3x^2 - 3 = 3(x^2 - 1) = 3(x+1)(x-1)$.  So the
critical numbers are $x = \pm 1$.

27.  $f(t) = 2t^3 + 3t^2 + 6t + 4$ ➔ $f'(t) = 6t^2 + 6t + 6$.  But $t^2 + t + 1 = 0$ has no
real solutions (by the quadratic formula).  No critical numbers.

29.  $s(t) = 2t^3 + 3t^2 - 6t + 4$ ➔ $s'(t) = 6t^2 + 6t - 6 = 6(t^2 + t - 1)$.  By the
quadratic formula, the critical numbers are $t = (-1 \pm \sqrt{5})/2$.

31.  $g(x) = \sqrt[9]{x} = x^{1/9}$ ➔ $g'(x) = \frac{1}{9}x^{-8/9} = 1/\sqrt[9]{x^8} \ne 0$, but $g'(0)$ does not
exist, so $x = 0$ is a critical number.

33.  $g(t) = 5t^{2/3} + t^{5/3}$ ➔ $g'(t) = (10/3)t^{-1/3} + (5/3)t^{2/3}$.  $g'(0)$ does
not exist, so $t = 0$ is a critical number.
$g'(t) = (5/3)t^{-1/3}(2+t) = 0$ ⟺ $t = -2$, so $t = -2$ is also a
critical number.

35. $f(r) = \dfrac{r}{r^2+1} \Rightarrow f'(r) = \dfrac{1(r^2+1)-r(2r)}{(r^2+1)^2} = \dfrac{-r^2+1}{(r^2+1)^2} = 0 \iff r^2 = 1 \iff$

 $r = \pm 1$, so these are the critical numbers.

37. $F(x) = x^{4/5}(x-4)^2 \Rightarrow F'(x) = (4/5)x^{-1/5}(x-4)^2+2x^{4/5}(x-4)$

 $= (x-4)(7x-8)/5x^{1/5} = 0$ when $x = 4$, $8/7$ and $F'(0)$ does not exist.

 Critical numbers are 0, 8/7, 4.

39. $V(x) = x\sqrt{x-2}$. $V'(x) = \sqrt{x-2} + x/2\sqrt{x-2} \Rightarrow V'(2)$ does not exist.

 For $x > 2$ (the domain of $V'(x)$), $V'(x) > 0$, so 2 is the only

 critical number.

41. $f(\theta) = \sin^2(2\theta) \Rightarrow f'(\theta) = 2\sin(2\theta)\cos(2\theta)(2) = 2\sin 4\theta = 0 \iff$

 $\sin(4\theta) = 0 \iff 4\theta = n\pi$, $n$ an integer. So $\theta = n\pi/4$ are the

 critical numbers.

43. $f(x) = x^2-2x+2$, $[0,3]$. $f'(x) = 2x-2 = 0 \iff x = 1$. $f(0) = 2$,

 $f(1) = 1$, $f(3) = 5$. So $f(3) = 5$ is the absolute maximum and $f(1)$

 $= 1$ the absolute minimum.

45. $f(x) = x^3-12x+1$, $[-3,5]$. $f'(x) = 3x^2-12 = 3(x^2-4) = 3(x+2)(x-2)$

 $= 0 \iff x = \pm 2$. $f(-3) = 10$, $f(-2) = 17$, $f(2) = -15$, $f(5) = 66$.

 So $f(2) = -15$ is the absolute minimum and $f(5) = 66$ is the absolute

 maximum.

47. $f(x) = 2x^3+3x^2+4$, $[-2,1]$. $f'(x) = 6x^2+6x = 6x(x+1) = 0 \iff$

 $x = -1,0$. $f(-2) = 0$, $f(-1) = 5$, $f(0) = 4$, $f(1) = 9$. So $f(1) = 9$

 is the absolute maximum and $f(-2) = 0$ is the absolute minimum.

49. $f(x) = x^4-4x^2+2$, $[-3,2]$. $f'(x) = 4x^3-8x = 4x(x^2-2) = 0 \iff$

 $x = 0$, $\pm\sqrt{2}$. $f(-3) = 47$, $f(-\sqrt{2}) = -2$, $f(0) = 2$, $f(\sqrt{2}) = -2$,

 $f(2) = 2$, so $f(\pm\sqrt{2}) = -2$ is the absolute minimum and $f(-3) = 47$ is

 the absolute maximum.

51. $f(x) = x^2+2/x$, $[1/2,2]$. $f'(x) = 2x-2/x^2 = 2(x^3-1)/x^2 = 0 \iff$

 $x = 1$. $f(1/2) = 17/4$, $f(1) = 3$, $f(2) = 5$. So $f(1) = 3$ is the

 absolute minimum and $f(2) = 5$ is the absolute maximum.

53. $f(x) = x^{4/5}$, $[-32,1]$. $f'(x) = (4/5)x^{-1/5} \Rightarrow f'(x) \neq 0$ but $f'(0)$

 does not exist, so 0 is the only critical number. $f(-32) = 16$,

 $f(0) = 0$, $f(1) = 1$. So $f(0) = 0$ is the absolute minimum and

 $f(-32) = 16$ is the absolute maximum.

55. $f(x) = |x-1|-1$, $[-1,2]$. $f'(x) = -1$ if $x < 1$, $f'(x) = 1$ if $x > 1$

 and $f'(1)$ does not exist, so 1 is the only critical number.

 $f(-1) = 1$, $f(1) = -1$, $f(2) = 0$. So $f(1) = -1$ is the absolute

 minimum and $f(-1) = 1$ is the absolute maximum.

57. $f(x) = \sin x + \cos x$, $[0, \pi/3]$. $f'(x) = \cos x - \sin x = 0$ $\Longleftrightarrow$ $x = \pi/4$. $f(0) = 1$, $f(\pi/4) = \sqrt{2}$, $f(\pi/3) = (\sqrt{3}+1)/2$. So $f(0) = 1$ is the absolute minimum and $f(\pi/4) = \sqrt{2}$ the absolute maximum.

59. $f(x) = x^5$. $f'(x) = 5x^4 \Rightarrow f'(0) = 0$ so 0 is a critical number. But $f(0) = 0$ and $f$ takes both positive and negative values in any open interval containing 0, so $f$ does not have a local extremum at 0.

61. $f(x) = x^{101}+x^{51}+x+1 \Rightarrow f'(x) = x^{100}+x^{50}+1 \geq 1$ for all $x$, so $f'(x) = 0$ has no solutions. Thus $f(x)$ has no critical numbers, so $f(x)$ can have no local extrema.

63.

65. If $f$ has a local minimum at $c$, then $g(x) = -f(x)$ has a local maximum at $c$, so $g'(c) = 0$ by the case of Fermat's Theorem proved in the text. Thus $f'(c) = -g'(c) = 0$.

## Section 4.2

## Exercises 4.2

1. $f(x) = x^3-x$, $[-1,1]$. $f$, being a polynomial, is continuous on $[-1,1]$ and differentiable on $(-1,1)$. Also $f(-1) = 0 = f(1)$. $f'(c) = 3c^2-1 = 0 \Rightarrow c = \pm 1/\sqrt{3}$

3. $f(x) = \cos 2x$, $[0,\pi]$. $f$ is continuous on $[0,\pi]$ and differentiable on $(0,\pi)$. Also $f(0) = 1 = f(\pi)$. $f'(c) = -2 \sin 2c = 0 \Rightarrow$ $\sin 2c = 0 \Rightarrow 2c = \pi \Rightarrow c = \pi/2$ (since $c \in (0,\pi)$)

5. $f(x) = 1-x^{2/3}$. $f(-1) = 1-(-1)^{2/3} = 1-1 = 0 = f(1)$. $f'(x) = -(2/3)x^{-1/3} \Rightarrow f'(c) = 0$ has no solutions. This does not contradict Rolle's Theorem since $f'(0)$ does not exist.

7. $f(x) = 1-x^2$, $[0,3]$. f, being a polynomial, is continuous on $[0,3]$ and differentiable on $(0,3)$. $\frac{f(3)-f(0)}{3-0} = \frac{-8-1}{3} = -3$ and

$-3 = f'(c) = -2c \Rightarrow c = 3/2$.

9. $f(x) = x^3-2x+1$, $[-2,3]$. f, being a polynomial, is continuous on $[-2,3]$ and differentiable on $(-2,3)$. $\frac{f(3)-f(-2)}{3-(-2)} = \frac{22-(-3)}{5} = 5$ and

$5 = f'(c) = 3c^2-2 \Rightarrow 3c^2 = 7 \Rightarrow c = \pm\sqrt{7/3}$.

11. $f(x) = 1/x$, $[1,2]$. f, being a rational function, is continuous on $[1,2]$ and differentiable on $(1,2)$. $\frac{f(2)-f(1)}{2-1} = \frac{(1/2)-1}{1} = -1/2$ and

$-1/2 = f'(c) = -1/c^2 \Rightarrow c^2 = 2 \Rightarrow c = \sqrt{2}$ (since c must lie in $[1,2]$).

13. $f(x) = |x-1|$. $f(3)-f(0) = |3-1|-|0-1| = 1$. Since $f'(c) = -1$ if $c < 1$ and $f'(c) = 1$ if $c > 1$, $f'(c)(3-0) = \pm 3$ and so is never $= 1$. This does not contradict the Mean Value Theorem since $f'(1)$ does not exist.

15. $f(x) = x^5+10x+3 = 0$. Since f is continuous and $f(-1) = -8$ and $f(0) = 3$, the equation has at least one root in $(-1,0)$ by the Intermediate Value Theorem. Suppose that the equation has more than one root, say a and b are both roots with $a < b$. Then $f(a) = 0 = f(b)$ so by Rolle's Theorem $f'(x) = 5x^4+10 = 0$ has a root in $(a,b)$. But this is impossible since clearly $f'(x) \geq 10 > 0$ for all real x.

17. $f(x) = x^5-6x+c = 0$. Suppose that $f(x)$ has two roots a and b with $-1 \leq a < b \leq 1$. Then $f(a) = 0 = f(b)$, so by Rolle's Theorem there is a number d in $(a,b)$ with $f'(d) = 0$. Now $0 = f'(d) = 5d^4-6 \Rightarrow d = \pm\sqrt[4]{6/5}$ neither of which is in the interval $[-1,1]$ and hence not in $(a,b)$. Thus $f(x)$ can have at most one root in $[-1,1]$.

19. (a) Suppose that a cubic polynomial $P(x)$ has roots $a_1 < a_2 < a_3 < a_4$, so $P(a_1) = P(a_2) = P(a_3) = P(a_4)$. By Rolle's Theorem there are numbers $c_1$, $c_2$, $c_3$ with $a_1 < c_1 < a_2$, $a_2 < c_2 < a_3$ and $a_3 < c_3 < a_4$ and $P'(c_1) = P'(c_2) = P'(c_3) = 0$. Thus the second degree polynomial $P'(x)$ has 3 distinct real roots, a contradiction.

    (b) We prove by induction that a polynomial of degree n has at most n real roots. This is certainly true for $n = 1$. Suppose that the result is true for all polynomials of degree n and let $P(x)$ be a polynomial of degree n+1. Suppose that $P(x)$ has more than n+1 real roots, say $a_1 < a_2 < a_3 < \cdots < a_{n+1} < a_{n+2}$. Then $P(a_1) = P(a_2) = \cdots = P(a_{n+2}) = 0$. By Rolle's Theorem there are real numbers $c_1, \cdots, c_{n+1}$ with $a_1 < c_1 < a_2, \cdots, a_{n+1} < c_{n+1} < a_{n+2}$ and $P'(c_1) = \cdots = P'(c_{n+1}) = 0$. Thus the nth degree polynomial $P'(x)$ has at least n+1 roots. This contradiction shows that $P(x)$ has at most n+1 real roots.

21. Suppose that such a function f exists. By the Mean Value theorem there is a number $0 < c < 2$ with $f'(c) = \dfrac{f(2)-f(0)}{2-0} = 5/2$. But this is impossible since $f'(x) \leq 2 < 5/2$ for all x, so no such function can exist.

23. Let $f(x) = \sin x$ and let $b < a$. Then $f(x)$ is continuous on $[b,a]$ and differentiable on $(b,a)$. By the Mean Value Theorem, there is a number $c \in (b,a)$ with $\sin a - \sin b = f(a)-f(b) = f'(c)(a-b) = (\cos c)(a-b)$. Thus $|\sin a - \sin b| \leq |\cos c||b-a| \leq |a-b|$. If $a < b$, then $|\sin a - \sin b| = |\sin b - \sin a| \leq |b-a| \leq |a-b|$. If $a = b$, both sides of the inequality collapse to 0.

25. For $x > 0$, $f(x) = g(x)$, so $f'(x) = g'(x)$. For $x < 0$, $f'(x) = (1/x)' = -1/x^2$ and $g'(x) = (1+1/x)' = -1/x^2$, so again $f'(x) = g'(x)$. However, the domain of $g(x)$ is not an interval (it is $(-\infty,0)\cup(0,\infty)$) so we cannot conclude that f-g is constant (in fact it is not).

27.  Let g(t) and h(t) be the position functions of the two runners and
     let f(t) = g(t)-h(t).  By hypothesis f(0) = g(0)-h(0) = 0 and f(b)
     = g(b)-h(b) = 0 where b is the finishing time.  Then by Rolle's
     Theorem, there is a time 0 < c < b with 0 = f'(c) = g'(c)-h'(c).
     Hence g'(c) = h'(c), so at time c, both runners have the same
     velocity g'(c) = h'(c).

## Section 4.3

Exercises 4.3

1.  $f(x) = 20-x-x^2$, $f'(x) = -1-2x = 0$ $\Rightarrow$ $x = -1/2$ (the only critical
    number)  (a)  $f'(x) > 0$ $\Longleftrightarrow$ $-1-2x > 0$ $\Longleftrightarrow$ $x < -1/2$, $f'(x) < 0$ $\Longleftrightarrow$
    $x > -1/2$, so f is increasing on $(-\infty,-1/2]$ and decreasing on
    $[-1/2,\infty)$.

    (b)  By the First Derivative Test,     (c)
    $f(-1/2) = 20.25$ is a local
    maximum.

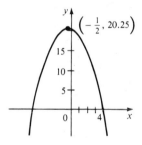

3.  $f(x) = x^3+x+1$.  $f'(x) = 3x^2+1 > 0$ for all $x \in$ R.
    (a) f is increasing on R.
    (b) f has no local extrema.     (c)

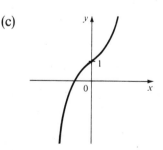

5. $f(x) = x^3-2x^2+x$. $f'(x) = 3x^2-4x+1 = (3x-1)(x-1)$. So the critical numbers are $x = 1/3, 1$. (a) $f'(x) > 0 \Leftrightarrow (3x-1)(x-1) > 0 \Leftrightarrow x < 1/3$ or $x > 1$ and $f'(x) < 0 \Leftrightarrow 1/3 < x < 1$. So f is increasing on $(-\infty, 1/3]$ and $[1,\infty)$ and f is decreasing on $[1/3,1]$. (b) The local maximum is $f(1/3)$ = 4/27 and the local minimum is $f(1) = 0$.

(c)

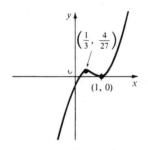

7. $f(x) = 1-3x+5x^2-x^3$. $f'(x) = -3+10x-3x^2 = -(3x-1)(x-3)$, so the critical numbers are $x = 1/3, 3$.
   (a) $f'(x) > 0 \Leftrightarrow -(3x-1)(x-3) > 0 \Leftrightarrow (3x-1)(x-3) < 0 \Leftrightarrow 1/3 < x < 3$. $f'(x) < 0 \Leftrightarrow x < 1/3$ or $x > 3$. So f is increasing on $[1/3,3]$ and decreasing on $(-\infty, 1/3]$ and $[3,\infty)$.
   (b) The local minimum is $f(1/3)$ = $\frac{14}{27}$ and the local maximum is $f(3) = 10$.

(c)

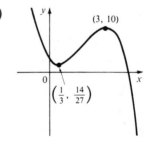

82

9. $f(x) = 2x^2-x^4$. $f'(x) = 4x-4x^3 = 4x(1-x^2) = 4x(1+x)(1-x)$, so the critical numbers are $x = -1,0,1$.

(a)

| Interval | 4x | 1+x | 1-x | f'(x) | f |
|----------|----|-----|-----|-------|---|
| x<-1 | - | - | + | + | increasing on $(-\infty,-1]$ |
| -1<x<0 | - | + | + | - | decreasing on $[-1,0]$ |
| 0<x<1 | + | + | + | + | increasing on $[0,1]$ |
| x>1 | + | + | - | - | decreasing on $[1,\infty)$ |

(b) Local maximum $f(-1) = 1$,
local minimum $f(0) = 0$,
local maximum $f(1) = 1$.

(c)
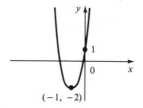

11. $f(x) = x^4+4x+1$   $f'(x) = 4x^3+4 = 4(x^3+1) = 4(x+1)(x^2-x+1)$ ⟹ $x = -1$ is the only critical number.   (a) $f'(x) > 0$ ⟺ $4(x^3+1) > 0$ ⟺ $x > -1$.   $f'(x) < 0$ ⟺ $x < -1$.   So f is increasing on $[-1,\infty)$ and decreasing on $(-\infty,-1]$.

(b) Local minimum $f(-1) = -2$.

(c)
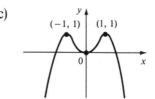

13. $f(x) = x^3(x-4)^4$.   $f'(x) = 3x^2(x-4)^4+x^3[4(x-4)^3] = x^2(x-4)^3(7x-12)$.
The critical numbers are $x = 0,4,12/7$.   (a) $x^2(x-4)^2 \geq 0$ so $f'(x) \geq 0$ ⟺ $(x-4)(7x-12) \geq 0$ ⟺ $x \leq 12/7$ or $x \geq 4$.   $f'(x) \leq 0$ ⟺ $12/7 \leq x \leq 4$.   So f is increasing on $(-\infty,12/7]$ and $[4,\infty)$ and decreasing on $[12/7,4]$.

(b) Local maximum $f(12/7)$
$= 12^3 \cdot 16^4/7^7 \approx 137.5$,
local minimum $f(4) = 0$.

(c)
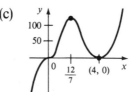

15. $f(x) = x\sqrt{6-x}$.  $f'(x) = \sqrt{6-x} + x(-1/2\sqrt{6-x}) = 3(4-x)/2\sqrt{6-x}$.  Critical numbers are $x = 4,6$.  (a) $f'(x) > 0 \Longleftrightarrow 4-x > 0$ (and $x < 6$) $\Longleftrightarrow$ $x < 4$ and $f'(x) < 0 \Longleftrightarrow 4-x < 0$ (and $x < 6$) $\Longleftrightarrow 4 < x < 6$.  So f is increasing on $(-\infty,4]$ and decreasing on $[4,6]$.

(b) Local maximum $f(4) = 4\sqrt{2}$.          (c)

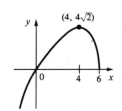

17. $f(x) = x^{1/5}(x+1)$.  $f'(x) = (1/5)x^{-4/5}(x+1)+x^{1/5} = (1/5)x^{-4/5}(6x+1)$. The critical numbers are $x = 0,-1/6$.  (a) $f'(x) > 0 \Longleftrightarrow 6x+1 > 0$ $(x \neq 0) \Longleftrightarrow x > -1/6$ $(x \neq 0)$ and $f'(x) < 0 \Longleftrightarrow x < -1/6$.  So f is increasing on $[-1/6,\infty)$ and decreasing on $(-\infty,-1/6]$.

(b) Local minimum $f(-1/6)$          (c)
$= -5/6^{6/5} \approx -.58$

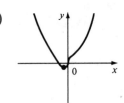

19. $f(x) = x\sqrt{x-x^2}$.  The domain of f is $\{x \mid x(1-x) \geq 0\} = [0,1]$. $f'(x) = \sqrt{x-x^2} + x(1-2x)/2\sqrt{x-x^2} = x(3-4x)/2\sqrt{x-x^2}$.  So the critical numbers are $x = 0,3/4,1$.  (a) $f'(x) > 0 \Longleftrightarrow 3-4x > 0 \Longleftrightarrow$ $0 < x < 3/4$.  $f'(x) < 0 \Longleftrightarrow 3/4 < x < 1$.  So f is increasing on $[0,3/4]$ and decreasing on $[3/4,1]$.
(b) Local maximum $f(3/4) = 3\sqrt{3}/16$.          (c)

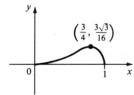

21. $f(x) = x - 2 \sin x$, $0 \leq x \leq 2\pi$. $f'(x) = 1 - 2 \cos x$. So $f'(x) = 0$
$\iff \cos x = 1/2 \iff x = \pi/3$ or $5\pi/3$. (a) $f'(x) > 0 \iff$
$1 - 2 \cos x > 0 \iff \frac{1}{2} > \cos x \iff \frac{\pi}{3} < x < \frac{5}{3}\pi$. $f'(x) < 0 \iff$
$0 \leq x < \pi/3$ or $5\pi/3 < x \leq 2\pi$. So $f$ is increasing on $[\pi/3, 5\pi/3]$ and
decreasing on $[0, \pi/3]$ and $[5\pi/3, 2\pi]$.
(b) Local minimum $f(\pi/3) = \pi/3 - \sqrt{3}$    (c)
$\approx -.68$, local maximum $f(5\pi/3)$
$= \sqrt{3} + 5\pi/3 \approx 6.97$.

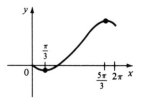

23. $f(x) = \sin^4 x + \cos^4 x$, $0 \leq x \leq 2\pi$. $f'(x) = 4 \sin^3 x \cos x$
$- 4 \cos^3 x \sin x = -4 \sin x \cos x (\cos^2 x - \sin^2 x) = -2 \sin 2x \cos 2x$
$= -\sin 4x$. $f'(x) = 0 \iff \sin 4x = 0 \iff 4x = n\pi \iff x = \pi n/4$.
So the critical numbers are $0, \pi/4, \pi/2, 3\pi/4, \pi, 5\pi/4, 3\pi/2, 7\pi/4, 2\pi$.
(a) $f'(x) > 0 \iff \sin 4x < 0 \iff \pi/4 < x < \pi/2$ or $3\pi/4 < x < \pi$ or
$5\pi/4 < x < 3\pi/2$ or $7\pi/4 < x < 2\pi$. $f$ is increasing on these
intervals. $f$ is decreasing on $[0, \pi/4]$, $[\pi/2, 3\pi/4]$, $[\pi, 5\pi/4]$,
$[3\pi/2, 7\pi/4]$.
(b) Local maxima $f(\pi/2) = f(\pi)$        (c)
$= f(3\pi/2) = 1$, local minima
$f(\pi/4) = f(3\pi/4) = f(5\pi/4)$
$= f(7\pi/4) = 1/2$.

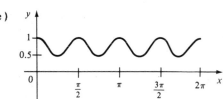

25. $f(x) = x^3 + 2x^2 - x + 1$. $f'(x) = 3x^2 + 4x - 1 = 0 \Rightarrow x = (-4 \pm \sqrt{28})/6$
$= (-2 \pm \sqrt{7})/3$. Now $f'(x) > 0$ for $x < (-2 - \sqrt{7})/3$ or $x > (-2 + \sqrt{7})/3$ and
$f'(x) < 0$ for $(-2 - \sqrt{7})/3 < x < (-2 + \sqrt{7})/3$. $f$ is increasing on
$(-\infty, (-2 - \sqrt{7})/3]$ and $[(-2 + \sqrt{7})/3, \infty)$ and decreasing on
$[(-2 - \sqrt{7})/3, (-2 + \sqrt{7})/3]$.

27. $f(x) = x^6 + 192x + 17$. $f'(x) = 6x^5 + 192 = 6(x^5 + 32)$. So $f'(x) > 0 \iff$
$x^5 > -32 \iff x > -2$ and $f'(x) < 0 \iff x < -2$. So $f$ is increasing
on $[-2, \infty)$ and decreasing on $(-\infty, -2]$.

29.  $f(x) = x^3-3x^2+6x-2$, $-1 \leq x \leq 1$.  $f'(x) = 3x^2-6x+6 = 3(x^2-2x+2)$.
Since $f'(x)$ has no real roots, $f'(x) > 0$ for all $x \in R$, so $f$ is
increasing and has no local extremum.  The absolute minimum is
$f(-1) = -12$ and the absolute maximum is $f(1) = 2$.

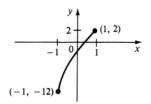

31.  $f(x) = x+\sqrt{1-x}$, $0 \leq x \leq 1$.  $f'(x) = 1-1/2\sqrt{1-x} = (2\sqrt{1-x}-1)/2\sqrt{1-x} = 0$
when $2\sqrt{1-x} - 1 = 0$ ⟹ $\sqrt{1-x} = 1/2$ ⟹ $1-x = 1/4$ ⟹ $x = 3/4$.  For
$0 < x < 3/4$, $f'(x) > 0$ and for $1 > x > 3/4$, $f'(x) < 0$.  So the
local maximum is $f(3/4) = 5/4$.  Also $f(0) = 1$ and $f(1) = 1$ are the
absolute minima and $f(3/4) = 5/4$ is the absolute maximum.

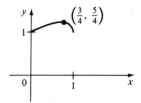

33.  $g(x) = \dfrac{x}{x^2+1}$, $-5 \leq x \leq 5$.  $g'(x) = \dfrac{(x^2+1)-x(2x)}{(x^2+1)^2} = \dfrac{1-x^2}{(x^2+1)^2}$.  The
critical numbers are $x = \pm 1$.  $g'(x) > 0$ ⟺ $x^2 < 1$ ⟺ $-1 < x < 1$
and $g'(x) < 0$ ⟺ $x < -1$ or $x > 1$.  So $g(-1) = -1/2$ is a local
minimum and $g(1) = 1/2$ is a local maximum.  Also $g(-5) = -5/26$ and
$g(5) = 5/26$.  So $g(-1) = -1/2$ is the absolute minimum and
$g(1) = 1/2$ is the absolute maximum.

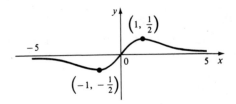

35. Let $f(x) = x+1/x$, so $f'(x) = 1-1/x^2 = (x^2-1)/x^2$. Thus $f'(x) > 0$ for $x > 1$ $\Rightarrow$ f is increasing on $[1,\infty)$. Hence for $1 < a < b$,
$$a + \frac{1}{a} = f(a) < f(b) = b + \frac{1}{b}$$

37. Let $f(x) = 2\sqrt{x}-3+\frac{1}{x}$. Then $f'(x) = \frac{1}{\sqrt{x}} - \frac{1}{x^2} > 0$ for $x > 1$ since for $x > 1$, $x^2 > x > \sqrt{x}$. Hence f is increasing, so for $x > 1$, $f(x) > f(1) = 0$ or $2\sqrt{x}-3+1/x > 0$ for $x > 1$. Hence $2\sqrt{x} > 3-1/x$ for $x > 1$.

39. Let $f(x) = \sin x - x + \frac{x^3}{6}$. Then $f'(x) = \cos x - 1 + \frac{x^2}{2}$. By Exercise 38, $f'(x) > 0$ for $x > 0$, so f is increasing for $x > 0$. Thus $f(x) > f(0) = 0$ for $x > 0$. Hence $\sin x - x + x^3/6 > 0$ or $\sin x > x - x^3/6$ for $x > 0$.

41. $f(x) = ax^3+bx^2+cx+d$ $\Rightarrow$ $f(1) = a+b+c+d = 0$ and $f(-2) = -8a+4b-2c+d = 3$. Also $f'(1) = 3a+2b+c = 0$ and $f'(-2) = 12a-4b+c = 0$ by Fermat's Theorem. Solving these four equations, we get $a = 2/9$, $b = 1/3$, $c = -4/3$, $d = 7/9$, so the function is
$f(x) = (1/9)(2x^3+3x^2-12x+7)$.

43. Let $x_1$ and $x_2$ be any two numbers in $[a,b]$ with $x_1 < x_2$. Then f is continuous on $[x_1,x_2]$ and differentiable on $(x_1,x_2)$, so by the Mean Value Theorem there is a number c between $x_1$ and $x_2$ such that $f(x_2)-f(x_1) = f'(c)(x_2-x_1)$. Now $f'(c) < 0$ by assumption and $x_2-x_1 > 0$ because $x_1 < x_2$. Thus $f(x_2)-f(x_1) = f'(c)(x_2-x_1)$ is negative, so $f(x_2)-f(x_1) < 0$ or $f(x_2) < f(x_1)$. This shows that f is decreasing on $[a,b]$.

**Exercises 4.4**

1. (a) $f(x) = x^3-x \Rightarrow f'(x) = 3x^2-1 = 0 \iff x^2 = 1/3 \iff x = \pm 1/\sqrt{3}$
   $f'(x) > 0 \iff x^2 > 1/3 \iff |x| > 1/\sqrt{3} \iff x > 1/\sqrt{3}$ or $x < -1/\sqrt{3}$
   $f'(x) < 0 \iff |x| < 1/\sqrt{3} \iff -1/\sqrt{3} < x < 1/\sqrt{3}$. So f is
   increasing on $(-\infty, -1/\sqrt{3}]$ and $[1/\sqrt{3}, \infty)$, decreasing on $[-1/\sqrt{3}, 1/\sqrt{3}]$.
   (b) Local maximum $f(-1/\sqrt{3}) = 2/3\sqrt{3} \approx .38$, local minimum $f(1/\sqrt{3}) =$
   $-2/3\sqrt{3} \approx -.38$.  (c) $f''(x) = 6x \Rightarrow f''(x) > 0 \iff x > 0$, so f is CU
   on $(0, \infty)$ and CD on $(-\infty, 0)$.  (d) Point of inflection at $x = 0$.

   (e) $\left(\frac{-1}{\sqrt{3}}, \frac{2}{3\sqrt{3}}\right)$
   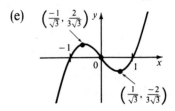
   $\left(\frac{1}{\sqrt{3}}, \frac{-2}{3\sqrt{3}}\right)$

3. (a) $f(x) = x^3-x^2-x+1 \Rightarrow f'(x) = 3x^2-2x-1 = (3x+1)(x-1) = 0 \iff$
   $x = -1/3$ or $1$.  $f'(x) > 0 \iff x < -1/3$ or $x > 1$; $f'(x) < 0 \iff$
   $-1/3 < x < 1$, so f is increasing on $(-\infty, -1/3]$ and $[1, \infty)$ and
   decreasing on $[-1/3, 1]$.  (b) Local maximum $f(-1/3) = 32/27$, local
   minimum $f(1) = 0$.  (c) $f''(x) = 6x-2 > 0 \iff x > 1/3$, so f is CU on
   $(1/3, \infty)$ and CD on $(-\infty, 1/3)$.  (d) Inflection point at $x = 1/3$.

   (e) $\left(-\frac{1}{3}, \frac{32}{27}\right)$
   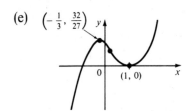
   $(1, 0)$

5. (a) $g(x) = x^4-3x^3+3x^2-x \Rightarrow g'(x) = 4x^3-9x^2+6x-1 = (x-1)^2(4x-1) = 0$
   when $x = 1$ or $1/4$.  $g'(x) \geq 0 \iff 4x-1 \geq 0 \iff x \geq 1/4$ and
   $g'(x) \leq 0 \iff x \leq 1/4$, so g is increasing on $[1/4, \infty)$ and
   decreasing on $(-\infty, 1/4]$.  (b) Local minimum $g(1/4) = -27/256$.
   (c) $g''(x) = 12x^2-18x+6 = 6(2x-1)(x-1) > 0 \iff x < 1/2$ or $x > 1$,

$g'(x) < 0 \iff 1/2 < x < 1$, so g is CU on $(-\infty, 1/2)$ and $(1, \infty)$ and CD on $(1/2, 1)$. (d) Inflection points at $x = 1/2$ and 1.

(e)

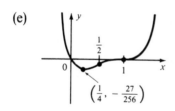

$\left(\frac{1}{4}, -\frac{27}{256}\right)$

7.  (a) $h(x) = 3x^5 - 5x^3 + 3 \Rightarrow h'(x) = 15x^4 - 15x^2 = 15x^2(x^2 - 1) = 0$ when $x = 0, \pm 1$. $h'(x) > 0 \iff x^2 > 1 \iff x > 1$ or $x < -1$, so h is increasing on $(-\infty, -1]$ and $[1, \infty)$ and decreasing on $[-1, 1]$.
(b) Local maximum $h(-1) = 5$, local minimum $h(1) = 1$.
(c) $h''(x) = 60x^3 - 30x = 30x(2x^2 - 1) = 60x(x + 1/\sqrt{2})(x - 1/\sqrt{2}) \Rightarrow$ $h''(x) > 0$ when $x > 1/\sqrt{2}$ or $-1/\sqrt{2} < x < 0$, so h is CU on $(1/\sqrt{2}, \infty)$ and $(-1/\sqrt{2}, 0)$ and CD on $(-\infty, -1/\sqrt{2})$ and $(0, 1/\sqrt{2})$.
(d) Inflection points at $x = \pm 1/\sqrt{2}$ and 0.

(e)

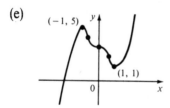

$(-1, 5)$

$(1, 1)$

9.  (a) $F(x) = (x^2 - x)^3 \Rightarrow F'(x) = 3(x^2 - x)^2(2x - 1) > 0 \iff x > 1/2$, so F is increasing on $[1/2, \infty)$ and decreasing on $(-\infty, 1/2]$.
(b) Local minimum $F(1/2) = -1/64$.
(c) $F''(x) = 6(x^2 - x)(2x - 1)^2 + 3(x^2 - x)^2(2) = 6(x^2 - x)(5x^2 - 5x + 1)$
The quadratic formula gives the roots of $5x^2 - 5x + 1 = 0$ as $(5 \pm \sqrt{5})/10$ so the roots of $F''(x) = 0$ are 0, 1, $\alpha = (5 - \sqrt{5})/10$, $\beta = (5 + \sqrt{5})/10$.

| Interval | x | x−1 | x−α | x−β | F''(x) | F |
|---|---|---|---|---|---|---|
| x<0 | − | − | − | − | + | CU on $(-\infty, 0)$ |
| 0<x<α | + | − | − | − | − | CD on $(0, \alpha)$ |
| α<x<β | + | − | + | − | + | CU on $(\alpha, \beta)$ |
| β<x<1 | + | − | + | + | − | CD on $(\beta, 1)$ |
| x>1 | + | + | + | + | + | CU on $(1, \infty)$ |

(d) Inflection points at x = 0, 1, $(5-\sqrt{5})/10$, $(5+\sqrt{5})/10$.

(e)

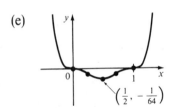

11. (a) $G(x) = 8-\sqrt[3]{x}$ ⇒ $G'(x) = -(1/3)x^{-2/3} < 0$ (x≠0) so G is decreasing on $(-\infty,\infty)$. (b) No extrema. (c) $G''(x) = (2/9)x^{-5/3} > 0$ if x > 0, so G is CU on $(0,\infty)$ and CD on $(-\infty,0)$. (d) IP at x = 0.

(e)

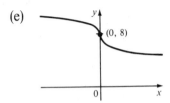

(0, 8)

13. (a) $P(x) = x\sqrt{x^2+1}$ ⇒ $P'(x) = \sqrt{x^2+1} + x^2/\sqrt{x^2+1} = (2x^2+1)/\sqrt{x^2+1} > 0$, so P is increasing on R. (b) No extrema.

(c) $P''(x) = \dfrac{4x\sqrt{x^2+1} - (2x^2+1)(x/\sqrt{x^2+1})}{x^2+1} = \dfrac{x(2x^2+3)}{(x^2+1)^{3/2}} > 0 \iff x > 0$

so P is CU on $(0,\infty)$ and CD on $(-\infty,0)$. (d) IP at x = 0.

(e)

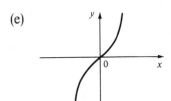

15. (a) $Q(x) = x^{1/3}(x+3)^{2/3}$ ⇒ $Q'(x) = \frac{1}{3}x^{-2/3}(x+3)^{2/3}+x^{1/3}\left[\frac{2}{3}\right](x+3)^{-1/3}$
$= (x+1)/x^{2/3}(x+3)^{1/3}$  The critical numbers are -3, -1, and 0.
Note that $x^{2/3} \geq 0$ for all x. So $Q'(x)>0$ when x<-3 or x>-1 and $Q'(x)<0$ when -3<x<-1 ⇒ Q is increasing on $(-\infty,-3]$ and $[-1,\infty)$ and decreasing on $[-3,-1]$. (b) $Q(-3) = 0$ is a local maximum and $Q(-1) = -4^{1/3} \approx -1.6$ is a local minimum.

(c) $Q''(x) = -2/x^{5/3}(x+3)^{4/3} \Rightarrow Q''(x)>0 \Leftrightarrow x<0$, so Q is CU on $(-\infty,-3)$ and $(-3,0)$ and CD on $(0,\infty)$.   (d) IP at $x = 0$.

(e)

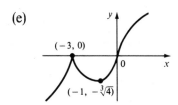

17.  (a) $f(\theta) = \sin^2\theta \Rightarrow f'(\theta) = 2\sin\theta\cos\theta = \sin 2\theta > 0 \Leftrightarrow 2\theta \in (2n\pi,(2n+1)\pi) \Leftrightarrow \theta \in (n\pi,n\pi+\pi/2)$, n an integer.  So f is increasing on $[n\pi,n\pi+\pi/2]$ and decreasing on $[n\pi+\pi/2,(n+1)\pi]$.
(b) Local minima $f(n\pi) = 0$, local maxima $f(n\pi+\pi/2) = 1$.
(c) $f''(\theta) = 2\cos 2\theta > 0 \Leftrightarrow 2\theta \in (2n\pi-\pi/2,2n\pi+\pi/2) \Leftrightarrow \theta \in (n\pi-\pi/4,n\pi+\pi/4)$, so f is CU on these intervals and CD on $(n\pi+\pi/4,n\pi+3\pi/4)$.   (d) IP at $\theta = n\pi\pm\pi/4$, n an integer.

(e)

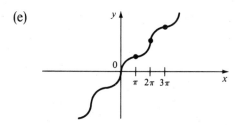

19.  (a) $f(t) = t + \sin t \Rightarrow f'(t) = 1 + \cos t \geq 0$ for all t $\Rightarrow$ f is increasing on R.   (b) Thus there is no local extremum.
(c) $f''(t) = -\sin t > 0$ when $\sin t < 0$  So f is CU on $((2n-1)\pi,2n\pi)$ and CD on $(2n\pi,(2n+1)\pi)$, n an integer.   (d) IP at $x = n\pi$.

(e)

**21.**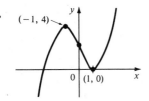

**23.**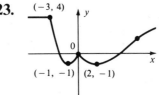

**25.** $f(x) = ax^3+bx^2+cx+d$, $a \neq 0$, $\Rightarrow$ $f'(x) = 3ax^2+2bx+c$ $\Rightarrow$ $f''(x) = 6ax+2b$

f is CU when $6ax+2b > 0$ and CD when $6ax+2b < 0$. So there is exactly one inflection point, namely at $x = -b/3a$.

**27.** By hypothesis $g = f'$ is differentiable on an open interval containing c. Since $(c,f(c))$ is a point of inflection, the concavity changes at $x = c$, so $f'(x)$ changes signs at $x = c$. Hence, by the First Derivative Test, $f'$ has a local extremum at $x = c$. Thus, by Fermat's Theorem, $f''(c) = 0$.

**29.** If f and g are CU on I, then $f'' > 0$ and $g'' > 0$ on I, so $(f+g)'' = f'' + g'' > 0$ on I $\Rightarrow$ f+g is CU on I.

**31.** Since f and g are positive, increasing, and CU on I, we have $f > 0$, $f' > 0$, $f'' > 0$, $g > 0$, $g' > 0$, $g'' > 0$ on I. Then $(fg)' = f'g+fg'$ $\Rightarrow$ $(fg)'' = f''g+2f'g'+fg'' > 0$ $\Rightarrow$ fg is CU on I.

Section 4.5

Exercises 4.5

**1.** $\displaystyle \lim_{x\to\infty} \frac{1}{x\sqrt{x}} = \lim_{x\to\infty} \frac{1}{x^{3/2}} = 0$ by Theorem 4.32.

**3.** $\displaystyle \lim_{x\to\infty} \frac{5+2x}{3-x} = \lim_{x\to\infty} \frac{\frac{5}{x} + 2}{\frac{3}{x} - 1} \underset{(5)}{=} \frac{\displaystyle\lim_{x\to\infty} \left[\frac{5}{x} + 2\right]}{\displaystyle\lim_{x\to\infty} \left[\frac{3}{x} - 1\right]} \underset{(1,2,3)}{=} \frac{5 \displaystyle\lim_{x\to\infty} \frac{1}{x} + \lim_{x\to\infty} 2}{3 \displaystyle\lim_{x\to\infty} \frac{1}{x} - \lim_{x\to\infty} 1}$

$= \dfrac{5(0)+2}{3(0)-1} = -2$ by (7) and Theorem 4.32.

5. $\lim\limits_{x\to-\infty} \dfrac{2x^2-x-1}{4x^2+7} = \lim\limits_{x\to-\infty} \dfrac{2 - \frac{1}{x} - \frac{1}{x^2}}{4 + \frac{7}{x^2}}$  $\underset{(5,1,2,3)}{=}$ $\dfrac{\lim\limits_{x\to-\infty} 2 - \lim\limits_{x\to-\infty}\frac{1}{x} - \lim\limits_{x\to-\infty}\frac{1}{x^2}}{\lim\limits_{x\to-\infty} 4 + 7\lim\limits_{x\to-\infty}\frac{1}{x^2}}$

$= \dfrac{2-0-0}{4+7(0)} = \dfrac{1}{2}$  by (7) and Theorem 4.32.

7. $\lim\limits_{x\to-\infty} \dfrac{(1-x)(2+x)}{(1+2x)(2-3x)} = \lim\limits_{x\to-\infty} \dfrac{\left[\frac{1}{x}-1\right]\left[\frac{2}{x}+1\right]}{\left[\frac{1}{x}+2\right]\left[\frac{2}{x}-3\right]} = \dfrac{\left[\lim\limits_{x\to-\infty}\frac{1}{x}-1\right]\left[\lim\limits_{x\to-\infty}\frac{2}{x}+1\right]}{\left[\lim\limits_{x\to-\infty}\frac{1}{x}+2\right]\left[\lim\limits_{x\to-\infty}\frac{2}{x}-3\right]}$

[by (5,4,1,2,7)] $= \dfrac{(0-1)(0+1)}{(0+2)(0-3)} = \dfrac{1}{6}$

9. $\lim\limits_{x\to\infty} \dfrac{1}{3+\sqrt{x}} = \lim\limits_{x\to\infty} \dfrac{1/\sqrt{x}}{(3/\sqrt{x})+1}$  $\underset{(5,1,3)}{=}$ $\dfrac{\lim\limits_{x\to\infty}(1/\sqrt{x})}{3\lim\limits_{x\to\infty}(1/\sqrt{x})+\lim\limits_{x\to\infty}1}$

$= \dfrac{0}{3(0)+1} = 0$  (by Theorem 4.32 with $r = \frac{1}{2}$)

[OR: Note that $0 < \dfrac{1}{3+\sqrt{x}} < \dfrac{1}{\sqrt{x}}$ and use the Squeeze Theorem.]

11. $\lim\limits_{r\to\infty} \dfrac{r^4-r^2+1}{r^5+r^3-r} = \lim\limits_{r\to\infty} \dfrac{\frac{1}{r}-\frac{1}{r^3}+\frac{1}{r^5}}{1+\frac{1}{r^2}-\frac{1}{r^4}} = \dfrac{\lim\limits_{r\to\infty}\frac{1}{r}-\lim\limits_{r\to\infty}\frac{1}{r^3}+\lim\limits_{r\to\infty}\frac{1}{r^5}}{\lim\limits_{r\to\infty}1+\lim\limits_{r\to\infty}\frac{1}{r^2}-\lim\limits_{r\to\infty}\frac{1}{r^4}} = \dfrac{0-0+0}{1+0-0} = 0$

13. $\lim\limits_{x\to\infty} \dfrac{\sqrt{1+4x^2}}{4+x} = \lim\limits_{x\to\infty} \dfrac{\sqrt{(1/x^2)+4}}{(4/x)+1} = \dfrac{\sqrt{0+4}}{0+1} = 2$

15. $\lim\limits_{x\to\infty} \dfrac{\sqrt[3]{x^2}+8}{x+2} = \lim\limits_{x\to\infty} \dfrac{\sqrt[3]{(1/x)+(8/x^3)}}{1+2/x} = \dfrac{\sqrt[3]{0+0}}{1+0} = 0$

17. $\lim\limits_{x\to\infty} \dfrac{1-\sqrt{x}}{1+\sqrt{x}} = \lim\limits_{x\to\infty} \dfrac{(1/\sqrt{x})-1}{(1/\sqrt{x})+1} = \dfrac{0-1}{0+1} = -1$

19. $\lim\limits_{x\to\infty} (\sqrt{x^2+1} - \sqrt{x^2-1}) = \lim\limits_{x\to\infty} (\sqrt{x^2+1} - \sqrt{x^2-1}) \dfrac{\sqrt{x^2+1}+\sqrt{x^2-1}}{\sqrt{x^2+1}+\sqrt{x^2-1}}$

$= \lim\limits_{x\to\infty} \dfrac{(x^2+1)-(x^2-1)}{\sqrt{x^2+1}+\sqrt{x^2-1}} = \lim\limits_{x\to\infty} \dfrac{2}{\sqrt{x^2+1}+\sqrt{x^2-1}} = \lim\limits_{x\to\infty} \dfrac{2/x}{\sqrt{1+(1/x^2)}+\sqrt{1-(1/x^2)}}$

$= \dfrac{0}{\sqrt{1+0}+\sqrt{1-0}} = 0$

21. $\lim\limits_{x\to\infty} (\sqrt{1+x}-\sqrt{x}) = \lim\limits_{x\to\infty} (\sqrt{1+x}-\sqrt{x}) \dfrac{\sqrt{1+x}+\sqrt{x}}{\sqrt{1+x}+\sqrt{x}} = \lim\limits_{x\to\infty} \dfrac{(1+x)-x}{\sqrt{1+x}+\sqrt{x}} = \lim\limits_{x\to\infty} \dfrac{1}{\sqrt{1+x}+\sqrt{x}}$

$= \lim\limits_{x\to\infty} \dfrac{1/\sqrt{x}}{\sqrt{(1/x)+1}+1} = \dfrac{0}{\sqrt{0+1}+1} = 0$

23. $\lim\limits_{x\to-\infty} (\sqrt{x^2+x+1} + x) = \lim\limits_{x\to-\infty} (\sqrt{x^2+x+1} + x)\dfrac{(\sqrt{x^2+x+1} - x)}{(\sqrt{x^2+x+1} - x)}$

$= \lim\limits_{x\to-\infty} \dfrac{x+1}{(\sqrt{x^2+x+1} - x)} = \lim\limits_{x\to-\infty} \dfrac{1+(1/x)}{-\sqrt{1+(1/x)+(1/x^2)} - 1} = \dfrac{1+0}{-\sqrt{1+0+0} - 1} = -\dfrac{1}{2}$

25. $\lim\limits_{x\to\infty} \cos x$ does not exist because, as x increases, $\cos x$ oscillates

between 1 and -1 infinitely often.

27. $\lim\limits_{x\to\pm\infty} \dfrac{1+2x}{2+x} = \lim\limits_{x\to\pm\infty} \dfrac{(1/x)+2}{(2/x)+1} = \dfrac{0+2}{0+1} = 2$, so y = 2 is the horizontal

asymptote.

29. $\lim\limits_{x\to\infty} \dfrac{x}{\sqrt[4]{x^4+1}} = \lim\limits_{x\to\infty} \dfrac{1}{\sqrt[4]{1+(1/x^4)}} = \dfrac{1}{\sqrt[4]{1+0}} = 1$ and $\lim\limits_{x\to-\infty} \dfrac{x}{\sqrt[4]{x^4+1}} =$

$\lim\limits_{x\to-\infty} \dfrac{1}{-\sqrt[4]{1+(1/x^4)}} = \dfrac{1}{-\sqrt[4]{1+0}} = -1$, so $y = \pm 1$ are horizontal asymptotes.

31. $\lim\limits_{x\to\pm\infty} \dfrac{x}{x^2+1} = \lim\limits_{x\to\pm\infty} \dfrac{1/x}{1+1/x^2} = \dfrac{0}{1+0} = 0$, so y = 0 is a horizontal

asymptote. $y' = \dfrac{x^2+1-x(2x)}{(x^2+1)^2} = \dfrac{1-x^2}{(x^2+1)^2} = 0$ when $x = \pm 1$ and y' > 0

$\Longleftrightarrow x^2 < 1 \Longleftrightarrow -1 < x < 1$, so y is increasing on [-1,1] and

decreasing on $(-\infty,-1]$ and $[1,\infty)$. $y" = \dfrac{(1+x^2)^2(-2x)-(1-x^2)2(x^2+1)2x}{(1+x^2)^4}$

$= \dfrac{2x(x^2-3)}{(1+x^2)^3} \Rightarrow y" > 0 \Longleftrightarrow x > \sqrt{3}$ or $-\sqrt{3} < x < 0$, so y is CU on

$(\sqrt{3},\infty)$ and $(-\sqrt{3},0)$ and CD on $(-\infty,-\sqrt{3})$ and $(0,\sqrt{3})$.

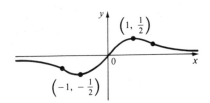

33. $\lim\limits_{x\to\pm\infty} (1 - 1/\sqrt{x^2+1}) = 1-0 = 1$, so y = 1 is a horizontal asymptote.

$y' = -(-1/2)(x^2+1)^{-3/2}(2x) = x/(x^2+1)^{3/2} > 0 \Longleftrightarrow x > 0$, so y is

increasing on $[0,\infty)$ and decreasing on $(-\infty,0]$.

$$y" = \frac{(x^2+1)^{3/2}-x(3/2)(x^2+1)^{1/2}(2x)}{(x^2+1)^3} = \frac{1-2x^2}{(x^2+1)^{5/2}} > 0 \iff x^2 < \frac{1}{2} \iff$$

$|x|<1/\sqrt{2}$, so y is CU on $(-1/\sqrt{2},1/\sqrt{2})$, CD on $(-\infty,-1/\sqrt{2})$ and $(1/\sqrt{2},\infty)$

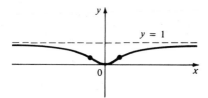

**35.** (a) If $t = \frac{1}{x}$ then $\lim\limits_{x\to\infty} x \sin\frac{1}{x} = \lim\limits_{t\to 0^+} \frac{1}{t} \sin t = \lim\limits_{t\to 0^+} \frac{\sin t}{t} = 1$

(b)

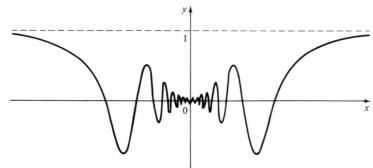

**37.**

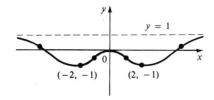

**39.** If $f(x) = x^2/2^x$, then a calculator gives $f(0) = 0$, $f(1) = 0.5$, $f(2) = 1$, $f(3) = 1.125$, $f(4) = 1$, $f(5) = 0.78125$, $f(6) = 0.5625$, $f(7) = 0.3828125$, $f(8) = 0.25$, $f(9) = 0.158203125$, $f(10) = 0.09765625$, $f(20) \approx 0.00038147$, $f(50) \approx 2.2204 \times 10^{-12}$, $f(100) \approx 7.8886 \times 10^{-27}$. It appears that $\lim\limits_{x\to\infty} \frac{x^2}{2^x} = 0$.

41. If $\epsilon > 0$ is given, then $1/x^2 < \epsilon \iff x^2 > 1/\epsilon \iff x > 1/\sqrt{\epsilon}$.
Let $N = 1/\sqrt{\epsilon}$. Then $x > N \Rightarrow x > 1/\sqrt{\epsilon} \Rightarrow |1/x^2 - 0| = 1/x^2 < \epsilon$
so $\lim\limits_{x \to \infty} \dfrac{1}{x^2} = 0$.

43. If $\epsilon > 0$ is given, then $1/\sqrt{x} < \epsilon \iff \sqrt{x} > 1/\epsilon \iff x > 1/\epsilon^2$.
Let $N = 1/\epsilon^2$. Then $x > N \Rightarrow x > 1/\epsilon^2 \Rightarrow |1/\sqrt{x} - 0| = 1/\sqrt{x} < \epsilon$
so $\lim\limits_{x \to \infty} \dfrac{1}{\sqrt{x}} = 0$.

45. Suppose that $\lim\limits_{x \to \infty} f(x) = L$ and let $\epsilon > 0$ be given. Then there

exists $N > 0$ such that $x > N \Rightarrow |f(x)-L| < \epsilon$. Let $\delta = 1/N$. Then
$0 < t < \delta \Rightarrow t < 1/N \Rightarrow 1/t > N \Rightarrow |f(1/t)-L| < \epsilon$. So
$\lim\limits_{t \to 0^+} f(1/t) = L = \lim\limits_{x \to \infty} f(x)$.

Now suppose that $\lim\limits_{x \to -\infty} f(x) = L$ and let $\epsilon > 0$ be given. Then there

exists $N < 0$ such that $x < N \Rightarrow |f(x)-L| < \epsilon$. Let $\delta = -1/N$. Then
$-\delta < t < 0 \Rightarrow t > 1/N \Rightarrow 1/t < N \Rightarrow |f(1/t)-L| < \epsilon$. So
$\lim\limits_{t \to 0^-} f(1/t) = L = \lim\limits_{x \to -\infty} f(x)$.

**Section 4.6**

**Exercises 4.6**

1. Vertical asymptotes: $x = -9$, $x = -4$, $x = 3$, $x = 7$
Horizontal asymptotes: $y = -2$, $y = 2$

3. $\lim\limits_{x \to -1} \dfrac{-2}{(x+1)^6} = -\infty$ since $(x+1)^6 \to 0$ as $x \to -1$ and $\dfrac{-2}{(x+1)^6} < 0$

5. $\lim\limits_{x \to 5^+} \dfrac{6}{x-5} = \infty$ since $x-5 \to 0$ as $x \to 5^+$ and $\dfrac{6}{x-5} > 0$ for $x > 5$

7. $\lim\limits_{x \to -1^-} \dfrac{x^2}{x+1} = -\infty$ since $x+1 \to 0$ as $x \to -1^-$ and $\dfrac{x^2}{x+1} < 0$ for $x < -1$

9. $\lim\limits_{t \to 3^-} \dfrac{t+3}{t^2-9} = \lim\limits_{t \to 3^-} \dfrac{1}{t-3} = -\infty$ since $t-3 \to 0$ as $t \to 3^-$ and $\dfrac{1}{t-3} < 0$ for $t < 3$

11. $\lim\limits_{x \to -3^-} \dfrac{x^2}{x^2-9} = \infty$ since $x^2-9 \to 0$ as $x \to -3^-$ and $\dfrac{x^2}{x^2-9} > 0$ for $x < -3$

13. $\lim\limits_{x \to -2^+} \dfrac{x-1}{x^2(x+2)} = -\infty$ since $x^2(x+2) \to 0$ as $x \to -2^+$ and $\dfrac{x-1}{x^2(x+2)} < 0$ if

$-2 < x < 0$

15. $\lim\limits_{x \to -3^+} \dfrac{x^3}{x^2+5x+6} = \lim\limits_{x \to -3^+} \dfrac{x^3}{(x+2)(x+3)} = \infty$ since $(x+2)(x+3) \to 0$ as

$x \to -3^+$ and, for $-3 < x < -2$, $x^3 < 0$, $x+3 > 0$, $x+2 < 0 \Rightarrow \dfrac{x^3}{(x+2)(x+3)} > 0$

17. $\lim\limits_{x \to 2^+} \dfrac{-3}{\sqrt[3]{x-2}} = -\infty$ since $\sqrt[3]{x-2} \to 0$ as $x \to 2^+$ and $\dfrac{-3}{\sqrt[3]{x-2}} < 0$ for $x > 2$

19. $\lim\limits_{x \to 0^-} \cot x = \lim\limits_{x \to 0^-} \dfrac{\cos x}{\sin x} = -\infty$ since $\sin x \to 0$ as $x \to 0$ and $\cot x < 0$

for $-\pi/2 < x < 0$

21. $\lim\limits_{x \to \pi^-} \csc x = \lim\limits_{x \to \pi^-} \dfrac{1}{\sin x} = \infty$ since $\sin x \to 0$ as $x \to \pi^-$ and $\csc x > 0$

for $0 < x < \pi$

23. $\lim\limits_{t \to -3\pi/2^-} \sec t = \lim\limits_{t \to -3\pi/2^-} \dfrac{1}{\cos t} = \infty$ since $\cos t \to 0$ as $t \to -3\pi/2^-$

and $\sec t > 0$ for $-5\pi/2 < t < -3\pi/2$

25. $\sqrt[3]{x}$ is large negative when $x$ is large negative, so $\lim\limits_{x \to -\infty} \sqrt[3]{x} = -\infty$

27. $\lim\limits_{x \to \infty} (x+\sqrt{x}) = \infty$ since $x \to \infty$ and $\sqrt{x} \to \infty$

29. $\lim\limits_{x \to \infty} (x^2-x^4) = \lim\limits_{x \to \infty} x^2(1-x^2) = -\infty$ since $x^2 \to \infty$ and $1-x^2 \to -\infty$

31. $\lim\limits_{x \to \infty} \dfrac{x^3-1}{x^4+1} = \lim\limits_{x \to \infty} \dfrac{(1/x) - (1/x^4)}{1 + (1/x^4)} = \dfrac{0-0}{1+0} = 0$

33. $\lim\limits_{x \to \infty} \dfrac{x}{\sqrt{x-1}} = \lim\limits_{x \to \infty} \dfrac{x/\sqrt{x}}{\sqrt{x-1}/\sqrt{x}} = \lim\limits_{x \to \infty} \dfrac{\sqrt{x}}{\sqrt{1-1/x}} = \infty$ since $\sqrt{x} \to \infty$ and $\sqrt{1-1/x} \to 1$

[OR: Divide numerator and denominator by $x$ instead of $\sqrt{x}$.]

35. $\lim\limits_{x \to -\infty} \dfrac{x^8+3x^4+2}{x^5+x^3} = \lim\limits_{x \to -\infty} \dfrac{1+(3/x^4)+(2/x^8)}{(1/x^3)+(1/x^5)} = -\infty$ since denominator $\to 0^-$

[OR: Divide numerator and denominator by $x^5$ instead of $x^8$.]

37.  Since $x^2-1 \to 0$ and $y < 0$ for $-1 < x < 1$ and $y > 0$ for $x < -1$ and $x > 1$, we have $\lim\limits_{x\to1^-} \dfrac{x^2+4}{x^2-1} = -\infty$, $\lim\limits_{x\to1^+} \dfrac{x^2+4}{x^2-1} = \infty$, $\lim\limits_{x\to-1^-} \dfrac{x^2+4}{x^2-1} = \infty$, and

$\lim\limits_{x\to-1^+} \dfrac{x^2+4}{x^2-1} = -\infty$, so $x = 1$ and $x = -1$ are vertical asymptotes.

Also $\lim\limits_{x\to\pm\infty} \dfrac{x^2+4}{x^2-1} = \lim\limits_{x\to\pm\infty} \dfrac{1+4/x^2}{1-1/x^2} = \dfrac{1+0}{1-0} = 1$, so $y = 1$ is a horizontal asymptote.

39.  Since $y = \dfrac{x^3+1}{x^3+x} = \dfrac{x^3+1}{x(x^2+1)} > 0$ for $x > 0$ and $y < 0$ for $-1 < x < 0$,

$\lim\limits_{x\to0^+} \dfrac{x^3+1}{x^3+x} = \infty$ and $\lim\limits_{x\to0^-} \dfrac{x^3+1}{x^3+x} = -\infty$, so $x = 0$ is a vertical asymptote.

$\lim\limits_{x\to\pm\infty} \dfrac{x^3+1}{x^3+x} = \lim\limits_{x\to\pm\infty} \dfrac{1+1/x^3}{1+1/x^2} = 1$, so $y = 1$ is a horizontal asymptote.

41.  $\lim\limits_{x\to4^+} \dfrac{4}{x-4} = \infty$ and $\lim\limits_{x\to4^-} \dfrac{4}{x-4} = -\infty$, so $x = 4$ is a vertical asymptote.

$\lim\limits_{x\to\pm\infty} \dfrac{4}{x-4} = \lim\limits_{x\to\pm\infty} \dfrac{4/x}{1-4/x} = 0$, so $y = 0$ is a horizontal asymptote.

$y' = -4/(x-4)^2 < 0$ $(x \neq 4)$ so $y$ is decreasing on $(-\infty,4)$ and $(4,\infty)$.

$y'' = 8/(x-4)^3 > 0$ for $x > 4$, so $y$ is CU on $(4,\infty)$ and CD on $(-\infty,4)$.

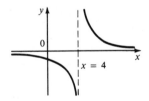

43.  $\lim\limits_{x\to1^-} \dfrac{1}{1-x^2} = \infty$, $\lim\limits_{x\to1^+} \dfrac{1}{1-x^2} = -\infty$, $\lim\limits_{x\to-1^-} \dfrac{1}{1-x^2} = -\infty$, $\lim\limits_{x\to-1^+} \dfrac{1}{1-x^2} = \infty$

so $x = 1$ and $x = -1$ are vertical asymptotes.

$\lim\limits_{x\to\pm\infty} \dfrac{1}{1-x^2} = 0$, so $y = 0$ is a horizontal asymptote.

$y' = 2x/(1-x^2)^2 > 0$ if $x > 0$ $(x \neq 1)$, so $y$ is increasing on $[0,1)$ and $(1,\infty)$ and decreasing on $(-\infty,-1)$ and $(-1,0]$.

$$y" = \frac{2(1-x^2)^2 - 2x[2(1-x^2)(-2x)]}{(1-x^2)^4} = \frac{2(3x^2+1)}{(1-x^2)^3} \Rightarrow y" > 0 \iff x^2 < 1$$

$\iff$ -1 < x < 1, so y is CU on (-1,1) and CD on (-∞,-1) and (1,∞).

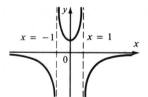

**45.** Divide numerator and denominator by the highest power of x in Q(x).
(a) If deg(P) < deg(Q), then numerator → 0 but denominator doesn't.
So $\lim_{x \to \infty} \frac{P(x)}{Q(x)} = 0$.   (b) If deg(P) > deg(Q), then numerator → ±∞ but

denominator doesn't, so $\lim_{x \to \infty} \frac{P(x)}{Q(x)} = \pm\infty$ (depending on the ratio of

the leading coefficients of P and Q).

**47.** $\frac{1}{(x+3)^4} > 10000 \iff (x+3)^4 < \frac{1}{10000} \iff |x-(-3)| = |x+3| < \frac{1}{10}$

**49.** Let N < 0 be given.  Then, for x < -1, we have
$\frac{5}{(x+1)^3} < N \iff \frac{5}{N} < (x+1)^3 \iff \sqrt[3]{5/N} < x+1$.  Let $\delta = -\sqrt[3]{5/N}$.  Then

-1-$\delta$ < x < -1 $\Rightarrow$ $\sqrt[3]{5/N}$ < x+1 < 0 $\Rightarrow$ $\frac{5}{(x+1)^3}$ < N, so $\lim_{x \to -1^-} \frac{5}{(x+1)^3} = -\infty$

**51.** Let M be given.  Since $\lim_{x \to a} f(x) = \infty$, there exists $\delta_1 > 0$ such that

0 < |x-a| < $\delta_1$ $\Rightarrow$ f(x) > M+1-c.  Since $\lim_{x \to a} g(x) = c$, there exists

$\delta_2 > 0$ such that 0 < |x-a| < $\delta_2$ $\Rightarrow$ |g(x)-c| < 1 $\Rightarrow$ g(x) > c-1.

Let $\delta$ be the smaller of $\delta_1$ and $\delta_2$.  Then 0 < |x-a| < $\delta$ $\Rightarrow$

f(x)+g(x) > (M+1-c)+(c-1) = M.  Thus $\lim_{x \to a} [f(x)+g(x)] = \infty$.

Exercises 4.7

*Abbreviations*:  D : the domain of f,  VA : vertical asymptote(s),
                  HA : horizontal asymptote,  IP : inflection point(s)

1.  $y = f(x) = 1/(x-1)$   A. $D = \{x \mid x \neq 1\} = (-\infty,1) \cup (1,\infty)$

    B. y-intercept $= f(0) = -1$, no x-intercept  C. No symmetry

    D. $\lim\limits_{x\to\pm\infty} \frac{1}{x-1} = 0$, so $y = 0$ is a HA.  $\lim\limits_{x\to1^+} \frac{1}{x-1} = \infty$ and $\lim\limits_{x\to1^-} \frac{1}{x-1} = -\infty$

    so $x = 1$ is a VA.   E. $f'(x) = -1/(x-1)^2 < 0$ $(x\neq1)$, so f is

    decreasing on $(-\infty,1)$ and $(1,\infty)$.     H.

    F. No extrema.

    G. $f''(x) = 2/(x-1)^3$ $\Rightarrow$ $f''(x) > 0$

    $\Leftrightarrow$ $x > 1$, so f is CU on $(1,\infty)$ and

    CD on $(-\infty,1)$.  No IP.

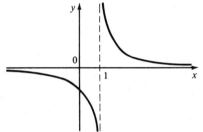

3.  $y = f(x) = 1/(x^2-9)$   A. $D = \{x \mid x \neq \pm3\} = (-\infty,-3) \cup (-3,3) \cup (3,\infty)$

    B. y-intercept $= f(0) = -1/9$, no x-intercept

    C. $f(-x) = f(x)$ $\Rightarrow$ f is even, the curve is symmetric about y-axis.

    D. $\lim\limits_{x\to\pm\infty} \frac{1}{x^2-9} = 0$, so $y = 0$ is a HA.  $\lim\limits_{x\to3^-} \frac{1}{x^2-9} = -\infty$,  $\lim\limits_{x\to3^+} \frac{1}{x^2-9} = \infty$

    $\lim\limits_{x\to-3^-} \frac{1}{x^2-9} = \infty$,  $\lim\limits_{x\to-3^+} \frac{1}{x^2-9} = -\infty$, so $x = 3$ and $x = -3$ are VA.

    E. $f'(x) = -2x/(x^2-9)^2 > 0$ $\Leftrightarrow$ $x < 0$ $(x\neq-3)$ so f is increasing on

    $(-\infty,-3)$ and $(-3,0]$ and decreasing on $[0,3)$ and $(3,\infty)$.

    F. Local maximum $f(0) = -1/9$      H.

    G. $y'' = \dfrac{-2(x^2-9)^2+(2x)2(x^2-9)(2x)}{(x^2-9)^4}$

    $= \dfrac{6(x^2+3)}{(x^2-9)^3} > 0$ $\Leftrightarrow$ $x^2 > 9$

    $\Leftrightarrow$ $x > 3$ or $x < -3$, so f is

    CU on $(-\infty,-3)$ and $(3,\infty)$ and

    CD on $(-3,3)$.  No IP.

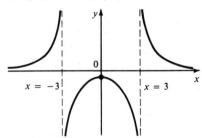

5.  $y = f(x) = x^3/(x^2-1)$   A. $D = \{x \mid x \neq \pm1\} = (-\infty,-1) \cup (-1,1) \cup (1,\infty)$

    B. x-intercept $= 0$, y-intercept $= 0$     C. $f(-x) = -f(x)$ $\Rightarrow$ f is

    odd, so the curve is symmetric about the origin.

D. $\lim\limits_{x\to\infty} \dfrac{x^3}{x^2-1} = \infty$ but long division gives $\dfrac{x^3}{x^2-1} = x + \dfrac{x}{x^2-1}$ so

$f(x) - x = \dfrac{x}{x^2-1} \to 0$ as $x \to \pm\infty$ $\Rightarrow$ $y = x$ is a slant asymptote.

$\lim\limits_{x\to 1^-} \dfrac{x^3}{x^2-1} = -\infty$, $\lim\limits_{x\to 1^+} \dfrac{x^3}{x^2-1} = \infty$, $\lim\limits_{x\to -1^-} \dfrac{x^3}{x^2-1} = -\infty$, $\lim\limits_{x\to -1^+} \dfrac{x^3}{x^2-1} = \infty$

so $x = 1$ and $x = -1$ are VA. E. $f'(x) = \dfrac{3x^2(x^2-1)-x^3(2x)}{(x^2-1)^2} = \dfrac{x^2(x^2-3)}{(x^2-1)^2}$

$\Rightarrow$ $f'(x) > 0$ $\Leftrightarrow$ $x^2 > 3$ $\Leftrightarrow$ $x > \sqrt{3}$ or $x < -\sqrt{3}$, so $f$ is increasing on

$(-\infty, -\sqrt{3}]$ and $[\sqrt{3}, \infty)$ and decreasing on $[-\sqrt{3}, -1)$, $(-1,1)$, and $(1, \sqrt{3}]$.

F. $f(-\sqrt{3}) = -3\sqrt{3}/2$ is a local

maximum and $f(\sqrt{3}) = 3\sqrt{3}/2$ is

a local minimum.

G. $y'' = \dfrac{2x(x^2+3)}{(x^2-1)^3} > 0$ $\Leftrightarrow$ $x > 1$ or

$-1 < x < 0$, so $f$ is CU on $(1, \infty)$ and

$(-1, 0)$ and CD on $(-\infty, -1)$ and

$(0, 1)$. IP is $(0,0)$.

H.

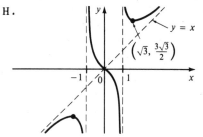

7. $y = f(x) = x/(2x-3)^2$ A. $D = \{x \mid x \neq 3/2\} = (-\infty, 3/2) \cup (3/2, \infty)$

B. Both intercepts are 0. C. No symmetry

D. $\lim\limits_{x\to \pm\infty} \dfrac{x}{(2x-3)^2} = 0$, so $y = 0$ is a HA. $\lim\limits_{x\to 3/2} \dfrac{x}{(2x-3)^2} = \infty$, so

$x = 3/2$ is a VA. E. $f'(x) = \dfrac{(2x-3)^2 - x[2(2x-3)(2)]}{(2x-3)^4} = -\dfrac{2x+3}{(2x-3)^3}$ $\Rightarrow$

$f'(x) > 0$ $\Leftrightarrow$ $-3/2 < x < 3/2$, so $f$ is increasing on $[-3/2, 3/2)$ and

decreasing on $(-\infty, -3/2]$ and $(3/2, \infty)$.

F. $f(-3/2) = -1/24$ is a local

minimum.

G. $y'' = \dfrac{-2(2x-3)^3 + (2x+3)6(2x-3)^2}{(2x-3)^6}$

$= \dfrac{8(x+3)}{(2x-3)^4} > 0$ $\Leftrightarrow$ $x > -3$ $(x \neq 3/2)$

So $f$ is CU on $(-3, 3/2)$ and $(3/2, \infty)$

and CD on $(-\infty, -3)$. IP $= (-3, -1/27)$

H.

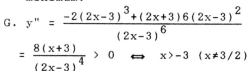

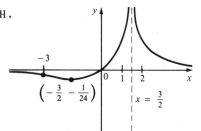

9.  $y = f(x) = (x^2+4)/x = x + 4/x$   A. $D = \{x \mid x \neq 0\} = (-\infty,0) \cup (0,\infty)$

B. No intercepts   C. $f(-x) = -f(x)$   $\Rightarrow$   symmetry about the origin

D. $\lim\limits_{x \to \infty} (x+\frac{4}{x}) = \infty$ but $f(x) - x = \frac{4}{x} \to 0$ as $x \to \pm\infty$, so $y = x$ is a

slant asymptote.  $\lim\limits_{x \to 0^+} (x+\frac{4}{x}) = \infty$ and $\lim\limits_{x \to 0^-} (x+\frac{4}{x}) = -\infty$, so $x = 0$ is a

VA.  E. $f'(x) = 1 - \frac{4}{x^2} > 0$  $\iff$  $x^2 > 4$  $\iff$  $x > 2$ or $x < -2$, so $f$ is

increasing on $(-\infty,-2]$ and $[2,\infty)$ and decreasing on $[-2,0)$ and $(0,2]$.

F. $f(-2) = -4$ is a local

maximum and $f(2) = 4$ is

a local minimum.

H.

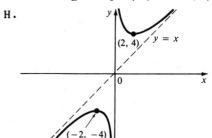

G. $f''(x) = \frac{8}{x^3} > 0$  $\iff$  $x > 0$

so $f$ is CU on $(0,\infty)$ and

CD on $(-\infty,0)$.  No IP.

11.  $y = f(x) = (x-3)/(x+3)$   A. $D = \{x \mid x \neq -3\} = (-\infty,-3) \cup (-3,\infty)$

B. x-intercept is 3, y-intercept $= f(0) = -1$   C. No symmetry

D. $\lim\limits_{x \to \pm\infty} \frac{x-3}{x+3} = \lim\limits_{x \to \pm\infty} \frac{1-3/x}{1+3/x} = 1$, so $y = 1$ is a HA.

$\lim\limits_{x \to -3^-} \frac{x-3}{x+3} = \infty$ and $\lim\limits_{x \to -3^+} \frac{x-3}{x+3} = -\infty$, so $x = -3$ is a VA.

E. $f'(x) = \frac{(x+3)-(x-3)}{(x+3)^2} = \frac{6}{(x+3)^2}$     H.

$\Rightarrow$  $f'(x) > 0$ $(x \neq -3)$ so $f$ is

increasing on $(-\infty,-3)$ and $(-3,\infty)$

F. No extrema

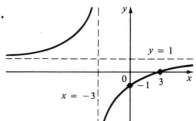

G. $f''(x) = \frac{-12}{(x+3)^3} > 0$  $\iff$

$x < -3$, so $f$ is CU on $(-\infty,-3)$

and CD on $(-3,\infty)$.  No IP.

13.  $y = f(x) = 1/(x-1)(x+2)$   A. $D = \{x \mid x \neq 1,-2\} = (-\infty,-2) \cup (-2,1) \cup (1,\infty)$

B. No x-intercept, y-intercept $= f(0) = -1/2$   C. No symmetry

D. $\lim\limits_{x \to \pm\infty} \frac{1}{(x-1)(x+2)} = 0$, so $y = 0$ is a HA.  $\lim\limits_{x \to 1^-} \frac{1}{(x-1)(x+2)} = -\infty$,

$\lim\limits_{x \to 1^+} \frac{1}{(x-1)(x+2)} = \infty$,  $\lim\limits_{x \to -2^-} \frac{1}{(x-1)(x+2)} = \infty$,  $\lim\limits_{x \to -2^+} \frac{1}{(x-1)(x+2)} = -\infty$

So x = 1 and x = -2 are VA.  E.  $f'(x) = -(2x+1)/[(x-1)(x+2)]^2$ ⇒
$f'(x)>0$ ⟺ x<-1/2 (x≠-2), so f is increasing on (-∞,-2) and
(-2,-1/2] and decreasing on [-1/2,1) and (1,∞).

F.  f(-1/2) = -4/9 is a local

   maximum

H.

G.  $f''(x) = \dfrac{6(x^2+x+1)}{[(x-1)(x+2)]^3}$

Now $x^2+x+1>0$ for all x, so
$f''(x)>0$ ⟺ (x-1)(x+2)>0 ⟺
x<-2 or x>1.  Thus f is CU on
(-∞,-2) and (1,∞) and CD on
(-2,1).  No IP.

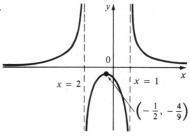

$\left(-\frac{1}{2}, -\frac{4}{9}\right)$

15.  $y = f(x) = \dfrac{1+x^2}{1-x^2} = -1 + \dfrac{2}{1-x^2}$　　A.  D = {x | x≠±1}

B.  No x-intercept, y-intercept = f(0) = 1　　C.  f(-x) = f(x), so f
is even and the curve is symmetric about the y-axis.

D.  $\lim\limits_{x\to\pm\infty} \dfrac{1+x^2}{1-x^2} = \lim\limits_{x\to\pm\infty} \dfrac{(1/x^2)+1}{(1/x^2)-1} = -1$, so y = -1 is a HA.

$\lim\limits_{x\to1^-} \dfrac{1+x^2}{1-x^2} = \infty$, 　$\lim\limits_{x\to1^+} \dfrac{1+x^2}{1-x^2} = -\infty$, 　$\lim\limits_{x\to-1^-} \dfrac{1+x^2}{1-x^2} = -\infty$, 　$\lim\limits_{x\to-1^+} \dfrac{1+x^2}{1-x^2} = \infty$

So x = 1 and x = -1 are VA.  E.  $f'(x) = 4x/(1-x^2)^2 > 0$ ⟺ x>0
(x≠1), so f increases on [0,1), (1,∞), decreases on (-∞,-1), (-1,0]

F.  f(0) = 1 is a local minimum　　H.

G.  $y'' = \dfrac{4(1-x^2)^2 - 4x\cdot2(1-x^2)(-2x)}{(1-x^2)^4}$

    $= \dfrac{4(1+3x^2)}{(1-x^2)^3} > 0$ ⟺ $x^2<1$

   ⟺ -1<x<1, so f is CU on (-1,1)
and CD on (-∞,-1) and (1,∞).
No IP.

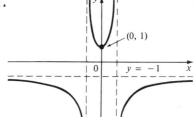

(0, 1)

y = -1

17.  $y = f(x) = (x^3-1)/(x^3+1)$　A.  D = {x | x ≠ -1} = (-∞,-1) ∪ (-1,∞)

B.  x-intercept = 1, y-intercept = f(0) = -1　C.  No symmetry

D.  $\lim\limits_{x\to\pm\infty} \dfrac{x^3-1}{x^3+1} = \lim\limits_{x\to\pm\infty} \dfrac{1-1/x^3}{1+1/x^3} = 1$, so y = 1 is a HA.

$\lim\limits_{x\to-1^-} \dfrac{x^3-1}{x^3+1} = \infty$ and $\lim\limits_{x\to-1^+} \dfrac{x^3-1}{x^3+1} = -\infty$, so x = -1 is a VA.

E. $f'(x) = \dfrac{(x^3+1)(3x^2)-(x^3-1)(3x^2)}{(x^3+1)^2} = \dfrac{6x^2}{(x^3+1)^2} > 0$ $(x \neq -1)$ so f is

increasing on $(-\infty,-1)$ and $(-1,\infty)$.    H.

F. No extrema

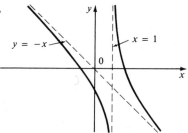

G. $y'' = \dfrac{12x(x^3+1)^2-6x^2 \cdot 2(x^3+1) \cdot 3x^2}{(x^3+1)^4}$

$= \dfrac{12x(1-2x^3)}{(x^3+1)^3} > 0 \iff x < -1$ or

$0 < x < 1/\sqrt[3]{2}$, so f is CU on $(-\infty,-1)$ and

$(0,1/\sqrt[3]{2})$ and CD on $(-1,0)$ and

$(1/\sqrt[3]{2},\infty)$. IP $(0,-1)$, $(1/\sqrt[3]{2},-1/3)$

19.   $y = \dfrac{1}{x-1} - x$    A. $D = \{x \mid x \neq 1\}$    B. $y = 0 \iff x = 1/(x-1) \iff$

$x^2-x-1 = 0 \Rightarrow x = (1\pm\sqrt{5})/2$ (x-intercepts), y-intercept $= f(0) = -1$

C. No symmetry    D. $y-(-x) = 1/(x-1) \to 0$ as $x \to \pm\infty$, so $y = -x$ is a

slant asymptote.   $\lim\limits_{x\to 1^+} \left[\dfrac{1}{x-1} - x\right] = \infty$ and $\lim\limits_{x\to 1^-} \left[\dfrac{1}{x-1} - x\right] = -\infty$, so

$x = 1$ is a VA.    E. $f'(x) = -1-1/(x-1)^2 < 0$ for all $x \neq 1$, so f is

decreasing on $(-\infty,1)$ and $(1,\infty)$.    H.

F. No local extrema.

G. $f''(x) = \dfrac{2}{(x-1)^3} > 0 \iff x > 1$

   so f is CU on $(1,\infty)$ and CD on

   $(-\infty,1)$.   No IP.

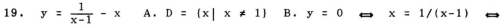

21.   $y = f(x) = 1/(x^3-x) = 1/x(x-1)(x+1)$    A. $D = \{x \mid x \neq 0,\pm 1\}$

B. No intercepts    C. $f(-x) = -f(x)$, symmetric about $(0,0)$

D. $\lim\limits_{x\to\pm\infty} \dfrac{1}{x^3-x} = 0$, so $y = 0$ is a HA.   $\lim\limits_{x\to 0^-} \dfrac{1}{x^3-x} = \infty$,   $\lim\limits_{x\to 0^+} \dfrac{1}{x^3-x} = -\infty$

$\lim\limits_{x\to 1^-} \dfrac{1}{x^3-x} = -\infty$,   $\lim\limits_{x\to 1^+} \dfrac{1}{x^3-x} = \infty$,   $\lim\limits_{x\to -1^-} \dfrac{1}{x^3-x} = -\infty$,   $\lim\limits_{x\to -1^+} \dfrac{1}{x^3-x} = \infty$

So $x = 0$, $x = 1$, and $x = -1$ are VA.

E. $f'(x) = (1-3x^2)/(x^3-x)^2 \Rightarrow f'(x) > 0 \iff x^2 < 1/3 \iff$

$-1/\sqrt{3} < x < 1/\sqrt{3}$ $(x \neq 0)$, so f is increasing on $[-1/\sqrt{3},0)$, $(0,1/\sqrt{3}]$

and decreasing on $(-\infty,-1)$, $(-1,-1/\sqrt{3}]$, $[1/\sqrt{3},1)$, and $(1,\infty)$.

F. Local minimum $f(-1/\sqrt{3}) = 3\sqrt{3}/2$   H.

local maximum $f(1/\sqrt{3}) = -3\sqrt{3}/2$

G. $f''(x) = \dfrac{2(6x^4-3x^2+1)}{(x^3-x)^3}$   Since

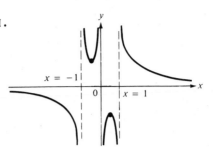

$6x^4-3x^2+1$ has negative discriminant

as a quadratic in $x^2$, it is $> 0 \Rightarrow$

$f''(x)>0 \iff x^3-x>0 \iff x>1$ or $-1<x<0$.

f is CU on $(-1,0)$ and $(1,\infty)$, and CD

on $(-\infty,-1)$ and $(0,1)$.   No IP.

23.  $y = f(x) = \sqrt{2-x}$  A. $D = \{x \mid x\leq 2\} = (-\infty,2]$  B. x-intercept $= 2$,

y-intercept $= f(0) = \sqrt{2}$   C. No symmetry  D. $\displaystyle\lim_{x\to-\infty} \sqrt{2-x} = \infty$,   no

asymptotes.  E. $f'(x) = -1/2\sqrt{2-x} < 0$, so f is decreasing on $(-\infty,2]$

F. No extrema                        H.

G. $f''(x) = -1/4(2-x)^{3/2} < 0$

so f is CD on $(-\infty,2)$

No IP.

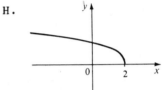

25.  $y = f(x) = x+\sqrt{x}$  A. $D = \{x \mid x\geq 0\} = [0,\infty)$  B. Both intercepts are 0.

C. No symmetry  D. $\displaystyle\lim_{x\to\infty} (x+\sqrt{x}) = \infty$, no asymptotes

E. $f'(x) = 1+1/2\sqrt{x} > 0$, so f          H.

is increasing on $[0,\infty)$.

F. No extrema

G. $f''(x) = -(1/4)x^{-3/2} < 0$, so

f is CD on $(0,\infty)$.   No IP.

27.  $y = f(x) = \sqrt{x^2+1} - x$  A. $D = R$  B. No x-intercept, y-intercept $= 1$

C.  No symmetry  D. $\displaystyle\lim_{x\to-\infty} (\sqrt{x^2+1} - x) = \infty$  and  $\displaystyle\lim_{x\to\infty} (\sqrt{x^2+1} - x)$

$= \displaystyle\lim_{x\to\infty} (\sqrt{x^2+1} - x) \dfrac{\sqrt{x^2+1} + x}{\sqrt{x^2+1} + x} = \displaystyle\lim_{x\to\infty} \dfrac{1}{\sqrt{x^2+1} + x} = 0$, so $y = 0$ is a HA.

E. $f'(x) = \dfrac{x}{\sqrt{x^2+1}} - 1 = \dfrac{x - \sqrt{x^2+1}}{\sqrt{x^2+1}}$   H.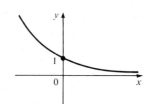

→ $f'(x)<0$, so f is decreasing on R.

F. No extrema

G. $f''(x) = 1/(x^2+1)^{3/2} > 0$, so f is
   CU on R.  No IP.

29. $y = f(x) = 1/x - \sqrt{x}$   A. $D = \{x \mid x>0\} = (0,\infty)$   B. x-intercept = 1,
no y-intercept   C. No symmetry   D. $\lim\limits_{x\to\infty} \left[\dfrac{1}{x} - \sqrt{x}\right] = -\infty$

$\lim\limits_{x\to 0^+} \left[\dfrac{1}{x} - \sqrt{x}\right] = \infty$, so x = 0 is a VA.

E. $f'(x) = (-1/x^2) - 1/2\sqrt{x} < 0$,   H.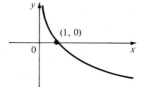
   so f decreases on $(0,\infty)$.

F. No extrema

G. $f''(x) = (2/x^3) + 1/4x^{3/2} > 0$
   so f is CU on $(0,\infty)$

31. $y = f(x) = \sqrt[4]{x^2-25}$   A. $D = \{x \mid x^2 \geq 25\} = (-\infty,-5] \cup [5,\infty)$
B. x-intercepts are ±5, no y-intercept   C. $f(-x) = f(x)$, so the
curve is symmetric about the y-axis.   D. $\lim\limits_{x\to\pm\infty} \sqrt[4]{x^2-25} = \infty$, no
asymptotes.   E. $f'(x) = (1/4)(x^2-25)^{-3/4}(2x) = x/2(x^2-25)^{3/4} > 0$ if
x>5, so f is increasing on $[5,\infty)$ and decreasing on $(-\infty,-5]$.
F. No local extrema   H.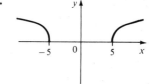

G. $y'' = \dfrac{2(x^2-25)^{3/4}-3x^2(x^2-25)^{-1/4}}{4(x^2-25)^{3/2}}$

$= - \dfrac{x^2+50}{4(x^2-25)^{7/4}} < 0$ so f is CD

on $(-\infty,-5)$ and $(5,\infty)$.  No IP.

33. $y = f(x) = x\sqrt{x^2-9}$   A. $D = \{x \mid x^2 \geq 9\} = (-\infty,-3] \cup [3,\infty)$
B. x-intercepts are ±3, no y-intercept   C. $f(-x) = -f(-x)$, so the
curve is symmetric about the origin.   D. $\lim\limits_{x\to\infty} x\sqrt{x^2-9} = \infty$,
$\lim\limits_{x\to-\infty} x\sqrt{x^2-9} = -\infty$, no asymptotes.   E. $f'(x) = \sqrt{x^2-9} + x^2/\sqrt{x^2-9} > 0$
for x∈D, so f is increasing on $(-\infty,-3]$ and $[3,\infty)$.   F. No extrema

G. $f''(x) =$

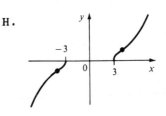

$$\frac{x}{\sqrt{x^2-9}} + \frac{2x\sqrt{x^2-9}-x^2(x/\sqrt{x^2-9})}{x^2-9}$$

$$= \frac{x(2x^2-27)}{(x^2-9)^{3/2}} > 0 \iff x>3\sqrt{3/2} \text{ or}$$

$-3\sqrt{3/2}<x<0$, so f is CU on $(3\sqrt{3/2},\infty)$
and $(-3\sqrt{3/2},-3)$ and CD on $(-\infty,-3\sqrt{3/2})$
and $(3,3\sqrt{3/2})$. IP $(\pm3\sqrt{3/2},\pm9\sqrt{3}/2)$

35. $y = f(x) = \sqrt{1-x^2}/x$  A. D = $\{x|\ |x|\leq1,\ x\neq0\}$ = $[-1,0) \cup (0,1]$
    B. x-intercepts $\pm1$, no y-intercept  C. $f(-x) = -f(x)$, so the curve

    is symmetric about $(0,0)$. D. $\lim\limits_{x\to0^+} \frac{\sqrt{1-x^2}}{x} = \infty$, $\lim\limits_{x\to0^-} \frac{\sqrt{1-x^2}}{x} = -\infty$, so

    x = 0 is a VA.  E. $f'(x) = \dfrac{(-x^2/\sqrt{1-x^2}) - \sqrt{1-x^2}}{x^2} = -1/x^2\sqrt{1-x^2} < 0$

    so f is decreasing on $[-1,0)$

    and $(0,1]$.  F. No extrema

    G. $f''(x) = \dfrac{2-3x^2}{x^3(1-x^2)^{3/2}} > 0 \iff$

    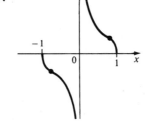

    $-1<x<-\sqrt{2/3}$ or $0<x<\sqrt{2/3}$, so f is
    CU on $(-1,-\sqrt{2/3})$ and $(0,\sqrt{2/3})$ and
    CD on $(-\sqrt{2/3},0)$ and $(\sqrt{2/3},1)$
    IP $(\pm\sqrt{2/3},\pm1/\sqrt{2})$

37. $y = f(x) = x+3x^{2/3}$  A. D = R  B. $y = x+3x^{2/3} = x^{2/3}(x^{1/3}+3) = 0$ if
    x = 0 or -27 (x-intercepts), y-intercept = f(0) = 0  C. No symmetry
    D. $\lim\limits_{x\to\infty} (x+3x^{2/3}) = \infty$,  $\lim\limits_{x\to-\infty} (x+3x^{2/3}) = \lim\limits_{x\to-\infty} x^{2/3}(x^{1/3}+3) = -\infty$, no
    asymptotes  E. $f'(x) = 1+2x^{-1/3} = (x^{1/3}+2)/x^{1/3} > 0 \iff x>0$ or
    $x<-8$, so f increases on $(-\infty,-8]$, $[0,\infty)$ and decreases on $[-8,0]$.
    F. Local maximum f(-8) = 4
       local minimum f(0) = 0
    G. $f''(x) = -(2/3)x^{-4/3} < 0$  $(x\neq0)$
       so f is CD on $(-\infty,0)$ and $(0,\infty)$
       No IP.

H.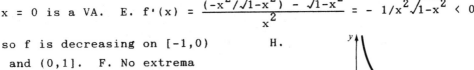

H.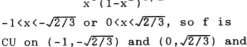

H.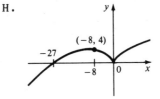

**39.** $y = f(x) = x + \sqrt{|x|}$   A. $D = R$   B. x-intercepts $= 0, -1$, y-intercept 0
C. No symmetry   D. $\lim\limits_{x \to \infty} (x + \sqrt{|x|}) = \infty$,   $\lim\limits_{x \to -\infty} (x + \sqrt{|x|}) = -\infty$. No

asymptotes   E. For $x > 0$, $f(x) = x + \sqrt{x}$ ⇒ $f'(x) = 1 + 1/2\sqrt{x} > 0$, so f
increases on $[0, \infty)$. For $x < 0$, $f(x) = x + \sqrt{-x}$ ⇒ $f'(x) = 1 - 1/2\sqrt{-x} > 0$
⟺ $2\sqrt{-x} > 1$ ⟺ $-x > 1/4$ ⟺ $x < -1/4$, so f increases on $(-\infty, -1/4]$ and
decreases on $[-1/4, 0]$.                    H.

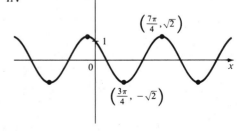

F. $f(-1/4) = 1/4$ is a local
   maximum, $f(0) = 0$ is a
   local minimum
G. For $x > 0$, $f''(x) = -(1/4)x^{-3/2}$
⇒ $f''(x) < 0$, so f is CD on $(0, \infty)$.
For $x < 0$, $f''(x) = -(1/4)(-x)^{-3/2}$
⇒ $f''(x) < 0$, so f is CD on $(-\infty, 0)$.

**41.**   $y = f(x) = \cos x - \sin x$   A. $D = R$   B. $y = 0$ ⟺ $\cos x = \sin x$
⟺ $x = n\pi + \pi/4$, n an integer, (x-intercepts), y-intercept $= f(0) = 1$
C. Periodic with period $2\pi$   D. No asymptotes
E. $f'(x) = -\sin x - \cos x = 0$ ⟺ $\cos x = -\sin x$ ⟺ $x = 2n\pi + 3\pi/4$
or $2n\pi + 7\pi/4$.   $f'(x) > 0$ ⟺ $\cos x < -\sin x$ ⟺ $2n\pi + 3\pi/4 < x < 2n\pi + 7\pi/4$
so f is increasing on $[2n\pi + 3\pi/4, 2n\pi + 7\pi/4]$ and decreasing on
$[2n\pi - \pi/4, 2n\pi + 3\pi/4]$.   F. Local maximum $f(2n\pi - \pi/4) = \sqrt{2}$, local
minimum $f(2n\pi + 3\pi/4) = -\sqrt{2}$.        H.

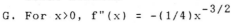
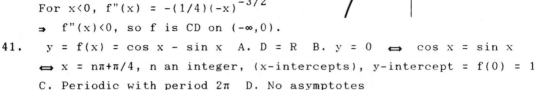

G. $f''(x) = -\cos x + \sin x > 0$
⟺ $\sin x > \cos x$ ⟺
$x \in (2n\pi + \pi/4, 2n\pi + 5\pi/4)$, so f is
CU on these intervals and CD on
$(2n\pi - 3\pi/4, 2n\pi + \pi/4)$
IP $(n\pi + \pi/4, 0)$

**43.**   $y = f(x) = \sin x - \tan x$   A. $D = \{x \mid x \neq (2n+1)\pi/2\}$   B. $y = 0$ ⟺
$\sin x = \tan x = \sin x / \cos x$ ⟺ $\sin x = 0$ or $\cos x = 1$ ⟺ $x = n\pi$
(x-intercepts), y-intercept $= f(0) = 0$   C. $f(-x) = -f(x)$, so the
curve is symmetric about $(0,0)$. Also periodic with period $2\pi$.
D. $\lim\limits_{x \to \pi/2^-} (\sin x - \tan x) = -\infty$ and $\lim\limits_{x \to \pi/2^+} (\sin x - \tan x) = \infty$, so
$x = n\pi + \pi/2$ are VA.   E. $f'(x) = \cos x - \sec^2 x \leq 0$, so f decreases
on each interval in its domain, i.e., on $((2n-1)\pi/2, (2n+1)\pi/2)$.

F. No extrema

G. $f''(x) = - \sin x - 2 \sec^2 x \tan x$

$= - \sin x (1 + 2 \sec^3 x)$   Note that

$1 + 2 \sec^3 x \neq 0$ since $\sec^3 x \neq -1/2$.

$f''(x) > 0$ for $-\pi/2 < x < 0$ and $3\pi/2 < x < 2\pi$

so f is CU on $(n\pi - \pi/2, n\pi)$ and CD on

$(n\pi, n\pi + \pi/2)$. IP $(n\pi, 0)$. Note also

that $f'(0) = 0$ but $f'(\pi) = -2$.

H.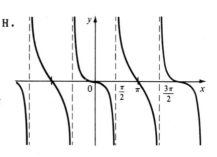

45.  $y = f(x) = x \tan x$, $-\pi/2 < x < \pi/2$   A. $D = (-\pi/2, \pi/2)$   B. Intercepts 0

C. $f(-x) = f(x)$, so the curve is symmetric about the y-axis.

D. $\lim\limits_{x \to \pi/2^-} x \tan x = \infty$ and $\lim\limits_{x \to -\pi/2^+} x \tan x = \infty$, so $x = \dfrac{\pi}{2}$ and $x = -\dfrac{\pi}{2}$

are VA.   E. $f'(x) = \tan x + x \sec^2 x > 0 \iff 0 < x < \pi/2$, so f

increases on $[0, \pi/2)$ and

decreases on $(-\pi/2, 0]$.

F. Absolute minimum $f(0) = 0$

G. $y'' = 2 \sec^2 x + 2x \tan x \sec^2 x$

   $>0$ for $-\pi/2 < x < \pi/2$, so f is

   CU on $(-\pi/2, \pi/2)$.

H.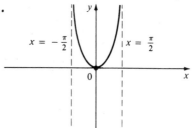

47.  $y = f(x) = x/2 - \sin x$, $0 < x < 3\pi$   A. $D = (0, 3\pi)$   B. No y-intercept

(x-intercept could be found approximately by Newton's Method, see

Exercise 3.10.11.)   C. No symmetry   D. No asymptotes

E. $f'(x) = 1/2 - \cos x > 0 \iff \cos x < 1/2 \iff \pi/3 < x < 5\pi/3$ or

$7\pi/3 < x < 3\pi$, so f is increasing on $[\pi/3, 5\pi/3]$ and $[7\pi/3, 3\pi)$ and

decreasing on $(0, \pi/3]$ and $[5\pi/3, 7\pi/3]$.

F. $f(\pi/3) = \pi/6 - \sqrt{3}/2$ is a local

minimum, $f(5\pi/3) = 5\pi/6 + \sqrt{3}/2$ is

a local maximum, $f(7\pi/3) = 7\pi/6 -$

$\sqrt{3}/2$ is a local minimum.

G. $f''(x) = \sin x > 0 \iff 0 < x < \pi$ or

$2\pi < x < 3\pi$, so f is CU on $(0, \pi)$ and

$(2\pi, 3\pi)$ and CD on $(\pi, 2\pi)$.

IP $(\pi, \pi/2)$ and $(2\pi, \pi)$

H.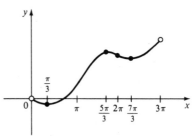

**49.** $y = f(x) = 2 \cos x + \sin^2 x$   A. $D = R$   B. y-intercept $= f(0) = 2$
C. $f(-x) = f(x)$, so the curve is symmetric about the y-axis.
Periodic with period $2\pi$   D. No asymptotes
E. $f'(x) = -2 \sin x + 2 \sin x \cos x = 2 \sin x (\cos x - 1) > 0 \iff$
$\sin x < 0 \iff (2n-1)\pi < x < 2n\pi$, so f is increasing on $[(2n-1)\pi, \ 2n\pi]$
and decreasing on $[2n\pi, (2n+1)\pi]$.   F. $f(2n\pi) = 2$ is a local
maximum.   $f((2n+1)\pi) = -2$ is a local minimum

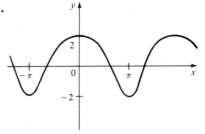

G. $f''(x) = -2 \cos x + 2 \cos 2x$   H.
$= 2(2 \cos^2 x - \cos x - 1)$
$= 2(2 \cos x + 1)(\cos x - 1) > 0$
$\iff \cos x < -1/2 \iff$
$x \in (2n\pi + 2\pi/3, 2n\pi + 4\pi/3)$, so f is
CU on these intervals and CD on
$(2n\pi - 2\pi/3, 2n\pi + 2\pi/3)$.
IP when $x = 2n\pi \pm 2\pi/3$

**51.** $y = f(x) = \sin 2x - 2 \sin x$   A. $D = R$   B. y-intercept $= f(0) = 0$
$y = 0 \iff 2 \sin x = \sin 2x = 2 \sin x \cos x \iff \sin x = 0$ or
$\cos x = 1 \iff x = n\pi$ (the x-intercepts)   C. $f(-x) = -f(x)$, so the
curve is symmetric about $(0,0)$.   f is periodic with period $2\pi$, so
we do parts D-G for $-\pi \le x \le \pi$.   D. No asymptotes.
E. $f'(x) = 2 \cos 2x - 2 \cos x$   As in Exercise 49G, we see that
$f'(x) > 0 \iff -\pi < x < -2\pi/3$ or $2\pi/3 < x < \pi$, so f is increasing on $[-\pi, -2\pi/3]$
and $[2\pi/3, \pi]$ and decreasing on $[-2\pi/3, 2\pi/3]$.   F. $f(-2\pi/3) = 3\sqrt{3}/2$
is a local maximum, $f(2\pi/3) = -3\sqrt{3}/2$ is a local minimum.

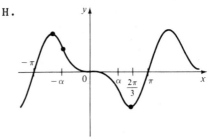

G. $f''(x) = -4 \sin 2x + 2 \sin x$   H.
$= 2 \sin x (1 - 4 \cos x) = 0$ when
$x = 0, \pm\pi$ or $\cos x = 1/4$.
If $\alpha = \cos^{-1}(1/4)$, then f is CU
on $(-\alpha, 0)$ and $(\alpha, \pi)$ and CD on
$(-\pi, -\alpha)$ and $(0, \alpha)$.   IP $(0,0)$,
$(\pi, 0)$, $(\alpha, -3\sqrt{15}/8)$, $(-\alpha, 3\sqrt{15}/8)$

Exercises 4.8

1.  If x is one number, the other is 100-x.  Maximize $f(x) = x(100-x)$
    $= 100x-x^2$.  $f'(x) = 100-2x = 0$ ⇒ $x = 50$.  Now $f''(x) = -2 < 0$, so
    there is an absolute maximum at $x = 50$.  The numbers are 50 and 50.

3.  The two numbers are x and 100/x where $x > 0$.  Minimize $f(x)$
    $= x+100/x$.  $f'(x) = 1-100/x^2 = (x^2-100)/x^2$.  The critical number is

    $x = 10$.  Since $f'(x) < 0$ for $0 < x < 10$ and $f'(x) > 0$ for $x > 10$,
    there is an absolute minimum at $x = 10$.  The numbers are 10 and 10.

5.  Let p be the perimeter and x and y the sides, so $p = 2x+2y$ ⇒ $y = \frac{1}{2}p-x$.  The area is $A(x) = x(\frac{1}{2}p-x) = \frac{1}{2}px-x^2$.  Now $0 = A'(x) = \frac{1}{2}p-2x$
    ⇒ $x = p/4$.  Since $A''(x) = -2 < 0$, there is an absolute maximum
    where $x = p/4$.  The sides of the rectangle are p/4 and p/2-p/4
    $= p/4$, so the rectangle is a square.

7.

Here $s^2 = h^2+b^2/4$ so $h^2 = s^2-b^2/4$.
The area is $A = \frac{1}{2}b\sqrt{s^2-b^2/4}$.  Let the
perimeter be p so $2s+b = p$ or $s = (p-b)/2$ ⇒ $A(b) = \frac{1}{2}b\sqrt{(p-b)^2/4-b^2/4}$
$= b\sqrt{p^2-2pb}/4$.  Now $A'(b) = \frac{1}{4}\sqrt{p^2-2pb}$

    $- \frac{1}{4}bp/\sqrt{p^2-2pb} = (-3pb+p^2)/4\sqrt{p^2-2pb}$.  Now $A'(b) = 0$ ⇒ $-3pb+p^2 = 0$
    ⇒ $b = p/3$.  Since $A'(b) > 0$ for $b < p/3$ and $A'(b) < 0$ for $b > p/3$,
    there is an absolute maximum when $b = p/3$.  But then $2s+p/3 = p$ so
    $s = p/3$ ⇒ $s = b$ ⇒ the triangle is equilateral.

9.

Here $5x+2y = 750$ so $y = (750-5x)/2$.
Maximize $A = xy = x(750-5x)/2$
$= 375x-(5/2)x^2$.  Now $A'(x) = 375-5x$
$= 0$ ⇒ $x = 75$.  Since $A''(x) = -5<0$
there is an absolute maximum when x
    $= 75$. Then $y = 375/2$. The largest area is $75(375/2) = 14,062.5$ ft$^2$.

11.  Let b be the base of the box and h the height.  The surface area is
    $1200 = b^2+4hb$ ⇒ $h = (1200-b^2)/4b$.  The volume is $V = b^2h =$

111

$b^2(1200-b^2)/4b = 300b-b^3/4$ ➙ $V'(b) = 300-(3/4)b^2$. $V'(b) = 0$ ➙ $b = \sqrt{400} = 20$. Since $V'(b) > 0$ for $0 < b < 20$ and $V'(b) < 0$ for $b > 20$, there is an absolute maximum when $b = 20$. Then $h = 10$, so the largest possible volume is $(20)^2(10) = 4000$ cm$^3$.

13.

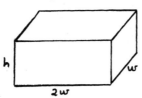

$10 = (2w)(w)h = 2w^2h$, so $h = 5/w^2$. The cost is $C(w) = 10(2w^2)+6[2(2wh)+2hw]+6(2w^2) = 32w^2+36wh$ $= 32w^2+180/w$. $C'(w) = 64w-180/w^2$ $= 4(16w^3-45)/w^2$ ➙ $w = \sqrt[3]{45/16}$ is the critical number. $C'(w) < 0$ for $0 < w < \sqrt[3]{45/16}$ and $C'(w) > 0$ for $w > \sqrt[3]{45/16}$. The minimum cost is $C(\sqrt[3]{45/16}) = 32(2.8125)^{2/3}$ $+180/\sqrt[3]{2.8125} \approx \$191.28$.

15. For $(x,y)$ on the line $y = 2x-3$, the distance to the origin is $\sqrt{(x-0)^2+(2x-3)^2}$. We minimize the square of the distance $= x^2+(2x-3)^3 = 5x^2-12x+9 = D(x)$. $D'(x) = 10x-12 = 0$ ➙ $x = 6/5$. Since there is a point closest to the origin, $x = 6/5$ and hence $y = -3/5$. So the point is $(6/5,-3/5)$.

17. By symmetry, the points are $(x,y)$ and $(x,-y)$, where $y > 0$. The square of the distance is $D(x) = (x-2)^2+y^2 = (x-2)^2+4+x^2$ $= 2x^2-4x+8$. So $D'(x) = 4x-4 = 0$ ➙ $x = 1$ and $y = \pm\sqrt{4+1} = \pm\sqrt{5}$. The points are $(1,\pm\sqrt{5})$.

19.

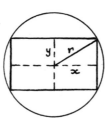

Area of rectangle is $4xy$. Also $r^2 = x^2+y^2$ so $y = \sqrt{r^2-x^2}$, so the area is $A(x) = 4x\sqrt{r^2-x^2}$. Now $A'(x) = 4\left[\sqrt{r^2-x^2} - x^2/\sqrt{r^2-x^2}\right]$ $= 4(r^2-2x^2)/\sqrt{r^2-x^2}$. The critical number is $x = r/\sqrt{2}$. Clearly this gives a maximum. The dimensions are $2x = \sqrt{2}r$ and $2y = \sqrt{2}r$.

21.

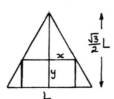

$\frac{(\sqrt{3}/2)L-y}{x} = \frac{(\sqrt{3}/2)L}{L/2} = \sqrt{3}$ (similar triangles) ➙ $\sqrt{3}x = (\sqrt{3}/2)L-y$ ➙ $y = (\sqrt{3}/2)(L-2x)$. The area of the inscribed rectangle is $A(x) = (2x)y = \sqrt{3}x(L-2x)$ where

$0 \le x \le L/2$. Now $0 = A'(x) = \sqrt{3}L - 4\sqrt{3}x \Rightarrow x = \sqrt{3}L/4\sqrt{3} = L/4$.
Since $A(0) = A(L/2) = 0$, the maximum occurs when $x = L/4$, and $y$
$= (\sqrt{3}/2)L - \sqrt{3}L/4 = \sqrt{3}L/4$, so the dimensions are $L/2$ and $\sqrt{3}L/4$.

23.

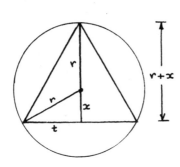

The area of the triangle is $A(x) =$

$\frac{1}{2}(2t)(r+x) = t(r+x) = \sqrt{r^2-x^2}\,(r+x)$.

Then $0 = A'(x) = r\dfrac{-2x}{2\sqrt{r^2-x^2}} + \sqrt{r^2-x^2}$

$+ x\dfrac{-2x}{2\sqrt{r^2-x^2}} = -\dfrac{x^2+rx}{\sqrt{r^2-x^2}} + \sqrt{r^2-x^2} \Rightarrow$

$\dfrac{x^2+rx}{\sqrt{r^2-x^2}} = \sqrt{r^2-x^2} \Rightarrow x^2+rx = r^2-x^2 \Rightarrow$

$0 = 2x^2+rx-r^2 = (2x-r)(x+r) \Rightarrow x = r/2$ or $x = -r$. Now $A(r) = 0$
$= A(-r) \Rightarrow$ the maximum occurs where $x = r/2$, so the triangle has
height $r+r/2 = 3r/2$ and base $2\sqrt{r^2-(r/2)^2} = 2\sqrt{3r^2/4} = \sqrt{3}r$.

25.

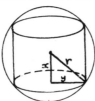

The cylinder has volume $V = \pi y^2(2x)$
Also $x^2+y^2 = r^2 \Rightarrow y^2 = r^2-x^2$, so
$V(x) = \pi(r^2-x^2)(2x) = 2\pi(r^2x-x^3)$,
where $0 \le x \le r$. $V'(x) = 2\pi(r^2-3x^2) = 0$
$\Rightarrow x = r/\sqrt{3}$. Now $V(0) = V(r) = 0$,

so there is a maximum when $x = r/\sqrt{3}$ and $V(r/\sqrt{3}) = 4\pi r^3/3\sqrt{3}$.

27.

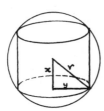

The cylinder has surface area
$2\pi y^2+2\pi y(2x)$. Now $x^2+y^2 = r^2 \Rightarrow y$
$= \sqrt{r^2-x^2}$, so the surface area is
$S(x) = 2\pi(r^2-x^2)+4\pi x\sqrt{r^2-x^2}$, $0 \le x \le r$.
$S'(x) = -4\pi x+4\pi\sqrt{r^2-x^2}-4\pi x^2/\sqrt{r^2-x^2}$

$= 4\pi(r^2-2x^2-x\sqrt{r^2-x^2})/\sqrt{r^2-x^2} = 0 \Rightarrow x\sqrt{r^2-x^2} = r^2-2x^2 \quad$ (i)
$\Rightarrow x^2(r^2-x^2) = r^4-4r^2x^2+4x^4 \Rightarrow 5x^4-5r^2x^2+r^4 = 0 \quad$ By the quadratic
formula, $x^2 = [(5\pm\sqrt{5})/10]r^2$, but we reject the root with the + sign
since it doesn't satisfy (i). So $x = \sqrt{(5-\sqrt{5})/10}\,r$. Since $S(0) =$
$S(r) = 0$, the maximum occurs at the critical number and $x^2 =$
$[(5-\sqrt{5})/10]r^2 \Rightarrow y^2 = [(5+\sqrt{5})/10]r^2 \Rightarrow$ the surface area is
$2\pi[(5+\sqrt{5})/10]r^2 + 4\pi\sqrt{(5-\sqrt{5})/10}\,\sqrt{(5+\sqrt{5})/10}\,r^2 = \pi r^2(1+\sqrt{5})$.

29.

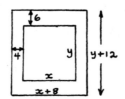

$xy = 384 \Rightarrow y = 384/x$. Total area
is $A(x) = (8+x)(12+384/x)$

$= 12(40+x+256/x)$

$A'(x) = 12(1-256/x^2) = 0 \Rightarrow x = 16$

There is an absolute minimum when
$x = 16$ since $A'(x)<0$ for $0<x<16$ and $A'(x)>0$ for $x>16$. When
$x = 16$, $y = 384/16 = 24$, so the dimensions are 24 cm and 36 cm.

31.

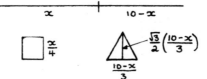

The total area is $A(x) =$
$x^2/16 + (\sqrt{3}/36)(10-x)^2$, $0 \leq x \leq 10$.

$A'(x) = x/8 - (\sqrt{3}/18)(10-x) = 0 \Rightarrow$

$x = 40\sqrt{3}/(9+4\sqrt{3})$. Now $A(0) =$

$(\sqrt{3}/36)100 \approx 4.81$, $A(10) = 100/16 =$

6.25 and $A(40\sqrt{3}/(9+4\sqrt{3})) \approx 2.72$, so the maximum occurs when
$x = 10$ m and the minimum when $x = 40\sqrt{3}/(9+4\sqrt{3}) \approx 4.35$ m.

33.

The volume is $V = \pi r^2 h$ and the
surface area is $S(r) = \pi r^2 + 2\pi rh =$
$\pi r^2 + 2\pi r(V/\pi r^2) = \pi r^2 + 2V/r$

$S'(r) = 2\pi r - 2V/r^2 = 0 \Rightarrow 2\pi r^3 = 2V$

$\Rightarrow r = \sqrt[3]{V/\pi}$. This gives an absolute
minimum since $S'(r)<0$ for $0<r<\sqrt[3]{V/\pi}$ and $S'(r)>0$ for $r>\sqrt[3]{V/\pi}$. When
$r = \sqrt[3]{V/\pi}$, $h = V/\pi r^2 = V/\pi(V/\pi)^{2/3} = \sqrt[3]{V/\pi}$.

35.

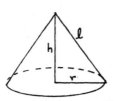

$h^2 + r^2 = \ell^2 \Rightarrow V = (\pi/3)r^2h$
$= (\pi/3)(\ell^2 - h^2)h = (\pi/3)(\ell^2 h - h^3)$

$V'(h) = (\pi/3)(\ell^2 - 3h^2) = 0$ when
$h = \ell/\sqrt{3}$. This gives an absolute
maximum since $V'(h)>0$ for $0<h<\ell/\sqrt{3}$
and $V'(h)<0$ for $h>\ell/\sqrt{3}$. Maximum volume is $V(\ell/\sqrt{3}) = 2\pi\ell^3/9\sqrt{3}$.

37. $E(n) = (2/3)n - (1/90)n^2 \Rightarrow E'(n) = 2/3 - n/45 = 0$ when $n = (2/3)45$
$= 30$. Since $E''(n) = -1/45 < 0$, the maximum effectiveness occurs
when the viewer watches the commercial 30 times.

39.

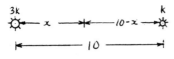

The total illumination is

$$I(x) = \frac{3k}{x^2} + \frac{k}{(10-x)^2} \qquad 0 < x < 10. \quad \text{Then}$$

$$I'(x) = \frac{-6k}{x^3} + \frac{2k}{(10-x)^3} = 0 \Rightarrow$$

$$6k(10-x)^3 = 2kx^3 \Rightarrow \sqrt[3]{3}(10-x) = x \Rightarrow$$

$x = 10\sqrt[3]{3}/(1+\sqrt[3]{3}) \approx 5.9$ ft. This gives a minimum since there is clearly no maximum.

41.

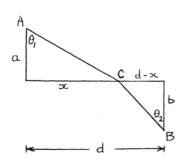

The total time T(x) = (time from A to C) + (time from C to B) =

$$\sqrt{a^2+x^2}/v_1 + \sqrt{b^2+(d-x)^2}/v_2 \qquad 0 < x < d.$$

$$T'(x) = \frac{x}{v_1 \sqrt{a^2+x^2}} - \frac{d-x}{v_2 \sqrt{b^2+(d-x)^2}}$$

$$= \frac{\sin\theta_1}{v_1} - \frac{\sin\theta_2}{v_2}$$

The minimum occurs when T'(x) = 0

$$\Rightarrow \frac{\sin\theta_1}{v_1} = \frac{\sin\theta_2}{v_2}$$

43.

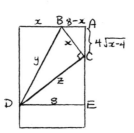

$y^2 = x^2+z^2$, but triangles CDE and BCA are similar, so $z/8 = x/4\sqrt{x-4}$. Thus we minimize $f(x) = y^2 =$

$x^2 + 4x^2/(x-4) = x^3/(x-4)$, $4 < x \le 8$.

$$f'(x) = \frac{3x^2(x-4)-x^3}{(x-4)^2} = \frac{2x^2(x-6)}{(x-4)^2} = 0$$

when x = 6. f'(x) < 0 when x < 6,

f'(x) > 0 when x < 6, so the minimum occurs when x = 6 in.

45. Let x = selling price of ticket. Then 12-x is the amount the ticket price has been lowered, so the number of tickets sold is 11000+1000(12-x) = 23000-1000x. The revenue is R(x) = x(23000-1000x) = 23000x-1000x$^2$, so R'(x) = 23000-2000x = 0 when x = 11.5. Since R"(x) = -2000 < 0, the maximum revenue occurs when the ticket prices are $11.50.

47.

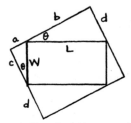

$a = W \sin \theta$, $c = W \cos \theta$, $b = L \cos \theta$, $d = L \sin \theta$, so the area of the circumscribed rectangle is $A(\theta) = (a+b)(c+d)$
$= (W \sin \theta + L \cos \theta)(W \cos \theta + L \sin \theta) = LW \sin^2\theta + LW \cos^2\theta$
$+ (L^2+W^2) \sin \theta \cos \theta = LW + (1/2)(L^2+W^2) \sin 2\theta$, $0 \le \theta \le \pi/2$. This expression shows, without calculus, that the maximum value of $A(\theta)$ occurs when $\sin 2\theta = 1 \iff 2\theta = \pi/2 \iff x = \pi/4$. So the maximum area is $A(\pi/4) = LW + (1/2)(L^2+W^2) = (L+W)^2/2$.

Section 4.9

Exercises 4.9

1.  (a) $C(x) = 10000+25x+x^2$, $C(1000) = \$1035000$. $c(x) = \dfrac{C(x)}{x} = \dfrac{10000}{x}$
    $+25+x$, $c(1000) = \$1035$. $C'(x) = 25+2x$, $C'(1000) = \$2025/\text{unit}$.
    (b) We must have $c(x) = C'(x) \Rightarrow 10000/x+25+x = 25+2x \Rightarrow 10000/x = x$
    $\Rightarrow x^2 = 10000 \Rightarrow x = 100$. This is a minimum since $c''(x) = 20000/x^3 > 0$.
    (c) The minimum average cost is $c(100) = \$225$.

3.  (a) $C(x) = 45+\dfrac{x}{2}+\dfrac{x^2}{560}$, $C(1000) = \$2330.71$. $c(x) = \dfrac{45}{x} + \dfrac{1}{2} + \dfrac{x}{560}$,
    $c(1000) = \$2.33$. $C'(x) = \dfrac{1}{2}+\dfrac{x}{280}$, $C'(1000) = \$4.07/\text{unit}$ (b) We must
    have $C'(x) = c(x) \Rightarrow \dfrac{1}{2}+\dfrac{x}{280} = \dfrac{45}{x}+\dfrac{1}{2}+\dfrac{x}{560} \Rightarrow \dfrac{45}{x} = \dfrac{x}{560} \Rightarrow x^2 = (45)(560) \Rightarrow$
    $x = \sqrt{25200} \approx 159$. This is a minimum since $c''(x) = 90/x^2 > 0$.
    (c) The minimum average cost is $c(159) = \$1.07$.

5.  (a) $C(x) = 2\sqrt{x}+\dfrac{x^2}{8000}$, $C(1000) = \$188.25$. $c(x) = \dfrac{2}{\sqrt{x}} + \dfrac{x}{8000}$,
    $c(1000) = \$.19$. $C'(x) = \dfrac{1}{\sqrt{x}} + \dfrac{x}{4000}$, $C'(1000) = \$.28/\text{unit}$.

(b)  We must have $C'(x) = c(x) \Rightarrow \dfrac{1}{\sqrt{x}} + \dfrac{x}{4000} = \dfrac{2}{\sqrt{x}} + \dfrac{x}{8000} \Rightarrow \dfrac{x}{8000}$

$= 1/\sqrt{x} \Rightarrow x^{3/2} = 8000 \Rightarrow x = (8000)^{2/3} = 400.$  This is a minimum
since $c''(x) = (3/2)x^{-5/2} > 0.$

(c)  The minimum average cost is $c(400) = \$.15.$

7.  $C(x) = 680+4x+0.01x^2$, $p(x) = 12 \Rightarrow R(x) = xp(x) = 12x.$ If the profit
is maximum, then $R'(x) = C'(x) \Rightarrow 12 = 4+0.02x \Rightarrow 0.02x = 8 \Rightarrow x =$
400.  Now $R''(x) = 0 < .02 = C''(x)$, so $x = 400$ gives a maximum.

9.  $C(x) = 1200+25x-.0001x^2$, $p(x) = 55-x/1000.$  Then $R(x) = xp(x)$
$= 55x-x^2/1000.$  If the profit is maximum, then $R'(x) = C'(x)$ or
$55-x/500 = 25-.0002x \Rightarrow 30 = 0.0018x \Rightarrow x = 30/0.0018 \approx 16667.$  Now
$R''(x) = -\dfrac{1}{500} < -.0002 = C''(x)$, so $x = 16667$ gives a maximum.

11.  $C(x) = 1450+36x-x^2+0.001x^3$, $p(x) = 60-0.01x.$  Then $R(x) = xp(x)$
$= 60x-.01x^2.$  If the profit is maximum, then $R'(x) = C'(x)$ or
$60-.02x = 36-2x+.003x^2 \Rightarrow .003x^2-1.98x-24 = 0.$  By the quadratic

formula, $x = \dfrac{1.98\pm\sqrt{(-1.98)^2+4(.003)(24)}}{2(.003)} = \dfrac{1.98\pm\sqrt{4.2084}}{.006}$  Since x>0,

$x \approx (1.98+2.05)/.006 \approx 672.$  Now $R''(x) = -.02$ and $C''(x) = -2+.006x,$
$\Rightarrow C''(672) = 2.032 \Rightarrow R''(672) < C''(672) \Rightarrow$ there is a maximum at
$x = 672.$

13.  $C(x) = .001x^3-0.3x^2+6x+900.$  The marginal cost is $C'(x) =$
$.003x^2-.6x+6.$  $C'(x)$ is increasing when $C''(x) > 0$ or $.006x-.6 > 0$
or $x > .6/.006 = 100.$  So $C'(x)$ starts to increase when $x = 100.$

15.  (a) We are given that the demand function p is linear and
$p(27000) = 10$, $p(33000) = 8$, so the slope is $\dfrac{10-8}{27000-33000} = -\dfrac{1}{3000}$
and the equation of the graph is $y-10 = (-1/3000)(x-27000) \Rightarrow$
$p(x) = 19 - x/3000.$
(b) The revenue is $R(x) = xp(x) = 19x-x^2/3000 \Rightarrow R'(x) = 19-x/1500 =$
0 when $x = 28500.$  Since $R''(x) = -1/1500<0$, the maximum revenue
occurs when $x = 28500 \Rightarrow$ the price is $p(28500) = \$9.50.$

17.  (a)  $p(x) = 450-(1/10)(x-1000) = 550-x/10.$  (b)  $R(x) = xp(x)$
$= 500x-x^2/10.$  $R'(x) = 550-x/5 = 0$ when $x = 5(550) = 2750.$
$p(2750) = 275$, so the rebate should be $450-275 = \$175.$
(c)  $P(x) = R(x)-C(x) = 550x-x^2/10-6800-150x = 400x-x^2/10-6800.$
$P'(x) = 400-x/5 = 0$ when $x = 2000.$  $p(2000) = 550-200 = 350.$
Therefore the rebate to maximize profits should be $450-350 = \$100.$

## Exercises 4.10

1. $f(x) = 12x^2 + 6x - 5$ ⇒ $F(x) = 12(x^3/3) + 6(x^2/2) - 5x + C = 4x^3 + 3x^2 - 5x + C$

3. $f(x) = 6x^9 - 4x^7 + 3x^2 + 1$ ⇒ $F(x) = 6(x^{10}/10) - 4(x^8/8) + 3(x^3/3) + x + C$
$= (3/5)x^{10} - (1/2)x^8 + x^3 + x + C$

5. $f(x) = \sqrt{x} + \sqrt[3]{x} = x^{1/2} + x^{1/3}$ ⇒ $F(x) = x^{3/2}/(3/2) + x^{4/3}/(4/3) + C$
$= (2/3)x^{3/2} + (3/4)x^{4/3} + C$

7. $f(x) = 6/x^5 = 6x^{-5}$ ⇒ $F(x) = 6x^{-4}/(-4) + C_1 = -3/2x^4 + C_1$, if $x > 0$;
$F(x) = -3/2x^4 + C_2$, if $x < 0$

9. $f(x) = \sqrt{x} + 1/\sqrt{x} = x^{1/2} + x^{-1/2}$ ⇒ $F(x) = x^{3/2}/(3/2) + x^{1/2}/(1/2) + C$
$= (2/3)x^{3/2} + 2x^{1/2} + C$

11. $g(t) = (t^3 + 2t^2)/\sqrt{t} = t^{5/2} + 2t^{3/2}$ ⇒ $G(t) = t^{7/2}/(7/2) + 2t^{5/2}/(5/2) + C$
$= (2/7)t^{7/2} + (4/5)t^{5/2} + C$

13. $h(x) = \sin x - 2\cos x$ ⇒ $H(x) = -\cos x - 2\sin x + C$

15. $f(t) = \sec^2 t + t^2$ ⇒ $F(t) = \tan t + t^3/3 + C_n$ on the interval
$((2n-1)\pi/2, (2n+1)\pi/2)$

17. $f'(x) = x^4 - 2x^2 + x - 1$ ⇒ $f(x) = x^5/5 - 2x^3/3 + x^2/2 - x + C$

19. $f''(x) = x^2 + x^3$ ⇒ $f'(x) = x^3/3 + x^4/4 + C$ ⇒
$f(x) = x^4/12 + x^5/20 + Cx + D$

21. $f''(x) = 1$ ⇒ $f'(x) = x + C$ ⇒ $f(x) = x^2/2 + Cx + D$

23. $f'''(x) = 24x$ ⇒ $f''(x) = 12x^2 + C$ ⇒ $f'(x) = 4x^3 + Cx + D$ ⇒
$f(x) = x^4 + Cx^2/2 + Dx + E$

25. $f'(x) = 4x + 3$ ⇒ $f(x) = 2x^2 + 3x + C$ ⇒ $-9 = f(0) = C$ ⇒ $f(x) = 2x^2 + 3x - 9$

27. $f'(x) = 3\sqrt{x} - 1/\sqrt{x} = 3x^{1/2} - x^{-1/2}$ ⇒ $f(x) = 3x^{3/2}/(3/2) - x^{1/2}/(1/2)$
$+ C$ ⇒ $2 = f(1) = 2 - 2 + C = C$ ⇒ $f(x) = 2x^{3/2} - 2x^{1/2} + 2$

29. $f'(x) = 3\cos x + 5\sin x$ ⇒ $f(x) = 3\sin x - 5\cos x + C$ ⇒
$4 = f(0) = -5 + C$ ⇒ $C = 9$ ⇒ $f(x) = 3\sin x - 5\cos x + 9$

31. $f'(x) = 2 + x^{3/5}$ ⇒ $f(x) = 2x + (5/8)x^{8/5} + C$ ⇒ $3 = f(1) = 2 + (5/8) + C$
⇒ $C = 3/8$ ⇒ $f(x) = 2x + (5/8)x^{8/5} + 3/8$

33. $f''(x) = -8$ ⇒ $f'(x) = -8x + C$ ⇒ $5 = f'(0) = C$ ⇒ $f'(x) = -8x + 5$ ⇒
$f(x) = -4x^2 + 5x + D$ ⇒ $6 = f(0) = D$ ⇒ $f(x) = -4x^2 + 5x + 6$

35. $f''(x) = 20x^3 - 10$ ⇒ $f'(x) = 5x^4 - 10x + C$ ⇒ $-5 = f'(1) = 5 - 10 + C$ ⇒
$C = 0$ ⇒ $f'(x) = 5x^4 - 10x$ ⇒ $f(x) = x^5 - 5x^2 + D$ ⇒ $1 = f(1) = 1 - 5 + D$
⇒ $D = 5$ ⇒ $f(x) = x^5 - 5x^2 + 5$

37. $f''(x) = x^2 + 3 \cos x \Rightarrow f'(x) = x^3/3 + 3 \sin x + C \Rightarrow 3 = f'(0)$
$= C \Rightarrow f'(x) = x^3/3 + 3 \sin x + 3 \Rightarrow f(x) = x^4/12 - 3 \cos x + 3x + D$
$\Rightarrow 2 = f(0) = -3+D \Rightarrow D = 5 \Rightarrow f(x) = x^4/12 - 3 \cos x + 3x + 5$

39. $f''(x) = 6x+6 \Rightarrow f'(x) = 3x^2+6x+C \Rightarrow f(x) = x^3+3x^2+Cx+D \Rightarrow$
$4 = f(0) = D$ and $3 = f(1) = 1+3+C+D = 4+C+4 \Rightarrow C = -5 \Rightarrow$
$f(x) = x^3+3x^2-5x+4$

41. $f''(x) = x^{-3} \Rightarrow f'(x) = -(1/2)x^{-2}+C \Rightarrow f(x) = (1/2)x^{-1}+Cx+D \Rightarrow$
$0 = f(1) = (1/2)+C+D$ and $0 = f(2) = (1/4)+2C+D$.  Solving these
equations, we get $C = 1/4$, $D = -3/4$, so $f(x) = 1/2x + x/4 - 3/4$.

43. $v(t) = s'(t) = 3-2t \Rightarrow s(t) = 3t-t^2+C \Rightarrow 4 = s(0) = C \Rightarrow$
$s(t) = 3t-t^2+4$

45. $a(t) = v'(t) = 3t+8 \Rightarrow v(t) = (3/2)t^2+8t+C \Rightarrow -2 = v(0) = C \Rightarrow$
$v(t) = (3/2)t^2+8t-2 \Rightarrow s(t) = (1/2)t^3+4t^2-2t+D \Rightarrow 1 = s(0) = D \Rightarrow$
$s(t) = (1/2)t^3+4t^2-2t+1$

47. $a(t) = v'(t) = t^2-t \Rightarrow v(t) = (1/3)t^3-(1/2)t^2+C \Rightarrow s(t) =$
$(1/12)t^4-(1/6)t^3+Ct+D \Rightarrow 0 = s(0) = D$ and $12 = s(6) = 108-36+6C+0$
$\Rightarrow C = -10 \Rightarrow s(t) = (1/12)t^4-(1/6)t^3-10t$

49. (a) $v'(t) = a(t) = -9.8 \Rightarrow v(t) = -9.8t+C$, but $C = v(0) = 0$, so
$v(t) = -9.8t \Rightarrow s(t) = -4.9t^2+D \Rightarrow D = s(0) = 450 \Rightarrow s(t) = 450-4.9t^2$
(b) It reaches the ground when $0 = s(t) = 450-4.9t^2 \Rightarrow t^2 = 450/4.9$
$\Rightarrow t = \sqrt{450/4.9} \approx 9.58$ s.  (c) $v = -9.8\sqrt{450/4.9} \approx -93.9$ m/s

51. (a) $v'(t) = -9.8 \Rightarrow v(t) = -9.8t+C \Rightarrow 5 = v(0) = C$, so
$v(t) = 5-9.8t \Rightarrow s(t) = 5t-4.9t^2+D \Rightarrow D = s(0) = 450 \Rightarrow$
$s(t) = 450+5t-4.9t^2$ (b) It reaches the ground when $450+5t-4.9t^2 = 0$
By the quadratic formula, the positive root of this equation is
$t = (5+\sqrt{8845})/9.8 \approx 10.1$ s.  (c) $v = 5-9.8(5+\sqrt{8845})/9.8 \approx -94.0$ m/s

53. By Exercise 52, $s(t) = -4.9t^2+v_0t+s_0$ and $v(t) = s'(t) = -9.8t+v_0$.
So $[v(t)]^2 = (9.8)^2t^2-19.6v_0t+v_0^2$ and $v_0^2-19.6[s(t)-s_0] =$
$v_0^2-19.6[-4.9t^2+v_0t] = v_0^2+(9.8)^2t^2-19.6v_0t = [v(t)]^2$.

55. $a(t) = a$ and the initial velocity is $30$ mi/h $= 30\cdot5280/(60)^2 = 44$
ft/s and final velocity $50$ mi/h $= 50\cdot5280/(60)^2 = 220/3$ ft/s.  So
$v(t) = at+44 \Rightarrow 220/3 = v(5) = 5a+44 \Rightarrow a = 88/15 \approx 5.87$ ft/s$^2$.

57. The height at time $t$ is $s(t) = -16t^2+h$, where $h = s(0)$ is the
height of the cliff.  $v(t) = -32t = -120$ when $t = 3.75$, so
$0 = s(3.75) = -16(3.75)^2+h \Rightarrow h = 16(3.75)^2 = 225$ ft.

## Review Exercises for Chapter 4

1. $f(x) = x^3 - 12x + 5$, $-5 \leq x \leq 3$. $f'(x) = 3x^2 - 12 = 0 \Rightarrow x^2 = 4 \Rightarrow x = \pm 2$. $f''(x) = 6x \Rightarrow f''(-2) = -12 < 0$, so $f(-2) = 21$ is a local maximum, and $f''(2) = 12 > 0$, so $f(2) = -11$ is a local minimum. Also $f(-5) = -60$ and $f(3) = -4$, so $f(-2) = 21$ is the absolute maximum and $f(-5) = -60$ is the absolute minimum.

3. $f(x) = \frac{x-2}{x+2}$ $0 \leq x \leq 4$. $f'(x) = \frac{(x+2)-(x-2)}{(x+2)^2} = \frac{4}{(x+2)^2} > 0 \Rightarrow f$ is increasing on $[0,4]$, so $f$ has no local extrema and $f(0) = -1$ is the absolute minimum and $f(4) = 1/3$ is the absolute maximum.

5. $f(x) = x - \sqrt{2} \sin x$, $0 \leq x \leq \pi$. $f'(x) = 1 - \sqrt{2} \cos x = 0 \Rightarrow \cos x = 1/\sqrt{2} \Rightarrow x = \pi/4$. $f''(\pi/4) = \sqrt{2} \sin(\pi/4) = 1 > 0$, so $f(\pi/4) = (\pi/4) - 1$ is a local minimum. Also $f(0) = 0$ and $f(\pi) = \pi$, so the absolute minimum is $f(\pi/4) = \frac{\pi}{4} - 1$, the absolute maximum is $f(\pi) = \pi$.

7. $\lim\limits_{t \to 6} \frac{17}{(t-6)^2} = \infty$ since $(t-6)^2 \to 0$ and $\frac{17}{(t-6)^2} > 0$

9. $\lim\limits_{x \to 2^-} \frac{3}{\sqrt{2-x}} = \infty$ since $\sqrt{2-x} \to 0$ and $\frac{3}{\sqrt{2-x}} > 0$

11. $\lim\limits_{x \to \infty} \frac{1+x}{1-x^2} = \lim\limits_{x \to \infty} \frac{1}{1-x} = \lim\limits_{x \to \infty} \frac{1/x}{(1/x)-1} = \frac{0}{0-1} = 0$

13. $\lim\limits_{x \to \infty} \frac{\sqrt{x^2-9}}{2x-6} = \lim\limits_{x \to \infty} \frac{\sqrt{1-9/x^2}}{2-6/x} = \frac{\sqrt{1-0}}{2-0} = \frac{1}{2}$

15. $\lim\limits_{x \to -\infty} \frac{4x^5+x^2+3}{2x^5+x^3+1} = \lim\limits_{x \to -\infty} \frac{4 + 1/x^3 + 3/x^5}{2 + 1/x^2 + 1/x^5} = \frac{4+0+0}{2+0+0} = 2$

17. $\lim\limits_{x \to \infty} \left[\sqrt[3]{x} - \frac{x}{3}\right] = \lim\limits_{x \to \infty} \sqrt[3]{x}(1 - \frac{1}{3}x^{2/3}) = -\infty$ since $\sqrt[3]{x} \to \infty$ and $1 - \frac{1}{3}x^{2/3} \to -\infty$

19. $\lim\limits_{x \to -2^+} \frac{x}{x^2-4} = \infty$ since $x^2-4 \to 0$ and $\frac{x}{x^2-4} > 0$ for $-2 < x < 0$

21. $y = f(x) = 1 + x + x^3$ A. $D = R$ B. y-intercept $= 1$, x-intercept requires the solution of a cubic, so we don't find it.

C. No symmetry  D. $\lim\limits_{x \to \infty} (1+x+x^3) = \infty$, $\lim\limits_{x \to -\infty} (1+x+x^3) = -\infty$, no

asymptotes.  E. $f'(x) = 1+3x^2 \Rightarrow$

$f'(x)>0$, so f is increasing on R.

F. No local extrema

G. $f''(x) = 6x \Rightarrow f''(x)>0$ if $x>0$

and $f''(x)<0$ if $x<0$, so f is CU on

$(0,\infty)$ and CD on $(-\infty,0)$. IP $(0,1)$

H.

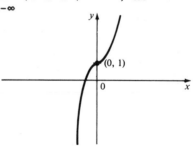

23.  $y = f(x) = 1/x(x-3)^2$  A. D = {x| x≠0,3} = $(-\infty,0) \cup (0,3) \cup (3,\infty)$

B. No intercepts  C. No symmetry  D. $\lim\limits_{x \to \pm\infty} \dfrac{1}{x(x-3)^2} = 0$, so y = 0 is

a HA.  $\lim\limits_{x \to 0^+} \dfrac{1}{x(x-3)^2} = \infty$, $\lim\limits_{x \to 0^-} \dfrac{1}{x(x-3)^2} = -\infty$, $\lim\limits_{x \to 3} \dfrac{1}{x(x-3)^2} = \infty$, so

x = 0 and x = 3 are VA.  E. $f'(x) = -\dfrac{(x-3)^2+2x(x-3)}{x^2(x-3)^4} = \dfrac{3(1-x)}{x^2(x-3)^3} \Rightarrow$

$f'(x)>0 \iff 1<x<3$, so f is increasing on [1,3) and decreasing on

$(-\infty,0)$, $(0,1]$, and $(3,\infty)$.  H.

F. $f(1) = 1/4$ is a local minimum

G. $f''(x) = \dfrac{6(2x^2-4x+3)}{x^3(x-3)^4}$  Note that

$2x^2-4x+3>0$ for all x since it has

negative discriminant.  So $f''(x)>0$

$\iff x>0 \Rightarrow$ f is CU on $(0,3)$ and

$(3,\infty)$ and CD on $(-\infty,0)$. No IP.

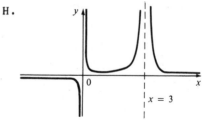

25.  $y = f(x) = x\sqrt{5-x}$  A. D = {x| x≤5} = $(-\infty,5]$  B. x-intercepts 0,5;

y-intercept = f(0) = 0  C. No symmetry  D. $\lim\limits_{x \to -\infty} x\sqrt{5-x} = -\infty$, no

asymptote  E. $f'(x) = \sqrt{5-x} - \dfrac{x}{2\sqrt{5-x}} = \dfrac{10-3x}{2\sqrt{5-x}} > 0 \iff x < \dfrac{10}{3}$ So f is

increasing on $(-\infty,10/3]$ and decreasing on $[10/3,5]$.

F. $f(10/3) = 10\sqrt{5}/3\sqrt{3}$ is a local and absolute maximum.

G. f"(x)

$$= \frac{-6\sqrt{5-x} - (10-3x)(-1/\sqrt{5-x})}{4(5-x)}$$

$$= \frac{3x-20}{4(5-x)^{3/2}} < 0 \text{ for all x in D}$$

so f is CD on $(-\infty, 5)$.

H.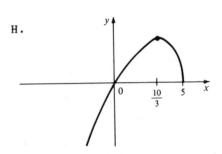

27. $y = f(x) = \frac{x^2}{x+8} = x-8 + \frac{64}{x+8}$   A. D = {x | x≠-8}   B. Intercepts are 0.

C. No symmetry   D. $\lim_{x\to\infty} \frac{x^2}{x+8} = \infty$, but $f(x) - (x-8) = \frac{64}{x+8} \to 0$ as $x \to \infty$

so $y = x-8$ is a slant asymptote. $\lim_{x\to-8^+} \frac{x^2}{x+8} = \infty$ and $\lim_{x\to-8^-} \frac{x^2}{x+8} = -\infty$

so $x = -8$ is a VA.   E. $f'(x) = 1 - \frac{64}{(x+8)^2} = \frac{x(x+16)}{(x+8)^2} > 0 \iff$ x>0 or

x<-16, so f is increasing on
$(-\infty,-16]$ and $[0,\infty)$ and decreasing   H.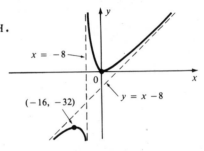
on $[-16,-8)$ and $(-8,0]$.
F. $f(-16) = -32$ is a local
maximum, $f(0) = 0$ is a local
minimum.   G. $f"(x) = 128/(x+8)^3$
$>0 \iff$ x>-8, so f is CU on $(-8,\infty)$
and CD on $(-\infty,-8)$.   No IP.

29. $y = f(x) = (x+3)^4(x-2)^5$   A. D = R   B. x-intercepts are -3, 2;
y-intercept = $f(0) = -2592$   C. No symmetry   D. $\lim_{x\to\infty} (x+3)^4(x-2)^5 = \infty$

and $\lim_{x\to-\infty} (x+3)^4(x-2)^5 = -\infty$, no asymptotes.

E. $f'(x) = 4(x+3)^3(x-2)^5 + 5(x+3)^4(x-2)^4 = (x+3)^3(x-2)^4(9x+7) > 0$
$\iff$ x<-3 or x>-7/9, so f is increasing on $(-\infty,-3]$ and $[-7/9,\infty)$ and
decreasing on $[-3,-7/9]$. F. $f(-3) = 0$ is a local maximum
$f(-7/9) \approx -4033$ is a local minimum

G. $y'' = 4(x+3)^2(x-2)^3(18x^2+28x-3)$   H.

The roots of $18x^2+28x-3 = 0$ are

$\alpha_1$, $\alpha_2 = (-14\pm5\sqrt{10})/18$, so $y''>0$

$\Longleftrightarrow \alpha_1<x<\alpha_2$ or $x>2$.   f is CU on

$(\alpha_1,\alpha_2)$, $(2,\infty)$ and CD on $(-\infty,\alpha_1)$ and

$(\alpha_2,2)$.   IP when $x = \alpha_1$, $\alpha_2$, 2.

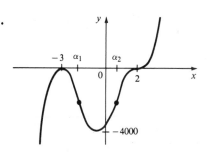

31.   $y = f(x) = x\sqrt[3]{x^2+5}$   A. $D = R$   B. Intercepts are 0   C. $f(-x) = -f(x)$

so f is symmetric about $(0,0)$.   D. $\lim\limits_{x\to\infty} x\sqrt[3]{x^2+5} = \infty$,  $\lim\limits_{x\to-\infty} x\sqrt[3]{x^2+5} = -\infty$

No asymptotes   E. $f'(x) = \sqrt[3]{x^2+5} + x\dfrac{2x}{3(x^2+5)^{2/3}} = \dfrac{5(x^2+3)}{3(x^2+5)^{2/3}} > 0$

so f is increasing on R.         H.

F. No extrema

G. $f''(x) =$

$\dfrac{3(x^2+5)^{2/3}(10x)-5(x^2+3)4x(x^2+5)^{-1/3}}{9(x^2+5)^{4/3}}$

$= \dfrac{10x(x^2+9)}{9(x^2+5)^{5/3}} > 0 \Longleftrightarrow x>0$ so f is

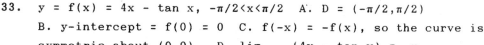

CU on $(0,\infty)$, CD on $(-\infty,0)$.   IP $(0,0)$

33.   $y = f(x) = 4x - \tan x$, $-\pi/2<x<\pi/2$   A. $D = (-\pi/2,\pi/2)$

B. y-intercept $= f(0) = 0$   C. $f(-x) = -f(x)$, so the curve is

symmetric about $(0,0)$.   D. $\lim\limits_{x\to\pi/2^-} (4x - \tan x) = -\infty$,

$\lim\limits_{x\to-\pi/2^+} (4x - \tan x) = \infty$, so $x = \pi/2$ and $x = -\pi/2$ are VA.

E. $f'(x) = 4 - \sec^2 x > 0 \Longleftrightarrow \sec x < 2 \Longleftrightarrow \cos x > 1/2 \Longleftrightarrow$

$-\pi/3<x<\pi/3$, so f is increasing on $[-\pi/3,\pi/3]$ and decreasing on

$(-\pi/2,-\pi/3]$ and $[\pi/3,\pi/2)$.         H.

F. $f(\pi/3) = 4\pi/3-\sqrt{3}$ is a local

maximum, $f(-\pi/3) = \sqrt{3}-4\pi/3$ is

a local minimum.

G. $f''(x) = -2\sec^2 x \tan x > 0 \Longleftrightarrow$

$\tan x < 0 \Longleftrightarrow -\pi/2<x<0$, so f is

CU on $(-\pi/2,0)$ and CD on $(0,\pi/2)$.

IP $(0,0)$

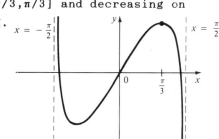

**35.** $f(x) = x^{101}+x^{51}+x-1 = 0$. Since f is continuous and $f(0) = -1$ and $f(1) = 2$, the equation has at least one root in $(0,1)$ by the Intermediate Value Theorem. Suppose the equation has two roots, a and b, with $a<b$. Then $f(a) = 0 = f(b)$ so by Rolle's Theorem $f'(x) = 101x^{100}+51x^{50}+1 = 0$ has a root in $(a,b)$. But this is impossible since $f'(x) \geq 1$ for all x.

**37.**

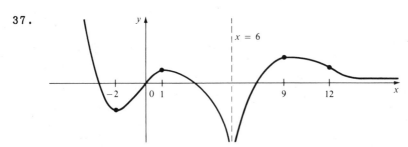

**39.** If $B = 0$, the line is vertical and the distance from $x = -C/A$ to $(x_1,y_1)$ is $|x_1+C/A| = |Ax_1+By_1+C|/\sqrt{A^2+B^2}$, so assume $B \neq 0$. The square of the distance from $(x_1,y_1)$ to the line is $f(x) = (x-x_1)^2 + (y-y_1)^2$ where $Ax+By+C = 0$, so we minimize

$$f(x) = (x-x_1)^2+(-\frac{A}{B}x-\frac{C}{B}-y_1)^2 \Rightarrow f'(x) = 2(x-x_1)+2(-\frac{A}{B}x-\frac{C}{B}-y_1)(-\frac{A}{B})$$

$f'(x) = 0 \Rightarrow x = (B^2x_1-ABy_1-AC)/(A^2+B^2)$ and this gives a minimum since $f''(x) = 2(1+A^2/B^2)>0$. Substituting this value of x and simplifying gives $f(x) = (Ax_1+By_1+C)^2/(A^2+B^2)$ so the minimum distance is $|Ax_1+By_1+C|/\sqrt{A^2+B^2}$.

**41.** By similar triangles, $\frac{y}{x} = r/\sqrt{x^2-2rx}$, so the area of the triangle is

$$A(x) = (1/2)(2y)x = xy = rx^2/\sqrt{x^2-2rx} \Rightarrow$$

$$A'(x) = \frac{2rx\sqrt{x^2-2rx} - rx^2(x-r)/\sqrt{x^2-2rx}}{x^2-2rx}$$

$$= \frac{rx^2(x-3r)}{(x^2-2rx)^{3/2}} = 0 \text{ when } x = 3r$$

$A'(x)<0$ when $2r<x<3r$, $A'(x)>0$ when $x>3r$. So $x = 3r$ gives a minimum and $A(3r) = r(9r^2)/\sqrt{3}r = 3\sqrt{3}r^2$.

**43.**

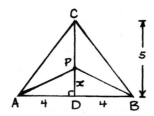

We minimize $L(x) = |PA| + |PB| + |PC|$

$= 2\sqrt{x^2+16} + (5-x)$, $0 \le x \le 5$.

$L'(x) = 2x/\sqrt{x^2+16} - 1 = 0 \Leftrightarrow 2x = \sqrt{x^2+16}$

$\Leftrightarrow 4x^2 = x^2+16 \Leftrightarrow x = 4/\sqrt{3}$    $L(0) = 13$,

$L(4/\sqrt{3}) \approx 11.9$, $L(5) \approx 12.8$, so the

minimum  occurs when $x = 4/\sqrt{3}$.

**45.**

The strength is $S(x) = k(2x)(2y)^2 = 8kxy^2$

$= 8kx(R^2-x^2) = 8k(R^2x-x^3)$, $0 \le x \le R$.

$S'(x) = 8k(R^2-3x^2) = 0 \Rightarrow x = R/\sqrt{3}$

$S(0) = 0 = S(R)$, so the maximum occurs

when $x = R/\sqrt{3}$.  Then the width is $2x =$

$2R/\sqrt{3}$ and the depth is $2y = 2\sqrt{R^2-R^2/3} = 2\sqrt{2}R/\sqrt{3}$.

**47.** $f'(x) = x - \sqrt[4]{x} = x - x^{1/4} \Rightarrow f(x) = (1/2)x^2 - (4/5)x^{5/4} + C$

**49.** $f'(x) = (1+x)/\sqrt{x} = x^{-1/2} + x^{1/2} \Rightarrow f(x) = 2x^{1/2} + (2/3)x^{3/2} + C \Rightarrow$

$0 = f(1) = 2 + 2/3 + C \Rightarrow C = -8/3 \Rightarrow f(x) = 2x^{1/2} + (2/3)x^{3/2} - 8/3$

**51.** $f''(x) = x^3 + x \Rightarrow f'(x) = x^4/4 + x^2/2 + C \Rightarrow 1 = f'(0) = C \Rightarrow$

$f'(x) = x^4/4 + x^2/2 + 1 \Rightarrow f(x) = x^5/20 + x^3/6 + x + D \Rightarrow$

$-1 = f(0) = D \Rightarrow f(x) = x^5/20 + x^3/6 + x - 1$

**53.** $v'(t) = a(t) = \sqrt{t} = t^{1/2} \Rightarrow v(t) = (2/3)t^{3/2} + C \Rightarrow 2 = v(0) = C \Rightarrow$

$v(t) = (2/3)t^{3/2} + 2 \Rightarrow s(t) = (4/15)t^{5/2} + 2t + D \Rightarrow 0 = s(0) = D \Rightarrow$

$s(t) = (4/15)t^{5/2} + 2t$

CHAPTER 5

Exercises 5.1

1. $\displaystyle\sum_{i=1}^{5} \sqrt{i} = \sqrt{1} + \sqrt{2} + \sqrt{3} + \sqrt{4} + \sqrt{5}$  3. $\displaystyle\sum_{i=4}^{6} 3^i = 3^4 + 3^5 + 3^6$

5. $\displaystyle\sum_{k=0}^{4} \frac{2k-1}{2k+1} = -1 + \frac{1}{3} + \frac{3}{5} + \frac{5}{7} + \frac{7}{9}$

7. $\displaystyle\sum_{i=1}^{n} i^{10} = 1^{10} + 2^{10} + 3^{10} + \cdots + n^{10}$

9. $\displaystyle\sum_{j=0}^{n-1} (-1)^j = 1 - 1 + 1 - 1 + \cdots + (-1)^{n-1}$

11. $1 + 2 + 3 + 4 + \cdots + 10 = \displaystyle\sum_{i=1}^{10} i$

13. $\dfrac{1}{2} + \dfrac{2}{3} + \dfrac{3}{4} + \dfrac{4}{5} + \cdots + \dfrac{19}{20} = \displaystyle\sum_{i=1}^{19} \dfrac{i}{i+1}$

15. $2 + 4 + 6 + 8 + \cdots + 2n = \displaystyle\sum_{i=1}^{n} 2i$

17. $1 + 2 + 4 + 8 + 16 + 32 = \displaystyle\sum_{i=0}^{5} 2^i$

19. $x + x^2 + x^3 + \cdots + x^n = \displaystyle\sum_{i=1}^{n} x^i$

21. $\displaystyle\sum_{i=4}^{8} (3i-2) = 10 + 13 + 16 + 19 + 22 = 80$

23. $\displaystyle\sum_{j=1}^{6} 3^{j+1} = 3^2 + 3^3 + 3^4 + 3^5 + 3^6 + 3^7 = 9 + 27 + 81 + 243 + 729$

$+ \ 2187 = 3276$   (For a more general method, see #49.)

25. $\displaystyle\sum_{n=1}^{20} (-1)^n = -1 + 1 - 1 + 1 - 1 + 1 - 1 + 1 - 1 + 1 - 1 + 1 - 1 + 1$

$- \ 1 + 1 - 1 + 1 - 1 + 1 = 0$

27. $\displaystyle\sum_{i=0}^{4} (2^i + i^2) = (1+0) + (2+1) + (4+4) + (8+9) + (16+16) = 61$

29. $\displaystyle\sum_{i=1}^{n} \left[\dfrac{1}{i} - \dfrac{1}{i+1}\right] = \left[1 - \dfrac{1}{2}\right] + \left[\dfrac{1}{2} - \dfrac{1}{3}\right] + \left[\dfrac{1}{3} - \dfrac{1}{4}\right] + \cdots$

$+ \left[\dfrac{1}{n-1} - \dfrac{1}{n}\right] + \left[\dfrac{1}{n} - \dfrac{1}{n+1}\right] = 1 - \dfrac{1}{n+1} = \dfrac{n}{n+1}$

31. $\displaystyle\sum_{i=1}^{n} 2i = 2 \sum_{i=1}^{n} i = n(n+1)$

33. $\displaystyle\sum_{i=1}^{n} (i^2+3i+4) = \sum_{i=1}^{n} i^2 + 3 \sum_{i=1}^{n} i + \sum_{i=1}^{n} 4$

$= \dfrac{n(n+1)(2n+1)}{6} + \dfrac{3n(n+1)}{2} + 4n = \dfrac{1}{6}[(2n^3+3n^2+n) + (9n^2+9n) + 24n]$

$= (1/6)[2n^3+12n^2+34n] = n(n^2+6n+17)/3$

35. $\displaystyle\sum_{i=1}^{n} (i+1)(i+2) = \sum_{i=1}^{n} (i^2+3i+2) = \sum_{i=1}^{n} i^2 + 3 \sum_{i=1}^{n} i + \sum_{i=1}^{n} 2$

$= \dfrac{n(n+1)(2n+1)}{6} + \dfrac{3n(n+1)}{2} + 2n = \dfrac{n(n+1)}{6} [(2n+1) + 9] + 2n$

$= \dfrac{n(n+1)}{3} (n+5) + 2n = \dfrac{n}{3} [(n+1)(n+5) + 6] = n(n^2+6n+11)/3$

37. $\displaystyle\sum_{i=1}^{n} (i^3-i-2) = \sum_{i=1}^{n} i^3 - \sum_{i=1}^{n} i - \sum_{i=1}^{n} 2 = \left[\dfrac{n(n+1)}{2}\right]^2 - \dfrac{n(n+1)}{2} - 2n$

$= \dfrac{n(n+1)}{4} [n(n+1) - 2] - 2n = \dfrac{n(n+1)(n+2)(n-1)}{4} - 2n$

$= \dfrac{n}{4}[(n+1)(n-1)(n+2) - 8] = \dfrac{n}{4} [(n^2-1)(n+2) - 8] = \dfrac{n(n^3+2n^2-n-10)}{4}$

39. By Theorem 5.2(a) and Example 3, $\displaystyle\sum_{i=1}^{n} c = c \sum_{i=1}^{n} 1 = cn$

41. (a) Let $S_n$ be the statement that $\displaystyle\sum_{i=1}^{n} i^3 = \left[\dfrac{n(n+1)}{2}\right]^2$

    1.    $S_1$ is true because $1^3 = \left[\dfrac{1 \cdot 2}{2}\right]^2$

    2.    Assume $S_k$ is true. Then $\displaystyle\sum_{i=1}^{k} i^3 = \left[\dfrac{k(k+1)}{2}\right]^2$, so

$\displaystyle\sum_{i=1}^{k+1} i^3 = \left[\dfrac{k(k+1)}{2}\right]^2 + (k+1)^3 = \dfrac{(k+1)^2}{4} [k^2+4(k+1)]$

$= \dfrac{(k+1)^2}{4} (k+2)^2 = \left[\dfrac{(k+1)\{(k+1)+1\}}{2}\right]^2$, showing that $S_{k+1}$ is true.

    Therefore $S_n$ is true for all $n$ by mathematical induction.

(b) $\displaystyle\sum_{i=1}^{n} [(i+1)^4 - i^4] = (2^4-1^4)+(3^4-2^4)+(4^4-3^4)+\cdots+[(n+1)^4 - n^4]$

$= (n+1)^4 - 1^4 = n^4 + 4n^3 + 6n^2 + 4n.$ On the other hand,

$$\sum_{i=1}^{n} [(i+1)^4 - i^4] = \sum_{i=1}^{n} (4i^3 + 6i^2 + 4i + 1) = 4 \sum_{i=1}^{n} i^3 + 6 \sum_{i=1}^{n} i^2 + 4 \sum_{i=1}^{n} i$$

$$+ \sum_{i=1}^{n} 1 = 4S + n(n+1)(2n+1) + 2n(n+1) + n \quad [\text{where } S = \sum_{i=1}^{n} i^3]$$

$$= 4S + 2n^3 + 3n^2 + n + 2n^2 + 2n + n = 4S + 2n^3 + 5n^2 + 4n \quad \text{Thus}$$

$n^4 + 4n^3 + 6n^2 + 4n = 4S + 2n^3 + 5n^2 + 4n$, from which it

follows that $4S = n^4 + 2n^3 + n^2 = n^2(n^2 + 2n + 1) = n^2(n+1)^2$ and

$$S = \left[\frac{n(n+1)}{2}\right]^2$$

(c) The area of $G_i$ is $\left(\sum_{k=1}^{i} k\right)^2 - \left(\sum_{k=1}^{i-1} k\right)^2 = \left[\frac{i(i+1)}{2}\right]^2 - \left[\frac{(i-1)i}{2}\right]^2$

$$= \frac{i^2}{4} [(i+1)^2 - (i-1)^2] = \frac{i^2}{4} [i^2 + 2i + 1) - (i^2 - 2i + 1)] = \frac{i^2}{4} [4i] = i^3$$

Thus the area of ABCD equals $\sum_{i=1}^{n} i^3$ and also equals $[n(n+1)/2]^2$

43. $\sum_{i=1}^{n} (a_i - a_{i-1}) = (a_1 - a_0) + (a_2 - a_1) + \cdots + (a_{n-1} - a_{n-2}) + (a_n - a_{n-1})$

$= a_n - a_0 \qquad$ [OR: Use mathematical induction.]

45. $\lim\limits_{n \to \infty} \sum\limits_{i=1}^{n} \frac{1}{n} \left(\frac{i}{n}\right)^2 = \lim\limits_{n \to \infty} \frac{1}{n^3} \sum\limits_{i=1}^{n} i^2 = \lim\limits_{n \to \infty} \frac{1}{n^3} \frac{n(n+1)(2n+1)}{6}$

$= \lim\limits_{n \to \infty} \frac{1}{6} \left[1 + \frac{1}{n}\right] \left[2 + \frac{1}{n}\right] = \frac{1}{6} \cdot 1 \cdot 2 = \frac{1}{3}$

47. $\lim\limits_{n \to \infty} \sum\limits_{i=1}^{n} \frac{2}{n} \left[\left(\frac{2i}{n}\right)^3 + 5\left(\frac{2i}{n}\right)\right] = \lim\limits_{n \to \infty} \sum\limits_{i=1}^{n} \left[\frac{16}{n^4} i^3 + \frac{20}{n^2} i\right]$

$= \lim\limits_{n \to \infty} \left[\frac{16}{n^4} \sum\limits_{i=1}^{n} i^3 + \frac{20}{n^2} \sum\limits_{i=1}^{n} i\right] = \lim\limits_{n \to \infty} \left[\frac{16}{n^4} \frac{n^2(n+1)^2}{4} + \frac{20}{n^2} \frac{n(n+1)}{2}\right]$

$= \lim\limits_{n \to \infty} \left[\frac{4(n+1)^2}{n^2} + \frac{10n(n+1)}{n^2}\right] = \lim\limits_{n \to \infty} \left[4\left(1 + \frac{1}{n}\right)^2 + 10\left(1 + \frac{1}{n}\right)\right]$

$= 4 \cdot 1 + 10 \cdot 1 = 14$

49. Let $S = \sum\limits_{i=1}^{n} ar^{i-1} = a + ar + ar^2 + \cdots + ar^{n-1}$. Then $rS = ar + ar^2$

$+ \cdots + ar^{n-1} + ar^n$. Subtracting the first relation from the

second, we find $(r-1)S = ar^n - a = a(r^n - 1)$, so $S = a(r^n - 1)/(r - 1)$.

51. By (18c) of Appendix B, $2 \sin u \sin v = \cos(u-v) - \cos(u+v)$ (*)
Taking $u = \frac{1}{2}x$ and $v = ix$, we get $2 \sin \frac{1}{2}x \sin ix = \cos(\frac{1}{2} - i)x - \cos(\frac{1}{2} + i)x = \cos(i - \frac{1}{2})x - \cos(i + \frac{1}{2})x$. Thus

$$2 \sin \frac{1}{2}x \sum_{i=1}^{n} \sin ix = \sum_{i=1}^{n} 2 \sin \frac{1}{2}x \sin ix = \sum_{i=1}^{n} [\cos(i - \frac{1}{2})x - \cos(i + \frac{1}{2})x] = -\sum_{i=1}^{n} [\cos(i + \frac{1}{2})x - \cos(i - \frac{1}{2})x]$$

$= -[\cos(n + \frac{1}{2})x - \cos \frac{1}{2}x]$  [by #43 with $a_i = \cos(i + \frac{1}{2})x$]

$= \cos[\frac{1}{2}(n+1)x - \frac{1}{2}nx] - \cos[\frac{1}{2}(n+1)x + \frac{1}{2}nx] = 2 \sin \frac{1}{2}(n+1)x \sin \frac{1}{2}nx$

[by (*) with $u = \frac{1}{2}(n+1)x$ and $v = \frac{1}{2}nx$]

If $x$ is not an integer multiple of $2\pi$, then $\sin \frac{1}{2}x \neq 0$, so we can

divide by $2 \sin \frac{1}{2}x$ and get $\sum_{i=1}^{n} \sin ix = \dfrac{\sin \frac{1}{2}nx \sin \frac{1}{2}(n+1)x}{\sin \frac{1}{2}x}$

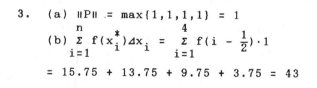

Section 5.2

## Exercises 5.2

1.  (a) $\|P\| = \max\{1,1,1,1\} = 1$
    (b) $\sum_{i=1}^{n} f(x_i^*)\Delta x_i = \sum_{i=1}^{4} f(i-1) \cdot 1$
    $= 16 + 15 + 12 + 7 = 50$

(c)
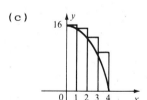

3.  (a) $\|P\| = \max\{1,1,1,1\} = 1$
    (b) $\sum_{i=1}^{n} f(x_i^*)\Delta x_i = \sum_{i=1}^{4} f(i - \frac{1}{2}) \cdot 1$
    $= 15.75 + 13.75 + 9.75 + 3.75 = 43$

(c)
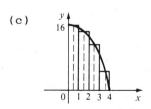

5.  (a) $\|P\| = \max\{1,1,1,1,1\} = 1$      (c)

     (b) $\sum_{i=1}^{n} f(x_i^*)\varDelta x_i = \sum_{i=1}^{5} f(i) \cdot 1$

     $= 3 + 5 + 7 + 9 + 11 = 35$

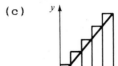

7.  (a) $\|P\| = \max\{.5,.5,.5,.5,.5,.5\} = .5$      (c)

     (b) $\sum_{i=1}^{n} f(x_i^*)\varDelta x_i = [f(-.5) + f(0) +$

     $f(.5) + f(1) + f(1.5) + f(2)](0.5) =$

     $\frac{1}{2}[1.875 + 2 + 2.125 + 3 + 5.375 + 10]$

     $= (1/2)(24.375) = 12.1875$

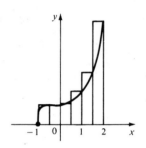

9.  (a) $\|P\| = \max\{\pi/4,\pi/4,\pi/4,\pi/4\} = \pi/4$      (c)

     (b) $\sum_{i=1}^{n} f(x_i^*)\varDelta x_i = \sum_{i=1}^{4} f(x_i^*) \frac{\pi}{4}$

     $= \frac{\pi}{4}\left[f(\frac{\pi}{6}) + f(\frac{\pi}{3}) + f(\frac{2\pi}{3}) + f(\frac{5\pi}{6})\right]$

     $= \frac{\pi}{4}[1 + \sqrt{3} + \sqrt{3} + 1] = \frac{\pi}{2}(1 + \sqrt{3})$

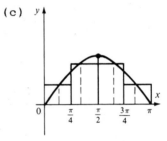

11. $f(x) = x^2 + 1$ on $[0,2]$ with partition points $x_i = 2i/n$ ($i = 0,1,$ $2,\cdots,n$), so $\varDelta x_1 = \varDelta x_2 = \cdots = \varDelta x_n = 2/n$. $\|P\| = \max\{\varDelta x_i\} = 2/n$, so $\|P\| \to 0$ is equivalent to $n \to \infty$. Taking $x_i^*$ to be the midpoint of $[x_{i-1},x_i] = [2(i-1)/n, 2i/n]$, we get $x_i^* = (2i-1)/n$. Thus

$$A = \lim_{\|P\|\to 0} \sum_{i=1}^{n} f(x_i^*)\varDelta x_i = \lim_{n\to\infty} \sum_{i=1}^{n} \left[\left[\frac{2i-1}{n}\right]^2 + 1\right]\frac{2}{n}$$

$$= \lim_{n\to\infty} \sum_{i=1}^{n} \left[\frac{8i^2}{n^3} - \frac{8i}{n^3} + \frac{2}{n^3} + \frac{2}{n}\right]$$

$$= \lim_{n\to\infty} \left[\frac{8}{n^3}\sum_{i=1}^{n} i^2 - \frac{8}{n^3}\sum_{i=1}^{n} i + \left[\frac{2}{n^3} + \frac{2}{n}\right]\sum_{i=1}^{n} 1\right]$$

$$= \lim_{n\to\infty} \left[\frac{8}{n^3}\frac{n(n+1)(2n+1)}{6} - \frac{8}{n^3}\frac{n(n+1)}{2} + \left[\frac{2}{n^3} + \frac{2}{n}\right]n\right]$$

$$= \lim_{n\to\infty} \left[\frac{4}{3}\cdot 1\cdot\left[1+\frac{1}{n}\right]\left[2+\frac{1}{n}\right] - \frac{4}{n}\cdot 1\cdot\left[1+\frac{1}{n}\right] + \frac{2}{n^2} + 2\right]$$

$$= \frac{4}{3}\cdot 1\cdot 1\cdot 2 - 0\cdot 1\cdot 1 + 0 + 2 = \frac{8}{3} + 2 = \frac{14}{3}$$

13.  $f(x) = 16 - x^2$ on $[-4, 4]$ with partition points

$x_i = -4 + \frac{8i}{n}$ $(i = 0, 1, \ldots, n)$, so $\Delta x_i = 8/n$ for all $i$.  $\|P\| = 8/n$,

so $\|P\| \to 0$ is equivalent to $n \to \infty$.  *ask about find*

(a)  $x_i^* = x_{i-1} = -4 + \frac{8(i-1)}{n}$

$A = \lim\limits_{n \to \infty} \sum\limits_{i=1}^{n} \left[ 16 - \left[ -4 + \frac{8(i-1)}{n} \right]^2 \right] \frac{8}{n} = \lim\limits_{n \to \infty} \sum\limits_{i=1}^{n} \left[ \frac{64(i-1)}{n} - \frac{64(i-1)^2}{n^2} \right] \frac{8}{n}$

*partition pts.*

$= \lim\limits_{n \to \infty} \sum\limits_{i=0}^{n-1} \left[ \frac{64i}{n} - \frac{64i^2}{n^2} \right] \frac{8}{n} = \lim\limits_{n \to \infty} \sum\limits_{i=1}^{n-1} \left[ \frac{64i}{n} - \frac{64}{n^2} i^2 \right] \frac{8}{n}$

$= \lim\limits_{n \to \infty} \left[ \frac{64 \cdot 8}{n^2} \sum\limits_{i=1}^{n-1} i - \frac{64 \cdot 8}{n^3} \sum\limits_{i=1}^{n-1} i^2 \right]$

$= \lim\limits_{n \to \infty} \left[ \frac{64 \cdot 8}{n^2} \cdot \frac{(n-1)n}{2} - \frac{64 \cdot 8}{n^3} \frac{(n-1)n(2n-1)}{6} \right]$

$= \lim\limits_{n \to \infty} \left[ 256 \cdot \left( 1 - \frac{1}{n} \right) - \frac{256}{3} \cdot \left( 1 - \frac{1}{n} \right) \cdot \left( 2 - \frac{1}{n} \right) \right] = 256 \cdot 1 - \frac{256}{3} \cdot 1 \cdot 2 = \frac{256}{3}$

(b)  $x_i^* = x_i = -4 + \frac{8i}{n}$    The only difference is that $i - 1$ is

replaced by $i$, so from (a) we see that

$A = \lim\limits_{n \to \infty} \sum\limits_{i=1}^{n} \left[ \frac{64i}{n} - \frac{64i^2}{n^2} \right] \frac{8}{n} = \lim\limits_{n \to \infty} \left[ \frac{64 \cdot 8}{n^2} \sum\limits_{i=1}^{n} i - \frac{64 \cdot 8}{n^3} \sum\limits_{i=1}^{n} i^2 \right]$

$= \lim\limits_{n \to \infty} \left[ \frac{64 \cdot 8}{n^2} \frac{n(n+1)}{2} - \frac{64 \cdot 8}{n^3} \frac{n(n+1)(2n+1)}{6} \right]$

$= \lim\limits_{n \to \infty} \left[ 256 \cdot \left( 1 + \frac{1}{n} \right) - \frac{256}{3} \left( 1 + \frac{1}{n} \right) \left( 2 + \frac{1}{n} \right) \right] = 256 - \frac{256}{3} \cdot 2 = \frac{256}{3}$

(c)  $x_i^* = -4 + \frac{8i-4}{n} = 4 \left[ \frac{2i-1}{n} - 1 \right] = 4 \left[ \frac{2i}{n} - \frac{1}{n} - 1 \right]$

$A = \lim\limits_{n \to \infty} \sum\limits_{i=1}^{n} \left[ 16 - 16 \left[ \frac{2i}{n} - \frac{1}{n} - 1 \right]^2 \right] \frac{8}{n}$

$= \lim\limits_{n \to \infty} \sum\limits_{i=1}^{n} \left[ 16 - 16 \left[ \frac{4i^2}{n^2} - \frac{4i}{n^2} - \frac{4i}{n} + \frac{1}{n^2} + \frac{2}{n} + 1 \right] \right] \frac{8}{n}$

$= \lim\limits_{n \to \infty} \left[ -64 \frac{8}{n^3} \sum\limits_{i=1}^{n} i^2 + 64 \frac{8}{n^3} \sum\limits_{i=1}^{n} i + 64 \frac{8}{n^2} \sum\limits_{i=1}^{n} i - 16 \cdot \frac{8}{n^3} - \frac{16^2}{n^2} - 16 \cdot \frac{8}{n} \right]$

$= \lim\limits_{n \to \infty} \left[ -64 \frac{8}{n^3} \frac{n(n+1)(2n+1)}{6} + 64 \frac{8}{n^3} \frac{n(n+1)}{2} + 64 \frac{8}{n^2} \frac{n(n+1)}{2} \right]$

(The last three terms all $\to 0$.)

$$= \lim_{n \to \infty} \left[ -64 \cdot \frac{4}{3} \cdot \left[1 + \frac{1}{n}\right]\left[2 + \frac{1}{n}\right] + 64 \cdot \frac{4}{n}\left[1 + \frac{1}{n}\right] + 64 \cdot 4\left[1 + \frac{1}{n}\right] \right]$$

$$= -64 \cdot \frac{4}{3} \cdot 1 \cdot 2 + 64 \cdot 0 \cdot 1 + 64 \cdot 4 \cdot 1 = -\frac{512}{3} + 256 = \frac{256}{3}$$

(a) 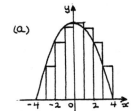   (b) 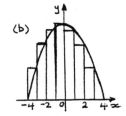   (c)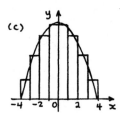

15. $f(x) = 5$ on $[-2,2]$ with partition points $x_i = -2 + \frac{4i}{n}$

$(i = 0, 1, \ldots, n)$ $\Delta x_i = \frac{4}{n}$ for $i = 1, \ldots, n$ $\Rightarrow$ $\|P\| = 4/n$

$$A = \lim_{n \to \infty} \sum_{i=1}^{n} f(x_i^*) \, \Delta x_i = \lim_{n \to \infty} \sum_{i=1}^{n} 5 \cdot \frac{4}{n}$$

$$= \lim_{n \to \infty} \frac{20}{n} \sum_{i=1}^{n} 1 = \lim_{n \to \infty} \frac{20}{n} \cdot n = 20$$

The approximation is exact in this instance.

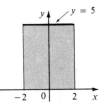

17. $f(x) = x^2 + 3x - 2$ on $[1,4]$. $x_i^* = x_i = 1 + \frac{3i}{n}$ for $i = 1, 2, \ldots, n$.

$\Delta x_i = \frac{3}{n}$ for all $i$. $\quad A = \lim_{n \to \infty} \sum_{i=1}^{n} \left[ \left[1 + \frac{3i}{n}\right]^2 + 3\left[1 + \frac{3i}{n}\right] - 2 \right] \frac{3}{n}$

$$= \lim_{n \to \infty} \sum_{i=1}^{n} \left[ \frac{9i^2}{n^2} + \frac{15i}{n} + 2 \right] \frac{3}{n} = \lim_{n \to \infty} \left[ \frac{27}{n^3} \sum_{i=1}^{n} i^2 + \frac{45}{n^2} \sum_{i=1}^{n} i + \frac{6}{n} \sum_{i=1}^{n} 1 \right]$$

$$= \lim_{n \to \infty} \left[ \frac{27}{n^3} \frac{n(n+1)(2n+1)}{6} + \frac{45}{n^2} \frac{n(n+1)}{2} + \frac{6}{n} n \right]$$

$$= \lim_{n \to \infty} \left[ \frac{9}{2} \cdot 1 \cdot \left[1 + \frac{1}{n}\right] \cdot \left[2 + \frac{1}{n}\right] + \frac{45}{2} \cdot 1 \cdot \left[1 + \frac{1}{n}\right] + 6 \right]$$

$$= \frac{9}{2} \cdot 1 \cdot 1 \cdot 2 + \frac{45}{2} \cdot 1 \cdot 1 + 6 = 9 + \frac{45}{2} + 6 = \frac{75}{2} = 37.5$$

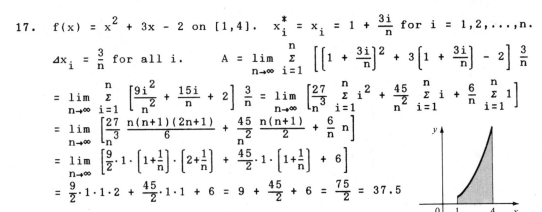

**19.** $f(x) = x^3 + 2x$ on $[0,2]$. $\quad x_i^* = x_i = \dfrac{2i}{n}$ for $i = 1, 2, \ldots, n$.

$$A = \lim_{n \to \infty} \sum_{i=1}^{n} \left[ \left(\frac{2i}{n}\right)^3 + \frac{4i}{n} \right] \frac{2}{n} = \lim_{n \to \infty} \left[ \frac{16}{n^4} \Sigma i^3 + \frac{8}{n^2} \Sigma i \right]$$

$$= \lim_{n \to \infty} \left[ \frac{16}{n^4} \frac{n^2(n+1)^2}{4} + \frac{8}{n^2} \frac{n(n+1)}{2} \right]$$

$$= \lim_{n \to \infty} \left[ 4 \cdot \left(1 + \frac{1}{n}\right)^2 + 4\left(1 + \frac{1}{n}\right) \right] = 4 \cdot 1 + 4 \cdot 1 = 8$$

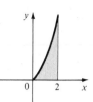

**21.** $f(x) = 1 - 2x^2 + x^4$ on $[-1, 1]$. $\quad x_i^* = x_i = -1 + \dfrac{2i}{n}, \quad \varDelta x_i = \dfrac{2}{n}$.

$$A = \lim_{n \to \infty} \sum_{i=1}^{n} \left[ 1 - 2\left(-1 + \frac{2i}{n}\right)^2 + \left(-1 + \frac{2i}{n}\right)^4 \right]\frac{2}{n}$$

$$= \lim_{n \to \infty} \sum_{i=1}^{n} \left[ 1 - 2 + \frac{8i}{n} - \frac{8i^2}{n^2} + 1 - \frac{8i}{n} + \frac{24i^2}{n^2} - \frac{32i^3}{n^3} + \frac{16i^4}{n^4} \right]\frac{2}{n}$$

$$= \lim_{n \to \infty} \left[ \frac{32}{n^5} \Sigma i^4 - \frac{64}{n^4} \Sigma i^3 + \frac{32}{n^3} \Sigma i^2 \right]$$

$$= \lim_{n \to \infty} \left[ \frac{32}{n^5} \frac{n(n+1)(2n+1)(3n^2+3n-1)}{30} - \frac{64}{n^2} \frac{n^2(n+1)^2}{4} + \frac{32}{n^3} \frac{n(n+1)(2n+1)}{6} \right]$$

$$= \lim_{n \to \infty} \left[ \frac{16}{15} \cdot 1 \cdot \left(1 + \frac{1}{n}\right)\left(2 + \frac{1}{n}\right)\left(3 + \frac{3}{n} - \frac{1}{n^2}\right) - 16 \cdot 1^2 \cdot \left(1 + \frac{1}{n}\right)^2 + \frac{16}{3} \cdot 1 \cdot \left(1 + \frac{1}{n}\right)\left(2 + \frac{1}{n}\right) \right]$$

$$= \frac{16}{15} \cdot 1^2 \cdot 2 \cdot 3 - 16 \cdot 1^4 + \frac{16}{3} \cdot 1^2 \cdot 2 = \frac{32}{5} - 16 + \frac{32}{3} = \frac{16}{15}$$

**21.**

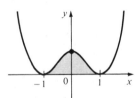

**23.**

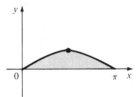

**23.** $f(x) = \sin x$ on $[0, \pi]$, $\quad x_i^* = x_i = \dfrac{\pi i}{n}, \quad \varDelta x_i = \dfrac{\pi}{n}$

$$A = \lim_{n \to \infty} \sum_{i=1}^{n} \left[ \sin \frac{\pi i}{n} \right] \frac{\pi}{n} = \lim_{n \to \infty} \frac{\pi}{n} \sum_{i=1}^{n} \left[ \sin i\frac{\pi}{n} \right]$$

$$= \lim_{n \to \infty} \frac{\pi}{n} \cdot \frac{\sin \frac{1}{2}n \frac{\pi}{n} \, \sin \frac{1}{2}(n+1)\frac{\pi}{n}}{\sin \frac{1}{2} \frac{\pi}{n}} \qquad \text{[by Exer. 5.1.51]}$$

$$= \lim_{n \to \infty} \frac{\pi}{n} \frac{\sin\left[\frac{\pi}{2} + \frac{\pi}{2n}\right]}{\sin\left[\frac{\pi}{2n}\right]} = \lim_{n \to \infty} 2 \frac{\frac{\pi}{2n}}{\sin \frac{\pi}{2n}} \lim_{n \to \infty} \cos \frac{\pi}{2n} = 2 \cdot 1 \cdot 1 = 2$$

$$\text{since } \sin\left[\frac{\pi}{2} + \frac{\pi}{2n}\right] = \cos \frac{\pi}{2n}$$

Exercises 5.3

1. $f(x) = 7 - 2x$      (a) $\|P\| = \max\{.6, .6, .8, 1.2, .8\} = 1.2$

(b) $\sum\limits_{i=1}^{5} f(x_i^*)\Delta x_i = f(1.3)(.6) + f(1.9)(.6) + f(2.6)(.8) +$

$f(3.6)(1.2) + f(4.6)(.8) = (4.4)(.6) + (3.2)(.6) + (1.8)(.8) +$

$(-.2)(1.2) + (-2.2)(.8) = 4$

3. $f(x) = 2 - x^2$      (a) $\|P\| = \max\{.6, .4, 1, .8, .6, .6\} = 1$

(b) $\sum\limits_{i=1}^{6} f(x_i^*)\Delta x_i = f(-1.4)(.6) + f(-1)(.4) + f(0)(1) + f(.8)(.8) +$

$f(1.4)(.6) + f(2)(.6) = (.04)(.6) + (1)(.4) + (2)(1) + (1.36)(.8) +$

$(.04)(.6) + (-2)(.6) = 2.336$

5. $f(x) = x^3$      (a) $\|P\| = \max\{.5, .5, .5, .5\} = .5$     (b) $\sum\limits_{i=1}^{n} f(x_i^*)\Delta x_i =$

$\frac{1}{2} \sum\limits_{i=1}^{4} f(x_i^*) = \frac{1}{2}[(-1)^3 + (-.4)^3 + (.2)^3 + 1^3] = -.028$

7. $\int_a^b c\, dx = \lim\limits_{n\to\infty} \frac{b-a}{n} \sum\limits_{i=1}^{n} c = \lim\limits_{n\to\infty} \frac{b-a}{n} nc$     [by Theorem 5.3]

$= \lim\limits_{n\to\infty} (b-a)c = (b-a)c$

9. $\int_1^4 (x^2-2)dx = \lim\limits_{n\to\infty} \frac{3}{n} \sum\limits_{i=1}^{n} \left[\left(1 + \frac{3i}{n}\right)^2 - 2\right] = \lim\limits_{n\to\infty} \frac{3}{n} \sum\limits_{i=1}^{n} \left[\frac{9i^2}{n^2} + \frac{6i}{n} - 1\right]$

$= \lim\limits_{n\to\infty} \left[\frac{27}{n^3} \sum i^2 + \frac{18}{n^2} \sum i - \frac{3}{n} \sum 1\right]$

$= \lim\limits_{n\to\infty} \left[\frac{27}{n^3} \frac{n(n+1)(2n+1)}{6} + \frac{18}{n^2} \frac{n(n+1)}{2} - \frac{3}{n}\cdot n\right]$

$= \lim\limits_{n\to\infty} \left[\frac{9}{2}\cdot 1\cdot\left[1+\frac{1}{n}\right]\left[2+\frac{1}{n}\right] + 9\cdot 1\cdot\left[1+\frac{1}{n}\right] - 3\right] = \frac{9}{2}\cdot 2 + 9 - 3 = 15$

11. $\int_{-3}^0 (2x^2-3x-4)dx = \lim\limits_{n\to\infty} \frac{3}{n} \sum\limits_{i=1}^{n} \left[2\left[-3 + \frac{3i}{n}\right]^2 - 3\left[-3 + \frac{3i}{n}\right] - 4\right]$

$= \lim\limits_{n\to\infty} \frac{3}{n} \sum\limits_{i=1}^{n} \left[\frac{18i^2}{n^2} - \frac{45i}{n} + 23\right] = \lim\limits_{n\to\infty} \left[\frac{54}{n^3} \sum i^2 - \frac{135}{n^2} \sum i + \frac{69}{n} \sum 1\right]$

$= \lim\limits_{n\to\infty} \left[\frac{54}{n^3} \frac{n(n+1)(2n+1)}{6} - \frac{135}{n^2} \frac{n(n+1)}{2} + \frac{69}{n}\cdot n\right]$

$= \lim\limits_{n\to\infty} \left[9\cdot 1\cdot\left[1+\frac{1}{n}\right]\left[2+\frac{1}{n}\right] - \frac{135}{2}\cdot 1\cdot\left[1+\frac{1}{n}\right] + 69\right] = 9\cdot 2 - \frac{135}{2} + 69 = 19.5$

13.  $\int_{-1}^{1} (t^3-t^2+1)dt = \lim_{n\to\infty} \frac{2}{n} \sum_{i=1}^{n} \left[ \left[-1 + \frac{2i}{n}\right]^3 - \left[-1 + \frac{2i}{n}\right]^2 + 1 \right]$

$= \lim_{n\to\infty} \frac{2}{n} \sum_{i=1}^{n} \left[ \left[\frac{8i^3}{n^3} - \frac{12i^2}{n^2} + \frac{6i}{n} - 1\right] - \left[\frac{4i^2}{n^2} - \frac{4i}{n} + 1\right] + 1 \right]$

$= \lim_{n\to\infty} \frac{2}{n} \sum_{i=1}^{n} \left[\frac{8i^3}{n^3} - \frac{16i^2}{n^2} + \frac{10i}{n} - 1\right]$

$= \lim_{n\to\infty} \left[\frac{16}{n^4} \Sigma i^3 - \frac{32}{n^3} \Sigma i^2 + \frac{20}{n^2} \Sigma i - \frac{2}{n} \Sigma 1\right]$

$= \lim_{n\to\infty} \left[\frac{16}{n^4} \frac{n^2(n+1)^2}{4} - \frac{32}{n^3} \frac{n(n+1)(2n+1)}{6} + \frac{20}{n^2} \frac{n(n+1)}{2} - \frac{2}{n}\cdot n\right]$

$= \lim_{n\to\infty} \left[4\cdot1^2\cdot\left[1+\frac{1}{n}\right]^2 - \frac{16}{3}\cdot1\cdot\left[1+\frac{1}{n}\right]\left[2+\frac{1}{n}\right] + 10\cdot1\cdot\left[1+\frac{1}{n}\right] - 2\right]$

$= 4 - \frac{32}{3} + 10 - 2 = \frac{4}{3}$

15.  $\int_0^b (x^3+4x)dx = \lim_{n\to\infty} \frac{b}{n} \sum_{i=1}^{n} \left[ \left[\frac{bi}{n}\right]^3 + 4\left[\frac{bi}{n}\right] \right] = \lim_{n\to\infty} \left[\frac{b^4}{n^4} \Sigma i^3 + 4\frac{b^2}{n^2} \Sigma i\right]$

$= \lim_{n\to\infty} \left[\frac{b^4}{n^4} \frac{n^2(n+1)^2}{4} + \frac{4b^2}{n^2} \frac{n(n+1)}{2}\right]$

$= \lim_{n\to\infty} \left[\frac{b^4}{4}\cdot1^2\cdot\left[1+\frac{1}{n}\right]^2 + 2b^2\cdot1\cdot\left[1+\frac{1}{n}\right]\right] = \frac{b^4}{4} + 2b^2$

17.  $\int_0^2 (x^4-x+1)dx = \lim_{n\to\infty} \frac{2}{n} \sum_{i=1}^{n} \left[ \left[\frac{2i}{n}\right]^4 - \left[\frac{2i}{n}\right] + 1 \right]$

$= \lim_{n\to\infty} \left[\frac{32}{n^5} \Sigma i^4 - \frac{4}{n^2} \Sigma i + \frac{2}{n} \Sigma 1\right]$

$= \lim_{n\to\infty} \left[\frac{32}{n^5} \frac{n(n+1)(2n+1)(3n^2+3n-1)}{30} - \frac{4}{n^2} \frac{n(n+1)}{2} + \frac{2}{n} n\right]$

$= \lim_{n\to\infty} \left[\frac{16}{15}\cdot1\cdot\left[1+\frac{1}{n}\right]\left[2+\frac{1}{n}\right]\left[3 + \frac{3}{n} - \frac{1}{n^2}\right] - 2\cdot1\cdot\left[1+\frac{1}{n}\right] + 2\right]$

$= \frac{16}{15}\cdot2\cdot3 - 2 + 2 = \frac{32}{5}$

19.  $\int_a^b x\,dx = \lim_{n\to\infty} \frac{b-a}{\cdot n} \sum_{i=1}^{n} \left[a + \frac{b-a}{n} i\right] = \lim_{n\to\infty}\left[\frac{a(b-a)}{n} \Sigma 1 + \frac{(b-a)^2}{n^2} \Sigma i\right]$

$= \lim_{n\to\infty} \left[\frac{a(b-a)}{n}\cdot n + \frac{(b-a)^2}{n^2}\cdot\frac{n(n+1)}{2}\right]$

$= a(b-a) + \lim_{n\to\infty} \frac{(b-a)^2}{2}\left[1+\frac{1}{n}\right] = a(b-a) + \frac{1}{2}(b-a)^2 = \frac{1}{2}(b^2-a^2)$

21. $\displaystyle\lim_{\|P\|\to 0}\ \sum_{i=1}^{n}\ [2(x_i^*)^2 - 5x_i^*]\ \Delta x_i = \int_0^1 (2x^2 - 5x)\,dx$

23. $\displaystyle\lim_{\|P\|\to 0}\ \sum_{i=1}^{n}\ \cos x_i\ \Delta x_i = \int_0^\pi \cos x\ dx$

25. $\displaystyle\lim_{n\to\infty}\ \sum_{i=1}^{n}\ \frac{i^4}{n^5} = \lim_{n\to\infty}\ \frac{1}{n}\ \sum_{i=1}^{n}\ \left[\frac{i}{n}\right]^4 = \int_0^1 x^4\,dx$

27. $\displaystyle\lim_{n\to\infty}\ \sum_{i=1}^{n}\ \left[3\left[1 + \frac{2i}{n}\right]^5 - 6\right]\frac{2}{n} = \int_1^3 (3x^5 - 6)\,dx$   (To get started, notice

    that $\dfrac{2}{n} = \Delta x = \dfrac{b-a}{n}$ and $1 + \dfrac{2i}{n} = a + \dfrac{b-a}{n}\,i$.)

29. By the definition in Note 6, $\displaystyle\int_9^4 \sqrt{t}\ dt = -\int_4^9 \sqrt{t}\ dt = -\frac{38}{3}$

31. f is bounded since $|f(x)| \le 1$ for all x in [a,b].  To see that f is

    not integrable on [a,b], notice that $\displaystyle\sum_{i=1}^{n} f(x_i^*)\Delta x_i = 0$ if $x_1^*,\ldots,x_n^*$

    are all chosen to be rational numbers, but $\displaystyle\sum_{i=1}^{n} f(x_i^*)\Delta x_i = \sum \Delta x_i =$

    b-a if $x_1^*,\ldots,x_n^*$ are all chosen to be irrational numbers.  This is

    true no matter how small $\|P\|$ is, since every interval $[x_{i-1},\ x_i]$

    with $x_{i-1} < x_i$ contains both rational and irrational numbers.

    $\displaystyle\sum_{i=1}^{n} f(x_i^*)\Delta x_i$ cannot simultaneously approach both 0 and b-a as

    $\|P\| \to 0$, so it has no limit as $\|P\| \to 0$.

33. Choose $x_i = 1 + \dfrac{i}{n}$ and $x_i^* = \sqrt{x_{i-1}\,x_i} = \sqrt{\left[1 + \dfrac{i-1}{n}\right]\left[1 + \dfrac{i}{n}\right]}$

    Then $\displaystyle\int_1^2 x^{-2}dx = \lim_{n\to\infty}\ \frac{1}{n}\ \sum_{i=1}^{n}\ \frac{1}{\left[1 + \dfrac{i-1}{n}\right]\left[1 + \dfrac{i}{n}\right]}$

    $\displaystyle = \lim_{n\to\infty} n\ \sum_{i=1}^{n}\ \frac{1}{(n+i-1)(n+i)} = \lim_{n\to\infty} n\ \sum_{i=1}^{n}\ \left[\frac{1}{n+i-1} - \frac{1}{n+1}\right]$ [by the hint]

    $\displaystyle = \lim_{n\to\infty} n\left[\sum_{i=0}^{n-1}\ \frac{1}{n+i} - \sum_{i=1}^{n}\ \frac{1}{n+i}\right] = \lim_{n\to\infty} n\left[\frac{1}{n} - \frac{1}{2n}\right] = \lim_{n\to\infty}\ [1 - \frac{1}{2}] = \frac{1}{2}$

Exercises 5.4

1. $\int_2^6 3 \; dx = 3(6-2) = 12$   [by Property 1]

3. $\int_{-4}^{-1} \sqrt{3} \; dx = \sqrt{3}(-1+4) = 3\sqrt{3}$

5. $\int_0^2 (5x+3)dx = \int_0^2 5x \; dx + \int_0^2 3 \; dx$   [Property 2]

   $= 5 \int_0^2 x \; dx + \int_0^2 3 \; dx$   [Property 3]

   $= 5 \cdot \frac{1}{2}(2^2 - 0^2) + 3(2-0) = 16$   [Property 1]

7. $\int_1^4 (2x^2 - 3x + 1)dx = 2 \int_1^4 x^2 dx - 3 \int_1^4 x \; dx + \int_1^4 1 \; dx$

   $= 2 \cdot \frac{1}{3}(4^3 - 1^3) - 3 \cdot \frac{1}{2}(4^2 - 1^2) + 1 \cdot (4-1) = \frac{45}{2}$

9. $\int_{-1}^1 (x-1)^2 dx = \int_{-1}^1 (x^2 - 2x + 1)dx = \int_{-1}^1 x^2 \; dx - 2 \int_{-1}^1 x \; dx + \int_{-1}^1 1 \; dx$

   $= \frac{1}{3}[1^3 - (-1)^3] - 2 \cdot \frac{1}{2}[1^2 - (-1)^2] + 1 \cdot [1-(-1)] = \frac{8}{3}$

11. $\int_0^{\pi/3} (1 - 2\cos x)dx = \int_0^{\pi/3} 1 \; dx - 2 \int_0^{\pi/3} \cos x \; dx$

    $= 1 \cdot (\frac{\pi}{3} - 0) - 2 \sin \frac{\pi}{3} = \frac{\pi}{3} - \sqrt{3}$

13. $\int_{-2}^2 |x+1|dx = \int_{-2}^{-1} |x + 1| \; dx + \int_{-1}^2 |x+1|dx$   [Property 5]

    $= \int_{-2}^{-1} (-x-1)dx + \int_{-1}^2 (x+1)dx$

    $= -\int_{-2}^{-1} x \; dx - \int_{-2}^{-1} 1 \; dx + \int_{-1}^2 x \; dx + \int_{-1}^2 1 \; dx$

    $= -\frac{1}{2}[(-1)^2 - (-2)^2] - 1 \cdot [-1-(-2)] + \frac{1}{2}[2^2 - (-1)^2] + 1[2-(-1)] = 5$

15. Using Property 5 we have

    $\int_{-2}^5 [\![x]\!]dx = \int_{-2}^{-1} [\![x]\!]dx + \int_{-1}^0 [\![x]\!]dx + \int_0^1 [\![x]\!]dx + \int_1^2 [\![x]\!]dx + \int_2^3 [\![x]\!]dx$

    $+ \int_3^4 [\![x]\!]dx + \int_4^5 [\![x]\!]dx = \int_{-2}^{-1} (-2)dx + \int_{-1}^0 (-1)dx + \int_0^1 0 \; dx + \int_1^2 1 \; dx$

    $+ \int_2^3 2 \; dx + \int_3^4 3 \; dx + \int_4^5 4 \; dx = (-2)(-1+2) + (-1)(0+1) + 0(1-0)$

    $+ 1(2-1) + 2(3-2) + 3(4-3) + 4(5-4) = 7$

17. $\int_{-1}^1 f(x)dx = \int_{-1}^0 f(x)dx + \int_0^1 f(x)dx = \int_{-1}^0 (-2x)dx + \int_0^1 3x^2 \; dx$

    $= -2 \int_{-1}^0 x \; dx + 3 \int_0^1 x^2 \; dx = -2 \cdot \frac{1}{2}[0^2 - (-1)^2] + 3 \cdot \frac{1}{3}[1^3 - 0^3] = 2$

19. $\int_{-2}^1 F(x)dx = \int_{-2}^0 F(x)dx + \int_0^1 F(x)dx = \int_{-2}^0 (1+x^2)dx + \int_0^1 (1+2x)dx$

    $= \int_{-2}^0 1 \; dx + \int_{-2}^0 x^2 \; dx + \int_0^1 1 \; dx + 2 \int_0^1 x \; dx$

    $= 1 \cdot [0-(-2)] + \frac{1}{3}[0^3 - (-2)^3] + 1 \cdot (1-0) + 2 \cdot \frac{1}{2}[1^2 - 0^2] = \frac{20}{3}$

21. $\int_0^4 x^2 \, dx + \int_4^{10} x^2 \, dx = \int_0^{10} x^2 \, dx = \frac{1}{3}[10^3 - 0^3] = \frac{1000}{3}$

23. $\int_1^3 f(x)dx + \int_3^6 f(x)dx + \int_6^{12} f(x)dx$

 $= \int_1^6 f(x)dx + \int_6^{12} f(x)dx = \int_1^{12} f(x)dx$

25. $\int_2^{10} f(x)dx - \int_2^7 f(x)dx = \int_2^7 f(x)dx + \int_7^{10} f(x)dx - \int_2^7 f(x)dx$

 $= \int_7^{10} f(x)dx$

27. $x \geq x^2$ on $[0,1]$, so $\int_0^1 x \, dx \geq \int_0^1 x^2 \, dx$    [Property 7]

29. $x^2 - 1 \geq 0$ on $[2,6]$, so $\int_2^6 (x^2-1)dx \geq 0$    [Property 6]

31. $0 \leq \sin x < 1$ on $[0, \pi/4]$, so $\sin^3 x \leq \sin^2 x$ on $[0, \pi/4]$.

 Hence $\int_0^{\pi/4} \sin^3 x \, dx \leq \int_0^{\pi/4} \sin^2 x \, dx$ [Property 7]

33. $5-x \geq 3 \geq x+1$ on $[1,2]$, so $\sqrt{5-x} \geq \sqrt{x+1}$ and $\int_1^2 \sqrt{5-x} \, dx \geq \int_1^2 \sqrt{x+1} \, dx$

35. $4 \leq x^2 \leq 16$ for $2 \leq x \leq 4$, so $4 \cdot (4-2) \leq \int_2^4 x^2 \, dx \leq 16 \cdot (4-2)$

 [Property 8]; that is, $8 \leq \int_2^4 x^2 \, dx \leq 32$

37. If $-1 \leq x \leq 1$, then $0 \leq x^2 \leq 1$ and $1 \leq 1+x^2 \leq 2$, so

 $1 \leq \sqrt{1+x^2} \leq \sqrt{2}$ and $1 \cdot [1-(-1)] \leq \int_{-1}^1 \sqrt{1+x^2} \, dx \leq \sqrt{2}\,[1-(-1)]$

 [Property 8]; that is, $2 \leq \int_{-1}^1 \sqrt{1+x^2} \, dx \leq 2\sqrt{2}$

39. $1 \leq x^3 \leq 27$ for $1 \leq x \leq 3$, so $1 \cdot (3-1) \leq \int_1^3 x^3 \, dx \leq 27 \cdot (3-1)$.

 Thus $2 \leq \int_1^3 x^3 \, dx \leq 54$

41. If $1 \leq x \leq 2$, then $\frac{1}{2} \leq \frac{1}{x} \leq 1$, so $\frac{1}{2}(2-1) \leq \int_1^2 \frac{1}{x} \, dx \leq 1 \cdot (2-1)$ or

 $\frac{1}{2} \leq \int_1^2 \frac{1}{x} \, dx \leq 1$

43. If $f(x) = x^2 + 2x$, $-3 \leq x \leq 0$, then $f'(x) = 2x+2 = 0$ when $x = -1$, and

 $f(-1) = -1$. At the endpoints, $f(-3) = 3$, $f(0) = 0$. Thus the

 absolute minimum is $m = -1$ and the absolute maximum is $M = 3$. Thus

 $-1 \cdot [0-(-3)] \leq \int_{-3}^0 (x^2+2x)dx \leq 3 \cdot [0-(-3)]$ or $-3 \leq \int_{-3}^0 (x^2+2x)dx \leq 9$

45. For $-1 \leq x \leq 1$, $0 \leq x^4 \leq 1$ and $1 \leq \sqrt{1+x^4} \leq \sqrt{2}$, so

 $1 \cdot [1-(-1)] \leq \int_{-1}^1 \sqrt{1+x^4} \, dx \leq \sqrt{2} \cdot [1-(-1)]$ or $2 \leq \int_{-1}^1 \sqrt{1+x^4} \, dx \leq 2\sqrt{2}$

47. $\sqrt{x^4+1} \geq \sqrt{x^4} = x^2$, so $\int_1^3 \sqrt{x^4+1} \, dx \geq \int_1^3 x^2 \, dx = \frac{1}{3}(3^3-1^3) = \frac{26}{3}$

49. $0 \le \sin x \le 1$ for $0 \le x \le \pi/2$, so $x \sin x \le x$ $\Rightarrow$

$\int_0^{\pi/2} x \sin x \, dx \le \int_0^{\pi/2} x \, dx = \frac{1}{2}[(\pi/2)^2 - 0^2] = \pi^2/8$

51. $\int_a^b [cf(x) + dg(x)]dx = \int_a^b cf(x)dx + \int_a^b dg(x)dx$     [Property 2]

$= c \int_a^b f(x)dx + d \int_a^b g(x)dx$     [Property 3]

53. By Property 7, $f(x) \le 0$ on $[a,b]$ implies that

$\int_a^b f(x)dx \le \int_a^b 0 \, dx = 0$     [Property 1].

55. By Property 7, the inequalities $-|f(x)| \le f(x) \le |f(x)|$ imply that

$\int_a^b (-|f(x)|)dx \le \int_a^b f(x)dx \le \int_a^b |f(x)|dx$. By Property 3, the

left-hand integral equals $-\int_a^b |f(x)|dx$. Thus $-M \le \int_a^b f(x)dx \le M$,

where $M = \int_a^b |f(x)|dx$. (Notice that $M \ge 0$ by Property 6.) It

follows that $|\int_a^b f(x)dx| \le M = \int_a^b |f(x)|dx$.

Section 5.5

Exercises 5.5

1. $g(x) = \int_1^x (t^2-1)^{20}dt$ $\Rightarrow$ $g'(x) = (x^2-1)^{20}$

3. $g(u) = \int_\pi^u \frac{1}{1+t^4} dt$ $\Rightarrow$ $g'(u) = \frac{1}{1+u^4}$

5. $F(x) = \int_x^2 \cos(t^2)dt = - \int_2^x \cos(t^2)dt$ $\Rightarrow$ $F'(x) = -\cos(x^2)$

7. Let $u = \frac{1}{x}$. Then $\frac{d}{dx} \int_2^{1/x} \sin^4 t \, dt = \frac{d}{du} \int_2^u \sin^4 t \, dt \cdot \frac{du}{dx}$

$= \sin^4 u \frac{du}{dx} = -\sin^4(1/x)/x^2$

9. Let $u = \tan x$. Then $\frac{d}{dx} \int_{\tan x}^{17} \sin(t^4)dt = - \frac{d}{dx} \int_{17}^{\tan x} \sin(t^4)dt$

$= - \frac{d}{du} \int_{17}^u \sin(t^4)dt \cdot \frac{du}{dx} = -\sin(u^4) \frac{du}{dx} = -\sin(\tan^4 x) \sec^2 x$

11. Let $t = 5x+1$. Then $\frac{d}{dx} \int_0^{5x+1} \frac{1}{u^2-5} du = \frac{d}{dt} \int_0^t \frac{1}{u^2-5} du \cdot \frac{dt}{dx}$

$= \frac{1}{t^2-5} \frac{dt}{dx} = \frac{5}{25x^2+10x-4}$

**13.** $g(x) = \int_{2x}^{3x} \frac{u-1}{u+1} \, du = \int_{2x}^{0} \frac{u-1}{u+1} \, du + \int_{0}^{3x} \frac{u-1}{u+1} \, du$

$= -\int_{0}^{2x} \frac{u-1}{u+1} \, du + \int_{0}^{3x} \frac{u-1}{u+1} \, du \quad \Rightarrow \quad g'(x) = -\frac{2x-1}{2x+1} \cdot 2 + \frac{3x-1}{3x+1} \cdot 3$

**15.** $y = \int_{\sqrt{x}}^{x^3} \sqrt{t} \, \sin t \, dt = \int_{\sqrt{x}}^{1} \sqrt{t} \, \sin t \, dt + \int_{1}^{x^3} \sqrt{t} \, \sin t \, dt$

$= -\int_{1}^{\sqrt{x}} \sqrt{t} \, \sin t \, dt + \int_{1}^{x^3} \sqrt{t} \, \sin t \, dt$

$y' = -\sqrt[4]{x} (\sin \sqrt{x})/(2\sqrt{x}) + x^{3/2} \sin(x^3)(3x^2)$

$= 3x^{7/2} \sin(x^3) - (\sin\sqrt{x})/(2 \sqrt[4]{x})$

**17.** $\int_{-1}^{8} 6 \, dx = 6x]_{-1}^{8} = 48 - (-6) = 54$

**19.** $\int_{0}^{1} (1-2x-3x^2) dx = x - 2\frac{x^2}{2} - 3\frac{x^3}{3}]_{0}^{1} = x-x^2-x^3]_{0}^{1} = (1-1-1) - 0 = -1$

**21.** $\int_{-3}^{0} (5y^4-6y^2+14) dy = 5\frac{y^5}{5} - 6\frac{y^3}{3} + 14y]_{-3}^{0} = y^5-2y^3+14y]_{-3}^{0}$

$= 0 - (-243+54-42) = 231$

**23.** $\int_{1}^{3} \left[\frac{1}{t^2} - \frac{1}{t^4}\right] dt = \int_{1}^{3} (t^{-2}-t^{-4}) dt = \frac{t^{-1}}{-1} - \frac{t^{-3}}{-3}]_{1}^{3} = \frac{1}{3t^3} - \frac{1}{t}]_{1}^{3}$

$= \left[\frac{1}{81} - \frac{1}{3}\right] - \left[\frac{1}{3} - 1\right] = \frac{28}{81}$

**25.** $\int_{1}^{2} \frac{x^2 + 1}{\sqrt{x}} \, dx = \int_{1}^{2} (x^{3/2}+x^{-1/2}) dx = \frac{x^{5/2}}{5/2} + \frac{x^{1/2}}{1/2}]_{1}^{2}$

$= \frac{2}{5} x^{5/2} + 2x^{1/2}]_{1}^{2} = (\frac{2}{5} \cdot 4\sqrt{2} + 2\sqrt{2}) - (\frac{2}{5} + 2) = \frac{6(3\sqrt{2}-2)}{5}$

**27.** $\int_{0}^{1} u(\sqrt{u} + \sqrt[3]{u}) du = \int_{0}^{1} (u^{3/2}+u^{4/3}) du = \frac{u^{5/2}}{5/2} + \frac{u^{7/3}}{7/3}]_{0}^{1}$

$= \frac{2}{5} u^{5/2} + \frac{3}{7} u^{7/3}]_{0}^{1} = \frac{2}{5} + \frac{3}{7} = \frac{29}{35}$

**29.** $\int_{-2}^{3} |x^2-1| dx = \int_{-2}^{-1} (x^2-1) dx + \int_{-1}^{1} (1-x^2) dx + \int_{1}^{3} (x^2-1) dx$

$= \left[\frac{x^3}{3} - x\right]]_{-2}^{-1} + \left[x - \frac{x^3}{3}\right]]_{-1}^{1} + \left[\frac{x^3}{3} - x\right]]_{1}^{3}$

$= (\frac{-1}{3} + 1) - (\frac{-8}{3} + 2) + (1 - \frac{1}{3}) - (-1 + \frac{1}{3}) + (9-3) - (\frac{1}{3} - 1) = \frac{28}{3}$

**31.** $\int_{3}^{3} \sqrt{x^5+2} \, dx = 0$

**33.** $\int_{-4}^{2} \frac{2}{x^6} \, dx$ does not exist since $f(x) = 2/x^6$ has an infinite discontinuity at 0.

**35.** $\int_{1}^{4} (\sqrt{t} - \frac{2}{\sqrt{t}}) dt = \int_{1}^{4} (t^{1/2}-2t^{-1/2}) dt = \frac{t^{3/2}}{3/2} - 2\frac{t^{1/2}}{1/2}]_{1}^{4}$

$= \frac{2}{3} t^{3/2} - 4t^{1/2}]_{1}^{4} = \left[\frac{2}{3} \cdot 8 - 4 \cdot 2\right] - \left[\frac{2}{3} - 4\right] = \frac{2}{3}$

37. $\int_{-1}^{0} (x+1)^3 dx = \int_{-1}^{0} (x^3+3x^2+3x+1) dx = \frac{x^4}{4} + 3\frac{x^3}{3} + 3\frac{x^2}{2} + x \Big]_{-1}^{0}$

$= \frac{x^4}{4} + x^3 + \frac{3}{2} x^2 + x \Big]_{-1}^{0} = 0 - \left[ \frac{1}{4} - 1 + \frac{3}{2} - 1 \right] = 2 - \frac{7}{4} = \frac{1}{4}$

39. $\int_{\pi/4}^{\pi/3} \sin t \, dt = - \cos t ]_{\pi/4}^{\pi/3} = - \cos \frac{\pi}{3} + \cos \frac{\pi}{4} = \frac{(\sqrt{2} - 1)}{2}$

41. $\int_{\pi/2}^{\pi} \sec x \tan x \, dx$ does not exist since $\sec x \tan x$ has an

infinite discontinuity at $\pi/2$.

43. $\int_{\pi/6}^{\pi/3} \csc^2 \theta \, d\theta = - \cot x ]_{\pi/6}^{\pi/3} = - \cot \frac{\pi}{3} + \cot \frac{\pi}{6} = - \frac{1}{3}\sqrt{3} + \sqrt{3} = \frac{2}{3}\sqrt{3}$

45. $\int_{0}^{1} \left[ \sqrt[4]{x^5} + \sqrt[5]{x^4} \right] dx = \int_{0}^{1} (x^{5/4} + x^{4/5}) dx = \frac{x^{9/4}}{9/4} + \frac{x^{9/5}}{9/5} \Big]_{0}^{1}$

$= \frac{4}{9} x^{9/4} + \frac{5}{9} x^{9/5} \Big]_{0}^{1} = \frac{4}{9} + \frac{5}{9} - 0 = 1$

47. $\int_{-1}^{2} (x - 2|x|) dx = \int_{-1}^{0} 3x \, dx + \int_{0}^{2} (-x) dx = 3\frac{x^2}{2} \Big]_{-1}^{0} - \frac{x^2}{2} \Big]_{0}^{2}$

$= [3 \cdot 0 - 3(1/2)] - (2-0) = -7/2$

49. $\int_{0}^{2} f(x) dx = \int_{0}^{1} x^4 \, dx + \int_{1}^{2} x^5 \, dx = \frac{x^5}{5} \Big]_{0}^{1} + \frac{x^6}{6} \Big]_{1}^{2}$

$= (\frac{1}{5} - 0) + (\frac{64}{6} - \frac{1}{6}) = 10.7$

51. area $= \int_{0}^{2} (4x^2 - 4x + 3) dx = 4\frac{x^3}{3} - 4\frac{x^2}{2} + 3x \Big]_{0}^{2}$

$= 4(8/3) - 4 \cdot 2 + 3 \cdot 2 - 0 = 26/3$

53. area $= \int_{0}^{27} x^{1/3} \, dx = \frac{3}{4} x^{4/3} \Big]_{0}^{27} = \frac{3}{4} \cdot 81 - 0 = \frac{243}{4}$

55. area $= \int_{0}^{\pi} \sin x \, dx = - \cos x ]_{0}^{\pi} = - \cos \pi + \cos 0 = - (-1) + 1 = 2$

57. area $= \int_{-1}^{2} x^{0.8} \, dx = \frac{x^{1.8}}{1.8} \Big]_{-1}^{2} = \frac{2^{1.8}}{1.8} - \frac{-1}{1.8} = \frac{1 + 2^{1.8}}{1.8}$ or $\frac{5}{9}(1 + 2^{9/5})$

59. $\frac{d}{dx} \left[ \frac{x}{a^2 \sqrt{a^2 - x^2}} + C \right] = \frac{1}{a^2} \frac{\sqrt{a^2 - x^2} - x(-x)/\sqrt{a^2 - x^2}}{(a^2 - x^2)} = \frac{1}{a^2} \frac{(a^2 - x^2) + x^2}{(a^2 - x^2)^{3/2}}$

$= \frac{1}{\sqrt{(a^2 - x^2)^3}}$

61. $\frac{d}{dx} (\frac{x}{2} - \frac{\sin 2x}{4} + C) = \frac{1}{2} - \frac{1}{4}(\cos 2x)(2) + 0$

$= \frac{1}{2} - \frac{1}{2} \cos 2x = \frac{1}{2} - \frac{1}{2}(1 - 2\sin^2 x) = \sin^2 x$

63. $\int x\sqrt{x} \, dx = \int x^{3/2} \, dx = \frac{2}{5} x^{5/2} + C$

65. $\int (2 - \sqrt{x})^2 dx = \int (4 - 4\sqrt{x} + x) dx = 4x - 4 \cdot \frac{x^{3/2}}{3/2} + \frac{x^2}{2} + C$

$= 4x - (8/3)x^{3/2} + (1/2)x^2 + C$

67. $\int (2x + \sec x \tan x)dx = x^2 + \sec x + C$

69. By (5.18), $\dfrac{g(x+h) - g(x)}{h} = \dfrac{1}{h} \int_x^{x+h} f$ for $h \neq 0$. Suppose $h < 0$.

Since $f$ is continuous on $[x+h,x]$, the Extreme Value Theorem says that there are numbers $u$ and $v$ in $[x+h,x]$ such that $f(u) = m$ and $f(v) = M$, where $m$ and $M$ are the absolute minimum and maximum values of $f$ on $[x+h,x]$. By Property 8 of integrals, $m(-h) \leq \int_{x+h}^x f \leq M(-h)$; that is, $f(u)(-h) \leq -\int_x^{x+h} f \leq f(v)(-h)$

Since $-h > 0$, we can divide this inequality by $-h$: $f(u) \leq \dfrac{1}{h} \int_x^{x+h} f \leq f(v)$ Using (5.18) to replace the middle part of the inequality, we obtain (5.19) in the case where $h < 0$.

71. $\int_4^8 \dfrac{1}{x}\, dx = \ln x\,\big]_4^8 = \ln 8 - \ln 4 = \ln \dfrac{8}{4} = \ln 2$

73. $\int_8^9 2^t dt = \dfrac{1}{\ln 2} 2^t\,\big]_8^9 = \dfrac{1}{\ln 2}(2^9 - 2^8) = \dfrac{2^8}{\ln 2}$

75. $\int_1^{\sqrt{3}} \dfrac{6}{1+x^2}\, dx = 6 \tan^{-1}x\,\big]_1^{\sqrt{3}} = 6 \tan^{-1}\sqrt{3} - 6 \tan^{-1}1 = 6\,\dfrac{\pi}{3} - 6\,\dfrac{\pi}{4} = \dfrac{\pi}{2}$

77. $\int_1^e \dfrac{x^2+x+1}{x}\, dx = \int_1^e \left[x + 1 + \dfrac{1}{x}\right]dx = \dfrac{x^2}{2} + x + \ln x\,\big]_1^e$

$= \left[\dfrac{e^2}{2} + e + \ln e\right] - \left[\dfrac{1}{2} + 1 + \ln 1\right] = \dfrac{e^2}{2} + e - \dfrac{1}{2}$

79. $\int \left[x^2 + 1 + \dfrac{1}{x^2+1}\right]dx = \dfrac{x^3}{3} + x + \tan^{-1}x + C$

## Section 5.6

Exercises 5.6

1. Let $u = x^2-1$. Then $du = 2x\, dx$, so $\int x(x^2-1)^{99}dx = \int u^{99}\, \dfrac{1}{2}\, du$

$= \dfrac{1}{2}\,\dfrac{u^{100}}{100} + C = \dfrac{(x^2-1)^{100}}{200} + C$

3. Let $u = 4x$. Then $du = 4\, dx$, so $\int \sin 4x\, dx = \int \sin u\, (1/4)du$
$= (1/4)(-\cos u) + C = -(1/4)\cos 4x + C$

5. Let $u = x^2+6x$. Then $du = 2(x+3)dx$, so $\int \dfrac{x+3}{(x^2+6x)^2}\, dx = \dfrac{1}{2} \int \dfrac{du}{u^2}$

$= (1/2) \int u^{-2}\, du = -(1/2)u^{-1} + C = -1/[2(x^2+6x)] + C$

7. Let $u = x^2+x+1$. Then $du = (2x+1)dx$, so $\int (2x+1)(x^2+x+1)^3 dx$

$= \int u^3 \, du = u^4/4 + C = (x^2+x+1)^4/4 + C$

9. Let $u = x-1$. Then $du = dx$, so $\int \sqrt{x-1} \, dx = \int u^{1/2} \, du$

$= \frac{2}{3} u^{3/2} + C = \frac{2}{3}(x-1)^{3/2} + C$

11. Let $u = 2+x^4$. Then $du = 4x^3 \, dx$, so $\int x^3 \sqrt{2+x^4} \, dx = \int u^{1/2} \frac{1}{4} \, du$

$= \frac{1}{4} \frac{u^{3/2}}{3/2} + C = \frac{1}{6}(2+x^4)^{3/2} + C$

13. Let $u = 3x^2-2x+1$. Then $du = 2(3x-1)dx$, so $\int \frac{3x-1}{(3x^2-2x+1)^4} \, dx$

$= \int u^{-4} \frac{1}{2} \, du = \frac{1}{2} \frac{u^{-3}}{-3} + C = -1/[6(3x^2-2x+1)^3] + C$

15. Let $u = t+1$. Then $du = dt$, so $\int \frac{2}{(t+1)^6} \, dt = 2 \int u^{-6} \, du$

$= -\frac{2}{5} u^{-5} + C = -2/5(t+1)^5 + C$

17. Let $u = 1-2y$. Then $du = -2 \, dy$, so $\int (1-2y)^{1.3} \, dy$

$= \int u^{1.3} (-1/2)du = -(1/2) \frac{u^{2.3}}{2.3} + C = -\frac{(1-2y)^{2.3}}{4.6} + C$

19. Let $u = 2\theta$. Then $du = 2 \, d\theta$, so $\int \cos 2\theta \, d\theta = \int \cos u \frac{1}{2} \, du$

$= \frac{1}{2} \sin u + C = \frac{1}{2} \sin 2\theta + C$

21. Let $u = x+2$. Then $du = dx$, so $\int \frac{x}{\sqrt[4]{x+2}} \, dx = \int \frac{u-2}{\sqrt[4]{u}} \, du$

$= \int (u^{3/4}-2u^{-1/4})du = (4/7)u^{7/4} - 2\cdot(4/3)u^{3/4} + C$

$= (4/7)(x+2)^{7/4} - (8/3)(x+2)^{3/4} + C$

23. Let $u = t^2$. Then $du = 2t \, dt$, so $\int t \sin(t^2)dt = \int \sin u \, (1/2)du$

$= -(1/2)\cos u + C = -(1/2) \cos(t^2) + C$

25. Let $u = 1-x^2$. Then $x^2 = 1-u$ and $2x \, dx = -du$, so

$\int x^3(1-x^2)^{3/2}dx = \int u^{3/2}(1-u)(-1/2)du = \frac{1}{2} \int (u^{5/2}-u^{3/2})du$

$= \frac{1}{2}\left[\frac{2}{7} u^{7/2} - \frac{2}{5} u^{5/2}\right] + C = \frac{1}{7}(1-x^2)^{7/2} - \frac{1}{5}(1-x^2)^{5/2} + C$

27. Let $u = \sin x$. Then $du = \cos x \, dx$, so $\int \sin^3 x \cos x \, dx$

$= \int u^3 \, du = \frac{u^4}{4} + C = \frac{1}{4} \sin^4 x + C$

29. Let $u = 1 + \sec x$. Then $du = \sec x \tan x \, dx$, so

$\int \sec x \tan x \sqrt{1 + \sec x} \, dx = \int u^{1/2} \, du = (2/3)u^{3/2} + C$

$= (2/3)(1 + \sec x)^{3/2} + C$

**31.** Let $u = ax^2+2bx+c$. Then $du = 2(ax+b)dx$, so $\int \dfrac{(ax+b)dx}{\sqrt{ax^2+2bx+c}}$

$= \int \dfrac{(1/2)du}{\sqrt{u}} = \frac{1}{2} \int u^{-1/2}\, du = u^{1/2} + C = \sqrt{ax^2+2bx+c} + C$

**33.** Let $u = 2x+3$. Then $du = 2\, dx$, so $\int \sin(2x+3)dx = \int \sin u\ (1/2)du$

$= -(1/2) \cos u + C = - (1/2) \cos(2x+3) + C$

**35.** Let $u = 3x$. Then $du = 3\, dx$, so $\int (\sin 3\alpha - \sin 3x)dx$

$= \int (\sin 3\alpha - \sin u)(1/3)du = (1/3)[(\sin 3\alpha)u + \cos u] + C$

$= (\sin 3\alpha)x + (1/3)\cos 3x + C$

**37.** Let $u = b+cx^{a+1}$. Then $du = (a+1)cx^a\, dx$, so $\int x^a \sqrt{b+cx^{a+1}}\, dx$

$= \int u^{1/2} \dfrac{1}{(a+1)c}\, du = \dfrac{1}{(a+1)c} \cdot \dfrac{2}{3} u^{3/2} + C = \dfrac{2}{3(a+1)c} (b+cx^{a+1})^{3/2} + C$

**39.** Let $u = 2x-1$. Then $du = 2\, dx$, so $\int_0^1 (2x-1)^{100}\, dx = \int_{-1}^1 u^{100}\, \frac{1}{2}\, du$

$= \int_0^1 u^{100}\, du$ [by 5.33 (a)] $= u^{101}/101 \Big]_0^1 = \dfrac{1}{101}$

**41.** Let $u = x^4+x$. Then $du = (4x^3+1)dx$, so $\int_0^1 (x^4+x)^5(4x^3+1)dx$

$= \int_0^2 u^5\, du = u^6/6 \Big]_0^2 = 2^6/6 = 32/3$

**43.** Let $u = x-1$. Then $du = dx$, so $\int_1^2 x\sqrt{x-1}\, dx = \int_0^1 (u+1)\sqrt{u}\, du$

$= \int_0^1 (u^{3/2}+u^{1/2})du = \left[\frac{2}{5} u^{5/2} + \frac{2}{3} u^{3/2}\right]_0^1 = \frac{2}{5} + \frac{2}{3} = \dfrac{16}{15}$

**45.** Let $u = \pi t$. Then $du = \pi\, dt$, so $\int_0^1 \cos \pi t\, dt = \int_0^\pi \cos u\ (\frac{1}{\pi}\, du)$

$= \frac{1}{\pi} \sin u \Big]_0^\pi = \frac{1}{\pi}(0-0) = 0$

**47.** Let $u = 1 + \dfrac{1}{x}$. Then $du = - \dfrac{dx}{x^2}$, so $\int_1^4 \dfrac{1}{x^2} \sqrt{1 + \dfrac{1}{x}}\, dx = \int_2^{5/4} u^{1/2}(-du)$

$= \int_{5/4}^2 u^{1/2}\, du = \frac{2}{3} u^{3/2} \Big]_{5/4}^2 = \frac{2}{3}\left[2\sqrt{2} - \dfrac{5\sqrt{5}}{8}\right] = \dfrac{4\sqrt{2}}{3} - \dfrac{5\sqrt{5}}{12}$

**49.** Let $u = \cos \theta$. Then $du = - \sin \theta\, d\theta$, so $\int_0^{\pi/3} \dfrac{\sin \theta}{\cos^2 \theta}\, d\theta$

$= \int_1^{1/2} \dfrac{-du}{u^2} = \int_{1/2}^1 u^{-2}\, du = \dfrac{-1}{u}\Big]_{1/2}^1 = -1+2 = 1$

**51.** Let $u = 1+2x$. Then $du = 2\, dx$, so $\int_0^{13} \dfrac{dx}{\sqrt[3]{(1+2x)^2}} = \int_1^{27} u^{-2/3}\, \frac{1}{2}\, du$

$= \frac{1}{2} \cdot 3u^{1/3} \Big]_1^{27} = \frac{3}{2}(3-1) = 3$

**53.** Let $u = a^2-x^2$. Then $du = -2x\, dx$, so $\int_0^a x\sqrt{a^2-x^2}\, dx$

$= \int_{a^2}^0 u^{1/2}(- \frac{1}{2}\, du) = \frac{1}{2} \int_0^{a^2} u^{1/2}\, du = \frac{1}{2} \cdot \frac{2}{3} u^{3/2} \Big]_0^{a^2} = \dfrac{a^3}{3}$

55. $\int_{-a}^{a} x\sqrt{x^2+a^2}\ dx = 0$ since $f(x) = x\sqrt{x^2+a^2}$ is an odd function.

57. Let $u = x+1$. Then $du = dx$, so area $= \int_{0}^{3} \sqrt{x+1}\ dx = \int_{1}^{4} u^{1/2}\ du$

    $= \frac{2}{3} u^{3/2}\Big]_{1}^{4} = \frac{2}{3}(8-1) = \frac{14}{3}$

59. Let $u = x/2$. Then $du = dx/2$, so area $= \int_{0}^{\pi/3} \sin(x/2)dx$

    $= \int_{0}^{\pi/6} \sin u \cdot 2\ du = 2(-\cos u)\Big]_{0}^{\pi/6} = -\sqrt{3} - (-2) = 2-\sqrt{3}$

61. Let $u = x+1$. Then $du = dx$, so area $= \int_{0}^{10} \frac{dx}{(x+1)^2} = \int_{1}^{11} u^{-2}\ du$

    $= -\frac{1}{u}\Big]_{1}^{11} = -\frac{1}{11} + 1 = \frac{10}{11}$

63. Let $u = -x$. Then $du = -dx$. When $x = a$, $u = -a$; when $x = b$,

    $u = -b$. So $\int_{a}^{b} f(-x)dx = \int_{-a}^{-b} f(u)(-du) = \int_{-b}^{-a} f(u)du = \int_{-b}^{-a} f(x)dx$

65. Let $u = 2x-1$. Then $du = 2\ dx$, so $\int \frac{dx}{2x-1} = \int \frac{(1/2)du}{u}$

    $= (1/2)\ln|u| + C = (1/2)\ln|2x-1| + C$

67. Let $u = \ln x$. Then $du = \frac{dx}{x}$ so $\int \frac{(\ln x)^2}{x}\ dx = \int u^2\ du$

    $= u^3/3 + C = (\ln x)^3/3 + C$

69. Let $u = 1 + e^x$. Then $du = e^x dx$, so $\int e^x(1+e^x)^{10}\ dx$

    $= \int u^{10}\ du = u^{11}/11 + C = (1+e^x)^{11}/11 + C$

71. Let $u = \ln x$. Then $du = \frac{dx}{x}$, so $\int \frac{dx}{x \ln x} = \int \frac{du}{u} = \ln|u| + C$

    $= \ln|\ln x| + C$

73. $\int \frac{e^x + 1}{e^x}\ dx = \int (1 + e^{-x})dx = x - e^{-x} + C$   (Substitute $u = -x$.)

75. Let $u = x^2+2x$. Then $du = 2(x+1)dx$, so $\int \frac{x+1}{x^2+2x}\ dx$

    $= \int \frac{(1/2)du}{u} = \frac{1}{2} \ln|u| + C = \frac{1}{2} \ln|x^2+2x| + C$

77. Let $u = e^x$. Then $du = e^x\ dx$, so $\int \frac{e^x}{e^{2x} + 1}\ dx = \int \frac{du}{u^2 + 1}$

    $= \tan^{-1}u + C = \tan^{-1}(e^x) + C$

79. $\int \frac{1 + x}{1 + x^2}\ dx = \int \frac{1}{1 + x^2}\ dx + \int \frac{x}{1 + x^2}\ dx$

    $= \tan^{-1}x + \frac{1}{2} \int \frac{2x\ dx}{1 + x^2} = \tan^{-1}x + \frac{1}{2} \ln(1+x^2) + C$

    [The second integral is $\int du/u$ where $u = 1 + x^2$.]

81. Let $u = x^2$. Then $du = 2x\ dx$, so $\int \frac{x}{1 + x^4}\ dx = \int \frac{(1/2)du}{1 + u^2}$

    $= (1/2) \tan^{-1}u + C = (1/2) \tan^{-1}(x^2) + C$

83. Let u = 2x+3. Then du = 2 dx, so $\int_0^3 \frac{dx}{2x+3} = \int_3^9 \frac{(1/2)du}{u}$

$= \frac{1}{2} \ln u \Big]_3^9 = \frac{1}{2}(\ln 9 - \ln 3) = \frac{1}{2} \ln 3$

85. Let u = ln x. Then du = $\frac{dx}{x}$ so $\int_e^{e^4} \frac{dx}{x\sqrt{\ln x}} = \int_1^4 u^{-1/2} du$

$= 2u^{1/2} \Big]_1^4 = 2 \cdot 2 - 2 \cdot 1 = 2$

87. Let u = - 2x. Then du = - 2 dx, so area = $\int_0^1 2e^{-2x} dx$

$= \int_0^{-2} e^u (-du) = \int_{-2}^0 e^u du = e^u \Big]_{-2}^0 = 1 - e^{-2}$

**Exercises 5.7**

1. area = $\int_{-1}^1 [(x^2+3) - x]dx = \int_{-1}^1 (x^2-x+3)dx = \frac{x^3}{3} - \frac{x^2}{2} + 3x \Big]_{-1}^1$

$= \left[\frac{1}{3} - \frac{1}{2} + 3\right] - \left[\frac{-1}{3} - \frac{1}{2} - 3\right] = \frac{20}{3}$

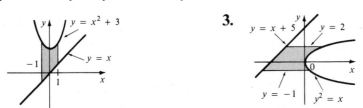

**1.** $y = x^2 + 3$, $y = x$, $-1$, $1$

**3.** $y = x + 5$, $y = 2$, $y = -1$, $y^2 = x$

3. area = $\int_{-1}^2 [y^2-(y-5)]dy = \frac{y^3}{3} - \frac{y^2}{2} + 5y \Big]_{-1}^2$

$= \left[\frac{8}{3} - 2 + 10\right] - \left[\frac{-1}{3} - \frac{1}{2} - 5\right] = 16.5$

5. area = $\int_0^6 [2x-(x^2-4x)]dx = \int_0^6 (6x-x^2)dx = 3x^2 - \frac{x^3}{3} \Big]_0^6 = 108-72 = 36$

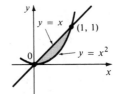

**5.** $y = 2x$, (6, 12), $y = x^2 - 4x$

**7.** $y = x$, (1, 1), $y = x^2$

7. area = $\int_0^1 (x-x^2)dx = \frac{x^2}{2} - \frac{x^3}{3} \Big]_0^1 = \frac{1}{2} - \frac{1}{3} = \frac{1}{6}$

9.  area $= \int_0^1 (\sqrt{x}-x^2)dx = \frac{2}{3}x^{3/2} - \frac{x^3}{3}\Big]_0^1 = \frac{2}{3} - \frac{1}{3} = \frac{1}{3}$

**9.**

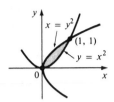

**11.**

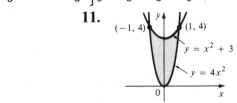

11.  area $= \int_{-1}^1 [(x^2+3)-4x^2]dx = 2\int_0^1 (3-3x^2)dx = 2(3x-x^3)]_0^1 = 2(3-1)-0 = 4$

13.  area $= \int_0^3 [(2x+5) - (x^2+2)]dx + \int_3^6 [(x^2+2) - (2x+5)]dx$

$= \int_0^3 (-x^2+2x+3)dx + \int_3^6 (x^2-2x-3)dx$

$= \left[-\frac{x^3}{3} + x^2 + 3x\right]_0^3 + \left[\frac{x^3}{3} - x^2 - 3x\right]_3^6$

$= (-9+9+9)-0+(72-36-18)-(9-9-9) = 36$

**13.**

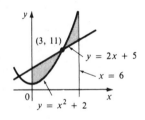

**15.**

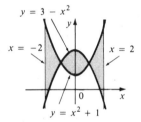

15.  area $= \int_{-2}^{-1} [(x^2+1) - (3-x^2)]dx + \int_{-1}^1 [(3-x^2) - (x^2+1)]dx$

$+ \int_1^2 [(x^2+1) - (3-x^2)]dx = \int_{-2}^{-1} (2x^2-2)dx + \int_{-1}^1 (2-2x^2)dx$

$+ \int_1^2 (2x^2-2)dx = 2\int_0^1 (2-2x^2)dx + 2\int_1^2 (2x^2-2)dx$   [by symmetry]

$= 2\left[2x-\frac{2}{3}x^3\right]\Big]_0^1 + 2\left[\frac{2}{3}x^3-2x\right]\Big]_1^2 = 2\left[2 - \frac{2}{3}\right] + 2\left[\frac{16}{3} - 4\right] - 2\left[\frac{2}{3} - 2\right] = 8$

17.  area $= \int_{-1}^3 (2y+3-y^2)dy = y^2 + 3y - \frac{y^3}{3}\Big]_{-1}^3 = (9+9-9) - (1-3+\frac{1}{3}) = \frac{32}{3}$

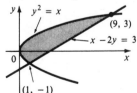

19.  area = $\int_{-1}^{1} [(1-y^2) - (y^2-1)]dy = \int_{-1}^{1} 2(1-y^2)dy = 4\int_{0}^{1} (1-y^2)dy$

$= 4\left[y - \frac{y^3}{3}\right]\Big]_{0}^{1} = 4\left[1 - \frac{1}{3}\right] = \frac{8}{3}$

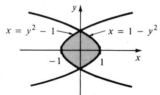

21.  area = $\int_{-2}^{0} [x^3 - (2x-x^2)]dx + \int_{0}^{1} [(2x-x^2) - x^3]dx$

$= \frac{x^4}{4} + \frac{x^3}{3} - x^2\Big]_{-2}^{0} + \left[-\frac{x^4}{4} - \frac{x^3}{3} + x^2\right]\Big]_{0}^{1} = 0-(4-\frac{8}{3}-4)+(-\frac{1}{4}-\frac{1}{3}+1)-0 = \frac{37}{12}$

**21.**

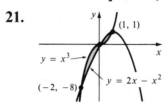

**23.**

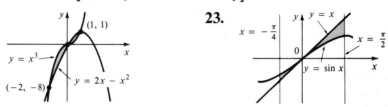

23.  area = $\int_{-\pi/4}^{0} (\sin x - x)dx + \int_{0}^{\pi/2} (x - \sin x)dx$

$= \left[-\cos x - \frac{x^2}{2}\right]\Big]_{-\pi/4}^{0} + \left[\frac{x^2}{2} + \cos x\right]\Big]_{0}^{\pi/2}$

$= -1 - (-1/\sqrt{2} - \pi^2/32) + \pi^2/8 - 1 = 5\pi^2/32 + 1/\sqrt{2} - 2$

25.  Notice that $\cos x = \sin 2x = 2 \sin x \cos x \Leftrightarrow 2 \sin x = 1$ or
$\cos x = 0 \Leftrightarrow x = \pi/6$ or $\pi/2$.
area = $\int_{0}^{\pi/6} (\cos x - \sin 2x)dx + \int_{\pi/6}^{\pi/2} (\sin 2x - \cos x)dx$

$= \sin x + \frac{1}{2} \cos 2x\Big]_{0}^{\pi/6} + \left[-\frac{1}{2} \cos 2x - \sin x\right]_{\pi/6}^{\pi/2}$

$= \frac{1}{2} + \frac{1}{2}\cdot\frac{1}{2} - \left[0 + \frac{1}{2}\cdot 1\right] + \left[\frac{1}{2} - 1\right] - \left[-\frac{1}{2}\cdot\frac{1}{2} - \frac{1}{2}\right] = \frac{1}{2}$

**25.**

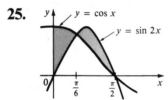

**27.**

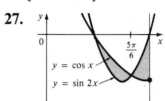

27.  $\cos x = \sin 2x = 2 \sin x \cos x \Leftrightarrow \cos x = 0$ or $\sin x = 1/2 \Leftrightarrow$
$x = \pi/2$ or $5\pi/6$.
area = $\int_{\pi/2}^{5\pi/6} (\cos x - \sin 2x)dx + \int_{5\pi/6}^{\pi} (\sin 2x - \cos x)dx$

$$= (\sin x + \tfrac{1}{2} \cos 2x)]_{\pi/2}^{5\pi/6} - (\sin x + \tfrac{1}{2} \cos 2x)]_{5\pi/6}^{\pi}$$

$$= \left[\tfrac{1}{2} + \tfrac{1}{2}\cdot\tfrac{1}{2}\right] - \left[1 - \tfrac{1}{2}\right] - \left[0 + \tfrac{1}{2}\right] + \left[\tfrac{1}{2} + \tfrac{1}{2}\cdot\tfrac{1}{2}\right] = \tfrac{1}{2}$$

29.  area $= \int_{-4}^{0} \{-x-[(x+1)^2-7]\}dx + \int_{0}^{2} \{x-[(x+1)^2-7]\}dx$

$$= \int_{-4}^{0} (-x^2-3x+6)dx + \int_{0}^{2}(-x^2-x+6)dx$$

$$= -\left.\frac{x^3}{3} - \frac{3x^2}{2} + 6x\right]_{-4}^{0} + \left[-\frac{x^3}{3} - \frac{x^2}{2} + 6x\right]_{0}^{2}$$

$$= 0 - (\tfrac{64}{3} - 24 - 24) + (-\tfrac{8}{3} - 2 + 12) - 0 = 34$$

**29.**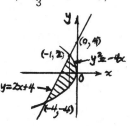

**31.**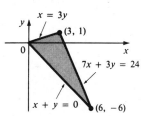

31.  area $= \int_{0}^{3}\left[\frac{x}{3} - (-x)\right]dx + \int_{3}^{6}\left[(8-\tfrac{7}{3}x) - (-x)\right]dx = \int_{0}^{3}\tfrac{4}{3}x\,dx + \int_{3}^{6}(-\tfrac{4}{3}x+8)dx$

$$= \left.\tfrac{2}{3}x^2\right]_{0}^{3} + (-\tfrac{2}{3}x^2+8x)\Big]_{3}^{6} = (6-0) + (24-18) = 12$$

33.  (a) area $= \int_{-4}^{-1}[(2x+4)+\sqrt{-4x}]dx + \int_{-1}^{0}2\sqrt{-4x}\,dx$

$$= x^2+4x]_{-4}^{-1} + 2\int_{-4}^{-1}\sqrt{-x}\,dx + 4\int_{-1}^{0}\sqrt{-x}\,dx$$

$$= (-3-0) + 2\int_{1}^{4}\sqrt{u}\,du + 4\int_{0}^{1}\sqrt{u}\,du \qquad [u=-x]$$

$$= -3 + \left[\tfrac{4}{3}u^{3/2}\right]_{1}^{4} + \left[\tfrac{8}{3}u^{3/2}\right]_{0}^{1} = -3 + \tfrac{28}{3} + \tfrac{8}{3} = 9$$

(b) area $= \int_{-4}^{2}\left[-\frac{y^2}{4} - \left[\frac{y}{2} - 2\right]\right]dy = -\frac{y^3}{12} - \frac{y^2}{4} + 2y\Big]_{-4}^{2}$

$$= (-\tfrac{2}{3} - 1 + 4) - (\tfrac{16}{3} - 4 - 8) = 9$$

**33.**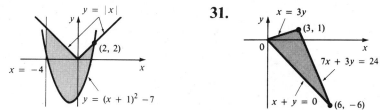

**35.**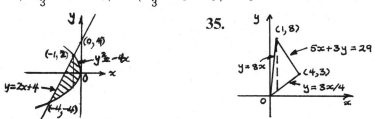

35.  area $= \int_{0}^{1}\left[8x - \tfrac{3}{4}x\right]dx + \int_{1}^{4}\left[\left[-\tfrac{5}{3}x + \tfrac{29}{3}\right] - \tfrac{3}{4}x\right]dx$

$$= \frac{29}{4}\int_{0}^{1}x\,dx + \int_{1}^{4}\left[-\frac{29}{12}x + \frac{29}{3}\right]dx = \frac{29}{4}\left.\frac{x^2}{2}\right]_{0}^{1} - \frac{29}{12}\left[\frac{x^2}{2} - 4x\right]\Big]_{1}^{4}$$

$$= \frac{29}{8} - \frac{29}{12}\left[-8 - \tfrac{1}{2} + 4\right] = 14.5$$

149

37. $\int_0^2 |x^2 - x^3| dx = \int_0^1 (x^2 - x^3) dx + \int_1^2 (x^3 - x^2) dx = \left[\frac{x^3}{3} - \frac{x^4}{4}\right]\Big]_0^1 + \left[\frac{x^4}{4} - \frac{x^3}{3}\right]\Big]_1^2$

$= \frac{1}{3} - \frac{1}{4} + \left[4 - \frac{8}{3}\right] - \left[\frac{1}{4} - \frac{1}{3}\right] = 1.5$

**37.**

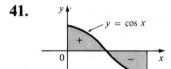

**39.**

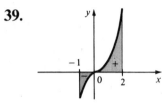

39. $\int_{-1}^2 x^3 \, dx = \frac{x^4}{4}\Big]_{-1}^2 = 4 - \frac{1}{4} = \frac{15}{4} = 3.75$

41. $\int_0^\pi \cos x \, dx = \sin x]_0^\pi = \sin \pi - \sin 0 = 0$

**41.**

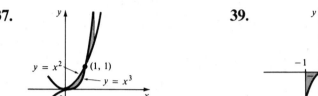

**43.**

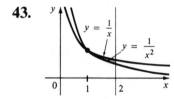

43. $A = \int_1^2 \left[\frac{1}{x} - \frac{1}{x^2}\right] dx = \ln x + \frac{1}{x}\Big]_1^2 = \left[\ln 2 + \frac{1}{2}\right] - \left[\ln 1 + 1\right] = \ln 2 - \frac{1}{2}$

45. $A = 2\int_0^1 \left[\frac{2}{x^2 + 1} - x^2\right] dx = 4 \tan^{-1} x - \frac{2}{3}x^3\Big]_0^1 = 4 \cdot \frac{\pi}{4} - \frac{2}{3} = \pi - \frac{2}{3}$

**45.**

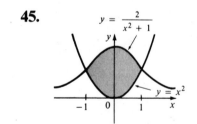

**47.**

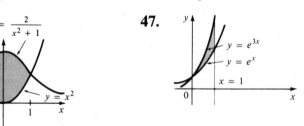

47. $A = \int_0^1 (e^{3x} - e^x) dx = \frac{1}{3}e^{3x} - e^x\Big]_0^1 = \left[\frac{e^3}{3} - e\right] - \left[\frac{1}{3} - 1\right] = \frac{e^3}{3} - e + \frac{2}{3}$

## Exercises 5.8

1. (a) displacement = $\int_0^2(15-2t^2)dt = 15t-\frac{2}{3}t^3\Big]_0^2 = 30 - \frac{16}{3} = \frac{74}{3}$ m

   (b) distance traveled = $\int_0^2|15-2t^2|dt = \int_0^2(15-2t^2)dt = \frac{74}{3}$ m

3. (a) displacement = $\int_1^6(t^2-2t-8)dt = \frac{t^3}{3}-t^2-8t\Big]_1^6 = (72-36-48)-(\frac{1}{3}-1-8)$

   $= -10/3$ m

   (b) distance traveled = $\int_1^6|t^2-2t-8|dt = \int_1^6|(t-4)(t+2)|dt$

   $= \int_1^4(-t^2+2t+8)dt + \int_4^6(t^2-2t-8)dt = \left[-\frac{t^3}{3}+t^2+8t\right]_1^4 + \left[\frac{t^3}{3}-t^2-8t\right]_4^6$

   $= \left[-\frac{64}{3}+16+32\right] - \left[-\frac{1}{3}+1+8\right] + \left[72-36-48\right] - \left[\frac{64}{3}-16-32\right] = \frac{98}{3}$ m

5. (a) displacement = $\int_1^{12}(3-\sqrt{t+4})dt = 3t - \frac{2}{3}(t+4)^{3/2}\Big]_1^{12}$

   $= \left[36 - \frac{2}{3}\cdot 64\right] - \left[3 - \frac{2}{3}\cdot 5^{3/2}\right] = \frac{(10\sqrt{5}-29)}{3}$ m

   (b) distance traveled = $\int_1^{12}|3-\sqrt{t+4}|dt = \int_1^5(3-\sqrt{t+4})dt +$

   $\int_5^{12}(\sqrt{t+4}-3)dt = \left[3t-\frac{2}{3}(t+4)^{3/2}\right]_1^5 + \left[\frac{2}{3}(t+4)^{3/2}-3t\right]_5^{12}$

   $= (15-\frac{2}{3}\cdot 27) - (3-\frac{2}{3}\cdot 5^{3/2}) + (\frac{2}{3}\cdot 64-36) - (\frac{2}{3}\cdot 27-15) = \frac{10\sqrt{5}-7}{3}$ m

7. (a) $v'(t) = a(t) = t+4 \Rightarrow v(t) = (1/2)t^2+4t+C \Rightarrow 5 = v(0) = C \Rightarrow$

   $v(t) = (1/2)t^2+4t+5$ m/s  [OR: $v(t)-v(0) = \int_0^t a(u)du = \int_0^t(u+4)du$

   $= \frac{u^2}{2} + 4u\Big]_0^t = \frac{t^2}{2} + 4t \Rightarrow v(t) = \frac{t^2}{2} + 4t + 5$ m/s ]

   (b) distance traveled = $\int_0^{10}|v(t)|dt = \int_0^{10}\left|\frac{t^2}{2} + 4t + 5\right|dt$

   $= \int_0^{10}\left[\frac{t^2}{2} + 4t + 5\right]dt = \frac{t^3}{6} + 2t^2 + 5t\Big]_0^{10} = \frac{500}{3} + 200 + 50 = \frac{1250}{3}$ m

9. (a) $v'(t) = 3t^2-2t-2$, so $v(t) = t^3-t^2-2t+C$, where $-2 = v(1)$

   $= 1-1-2+C$. $C = 0$, so $v(t) = t^3-t^2-2t$ m/s

   [OR: $v(t)-v(1) = \int_1^t(3u^2-2u-2)du = u^3-u^2-2u\Big]_1^t = t^3-t^2-2t-(-2)$, so

   $v(t) = t^3-t^2-2t+2+v(1) = t^3-t^2-2t$ m/s ]

   (b) $\int_1^3|t^3-t^2-2t|dt = \int_1^3|t(t-2)(t+1)|dt$

   $= \int_1^2(-t^3+t^2+2t)dt + \int_2^3(t^3-t^2-2t)dt = -\frac{t^4}{4} + \frac{t^3}{3} + t^2\Big]_1^2$

   $+ \left[\frac{t^4}{4} - \frac{t^3}{3} - t^2\right]_2^3 = \left[-4+\frac{8}{3}+4\right]-\left[-\frac{1}{4}+\frac{1}{3}+1\right]+\left[\frac{81}{4}-9-9\right]-\left[4-\frac{8}{3}-4\right] = 6.5$ m

11.  Mass $= \int_0^5 \rho(x)dx = \int_0^5 (9 + 2\sqrt{x+2})dx = 9x + 2(2/3)(x+2)^{3/2}\Big]_0^5$

$= 9(5) + (4/3)(7^{3/2}-2^{3/2}) = 45 + (4/3)(7\sqrt{7} - 2\sqrt{2})$ kg

13.  $C(5000)-C(3000) = \int_{3000}^{5000}(140-0.5x+0.012x^2)dx$

$= 140x-0.25x^2+0.004x^3\Big]_{3000}^{5000}$

$= 494,450,000 - 106,170,000 = \$388,280,000$

15.  $f(8)-f(4) = \int_4^8 f'(t)dt = \int_4^8 \sqrt{t}\, dt = \frac{2}{3}t^{3/2}\Big]_4^8 = \frac{2}{3}(16\sqrt{2}-8)$

$= \frac{16(2\sqrt{2}-1)}{3} \approx \$9.75$ million

17.  The volume of inhaled air in the lungs at time t is

$V(t) = \int_0^t f(u)du = \int_0^t (1/2) \sin(2\pi u/5)du$

$= \int_0^{2\pi t/5} \frac{1}{2} \sin v \cdot (5/2\pi)dv,\ [v = 2\pi u/5 \Rightarrow dv = (2\pi/5)du]$

$= (5/4\pi)(- \cos v)\Big]_0^{2\pi t/5} = \frac{5}{4\pi}[- \cos(2\pi t/5) + 1]$

$= (5/4\pi)[1 - \cos(2\pi t/5)]$ liters

Section 5.9

Exercises 5.9

1.  $f_{ave} = \frac{1}{3-0} \int_0^3 (1-2x)dx = \frac{1}{3}(x-x^2)\Big]_0^3 = \frac{1}{3}(3-9) = -2$

3.  $f_{ave} = \frac{1}{2-(-2)} \int_{-2}^2 (x^2+2x-5)dx = \frac{1}{4}\left[\frac{x^3}{3} + x^2 - 5x\right]\Big]_{-2}^2$

$= \frac{1}{4}\left[\left[\frac{8}{3}+4-10\right] - \left[-\frac{8}{3}+4+10\right]\right] = -\frac{11}{3}$

5.  $f_{ave} = \frac{1}{1-(-1)} \int_{-1}^1 x^4\, dx = \frac{1}{2}\cdot 2\int_0^1 x^4\, dx = \frac{x^5}{5}\Big]_0^1 = \frac{1}{5}$

7.  $f_{ave} = \frac{1}{(\pi/4)-(-\pi/2)} \int_{-\pi/2}^{\pi/4} \sin^2 x \cos x\, dx = \frac{4}{3\pi} \int_{-\pi/2}^{\pi/4} \sin^2 x \cos x\, dx$

$= \frac{4}{3\pi} \int_{-1}^{1/\sqrt{2}} u^2\, du \qquad [u = \sin x \Rightarrow du = \cos x\, dx]$

$= \frac{4}{3\pi}\cdot\frac{u^3}{3}\Big]_{-1}^{1/\sqrt{2}} = \frac{4}{9\pi}\left[\frac{1}{2\sqrt{2}} + 1\right] = \frac{4}{9\pi}\left[\frac{\sqrt{2}}{4} + 1\right] = \frac{\sqrt{2} + 4}{9\pi}$

9.   (a) $f_{ave} = \frac{1}{3-0} \int_0^3 2x\ dx = \frac{1}{3}x^2 \Big]_0^3 = \frac{1}{3}(9-0) = 3$

(b) $f_{ave} = f(c)$ when $3 = 2c$, that is when $c = \frac{3}{2}$

(c)

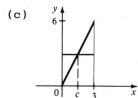

11.  (a) $f_{ave} = \frac{1}{2-0} \int_0^2 (4-x^2)dx = \frac{1}{2}\left[4x - \frac{x^3}{3}\right]_0^2 = \frac{1}{2}\left[\left[8-\frac{8}{3}\right]-0\right] = \frac{8}{3}$

(b) $f_{ave} = f(c) \iff \frac{8}{3} = 4-c^2 \iff c^2 = \frac{4}{3} \iff c = 2/\sqrt{3}$

(c)

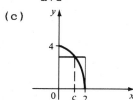

13.  $T_{ave} = \frac{1}{12} \int_0^{12}\left[50 + 14\sin\frac{\pi t}{12}\right]dt = \frac{1}{12}\left[50t - 14\cdot\frac{12}{\pi}\cos\frac{\pi t}{12}\right]_0^{12}$

$= \frac{1}{12}\left[50\cdot12 + 14\cdot\frac{12}{\pi} + 14\cdot\frac{12}{\pi}\right] = (50 + \frac{28}{\pi})°F \approx 59°F$

15.  $\rho_{ave} = \frac{1}{8} \int_0^8 \frac{12}{\sqrt{x+1}}\ dx = \frac{3}{2} \int_0^8 (x+1)^{-1/2}dx = 3\sqrt{x+1}\Big]_0^8 = 9-3 = 6$ kg/m

17.  $V_{ave} = \frac{1}{5} \int_0^5 V(t)dt = \frac{1}{5} \int_0^5 \frac{5}{4\pi}[1-\cos(2\pi t/5)]dt = \frac{1}{4\pi} \int_0^5 \left[1-\cos\left[\frac{2\pi t}{5}\right]\right]dt$

$= \frac{1}{4\pi}\left[t - \frac{5}{2\pi}\sin\left[\frac{2\pi t}{5}\right]\right]_0^5 = \frac{1}{4\pi}[(5-0)-0] = \frac{5}{4\pi} \approx 0.4$ L

19.  Let $F(x) = \int_a^x f(t)dt$ for x in [a,b]. Then F is continuous on [a,b]

and differentiable on (a,b), so by the Mean Value Theorem there is
a number c in (a,b) such that $F(b)-F(a) = F'(c)(b-a)$. But
$F'(x) = f(x)$ by the Fundamental Theorem of Calculus. Therefore
$$\int_a^b f(t)dt - 0 = f(c)(b-a).$$

Chapter 5 Review Exercises

1. $\displaystyle\sum_{i=2}^{5} \frac{1}{(-10)^i} = \frac{1}{(-10)^2} + \frac{1}{(-10)^3} + \frac{1}{(-10)^4} + \frac{1}{(-10)^5}$

   $= .01 - .001 + .0001 - .00001 = 0.00909$

3. $\displaystyle\sum_{k=1}^{n}(2+k^3) = \sum_{k=1}^{n} 2 + \sum_{k=1}^{n} k^3 = 2n + n^2(n+1)^2/4$

5. $\displaystyle\sum_{i=1}^{n} f(x_i^*)\Delta x_i = \sum_{i=1}^{4} f\left[\frac{i-1}{2}\right]\cdot\frac{1}{2} = \frac{1}{2}[f(0) + f(1/2) + f(1) + f(3/2)]$

   Since $f(x) = 2 + (x-2)^2$, we have $f(0) = 6$, $f(1/2) = 4.25$, $f(1) = 3$,

   and $f(3/2) = 2.25$. Thus $\displaystyle\sum_{i=1}^{n} f(x_i^*)\Delta x_i = \frac{1}{2}(15.5) = 7.75$

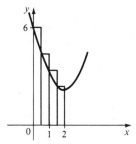

7. By (5.12), $\displaystyle\int_{2}^{4}(3-4x)dx = \lim_{n\to\infty}\frac{2}{n}\sum_{i=1}^{n}\left[3 - 4\left[2+\frac{2i}{n}\right]\right] = \lim_{n\to\infty}\frac{2}{n}\sum_{i=1}^{n}\left[-5 - \frac{8}{n}i\right]$

   $= \lim_{n\to\infty}\frac{2}{n}\left[-5n - \frac{8}{n}\frac{n(n+1)}{2}\right] = \lim_{n\to\infty}\left[-10 - 8\cdot 1\cdot\left[1+\frac{1}{n}\right]\right] = -10-8 = -18$

9. $\displaystyle\int_{0}^{5}(x^3-2x^2)dx = \lim_{n\to\infty}\frac{5}{n}\sum_{i=1}^{n}\left[\left[\frac{5i}{n}\right]^3 - 2\left[\frac{5i}{n}\right]^2\right]$  [by (5.12)]

   $= \lim_{n\to\infty}\left[\frac{625}{n^4}\Sigma i^3 - \frac{250}{n^3}\Sigma i^2\right] = \lim_{n\to\infty}\left[\frac{625}{n^4}\frac{n^2(n+1)^2}{4} - \frac{250}{n^3}\frac{n(n+1)(2n+1)}{6}\right]$

   $= \lim_{n\to\infty}\left[\frac{625}{4}\left[1+\frac{1}{n}\right]^2 - \frac{125}{3}\left[1+\frac{1}{n}\right]\left[2+\frac{1}{n}\right]\right] = \frac{625}{4} - \frac{250}{3} = \frac{875}{12}$

11. $\displaystyle\int_{0}^{5}(x^3-2x^2)dx = \frac{x^4}{4} - \frac{2}{3}x^3\bigg]_0^5 = \frac{625}{4} - \frac{250}{3} = \frac{875}{12}$

   NOTE: We evaluated this integral in #9.

13. $\displaystyle\int_{0}^{1}(1-x^9)dx = x - \frac{x^{10}}{10}\bigg]_0^1 = 1 - \frac{1}{10} = \frac{9}{10}$

**15.** $\int_1^8 \sqrt[3]{x}\,(x-1)dx = \int_1^8 (x^{4/3}-x^{1/3})dx = \frac{3}{7}x^{7/3} - \frac{3}{4}x^{4/3}\Big]_1^8$

$= \left[\frac{3}{7}\cdot 128 - \frac{3}{4}\cdot 16\right] - \left[\frac{3}{7}-\frac{3}{4}\right] = \frac{1209}{28}$

**17.** Let $u = 1+2x^3$. Then $du = 6x^2\,dx$, so $\int_0^2 x^2(1+2x^3)^3 dx$

$= \int_1^{17} u^3\,\frac{1}{6}\,du = \frac{u^4}{24}\Big]_1^{17} = \frac{17^4-1}{24} = 3480$

**19.** Let $u = 2x+3$. Then $du = 2\,dx$, so $\int_3^{11} \frac{dx}{\sqrt{2x+3}} = \int_9^{25} u^{-1/2}\,\frac{1}{2}\,du$

$= u^{1/2}\Big]_9^{25} = 5-3 = 2$

**21.** $\int_{-2}^{-1} \frac{dx}{(2x+3)^4}$ does not exist since the integrand has an infinite

discontinuity at $-3/2$.

**23.** Let $u = 2+x^5$. Then $du = 5x^4\,dx$, so $\int \frac{x^4\,dx}{(2+x^5)^6} = \int u^{-6}\,\frac{1}{5}\,du$

$= \frac{1}{5}\,\frac{u^{-5}}{-5} + C = \frac{-1}{25u^5} + C = \frac{-1}{25(2+x^5)^5} + C$

**25.** Let $u = \pi x$. Then $du = \pi\,dx$, so $\int \sin \pi x\,dx = \int \sin u\,\frac{1}{\pi}\,du$

$= \frac{1}{\pi}(-\cos u) + C = \frac{-1}{\pi}\cos \pi x + C$

**27.** Let $u = 1/t$. Then $du = -\,dt/t^2$, so $\int \cos(1/t)\,t^{-2}\,dt$

$= \int \cos u\,(-du) = -\sin u + C = -\sin(1/t) + C$

**29.** Let $u = \cos x$. Then $du = -\sin x\,dx$, so $\int \sin x\,\sec^2(\cos x)\,dx$

$= \int \sec^2 u\,(-du) = -\tan u + C = -\tan(\cos x) + C$

**31.** $\int_0^{2\pi} |\sin x|dx = \int_0^{\pi} \sin x\,dx - \int_{\pi}^{2\pi} \sin x\,dx$

$= 2\int_0^{\pi} \sin x\,dx = -2\cos x\Big]_0^{\pi} = -2[(-1)-1] = 4$

**33.** $F(x) = \int_1^x \sqrt{1+t^2+t^4}\,dt \Rightarrow F'(x) = \sqrt{1+x^2+x^4}$

**35.** $g(x) = \int_0^{x^3} \frac{t\,dt}{\sqrt{1+t^3}}$. Let $y = g(x)$ and $u = x^3$. Then $g'(x) = \frac{dy}{dx}$

$= \frac{dy}{du}\,\frac{du}{dx} = \frac{u}{\sqrt{1+u^3}}\,3x^2 = \frac{x^3}{\sqrt{1+x^9}}\,3x^2 = \frac{3x^5}{\sqrt{1+x^9}}$

**37.** $y = \int_{\sqrt{x}}^{x} \frac{\cos \theta}{\theta}\,d\theta = \int_1^x \frac{\cos \theta}{\theta}\,d\theta + \int_{\sqrt{x}}^1 \frac{\cos \theta}{\theta}\,d\theta = \int_1^x \frac{\cos \theta}{\theta}\,d\theta$

$- \int_1^{\sqrt{x}} \frac{\cos \theta}{\theta}\,d\theta \Rightarrow y' = \frac{\cos x}{x} - \frac{\cos \sqrt{x}}{\sqrt{x}}\,\frac{1}{2\sqrt{x}} = \frac{2\cos x - \cos \sqrt{x}}{2x}$

39. area $= \int_1^3 [0-(x^2-4x+3)]dx = \int_1^3(-x^2+4x-3)dx = -\frac{x^3}{3} + 2x^2 - 3x\big]_1^3$

    $= (-9+18-9) - (-1/3 +2-3) = 4/3$

41. area $= \int_0^6 [(12x-2x^2) - (x^2-6x)]dx = \int_0^6(18x-3x^2)dx$

    $= 9x^2-x^3\big]_0^6 = 9\cdot36 - 216 = 108$

43. By symmetry, area $= 2\int_0^1(x^{1/3}-x^3)dx = 2\left[\frac{3}{4}x^{4/3}-\frac{x^4}{4}\right]\big]_0^1 = 2\left[\frac{3}{4}-\frac{1}{4}\right] = 1$

45. area $= \int_0^\pi |\sin x - (-\cos x)|dx$

    $= \int_0^{3\pi/4}(\sin x + \cos x)dx - \int_{3\pi/4}^\pi(\sin x + \cos x)dx$

    $= (\sin x - \cos x)\big]_0^{3\pi/4} - (-\cos x + \sin x)\big]_{3\pi/4}^\pi$

    $= (1/\sqrt{2} + 1/\sqrt{2}) - (0-1) - (1+0) + (1/\sqrt{2} + 1/\sqrt{2}) = \sqrt{2}+1-1+\sqrt{2} = 2\sqrt{2}$

47. If $1 \le x \le 3$, then $2 \le \sqrt{x^2+3} \le 2\sqrt{3}$, so

    $2\cdot(3-1) \le \int_1^3 \sqrt{x^2+3}\, dx \le 2\sqrt{3}\cdot(3-1)$; that is, $4 \le \int_1^3 \sqrt{x^2+3}\, dx \le 4\sqrt{3}$

49. $|\cos x| \le 1 \Rightarrow \cos^8 x \le \cos^6 x \Rightarrow \int_0^\pi \cos^8 x\, dx \le \int_0^\pi \cos^6 x\, dx$

51. $0 \le x \le 1 \Rightarrow 0 \le \cos x \le 1 \Rightarrow x^2\cos x \le x^2 \Rightarrow$

    $\int_0^1 x^2\cos x\, dx \le \int_0^1 x^2\, dx = \frac{1}{3}x^3\big]_0^1 = \frac{1}{3}$    [Property 7]

53. $f_{ave} = \frac{1}{4-2} \int_2^4 x^3 dx = \frac{1}{2}\cdot\frac{x^4}{4}\big]_2^4 = \frac{1}{2}(64-4) = 30$

55. Let $u = 1-x$. Then $du = -dx$, so $\int_0^1 f(1-x)dx = \int_1^0 f(u)(-du)$

    $= \int_0^1 f(u)du = \int_0^1 f(x)dx$

57. Let $u = \sqrt{x}$. Then $du = \frac{dx}{2\sqrt{x}}$ so $\int \frac{e^{\sqrt{x}}}{\sqrt{x}} dx = \int e^u\, 2\, du = 2e^u + C$

    $= 2\, e^{\sqrt{x}} + C$

59. Let $u = \ln(\cos x)$. Then $du = -\tan x\, dx$, so $\int \tan x \ln(\cos x)dx$

    $= -\int u\, du = -u^2/2 + C = -\frac{1}{2}[\ln(\cos x)]^2 + C$

61. Let $u = 1+x^4$. Then $du = 4x^3\, dx$, so $\int \frac{x^3\, dx}{1+x^4} = \int \frac{(1/4)du}{u}$

    $= \frac{1}{4}\ln|u| + C = \frac{1}{4}\ln(1+x^4) + C$

# CHAPTER SIX

## Exercises 6.1

1.

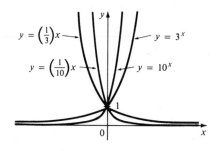

3.  $y = 100^x$

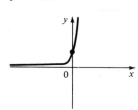

5.  $y = (0.9)^x$

7.  $y = 2^x$

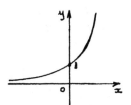

$y = 2^x + 1$

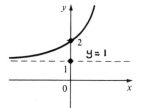

9.  $y = 3^x$    $y = 3^{-x}$

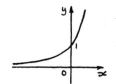

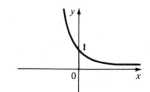

11.  $y = -3^{-x}$

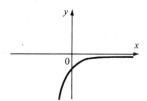

13.  $y = 5^{3x} = (5^3)^x = 125^x$      $y = 5^{-3x} = 125^{-x}$

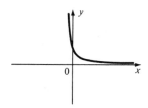

15.  $y = 2^{x-1}$                      $y = -2^{x-1}$                      $y = 3 - 2^{x-1}$

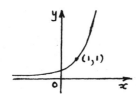     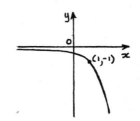     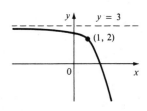

17.  $\lim\limits_{x\to\infty} (1.1)^x = \infty$ by (6.3) since $1.1 > 1$

19.  $\lim\limits_{x\to\infty} \pi^{-x} = \lim\limits_{x\to\infty} \left[\frac{1}{\pi}\right]^x = 0$ by (6.3) since $0 < \frac{1}{\pi} < 1$

    [OR: $\lim\limits_{x\to\infty} \pi^{-x} = 0$ since $\pi > 1$ and $-x \to -\infty$ as $x \to \infty$]

21.  $\lim\limits_{x\to\infty} 5^{2x+5} = \infty$ since $2x+5 \to \infty$ as $x \to \infty$

23.  $\lim\limits_{x\to\infty} \left[2^{-0.8x} + \frac{1}{x}\right] = \lim\limits_{x\to\infty} 2^{-0.8x} + \lim\limits_{x\to\infty} \frac{1}{x} = 0 + 0 = 0$   since $-0.8x \to -\infty$

25.  $\lim\limits_{x\to-\infty} \left[\frac{\pi}{4}\right]^x = \infty$ since $0 < \frac{\pi}{4} < 1$

27.  $\lim\limits_{x\to\pi/2^-} 2^{\tan x} = \infty$ since $\tan x \to \infty$ as $x \to \pi/2^-$

29.  $\lim\limits_{x\to\infty} 3^{1/x} = 3^0 = 1$ since $\frac{1}{x} \to 0$ as $x \to \infty$

31.  $\lim\limits_{x\to 0^+} 3^{1/x} = \infty$ since $\frac{1}{x} \to \infty$ as $x \to 0^+$

33.  Divide numerator and denominator by $10^x$:

    $\lim\limits_{x\to\infty} \dfrac{10^x}{10^x+1} = \lim\limits_{x\to\infty} \dfrac{1}{1+10^{-x}} = \dfrac{1}{1+0} = 1$

35.  $\lim\limits_{t\to 0^+} \pi^{\csc t} = \infty$ since $\csc t \to \infty$ as $t \to 0^+$

**Section 6.2**

**Exercises 6.2**

1.  Not one-to-one        3.  One-to-one            5.  Not one-to-one

7.  $x_1 \neq x_2 \Rightarrow 7x_1 \neq 7x_2 \Rightarrow 7x_1 - 3 \neq 7x_2 - 3 \Rightarrow f(x_1) \neq f(x_2)$, so $f$ is 1-1.

9. $x_1 \neq x_2 \Rightarrow \sqrt{x_1} \neq \sqrt{x_2} \Rightarrow g(x_1) \neq g(x_2)$, so g is 1-1.

11. $h(x) = x^4 + 5 \Rightarrow h(1) = 6 = h(-1)$, so h is not 1-1.

13. $x_1 \neq x_2 \Rightarrow 4x_1 \neq 4x_2 \Rightarrow 4x_1 + 7 \neq 4x_2 + 7 \Rightarrow f(x_1) \neq f(x_2)$, so f is 1-1.
    $y = 4x + 7 \Rightarrow 4x = y - 7 \Rightarrow x = (y-7)/4$. Interchange x and y:
    $y = (x-7)/4$. So $f^{-1}(x) = (x-7)/4$.

15. $f(x) = \dfrac{1+3x}{5-2x}$ If $f(x_1) = f(x_2)$, then $\dfrac{1+3x_1}{5-2x_1} = \dfrac{1+3x_2}{5-2x_2} \Rightarrow$

    $5 + 15x_1 - 2x_2 - 6x_1x_2 = 5 - 2x_1 + 15x_2 - 6x_1x_2 \Rightarrow 17x_1 = 17x_2 \Rightarrow x_1 = x_2$, so

    f is one-to-one. $y = \dfrac{1+3x}{5-2x} \Rightarrow 5y - 2xy = 1 + 3x \Rightarrow x(3+2y) = 5y - 1 \Rightarrow$

    $x = \dfrac{5y-1}{2y+3}$ Interchange x and y: $y = \dfrac{5x-1}{2x+3}$ So $f^{-1}(x) = \dfrac{5x-1}{2x+3}$

17. $x_1 \neq x_2 \Rightarrow 5x_1 \neq 5x_2 \Rightarrow 2 + 5x_1 \neq 2 + 5x_2 \Rightarrow \sqrt{2+5x_1} \neq \sqrt{2+5x_2} \Rightarrow$
    $f(x_1) \neq f(x_2)$, so f is 1-1. $y = \sqrt{2+5x} \Rightarrow y^2 = 2 + 5x$ and $y \geq 0 \Rightarrow$
    $5x = y^2 - 2 \Rightarrow x = (y^2-2)/5$, $y \geq 0$. Interchange x and y:
    $y = (x^2-2)/5$, $x \geq 0$. So $f^{-1}(x) = (x^2-2)/5$, $x \geq 0$.

19. $f(x) = 2x + 1$ (a) $x_1 \neq x_2 \Rightarrow 2x_1 \neq 2x_2 \Rightarrow 2x_1 + 1 \neq 2x_2 + 1 \Rightarrow$
    $f(x_1) \neq f(x_2)$, so f is 1-1. (b) $f(1) = 3 \Rightarrow g(3) = 1$. Also
    $f'(x) = 2$, so $g'(3) = 1/f'(1) = 1/2$ (c) $y = 2x + 1 \Rightarrow x = (y-1)/2$.
    Interchanging x and y gives
    $y = (x-1)/2$, so $f^{-1}(x) = (x-1)/2$
    Domain(g) = range(f) = R.
    Range(g) = dom(f) = R.
    (d) $g(x) = (x-1)/2 \Rightarrow g'(x) = 1/2$
    $\Rightarrow g'(3) = 1/2$ as in (b).

    (e)

21. (a) $x_1 \neq x_2 \Rightarrow x_1^3 \neq x_2^3 \Rightarrow f(x_1) \neq f(x_2)$, so f is one-to-one.
    (b) $f'(x) = 3x^2$ and $f(2) = 8 \Rightarrow g(8) = 2$, so $g'(8) = 1/f'(g(8))$
    $= 1/f'(2) = 1/12$ (c) $y = x^3 \Rightarrow x = \sqrt[3]{y}$. Interchanging x and y
    gives $y = \sqrt[3]{x}$, so $f^{-1}(x) = \sqrt[3]{x}$. (e)
    Domain(g) = range(f) = R.
    Range(g) = domain(f) = R.
    (d) $g(x) = \sqrt[3]{x} \Rightarrow$
    $g'(x) = (1/3)x^{-2/3} \Rightarrow g'(8) =$
    $(1/3)(1/4) = 1/12$ as in (b).

23. $f(x) = 9 - x^2$, $0 \leq x \leq 3$. (a) Since $x \geq 0$, $x_1 \neq x_2 \Rightarrow x_1^2 \neq x_2^2 \Rightarrow$
    $9 - x_1^2 \neq 9 - x_2^2 \Rightarrow f(x_1) \neq f(x_2)$, so f is 1-1. (b) $f'(x) = -2x$ and
    $f(1) = 8 \Rightarrow g(8) = 1$, so $g'(8) = 1/f'(g(8)) = 1/f'(1) = 1/(-2) = -1/2$.

(c) $y = 9-x^2$ ⇒ $x^2 = 9-y$       (e)

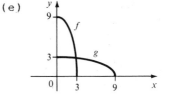

⇒ $x = \sqrt{9-y}$. Interchange x and y:

$y = \sqrt{9-x}$, so $f^{-1}(x) = \sqrt{9-x}$.

Domain(g) = range(f) = [0,9]

Range(g) = domain(f) = [0,3]

(d) $g'(x) = -1/2\sqrt{9-x}$ ⇒

$g'(8) = -1/2$ as in (b).

25.   $y = \sqrt[n]{x}$ ⇒ $y^n = x$ ⇒ $ny^{n-1}y' = 1$ ⇒ $y' = \dfrac{1}{ny^{n-1}} = \dfrac{1}{n\sqrt[n]{x}^{n-1}} = \dfrac{1}{n}x^{1/n-1}$

27.   Suppose that f is increasing. If $x_1 \neq x_2$, then either $x_1 < x_2$ or $x_2 < x_1$. If $x_1 < x_2$, then $f(x_1) < f(x_2)$. If $x_2 < x_1$, then $f(x_2) < f(x_1)$. In either case, $x_1 \neq x_2$ ⇒ $f(x_1) \neq f(x_2)$, so f is one-to-one.

**Section 6.3**

**Exercises 6.3**

1.   $\log_2 64 = 6$ since $2^6 = 64$      3.   $\log_8 2 = 1/3$ since $8^{1/3} = 2$

5.   $\log_3 \dfrac{1}{27} = -3$ since $3^{-3} = \dfrac{1}{27}$    7.   $\ln e^{\sqrt{2}} = \sqrt{2}$

9.   $\log_{10} 1.25 + \log_{10} 80 = \log_{10} (1.25 \cdot 80) = \log_{10} 100 = 2$

11.  $\log_8 6 - \log_8 3 + \log_8 4 = \log_8 \dfrac{6 \cdot 4}{3} = \log_8 8 = 1$

13.  $2^{(\log_2 3 + \log_2 5)} = 2^{\log_2 15} = 15$

15.  $\log_5 a + \log_5 b - \log_5 c = \log_5 \dfrac{ab}{c}$

17.  $(1/3)\ln x - 4 \ln(2x+3) = \ln(x^{1/3}) - \ln(2x+3)^4 = \ln[\sqrt[3]{x}/(2x+3)^4]$

**19.**

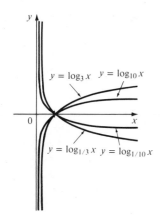

**21.** $y = \log_{1.1} x$

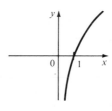

**23.** $y = \log_{0.1} x$

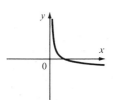

**25.** $y = \log_{10} x$ $\qquad$ $y = \log_{10}(x+5)$

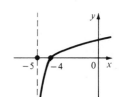

**27.** $y = \ln x$ $\qquad$ $y = -\ln x$

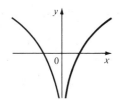

**29.** $y = \ln(-x)$ $\qquad$ $y = -\ln(-x)$

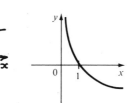

**31.** $y = \ln(x^2) = 2\ln|x|$

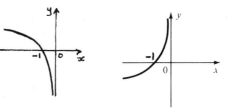

**33.** $y = \ln x$ $\qquad$ $y = \ln(x+3)$

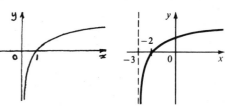

35. $y = 2e^{-3x}$

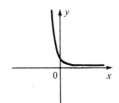

37. $y = -e^{2x}$     $y = 1-e^{-2x}$

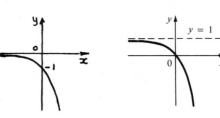

39. $\log_2 x = 3 \;\Rightarrow\; x = 2^3 = 8$

41. $y = \log_5(x-1) \;\Rightarrow\; 5^y = x-1 \;\Rightarrow\; x = 1+5^y$

43. $3^{x+2} = m \;\Rightarrow\; \log_3 m = x+2 \;\Rightarrow\; x = \log_3 m - 2$

45. $\ln x = 2 \;\Rightarrow\; x = e^{\ln x} = e^2$

47. $\ln x = \ln 5 + \ln 8 = \ln 40 \;\Rightarrow\; x = 40$

49. $5 = \ln(e^{2x-1}) = 2x-1 \;\Rightarrow\; x = 3$

51. $y = \ln(\ln x) \;\Rightarrow\; e^y = e^{\ln(\ln x)} = \ln x \;\Rightarrow\; e^{e^y} = e^{\ln x} = x$

53. Let $\log_a x = r$ and $\log_a y = s$. Then $a^r = x$ and $a^s = y$.

    (a) $xy = a^r a^s = a^{r+s} \;\Rightarrow\; \log_a(xy) = r+s = \log_a x + \log_a y$

    (b) $\dfrac{x}{y} = \dfrac{a^r}{a^s} = a^{r-s} \;\Rightarrow\; \log_a\left[\dfrac{x}{y}\right] = r-s = \log_a x - \log_a y$

    (c) $x^y = (a^r)^y = a^{ry} \;\Rightarrow\; \log_a(x^y) = ry = y \log_a x$

55. Take $a = e$ in Exercise 54(a): $\log_e b \, \log_b c = \log_e c \;\Rightarrow\;$
    $\ln b \, \log_b c = \ln c \;\Rightarrow\; \log_b c = (\ln c)/(\ln b)$

57. If $I$ is the intensity of the San Francisco earthquake, then
    $\log_{10}(I/S) = 8.3 \;\Rightarrow\; \log_{10}(4I/S) = \log_{10} 4 + \log_{10}(I/S)$
    $= \log_{10} 4 + 8.3 \approx 8.9$

59. $\displaystyle\lim_{x\to 5^+} \ln(x-5) = -\infty$ since $x-5 \to 0^+$ as $x \to 5^+$

61. $\displaystyle\lim_{x\to\infty} \log_2(x^2-x) = \infty$ since $x^2-x \to \infty$ as $x \to \infty$

63. Divide numerator and denominator by $e^{3x}$:
    $\displaystyle\lim_{x\to\infty} \frac{e^{3x}-e^{-3x}}{e^{3x}+e^{-3x}} = \lim_{x\to\infty} \frac{1-e^{-6x}}{1+e^{-6x}} = \frac{1-0}{1+0} = 1$

65. $\displaystyle\lim_{x\to\pi/2^-} \log_{10}(\cos x) = -\infty$ since $\cos x \to 0^+$ as $x \to \pi/2^-$

67. $\displaystyle\lim_{x\to 1^-} e^{2/(x-1)} = 0$ since $\dfrac{2}{x-1} \to -\infty$ as $x \to 1^-$

69. $\displaystyle\lim_{x\to \pi/2^-} \dfrac{2}{1 + e^{\tan x}} = 0$ since $\tan x \to \infty \Rightarrow e^{\tan x} \to \infty$

71. $\displaystyle\lim_{x\to\infty} [\ln(2+x)-\ln(1+x)] = \lim_{x\to\infty} \ln\dfrac{2+x}{1+x} = \lim_{x\to\infty} \ln\dfrac{2/x + 1}{1/x + 1} = \ln 1 = 0$

73. $\displaystyle\lim_{x\to\infty} \dfrac{e^x}{e^{2x}+e^{-x}} = \lim_{x\to\infty} \dfrac{e^{-x}}{1+e^{-3x}} = \dfrac{0}{1+0} = 0$

75. $f(x) = \log_{10}(1-x)$   Domain = $\{x\,|\,1-x>0\} = \{x\,|\,x<1\} = (-\infty,1)$   Range = R

77. $F(t) = \sqrt{t}\ln(t^2-1)$   Domain = $\{t\,|\ t\geq 0$ and $t^2-1>0\} = \{t\,|\ t>1\} = (1,\infty)$
    Range = R

79. $y = \ln(x+3) \Rightarrow e^y = e^{\ln(x+3)} = x+3 \Rightarrow x = e^y-3$.   Interchange x
    and y: the inverse function is $y = e^x-3$.

81. $y = e^{\sqrt{x}} \Rightarrow \ln y = \ln e^{\sqrt{x}} = \sqrt{x} \Rightarrow x = (\ln y)^2$.   Interchange x and
    y: the inverse function is $y = (\ln x)^2$.

83. $y = \dfrac{10^x}{10^x+1} \Rightarrow 10^x y+y = 10^x \Rightarrow 10^x(1-y) = y \Rightarrow 10^x = \dfrac{y}{1-y} \Rightarrow$

    $x = \log_{10}\left[\dfrac{y}{1-y}\right]$   Interchange x and y: $y = \log_{10}\left[\dfrac{x}{1-x}\right]$

85. (a) Let $\epsilon > 0$ be given. We need N such that $|a^x-0| < \epsilon$ when $x < N$.
    But $a^x < \epsilon \iff x < \log_a \epsilon$. Let $N = \log_a \epsilon$. Then $x < N \Rightarrow$

    $x < \log_a \epsilon \Rightarrow |a^x-0| = a^x < \epsilon$, so $\displaystyle\lim_{x\to-\infty} a^x = 0$.

    (b) Let $M > 0$ be given. We need N such that $a^x > M$ when $x > N$.
    But $a^x > M \iff x > \log_a M$. Let $N = \log_a M$. Then $x > N \Rightarrow$

    $x > \log_a M \Rightarrow a^x > M$, so $\displaystyle\lim_{x\to\infty} a^x = \infty$.

87. Let $t = 5x$. Then $x = t/5$, so $\displaystyle\lim_{x\to 0} (1+5x)^{1/x} = \lim_{t\to 0} (1+t)^{5/t}$

    $= \displaystyle\lim_{t\to 0} [(1+t)^{1/t}]^5 = \left[\lim_{t\to 0} (1+t)^{1/t}\right]^5 = e^5$   by the definition of e.

## Exercises 6.4

1. $f(x) = \ln(x+1) \Rightarrow f'(x) = 1/(x+1)$. $\text{Dom}(f) = \text{dom}(f') = \{x \mid x+1>0\}$
$= \{x \mid x>-1\} = (-1,\infty)$. [Note that, in general, $\text{dom}(f') \subset \text{dom}(f)$.]

3. $f(x) = \log_3(x^2-4) \Rightarrow f'(x) = \dfrac{1}{x^2-4}(\log_3 e)(2x) = \dfrac{2x}{(\ln 3)(x^2-4)}$
$\text{Dom}(f) = \text{dom}(f') = \{x \mid x^2-4>0\} = \{x \mid |x|>2\} = (-\infty,-2) \cup (2,\infty)$

5. $f(x) = \ln(\cos x) \Rightarrow f'(x) = \dfrac{1}{\cos x}(- \sin x) = - \tan x$
$\text{Dom}(f) = \text{dom}(f') = \{x \mid \cos x > 0\} = (-\pi/2,\pi/2) \cup (3\pi/2,5\pi/2) \cup \cdots$
$= \{x \mid (4n-1)\pi/2<x<(4n+1)\pi/2, \ n = 0,\pm1,\pm2,\ldots\}$

7. $f(x) = \ln(2-x-x^2) \Rightarrow f'(x) = (-1-2x)/(2-x-x^2)$ $\text{Dom}(f) = \text{dom}(f') =$
$\{x \mid 2-x-x^2>0\} = \{x \mid (2+x)(1-x)>0\} = \{x \mid -2<x<1\} = (-2,1)$

9. $f(x) = x^2\ln(1-x^2) \Rightarrow f'(x) = 2x \ln(1-x^2) + x^2[1/(1-x^2)](-2x)$
$= 2x \ln(1-x^2) - 2x^3/(1-x^2)$ $\text{Dom}(f) = \text{dom}(f') = \{x \mid 1-x^2>0\}$
$= \{x \mid |x|<1\} = (-1,1)$

11. $y = \log_{10} x \Rightarrow y' = \dfrac{1}{x} \log_{10} e \Rightarrow y'' = - \dfrac{1}{x^2} \log_{10} e$

[OR: $y' = 1/x \ln 10, \quad y'' = -1/x^2 \ln 10$]

13. $y = x \ln x \Rightarrow y' = \ln x + x(1/x) = \ln x + 1 \Rightarrow y'' = 1/x$

15. $f(x) = \sqrt{x} \ln x \Rightarrow f'(x) = (1/2\sqrt{x}) \ln x + \sqrt{x}(1/x) = (\ln x + 2)/2\sqrt{x}$

17. $g(x) = \ln \dfrac{a-x}{a+x} = \ln(a-x) - \ln(a+x) \Rightarrow g'(x) = \dfrac{-1}{a-x} - \dfrac{1}{a+x} = \dfrac{-2a}{a^2-x^2}$

19. $F(x) = \ln\sqrt{x} = (1/2) \ln x \Rightarrow F'(x) = (1/2)(1/x) = 1/2x$

21. $f(t) = \log_2(t^4-t^2+1) \Rightarrow f'(t) = [1/(t^4-t^2+1)](\log_2 e)(4t^3-2t)$
$= (4t^3-2t)/(\ln 2)(t^4-t^2+1)$

23. $h(y) = \ln(y^3\sin y) = 3 \ln y + \ln \sin y \Rightarrow$
$h'(y) = 3/y + (1/\sin y)(\cos y) = 3/y + \cot y$

25. $g(u) = \dfrac{1 - \ln u}{1 + \ln u} \Rightarrow g'(u) = \dfrac{(1 + \ln u)(-1/u) - (1 - \ln u)(1/u)}{(1 + \ln u)^2}$
$= -2/u(1 + \ln u)^2$

27. $y = (\ln \sin x)^3 \Rightarrow y' = 3(\ln \sin x)^2 \dfrac{\cos x}{\sin x} = 3(\ln \sin x)^2 \cot x$

29. $y = \dfrac{\ln x}{1+x^2} \Rightarrow y' = \dfrac{(1+x^2)(1/x) - 2x \ln x}{(1+x^2)^2} = \dfrac{1+x^2-2x^2\ln x}{x(1+x^2)^2}$

31. $y = \ln\left[\dfrac{x+1}{x-1}\right]^{3/5} = \dfrac{3}{5}[\ln(x+1)-\ln(x-1)] \Rightarrow y' = \dfrac{3}{5}\left[\dfrac{1}{x+1} - \dfrac{1}{x-1}\right] = \dfrac{-6}{5(x^2-1)}$

33. $y = \ln|x^3 - x^2| \quad \Rightarrow \quad y' = \dfrac{1}{x^3 - x^2}(3x^2 - 2x) = \dfrac{x(3x-2)}{x^2(x-1)} = \dfrac{3x-2}{x(x-1)}$

35. $y = \ln(x + \ln x) \quad \Rightarrow \quad y' = \dfrac{1}{x + \ln x}\left(1 + \dfrac{1}{x}\right)$

37. $\displaystyle\int_4^8 \dfrac{1}{x}\,dx = \ln x\Big]_4^8 = \ln 8 - \ln 4 = \ln \dfrac{8}{4} = \ln 2$

39. $\displaystyle\int_1^e \dfrac{x^2 + x + 1}{x}\,dx = \int_1^e \left(x + 1 + \dfrac{1}{x}\right)dx = \dfrac{x^2}{2} + x + \ln x\Big]_1^e$

$= \left(\dfrac{1}{2}e^2 + e + 1\right) - \left(\dfrac{1}{2} + 1 + 0\right) = \dfrac{1}{2}e^2 + e - \dfrac{1}{2}$

41. Let $u = 2x-1$. Then $du = 2\,dx$, so $\displaystyle\int \dfrac{dx}{2x-1} = \dfrac{1}{2}\int \dfrac{du}{u} = \dfrac{1}{2}\ln|u| + C$

$= (1/2)\ln|2x-1| + C$

43. Let $u = x^2 + 2x$. Then $du = 2(x+1)dx$, so $\displaystyle\int \dfrac{x+1}{x^2 + 2x}\,dx = \dfrac{1}{2}\int \dfrac{du}{u}$

$= (1/2)\ln|u| + C = (1/2)\ln|x^2 + 2x| + C$

45. Let $u = \ln x$. Then $du = \dfrac{dx}{x} \quad \Rightarrow \quad \displaystyle\int \dfrac{(\ln x)^2}{x}\,dx = \int u^2\,du = \dfrac{1}{3}u^3 + C$

$= (1/3)(\ln x)^3 + C$

47. Let $u = 1 + \cos x$. Then $du = -\sin x\,dx$, so
$\displaystyle\int \dfrac{\sin x}{1 + \cos x}\,dx = -\int \dfrac{du}{u} = -\ln|u| + C = -\ln(1 + \cos x) + C$

49. (a) $\dfrac{d}{dx}(\ln|\sin x| + C) = \dfrac{1}{\sin x}\cos x = \cot x$

(b) Let $u = \sin x$. Then $du = \cos x\,dx$, so $\displaystyle\int \cot x\,dx = \int \dfrac{\cos x}{\sin x}\,dx$

$= \displaystyle\int \dfrac{du}{u} = \ln|u| + C = \ln|\sin x| + C$

51. $y = (3x-7)^4(8x^2-1)^3 \quad \Rightarrow \quad \ln|y| = 4\ln|3x-7| + 3\ln|8x^2-1| \quad \Rightarrow$

$\dfrac{y'}{y} = \dfrac{12}{3x-7} + \dfrac{48x}{8x^2-1} \quad \Rightarrow \quad y' = (3x-7)^4(8x^2-1)^3\left[\dfrac{12}{3x-7} + \dfrac{48x}{8x^2-1}\right]$

53. $y = (x+1)^4(x-5)^3/(x-3)^8 \quad \Rightarrow \quad \ln|y| = 4\ln|x+1| + 3\ln|x-5| - 8\ln|x-3|$

$\Rightarrow \quad \dfrac{y'}{y} = \dfrac{4}{x+1} + \dfrac{3}{x-5} - \dfrac{8}{x-3} \quad \Rightarrow \quad y' = \dfrac{(x+1)^4(x-5)^3}{(x-3)^8}\left[\dfrac{4}{x+1} + \dfrac{3}{x-5} - \dfrac{8}{x-3}\right]$

55. $y = \dfrac{e^x\sqrt{x^5+2}}{(x+1)^4(x^2+3)^2} \quad \Rightarrow \quad \ln y = x + \dfrac{1}{2}\ln(x^5+2) - 4\ln|x+1| - 2\ln(x^2+3)$

$\Rightarrow \quad \dfrac{y'}{y} = 1 + \dfrac{5x^4}{2(x^5+2)} - \dfrac{4}{x+1} - \dfrac{4x}{x^2+3} \quad \Rightarrow$

$y' = \dfrac{e^x\sqrt{x^5+2}}{(x+1)^4(x^2+3)^2}\left[1 + \dfrac{5x^4}{2(x^5+2)} - \dfrac{4}{x+1} - \dfrac{4x}{x^2+3}\right]$

**57.** $y = \ln(x^2+y^2) \Rightarrow y' = \dfrac{2x+2yy'}{x^2+y^2} \Rightarrow x^2y'+y^2y' = 2x+2yy' \Rightarrow y' = \dfrac{2x}{x^2+y^2-2y}$

**59.** $f(x) = \dfrac{x}{\ln x} \Rightarrow f'(x) = \dfrac{\ln x - x(1/x)}{(\ln x)^2} = \dfrac{\ln x - 1}{(\ln x)^2} \Rightarrow f'(e) = \dfrac{1-1}{1^2} = 0$

**61.** $y = f(x) = \ln \ln x \Rightarrow f'(x) = (1/\ln x)(1/x) \Rightarrow f'(e) = 1/e$, so the equation of the tangent at $(e,0)$ is $y-0 = \dfrac{1}{e}(x-e)$ or $x-ey = e$.

**63.** $f(x) = \ln(x-1) \Rightarrow f'(x) = 1/(x-1) = (x-1)^{-1} \Rightarrow f''(x) = -(x-1)^{-2}$
$\Rightarrow f'''(x) = 2(x-1)^{-3} \Rightarrow f^{(4)}(x) = -2\cdot 3(x-1)^{-4} \Rightarrow \cdots \Rightarrow$
$f^{(n)}(x) = (-1)^{n-1}2\cdot 3\cdot 4\cdots(n-1)(x-1)^{-n} = (-1)^{n-1}(n-1)!/(x-1)^n$

**65.** $f(x) = 2x + \ln x \Rightarrow f'(x) = 2 + 1/x$. If $g = f^{-1}$, then $f(1) = 2 \Rightarrow$
$g(2) = 1$, so $g'(2) = 1/f'(g(2)) = 1/f'(1) = 1/3$.

**67.** $f(x) = (\ln x)/\sqrt{x} \Rightarrow f'(x) = \dfrac{\sqrt{x}(1/x)-(\ln x)(1/2\sqrt{x})}{x} = \dfrac{2 - \ln x}{2x^{3/2}} \Rightarrow$

$f''(x) = \dfrac{2x^{3/2}(-1/x) - (2 - \ln x)(3x^{1/2})}{4x^3} = \dfrac{3 \ln x - 8}{4x^{5/2}} > 0 \iff$

$\ln x > 8/3 \iff x > e^{8/3}$, so $f$ is CU on $(e^{8/3},\infty)$ and CD on $(0,e^{8/3})$
The inflection point is $(e^{8/3},(8/3)e^{-4/3})$.

**69.** $y = f(x) = \ln(\cos x)$  A. $D = \{x\mid \cos x > 0\} = (-\pi/2,\pi/2) \cup$
$(3\pi/2,5\pi/2) \cup \cdots = \{x\mid 2n\pi-\pi/2 < x < 2n\pi+\pi/2, n = 0,\pm 1,\pm 2,\cdots\}$
B. x-intercepts occur when $\ln(\cos x) = 0 \iff \cos x = 1 \iff x = 2n\pi$,
y-intercept $= f(0) = 0$  C. $f(-x) = f(x)$, so the curve is symmetric
about the y-axis. $f(x+2\pi) = f(x)$, $f$ has period $2\pi$, so in D-G we
consider only $-\pi/2<x<\pi/2$.  D. $\lim\limits_{x\to\pi/2^-} \ln(\cos x) = -\infty,$

$\lim\limits_{x\to-\pi/2^+} \ln(\cos x) = -\infty$, so $x = \pi/2$ and $x = -\pi/2$ are VA.  No HA.

E. $f'(x) = (1/\cos x)(-\sin x)$  
$= -\tan x > 0 \iff -\pi/2<x<0$  
so $f$ is increasing on $(-\pi/2,0]$  
and decreasing on $[0,\pi/2)$.  
F. $f(0) = 0$ is a local maximum  
G. $f''(x) = -\sec^2 x < 0 \Rightarrow f$ is  
CD on $(-\pi/2,\pi/2)$. No IP.

H.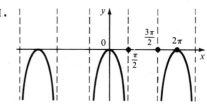

**71.** $y = f(x) = x + \ln x$  A. $D = \{x\mid x>0\} = (0,\infty)$  B. No y-intercept
C. No symmetry  D. $\lim\limits_{x\to\infty} (x + \ln x) = \infty$, no HA.  $\lim\limits_{x\to 0^+} (x + \ln x) = -\infty,$
so $x = 0$ is a VA.

E. $f'(x) = 1 + 1/x > 0$, so $f$
is increasing on $(0,\infty)$
F. No extrema
G. $f''(x) = -1/x^2 < 0$, so $f$ is
CD on $(0,\infty)$.  No IP.

H.

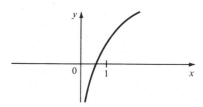

73.  $y = f(x) = \ln(x+\sqrt{1+x^2})$  A. $x+\sqrt{1+x^2} > 0$ for all $x$ since $1+x^2 > x^2$ $\Rightarrow$ $\sqrt{1+x^2} > |x|$, so $D = R$.  B. $y$-intercept $= f(0) = 0$, $x$-intercept occurs when $x+\sqrt{1+x^2} = 1$ $\Rightarrow$ $\sqrt{1+x^2} = 1-x$ $\Rightarrow$ $1+x^2 = 1-2x+x^2$ $\Rightarrow$ $x = 0$

C. $\ln(-x+\sqrt{1+x^2}) = -\ln(x+\sqrt{1+x^2})$ since $(\sqrt{1+x^2}-x)(\sqrt{1+x^2}+x) = 1$, so the curve is symmetric about the origin.  D. $\lim\limits_{x\to\infty} \ln(x+\sqrt{1+x^2}) = \infty$,

$\lim\limits_{x\to-\infty} \ln(x+\sqrt{1+x^2}) = \lim\limits_{x\to-\infty} \ln \dfrac{1}{\sqrt{1+x^2} - x} = -\infty$, no HA.

E. $f'(x) = \dfrac{1}{x + \sqrt{1+x^2}} \left[ 1 + \dfrac{x}{\sqrt{1+x^2}} \right] = \dfrac{1}{\sqrt{1+x^2}} > 0$, so $f$ is increasing on

$(-\infty,\infty)$.  F. No extrema

G. $f''(x) = -\dfrac{x}{(1+x^2)^{3/2}}$

$\Rightarrow$ $f''(x) > 0$ $\Leftrightarrow$ $x < 0$, so
$f$ is CU on $(-\infty,0)$, CD on $(0,\infty)$
and the IP is $(0,0)$.

H.

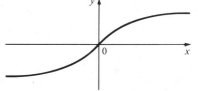

75.  $y = f(x) = \ln(1+x^2)$  A. $D = R$  B. Both intercepts are 0.
C. $f(-x) = f(x)$, so the curve is symmetric about the $y$-axis.
D. $\lim\limits_{x\to\pm\infty} \ln(1+x^2) = \infty$, no asymptotes  E. $f'(x) = \dfrac{2x}{1+x^2} > 0$ $\Leftrightarrow$ $x > 0$

so $f$ is increasing on $[0,\infty)$ and decreasing on $(-\infty,0]$.  F. $f(0) = 0$
is a local and absolute minimum.

G. $f''(x) = \dfrac{2(1+x^2)-2x(2x)}{(1+x^2)^2}$

$= \dfrac{2(1-x^2)}{(1+x^2)^2} > 0$ $\Leftrightarrow$ $|x| < 1$, so $f$ is

CU on $(-1,1)$, CD on $(-\infty,-1)$ and
$(1,\infty)$.  IP $(1,\ln 2)$ and $(-1,\ln 2)$

H.

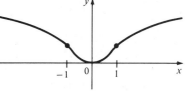

77. The domain of $f(x) = 1/x$ is $(-\infty,0) \cup (0,\infty)$, so its general
antiderivative is $F(x) = \begin{cases} \ln x + C_1 & \text{if } x > 0 \\ \ln|x| + C_2 & \text{if } x < 0 \end{cases}$

79. Let $f(x) = \ln(1+x) - x + x^2/2$. Then $f'(x) = \dfrac{1}{1+x} - 1 + x = \dfrac{x^2}{1+x} > 0$
for $x > 0$, so f is increasing on $(0,\infty)$. Thus, for $x > 0$, we have
$f(x) > f(0) = 0 \Rightarrow \ln(1+x) > x - x^2/2$.

Section 6.5

Exercises 6.5

1. $f(x) = e^{\sqrt{x}} \Rightarrow f'(x) = e^{\sqrt{x}}/2\sqrt{x}$  $\text{Dom}(f) = \{x \mid x \geq 0\} = [0,\infty)$
$\text{Dom}(f') = \{x \mid x > 0\} = (0,\infty)$

3. $f(x) = 7^{x^4} \Rightarrow f'(x) = 7^{x^4}(\ln 7)(4x^3) = (4 \ln 7)x^3 7^{x^4}$
$\text{Dom}(f) = \text{dom}(f') = R$

5. $f(x) = \sqrt{2}^{\sin x} \Rightarrow f'(x) = \sqrt{2}^{\sin x}(\ln\sqrt{2}) \cos x$  $\text{Dom}(f) = \text{dom}(f') = R$

7. $f(x) = (\ln x)^{\sqrt{5}} \Rightarrow f'(x) = \sqrt{5}(\ln x)^{\sqrt{5}-1}/x$
$\text{Dom}(f) = \text{dom}(f') = \{x \mid \ln x > 0\} = \{x \mid x > 1\} = (1,\infty)$

9. $y = e^{-x} \Rightarrow y' = -e^{-x} \Rightarrow y'' = e^{-x}$

11. $y = e^{\tan x} \Rightarrow y' = e^{\tan x} \sec^2 x \Rightarrow y'' = e^{\tan x} \sec^2 x \sec^2 x +$
$e^{\tan x}(2 \sec x)(\sec x \tan x) = e^{\tan x} \sec^2 x (2 \tan x + \sec^2 x)$

13. $f(x) = 2^{x^2} \Rightarrow f'(x) = 2^{x^2}(\ln 2)(2x) = (\ln 2)x2^{x^2+1}$

15. $F(x) = e^x \ln x \Rightarrow F'(x) = e^x \ln x + e^x(1/x) = e^x(\ln x + 1/x)$

17. $f(t) = \pi^{-t} \Rightarrow f'(t) = \pi^{-t}(\ln \pi)(-1) = -\pi^{-t} \ln \pi$

19. $h(t) = t^3 - 3^t \Rightarrow h'(t) = 3t^2 - 3^t \ln 3$

21. $y = xe^{2x} \Rightarrow y' = e^{2x} + xe^{2x}(2) = e^{2x}(1+2x)$

23. $y = 2^{3^x} \Rightarrow y' = 2^{3^x}(\ln 2) 3^x \ln 3 = (\ln 2)(\ln 3) 3^x 2^{3^x}$

25. $y = \sin(3^x) \Rightarrow y' = \cos(3^x) 3^x \ln 3$

27. $y = \ln|e^{2x}-2| \Rightarrow y' = [1/(e^{2x}-2)]e^{2x}(2) = 2e^{2x}/(e^{2x}-2)$

29. $y = e^{-1/x} \Rightarrow y' = e^{-1/x}/x^2$

31. $y = e^{x+e^x} \Rightarrow y' = e^{x+e^x}(1+e^x)$

33. $y = \ln[e^{-x}(1+x)] = \ln(e^{-x}) + \ln(1+x) = -x + \ln(1+x) \Rightarrow$

$y' = -1 + 1/(1+x) = -x/(1+x)$

35. $y = x^{\sin x} \Rightarrow \ln y = \sin x \ln x \Rightarrow y'/y = \cos x \ln x + (\sin x)/x$

$\Rightarrow y' = x^{\sin x}[\cos x \ln x + (\sin x)/x]$

37. $y = x^{e^x} \Rightarrow \ln y = e^x \ln x \Rightarrow y'/y = e^x \ln x + e^x/x \Rightarrow$

$y' = x^{e^x} e^x(\ln x + 1/x)$

39. $y = (\ln x)^x \Rightarrow \ln y = x \ln \ln x \Rightarrow y'/y = \ln \ln x + x \cdot \dfrac{1}{\ln x} \cdot \dfrac{1}{x} \Rightarrow$

$y' = (\ln x)^x(\ln \ln x + 1/\ln x)$

41. $y = x^{1/\ln x} \Rightarrow \ln y = (1/\ln x) \ln x = 1 \Rightarrow y = e \Rightarrow y' = 0$

43. $y = x^{1/x} \Rightarrow \ln y = (1/x) \ln x \Rightarrow y'/y = (-1/x^2)\ln x + (1/x)(1/x)$

$\Rightarrow y' = x^{1/x}(1 - \ln x)/x^2$

45. $y = (x^x)^x = x^{x^2} \Rightarrow \ln y = x^2 \ln x \Rightarrow y'/y = 2x \ln x + x^2(1/x) \Rightarrow$

$y' = x^{x^2} x(2 \ln x + 1) = x^{x^2+1}(2 \ln x + 1)$

47. $\cos(x-y) = xe^x \Rightarrow -\sin(x-y)(1-y') = e^x+xe^x \Rightarrow y' = 1 + \dfrac{e^x(1+x)}{\sin(x-y)}$

49. $g(t) = \sin(1-e^{2t}) \Rightarrow g'(t) = \cos(1-e^{2t})(-2e^{2t}) \Rightarrow g'(0) = 1(-2) = -2$

51. $y = f(x) = e^{-x}\sin x \Rightarrow f'(x) = -e^{-x}\sin x + e^{-x}\cos x \Rightarrow f'(\pi) =$

$e^{-\pi}(\cos \pi - \sin \pi) = -e^{-\pi}$, so the equation of the tangent at $(\pi,0)$

is $y-0 = -e^{-\pi}(x-\pi)$ or $x+e^{\pi}y = \pi$.

53. $y = f(x) = \ln(1+e^x) \Rightarrow f'(x) = e^x/(1+e^x) \Rightarrow f'(0) = 1/(1+1) = 1/2$

and so the equation of the tangent at $(0,\ln 2)$ is $y - \ln 2 = (1/2)x$

or $x - 2y + 2 \ln 2 = 0$.

55. $y = e^{2x}+e^{-3x} \Rightarrow y' = 2e^{2x}-3e^{-3x} \Rightarrow y'' = 4e^{2x}+9e^{-3x}$, so

$y''+y'-6y = (4e^{2x}+9e^{-3x}) + (2e^{2x}-3e^{-3x}) - 6(e^{2x}+e^{-3x}) = 0$

57. $f(x) = e^{-2x} \Rightarrow f'(x) = -2e^{-2x} \Rightarrow f''(x) = (-2)^2e^{-2x} \Rightarrow$

$f'''(x) = (-2)^3e^{-2x} \Rightarrow \cdots \Rightarrow f^{(8)}(x) = (-2)^8e^{-2x} = 256e^{-2x}$

59. (a) $v(t) = ce^{-kt} \Rightarrow a(t) = v'(t) = -kce^{-kt} = -kv(t)$

(b) $v(0) = ce^0 = c$, so c is the initial velocity.

(c) $v(t) = ce^{-kt} = \dfrac{c}{2} \Rightarrow e^{-kt} = \dfrac{1}{2} \Rightarrow -kt = \ln\dfrac{1}{2} = -\ln 2 \Rightarrow t = (\ln 2)/k$

61. (a) If the initial population is A, then $f(0) = A$, $f(3) = 2A$,

$f(6) = 4A = 10000 \Rightarrow A = 2500$. $f(3n) = 2^nA \Rightarrow f(t) = 2^{t/3}A$

$= 2500 \cdot 2^{t/3}$ (b) $f'(t) = 2500 \cdot 2^{t/3}(\ln 2)(1/3) = (2500/3)(\ln 2)2^{t/3}$

(c) $f(t) = 2500 \cdot 2^{t/3} = 15000 \Rightarrow 2^{t/3} = 6 \Rightarrow (t/3) \ln 2 = \ln 6 \Rightarrow$

$t = 3(\ln 6)/\ln 2 \approx 7.75$ h.

**63.** (a) $\lim\limits_{t\to\infty} C(t) = \lim\limits_{t\to\infty} K(e^{-at}-e^{-bt}) = K\lim\limits_{t\to\infty} e^{-at} - K\lim\limits_{t\to\infty} e^{-bt} = 0-0 = 0$

since $a>0$ and $b>0$.    (b) $C'(t) = K(-ae^{-at}+be^{-bt})$

(c) $C'(t) = 0 \Rightarrow ae^{-at} = be^{-bt} \Rightarrow b/a = e^{-at}/e^{-bt} = e^{(b-a)t} \Rightarrow$

$\ln(b/a) = \ln e^{(b-a)t} = (b-a)t \Rightarrow t = \ln(b/a)/(b-a)$

**65.** $f(x) = e^x+x$ is continuous on R and $f(-1) = e^{-1}-1 < 0 < 1 = f(0)$, so

by the Intermediate Value Theorem $e^x+x = 0$ has a root in $(-1,0)$.

**67.** $\int_3^4 5^t\,dt = \dfrac{5^t}{\ln 5}\Big]_3^4 = \dfrac{5^4-5^3}{\ln 5} = \dfrac{500}{\ln 5}$

**69.** $u = -6x \Rightarrow du = -6\,dx$, so $\int e^{-6x}dx = -\frac{1}{6}\int e^u du = -\frac{1}{6}e^u+C = -\frac{1}{6}e^{-6x} + C$

**71.** Let $u = 1+e^x$. Then $du = e^x dx$, so $\int e^x(1+e^x)^{10}dx = \int u^{10}du$

$= u^{11}/11 + C = (1+e^x)^{11}/11 + C$

**73.** $\int \dfrac{e^x+1}{e^x}\,dx = \int (1+e^{-x})dx = x-e^{-x}+C$

**75.** Let $u = -x^2$. Then $du = -2x\,dx$, so $\int xe^{-x^2}dx = -(1/2)\int e^u du$

$= -(1/2)e^u + C = -(1/2)e^{-x^2} + C$

**77.** Let $u = x^2-4x-3$. Then $du = 2(x-2)dx$, so $\int (x-2)e^{x^2-4x-3}dx$

$= (1/2)\int e^u du = (1/2)e^u + C = (1/2)e^{x^2-4x-3} + C$

**79.** Area $= \int_0^1 2e^{-2x}dx = -e^{-2x}\Big]_0^1 = -e^{-2}+1 \approx 0.865$

**81.** Area $= \int_0^1 (e^{3x}-e^x)dx = \frac{1}{3}e^{3x}-e^x\Big]_0^1 = (\frac{1}{3}e^3-e)-(\frac{1}{3}-1) = \frac{1}{3}e^3-e+\frac{2}{3} \approx 4.644$

**83.** (a) $f(x) = xe^x \Rightarrow f'(x) = e^x+xe^x = e^x(1+x)>0 \Longleftrightarrow 1+x>0 \Longleftrightarrow x>-1$, so

$f$ is increasing on $[-1,\infty)$ and decreasing on $(-\infty,-1]$.

(b) $f''(x) = e^x(1+x)+e^x = e^x(2+x)>0 \Longleftrightarrow 2+x>0 \Longleftrightarrow x>-2$, so $f$ is CU on

$(-2,\infty)$ and CD on $(-\infty,-2)$.    (c) IP is $(-2,-2e^{-2})$.

**85.** $y = f(x) = e^x-2e^{-x}$    A. D = R    B. x-intercept occurs when $e^x = 2e^{-x}$

$\Longleftrightarrow e^{2x} = 2 \Longleftrightarrow 2x = \ln 2 \Longleftrightarrow x = (\ln 2)/2$, y-intercept $= f(0) = -1$

C. No symmetry    D. $\lim\limits_{x\to\infty} (e^x-2e^{-x}) = \infty$, $\lim\limits_{x\to-\infty} (e^x-2e^{-x}) = -\infty$, no HA.

E. $f'(x) = e^x+2e^{-x} > 0$ for all x    H.

$\Rightarrow f$ is increasing on R.

F. No extrema

G. $f''(x) = e^x-2e^{-x} = f(x)$

As in B, $f''(x) > 0 \Longleftrightarrow x > (\ln 2)/2$

so $f$ is CU on $((\ln 2)/2,\infty)$ and

CD on $(-\infty,(\ln 2)/2)$.

IP $((\ln 2)/2,0)$

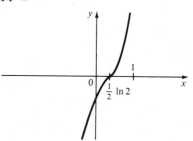

**87.** $y = f(x) = e^{-1/(x+1)}$   A. $D = \{x \mid x \neq -1\} = (-\infty,-1) \cup (-1,\infty)$
B. No x-intercept, y-intercept $= f(0) = e^{-1}$   C. No symmetry
D. $\lim\limits_{x\to\pm\infty} e^{-1/(x+1)} = 1$ since $-1/(x+1) \to 0$, so $y = 1$ is a HA.

$\lim\limits_{x\to-1^+} e^{-1/(x+1)} = 0$ since $-1/(x+1) \to -\infty$,   $\lim\limits_{x\to-1^-} e^{-1/(x+1)} = \infty$

since $-1/(x+1) \to \infty$, so $x = -1$ is a VA   E. $f'(x) = e^{-1/(x+1)}/(x+1)^2$

$\Rightarrow f'(x)>0$ for all $x$ $(\neq -1)$, so $f$ is increasing on $(-\infty,-1)$ and $(-1,\infty)$

F. No extrema                              H.

G. $f''(x) = e^{-1/(x+1)}/(x+1)^4$
$+ e^{-1/(x+1)}(-2)/(x+1)^3$
$= - e^{-1/(x+1)}(2x+1)/(x+1)^4$

$\Rightarrow f''(x)>0 \iff 2x+1<0 \iff x<-1/2$
So $f$ is CU on $(-\infty,-1)$, $(-1,-1/2)$,
and CD on $(-1/2,\infty)$. IP $(-1/2,e^{-2})$

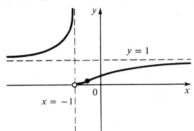

**89.** $y = f(x) = e^{-x^2}$   A. $D = R$   B. No x-intercept, y-intercept $= 1$
C. $f(-x) = f(x)$, so the curve is symmetric about the y-axis.
D. $\lim\limits_{x\to\pm\infty} e^{-x^2} = 0$, so $y = 0$ is a HA.   No VA.   E. $f'(x) = -2xe^{-x^2}$ $\Rightarrow$

$f'(x)>0 \iff x<0$, so $f$ is increasing on $(-\infty,0]$, decreasing on $[0,\infty)$.

F. $f(0) = 1$ is a local and           H.
absolute maximum

G. $f''(x) = 2e^{-x^2}(2x^2-1) > 0 \iff$
$x^2>1/2 \iff |x|>1/\sqrt{2}$, so $f$ is CU
on $(-\infty,-1/\sqrt{2})$ and $(1/\sqrt{2},\infty)$ and
CD on $(-1/\sqrt{2},1/\sqrt{2})$. IP $(\pm 1/\sqrt{2},1/\sqrt{e})$

**91.** $y = f(x) = xe^{x^2}$   A. $D = R$   B. Both intercepts are 0
C. $f(-x) = -f(x)$, so the curve is symmetric about the origin.
D. $\lim\limits_{x\to\infty} xe^{x^2} = \infty$,   $\lim\limits_{x\to-\infty} xe^{x^2} = -\infty$,   no asymptotes.

E. $f'(x) = e^{x^2} + xe^{x^2}(2x) = e^{x^2}(1+2x^2) > 0$, so $f$ is increasing on R.

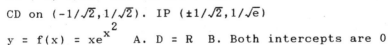

F. No extrema

G. $f''(x) = e^{x^2}(2x)(1+2x^2) +$
$e^{x^2}(4x) = e^{x^2}(2x)(3+2x^2) > 0$
$\Leftrightarrow x>0$, so f is CU on $(0,\infty)$
and CD on $(-\infty,0)$. IP $(0,0)$

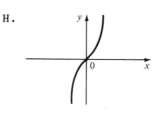

**93.** $y = f(x) = x^{\ln x}$  A. D = $(0,\infty)$  B. No intercepts  C. No symmetry

D. $\lim\limits_{x\to\infty} x^{\ln x} = \infty$, no HA. $\lim\limits_{x\to 0^+} x^{\ln x} = \lim\limits_{x\to 0^+} e^{(\ln x)^2} = \infty$, so x = 0

is a VA.  E. $\ln y = (\ln x)^2$ $\Rightarrow$ $y'/y = 2(\ln x)/x$ $\Rightarrow$
$f'(x) = x^{\ln x} 2(\ln x)/x > 0 \Leftrightarrow \ln x > 0 \Leftrightarrow x > 1$, so f is
increasing on $[1,\infty)$ and decreasing on $(0,1]$.  F. $f(1) = 1$ is a
local and absolute minimum.

G. $f''(x) = x^{\ln x}\left[\dfrac{2\ln x}{x}\right]^2$

$+ x^{\ln x}(2 - 2\ln x)/x^2$

$= 2x^{\ln x}[2(\ln x)^2 - \ln x + 1]/x^2$

$> 0$ for all x since $2t^2-t+1$ has
negative discriminant.  So f is
CU on $(0,\infty)$.

H.

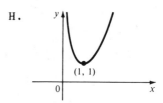

(1, 1)

**95.** $y = f(x) = e^{1/x^2}$  A. D = $\{x \mid x\neq 0\}$ = $(-\infty,0) \cup (0,\infty)$  B. No intercepts
C. $f(-x) = f(x)$, so the curve is symmetric about the y-axis.

D. $\lim\limits_{x\to\pm\infty} e^{1/x^2} = e^0 = 1$, so y = 1 is a HA. $\lim\limits_{x\to 0} e^{1/x^2} = \infty$, so x = 0

is a VA.  E. $f'(x) = e^{1/x^2}(-2/x^3) > 0 \Leftrightarrow x < 0$, so f is increasing
on $(-\infty,0)$ and decreasing on $(0,\infty)$.

F. No extrema

G. $f''(x) = e^{1/x^2}(4/x^6) +$
$e^{1/x^2}(6/x^4) > 0$, so f is CU
on $(-\infty,0)$ and $(0,\infty)$.

H.

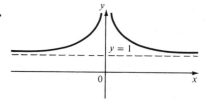

y = 1

**97.** $f(x) = 2e^x+3x^{1/2}$ $\Rightarrow$ $F(x) = 2e^x + 3x^{3/2}/(3/2) + C = 2(e^x + x^{3/2}) + C$

**99.** $f(x) = x+x^2+e^x$ $\Rightarrow$ $f'(x) = 1+2x+e^x$ and $f(0) = 1$ $\Rightarrow$ $g(1) = 0$, so
$g'(1) = 1/f'(g(1)) = 1/f'(0) = 1/2$

**101.** Let $f(x) = e^x-1-x$.  Then $f'(x) = e^x-1 > 0$ for $x > 0$, so f is
increasing on $(0,\infty)$.  Thus, for $x > 0$, $f(x) > f(0) = 0$. So $e^x > 1+x$.

## Exercises 6.6

1.

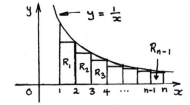

The area of $R_i$ is $\frac{1}{i+1}$ and so $\frac{1}{2} + \frac{1}{3} + \cdots + \frac{1}{n} < \int_1^n \frac{1}{t}\, dt = \ln n$.

The area of $S_i$ is $\frac{1}{i}$ and so $1 + \frac{1}{2} + \cdots + \frac{1}{n-1} > \int_1^n \frac{1}{t}\, dt = \ln n$.

3.  If $f(x) = \ln(x^r)$, then $f'(x) = (1/x^r)(rx^{r-1}) = r/x$. But if $g(x) = r \ln x$, then $g'(x) = r/x$. So $f$ and $g$ must differ by a constant: $\ln(x^r) = r \ln x + C$. Put $x = 1$: $\ln(1^r) = r \ln 1 + C \Rightarrow C = 0$, so $\ln(x^r) = r \ln x$.

5.  From (6.40c) and (6.48) we have $\ln[(e^x)^r] = r \ln(e^x) = rx = \ln(e^{rx})$ and, since $\ln$ is one-to-one, it follows that $(e^x)^r = e^{rx}$.

7.  Using (6.52), (6.40), and (6.50), we have $(ab)^x = e^{x \ln ab}$ $= e^{x(\ln a + \ln b)} = e^{x \ln a + x \ln b} = e^{x \ln a} e^{x \ln b} = a^x b^x$.

9.  Let $\log_a x = r$ and $\log_a y = s$. Then $a^r = x$ and $a^s = y$.

   (a)  $xy = a^r a^s = a^{r+s} \Rightarrow \log_a(xy) = r+s = \log_a x + \log_a y$

   (b)  $\dfrac{x}{y} = \dfrac{a^r}{a^s} = a^{r-s} \Rightarrow \log_a\left[\dfrac{x}{y}\right] = r-s = \log_a x - \log_a y$

   (c)  $x^y = (a^r)^y = a^{ry} \Rightarrow \log_a(x^y) = ry = y \log_a x$

## Section 6.7

## Exercises 6.7

1.  (a) By Theorem 6.59, $y(t) = y(0)e^{kt} = 100e^{kt} \Rightarrow y(1/3) = 100e^{k/3}$ $= 200 \Rightarrow k/3 = \ln(200/100) = \ln 2 \Rightarrow k = 3 \ln 2$. So $y(t) = 100e^{(3 \ln 2)t} = 100 \cdot 2^{3t}$ (since $e^{\ln 2} = 2$).

(b) $y(10) = 100 \cdot 2^{30} \approx 1.07 \times 10^{11}$ cells

(c) $y(t) = 100 \cdot 2^{3t} = 10000$ ➔ $2^{3t} = 100$ ➔ $3t \ln 2 = \ln 100$ ➔
$t = (\ln 100)/(3 \ln 2) \approx 2.2$ h

3. (a) $y(t) = y(0)e^{kt} = 500e^{kt}$ ➔ $y(3) = 500e^{3k} = 8000$ ➔ $e^{3k} = 16$
➔ $3k = \ln 16$ ➔ $y(t) = 500e^{(\ln 16)t/3} = 500 \cdot 16^{t/3}$

(b) $y(4) = 500 \cdot 16^{4/3} \approx 20159$ (c) $y(t) = 500 \cdot 16^{t/3} = 30000$ ➔
$16^{t/3} = 60$ ➔ $(t/3) \ln 16 = \ln 60$ ➔ $t = 3 \ln 60 / \ln 16 \approx 4.4$ h

5. (a) We are given that $k = 0.05$ in (6.58) and so, by Theorem 6.59,
$y(t) = y(0)e^{.05t} = 307000e^{.05t}$, where t is the number of years
after 1985. So $y(10) = 307000e^{10(.05)} \approx 506157$.
(b) $y(16) = 307000e^{16(.05)} \approx 683241$.

7. (a) If $y = [N_2O_5]$ then $\frac{dy}{dt} = -0.0005y$ ➔ $y(t) = y(0)e^{-.0005t}$
$= Ce^{-.0005t}$ (b) $y(t) = Ce^{-.0005t} = 0.9C$ ➔ $e^{-.0005t} = 0.9$ ➔
$-.0005t = \ln(.9)$ ➔ $t = -2000 \ln(.9) \approx 211$ s.

9. (a) If $y(t)$ is the mass remaining after t days, then $y(t) = y(0)e^{kt}$
$= 50e^{kt}$ ➔ $y(.00014) = 50e^{.00014k} = 25$ ➔ $e^{.00014k} = 1/2$ ➔
$k = - \ln 2 /0.00014$ ➔ $y(t) = 50e^{-(\ln 2)t/.00014} = 50 \cdot 2^{-t/.00014}$
(b) $y(.01) = 50 \cdot 2^{-.01/.00014} \approx 1.57 \times 10^{-20}$ mg
(c) $50e^{-(\ln 2)t/.00014} = 40$ ➔ $-(\ln 2)t/.00014 = \ln(.8)$ ➔
$t = - \ln(.8)(.00014)/(\ln 2) \approx 4.5 \times 10^{-5}$ s.

11. Let $y(t) =$ temperature after t minutes. Then $\frac{dy}{dt} = -\frac{1}{10}(y(t)-21)$.
If $u(t) = y(t)-21$, then $\frac{du}{dt} = -\frac{1}{10}u$ ➔ $u(t) = u(0)e^{-t/10} = 12e^{-t/10}$
➔ $y(t) = 21 + u(t) = 21 + 12e^{-t/10}$.

13. (a) Let $P(h)$ be the pressure at altitude h. Then $dP/dh = kP$ ➔
$P(h) = P(0)e^{kh} = 101.3e^{kh}$ ➔ $P(1000) = 101.3e^{1000k} = 87.14$ ➔
$1000k = \ln(87.14/101.3)$ ➔ $P(h) = 101.3e^{\ln(87.14/101.3)h/1000}$
$P(3000) = 101.3e^{\ln(87.14/101.3)(3)} \approx 64.5$ kPa
(b) $P(6187) = 101.3e^{\ln(87.14/101.3)(6187/1000)} \approx 39.9$ kPa

15. With the notation of Example 4, $A_0 = 3000$, $i = .09$, and $t = 5$.

(a) $n = 1$: $A = 3000(1.09)^5 = \$4615.87$
(b) $n = 2$: $A = 3000(1+.09/2)^{10} = \$4658.91$
(c) $n = 12$: $A = 3000(1+.09/12)^{60} = \$4697.04$
(d) $n = 52$: $A = 3000(1+.09/52)^{5(52)} = \$4703.11$
(e) $n = 365$: $A = 3000(1+.09/365)^{5(365)} = \$4704.67$
(f) continuously: $A = 3000e^{(.09)5} = \$4704.94$

17. (a) If y(t) is the amount of salt at time t, then y(0) = 1500(.3)
    = 450 kg. The rate of change of y is
    $$\frac{dy}{dt} = -\left[\frac{y(t)}{1500}\,\frac{kg}{L}\right]\left[20\,\frac{L}{min}\right] = -\frac{1}{75}\,y(t)\,\frac{kg}{min}$$
    so $y(t) = y(0)e^{-t/75} = 450e^{-t/75}$ ➡ $y(30) = 450e^{-0.4} \approx 301.6$ kg
    (b) When the concentration is .2 kg/L, the amount of salt is
    (0.2)(1500) = 300 kg. So $y(t) = 450e^{-t/75} = 300$ ➡ $e^{-t/75} = 2/3$
    ➡ $-t/75 = \ln(2/3)$ ➡ $t = -75\ln(2/3) \approx 30.41$ min.

## Section 6.8

## Exercises 6.8

1. $\cos^{-1}(-1) = \pi$ since $\cos \pi = -1$

3. $\tan^{-1}\sqrt{3} = \pi/3$ since $\tan \pi/3 = \sqrt{3}$

5. $\csc^{-1}\sqrt{2} = \pi/4$ since $\csc \pi/4 = \sqrt{2}$

7. $\cot^{-1}(-\sqrt{3}) = 5\pi/6$ since $\cot 5\pi/6 = -\sqrt{3}$

9. $\sin^{-1}(-1/\sqrt{2}) = -\pi/4$ since $\sin(-\pi/4) = -1/\sqrt{2}$

11. $\arctan(-\sqrt{3}/3) = -\pi/6$ since $\tan(-\pi/6) = -\sqrt{3}/3$

13. $\sin(\sin^{-1}0.7) = 0.7$          15. $\tan(\tan^{-1}10) = 10$

17. $\cos(\sin^{-1}\sqrt{3}/2) = \cos \pi/3 = 1/2$

19. Let $\theta = \cos^{-1}4/5$, so $\cos \theta = 4/5$. Then $\sin(\cos^{-1}4/5) = \sin \theta$
    $= \sqrt{1-(4/5)^2} = \sqrt{9/25} = 3/5$.

21. $\arcsin(\sin 5\pi/4) = \arcsin(-1/\sqrt{2}) = -\pi/4$

23. Let $\theta = \sin^{-1}(5/13)$. Then $\sin \theta = 5/13$, so $\cos(2\sin^{-1}(5/13)) =$
    $\cos 2\theta = 1 - 2\sin^2\theta = 1 - 2(5/13)^2 = 119/169$

25. Let $x = \sin^{-1}(1/3)$ and $y = \sin^{-1}(2/3)$. Then $\sin x = 1/3$, $\cos x =$
    $\sqrt{1-(1/3)^2} = 2\sqrt{2}/3$, $\sin y = 2/3$, $\cos y = \sqrt{1-(2/3)^2} = \sqrt{5}/3$, so
    $\sin[\sin^{-1}(1/3)+\sin^{-1}(2/3)] = \sin(x+y) = \sin x \cos y + \cos x \sin y =$
    $(1/3)(\sqrt{5}/3) + (2\sqrt{2}/3)(2/3) = (\sqrt{5}+4\sqrt{2})/9$.

27. Let $y = \sin^{-1}x$. Then $-\pi/2 \le y \le \pi/2$ ➡ $\cos y \ge 0$, so $\cos(\sin^{-1}x) =$
    $\cos y = \sqrt{1 - \sin^2 y} = \sqrt{1-x^2}$

29. Let $y = \tan^{-1}x$. Then $\tan y = x$, so from the triangle we see that
$$\sin(\tan^{-1}x) = \sin y = \frac{x}{\sqrt{1+x^2}}$$

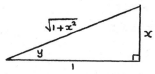

31. Let $y = \cos^{-1}x$. Then $\cos y = x$ and $0 \leq y \leq \pi$ $\Rightarrow$ $-\sin y \frac{dy}{dx} = 1$ $\Rightarrow$
$$\frac{dy}{dx} = -\frac{1}{\sin y} = -\frac{1}{\sqrt{1-\cos^2 y}} = -\frac{1}{\sqrt{1-x^2}} \qquad \left[\begin{array}{l}\text{Note that } \sin y \geq 0 \\ \text{for } 0 \leq y \leq \pi\end{array}\right]$$

33. Let $y = \cot^{-1}x$. Then $\cot y = x$ $\Rightarrow$ $-\csc^2 y\, dy/dx = 1$ $\Rightarrow$
$$\frac{dy}{dx} = -\frac{1}{\csc^2 y} = -\frac{1}{1+\cot^2 y} = -\frac{1}{1+x^2}$$

35. Let $y = \csc^{-1}x$. Then $\csc y = x$ $\Rightarrow$ $-\csc y \cot y\, dy/dx = 1$ $\Rightarrow$
$$\frac{dy}{dx} = -\frac{1}{\csc y \cot y} = -\frac{1}{\csc y \sqrt{\csc^2 y - 1}} = -\frac{1}{x\sqrt{x^2-1}}$$

[Note that $\cot y \geq 0$ on the domain of $\csc^{-1}x$.]

37. $g(x) = \tan^{-1}(x^3)$ $\Rightarrow$ $g'(x) = \frac{1}{1+(x^3)^2}(3x^2) = \frac{3x^2}{1+x^6}$

39. $y = \sin^{-1}(x^2)$ $\Rightarrow$ $y' = (1/\sqrt{1-(x^2)^2})(2x) = 2x/\sqrt{1-x^4}$

41. $F(x) = \tan^{-1}(x/a)$ $\Rightarrow$ $F'(x) = \frac{1}{1+(x/a)^2}\left[\frac{1}{a}\right] = \frac{a}{a^2+x^2}$

43. $H(x) = (1+x^2)\arctan x$ $\Rightarrow$ $H'(x) = (2x)\arctan x + (1+x^2)[1/(1+x^2)]$
$= 1 + 2x \arctan x$

45. $g(t) = \sin^{-1}(4/t)$ $\Rightarrow$ $g'(t) = (1/\sqrt{1-(4/t)^2})(-4/t^2) = -4/\sqrt{t^4-16t^2}$

47. $G(t) = \cos^{-1}\sqrt{2t-1}$ $\Rightarrow$ $G'(t) = -\frac{1}{\sqrt{1-(2t-1)}}\frac{2}{2\sqrt{2t-1}} = -\frac{1}{\sqrt{2(-2t^2+3t-1)}}$

49. $y = \sec^{-1}\sqrt{1+x^2}$ $\Rightarrow$ $y' = \frac{1}{\sqrt{1+x^2}\sqrt{(1+x^2)-1}}\frac{2x}{2\sqrt{1+x^2}} = \frac{x}{(1+x^2)\sqrt{x^2}} = \frac{x}{(1+x^2)|x|}$

51. $y = \tan^{-1}(\sin x)$ $\Rightarrow$ $y' = \frac{\cos x}{1+\sin^2 x}$

53. $y = \dfrac{\sin^{-1}x}{\cos^{-1}x}$ $\Rightarrow$ $y' = \dfrac{\cos^{-1}x/\sqrt{1-x^2} + \sin^{-1}x/\sqrt{1-x^2}}{(\cos^{-1}x)^2} = \dfrac{\cos^{-1}x + \sin^{-1}x}{\sqrt{1-x^2}(\cos^{-1}x)^2}$

$= \pi/2\sqrt{1-x^2}(\cos^{-1}x)^2$

55. $y = (\tan^{-1}x)^{-1}$ $\Rightarrow$ $y' = -(\tan^{-1}x)^{-2}[1/(1+x^2)] = -1/(1+x^2)(\tan^{-1}x)^2$

57. $y = x^2\cot^{-1}(3x)$ $\Rightarrow$ $y' = 2x \cot^{-1}(3x) + x^2[-1/(1+(3x)^2)](3)$
$= 2x \cot^{-1}(3x) - 3x^2/(1+9x^2)$

59.  $y = \arccos\left[\dfrac{b + a\cos x}{a + b\cos x}\right] \Rightarrow y' =$

$-\dfrac{1}{\sqrt{1-[(b+a\cos x)/(a+b\cos x)]^2}} \cdot \dfrac{(a+b\cos x)(-a\sin x)-(b+a\cos x)(-b\sin x)}{(a+b\cos x)^2}$

$= \dfrac{1}{\sqrt{a^2+b^2\cos^2 x-b^2-a^2\cos^2 x}} \dfrac{(a^2-b^2)\sin x}{|a + b\cos x|}$

$= \dfrac{1}{\sqrt{a^2-b^2}\,\sqrt{1-\cos^2 x}} \dfrac{(a^2-b^2)\sin x}{|a + b\cos x|} = \dfrac{\sqrt{a^2-b^2}}{|a + b\cos x|} \dfrac{\sin x}{|\sin x|}$

61.  $g(x) = \sin^{-1}(3x+1) \Rightarrow g'(x) = 3/\sqrt{1-(3x+1)^2} = 3/\sqrt{-9x^2-6x}$

$\text{Dom}(g) = \{x\mid -1\le 3x+1\le 1\} = \{x\mid -2/3\le x\le 0\} = [-2/3, 0]$

$\text{Dom}(g') = \{x\mid -1< 3x+1< 1\} = (-2/3, 0)$

63.  $S(x) = \sin^{-1}(\tan^{-1}x) \Rightarrow S'(x) = 1/\sqrt{1-(\tan^{-1}x)^2}(1+x^2)$

$\text{Dom}(S) = \{x\mid -1 \le \tan^{-1}x \le 1\} = \{x\mid \tan(-1) \le x \le \tan 1\}$

$= [-\tan 1, \tan 1] \quad \text{dom}(S') = \{x\mid -1 < \tan^{-1}x < 1\} = (-\tan 1, \tan 1)$

65.  $G(x) = \sqrt{\csc^{-1}x} \Rightarrow G'(x) = (1/2\sqrt{\csc^{-1}x})(-1/x\sqrt{x^2-1})$

$= -1/2x\sqrt{(x^2-1)\csc^{-1}x} \quad \text{Dom}(G) = \{x\mid |x|\ge 1\} = (-\infty,-1] \cup [1,\infty)$

$[\text{Note: } \csc^{-1}x > 0 ] \quad \text{dom}(G') = \{x\mid |x|>1\} = (-\infty,-1) \cup (1,\infty)$

67.  $U(t) = 2^{\arctan t} \Rightarrow U'(t) = 2^{\arctan t}(\ln 2)/(1+t^2)$

$\text{Dom}(U) = \text{dom}(U') = R$

69.  $g(x) = x\sin^{-1}(x/4) + \sqrt{16-x^2} \Rightarrow g'(x) = \sin^{-1}(x/4) + x/4\sqrt{1-(x/4)^2}$

$- x/\sqrt{16-x^2} = \sin^{-1}(x/4) \Rightarrow g'(2) = \sin^{-1}(1/2) = \pi/6$

71.  $\lim\limits_{x\to -1^+} \sin^{-1}x = \sin^{-1}(-1) = -\pi/2$

73.  $\lim\limits_{x\to -\infty} \cot^{-1}x = \pi$ since $\cot x \to -\infty$ as $x \to \pi^-$ (or see Figure 6.42)

75.  $\lim\limits_{x\to -\infty} \csc^{-1}x = \pi$ since $\csc x \to -\infty$ as $x \to \pi^+$ (or see Figure 6.40)

77.  $\lim\limits_{x\to\infty} (\tan^{-1}x)^2 = (\lim\limits_{x\to\infty} \tan^{-1}x)^2 = (\pi/2)^2 = \pi^2/4$

79.  $\lim\limits_{x\to 1^-} \dfrac{\arcsin x}{\tan(\pi x/2)} = 0$ since $\lim\limits_{x\to 1^-} \arcsin x = \dfrac{\pi}{2}$ and $\lim\limits_{x\to 1^-} \tan(\pi x/2) = \infty$

81.

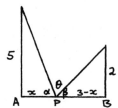

$x = 5 \cot \alpha$, $3-x = 2 \cot \beta$ $\Rightarrow$

$\theta = \pi - \cot^{-1}(x/5) - \cot^{-1}((3-x)/2)$

$\dfrac{d\theta}{dx} = \dfrac{1}{1+(x/5)^2}\left[\dfrac{1}{5}\right] + \dfrac{1}{1+((3-x)/2)^2}\left[-\dfrac{1}{2}\right] = 0$

$\Rightarrow 5(1+x^2/25) = 2[1+(9-6x+x^2)/4]$ $\Rightarrow$

$50+2x^2 = 65-30x+5x^2$ $\Rightarrow$ $x^2-10x+5 = 0$ $\Rightarrow$

$x = 5\pm2\sqrt{5}$. We reject the root with the + sign, since it is > 3.

$d\theta/dx > 0$ for $x < 5-2\sqrt{5}$ and $d\theta/dx < 0$ for $x > 5-2\sqrt{5}$, so $\theta$ is

maximized when $|AP| = x = 5-2\sqrt{5}$.

83. $y = f(x) = \sin^{-1}(x/(x+1))$  A. $D = \{x \mid -1 \le x/(x+1) \le 1\}$  For $x > -1$

we have $-x-1 \le x \le x+1$ $\Longleftrightarrow$ $2x \ge -1$ $\Longleftrightarrow$ $x \ge -1/2$, so $D = [-1/2,\infty)$.

B. Intercepts are 0.  C. No symmetry  D. $\lim\limits_{x\to\infty} \sin^{-1}\left[\dfrac{x}{x+1}\right] =$

$\lim\limits_{x\to\infty} \sin^{-1}\left[\dfrac{1}{1+1/x}\right] = \sin^{-1}1 = \pi/2$, so $y = \pi/2$ is a HA.

E. $f'(x) = \dfrac{1}{\sqrt{1-(x/(x+1))^2}}\dfrac{(x+1)-x}{(x+1)^2} = \dfrac{1}{(x+1)\sqrt{2x+1}} > 0$, so $f$ is

increasing on $[-1/2,\infty)$.  H.

F. $f(-1/2) = \sin^{-1}(-1) = -\pi/2$

is an absolute minimum.

G. $f''(x) = -\dfrac{\sqrt{2x+1}+(x+1)/\sqrt{2x+1}}{(x+1)^2(2x+1)}$

$= -\dfrac{3x+2}{(x+1)^2(2x+1)^{3/2}} < 0$ on $D$, so

f is CD on $(-1/2,\infty)$.

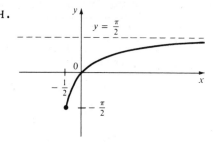

85. $y = f(x) = \cos^{-1}(1+x^2)$

$D = \{x \mid -1 \le 1+x^2 \le 1\}$

$= \{x \mid -2 \le x^2 \le 0\} = \{0\}$

So the graph of f consists

of a single point, $(0,0)$.

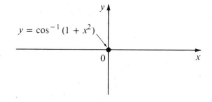

87. $y = f(x) = \arctan x + \arctan(1/x)$  A. $D = \{x \mid x \ne 0\} = (-\infty,0) \cup (0,\infty)$

Note that $f'(x) = \dfrac{1}{1+x^2} + \dfrac{1}{1+1/x^2}\left[-\dfrac{1}{x^2}\right] = \dfrac{1}{1+x^2} - \dfrac{1}{x^2+1} = 0$, so f is

constant on each of the

intervals $(-\infty,0)$ and $(0,\infty)$.

Now $f(1) = 2 \arctan 1 = \pi/2$

and $f(-1) = -f(-1) = -\pi/2$, so

$f(x) = \begin{cases} \pi/2 & \text{if } x>0 \\ -\pi/2 & \text{if } x<0 \end{cases}$

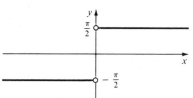

**89.** $y = f(x) = x - \tan^{-1}x$   A. D = R   B. Intercepts are 0

C. $f(-x) = -f(x)$, so the curve is symmetric about the origin.

D. $\lim\limits_{x\to\infty} (x - \tan^{-1}x) = \infty$ and $\lim\limits_{x\to-\infty} (x - \tan^{-1}x) = -\infty$, no HA.

But $f(x)-(x-\pi/2) = -\tan^{-1}x + \pi/2 \to 0$ as $x \to \infty$ and $f(x)-(x+\pi/2) = -\tan^{-1}x - \pi/2 \to 0$ as $x \to -\infty$, so $y = x\pm\pi/2$ are slant asymptotes.

E. $f'(x) = 1 - 1/(x^2+1) = x^2/(x^2+1) > 0$, so f is increasing on R.

F. No extrema

H.

G. $f''(x) = \dfrac{(1+x^2)(2x)-x^2(2x)}{(1+x^2)^2}$

$= \dfrac{2x}{(1+x^2)^2} > 0 \Leftrightarrow x > 0$, so f is

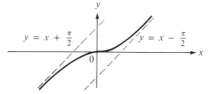

CU on $(0,\infty)$, CD on $(-\infty,0)$. IP $(0,0)$

**91.** $f(x) = 2x + 5/\sqrt{1-x^2} \Rightarrow F(x) = x^2 + 5\sin^{-1}x + C$

**93.** $\displaystyle\int_1^{\sqrt{3}} \frac{6}{1+x^2}\, dx = 6\tan^{-1}x\Big]_1^{\sqrt{3}} = 6(\tan^{-1}\sqrt{3} - \tan^{-1}1) = 6(\frac{\pi}{3} - \frac{\pi}{4}) = \frac{\pi}{2}$

**95.** Let $u = x^3$. Then $du = 3x^2dx$, so $\displaystyle\int \frac{x^2}{\sqrt{1-x^6}}\, dx = \frac{1}{3}\int \frac{1}{\sqrt{1-u^2}}\, du$

$= (1/3)\sin^{-1}u + C = (1/3)\sin^{-1}(x^3) + C$

**97.** $\displaystyle\int \frac{x+9}{x^2+9}\, dx = \int \frac{x}{x^2+9}\, dx + 9\int \frac{1}{x^2+9}\, dx = \frac{1}{2}\ln(x^2+9) + 9\cdot\frac{1}{3}\tan^{-1}\frac{x}{3} + C$

(Let $u = x^2+9$ in the first integral; use (6.75) in the second.)

**99.** Let $u = 3x$. Then $du = 3\, dx$, so $\displaystyle\int \frac{dx}{1+9x^2} = \frac{1}{3}\int \frac{du}{1+u^2} = \frac{1}{3}\tan^{-1}u + C$

$= (1/3)\tan^{-1}(3x) + C$

**101.** Let $u = e^x$. Then $du = e^xdx$, so $\displaystyle\int \frac{e^x}{e^{2x}+1}\, dx = \int \frac{1}{u^2+1}\, du = \tan^{-1}u + C$

$= \tan^{-1}(e^x) + C$

**103.** Let $u = \sin^{-1}x$. Then $du = \dfrac{1}{\sqrt{1-x^2}}\, dx$, so $\displaystyle\int_0^{1/2} \frac{\sin^{-1}x}{\sqrt{1-x^2}}\, dx = \int_0^{\pi/6} u\, du$

$= \frac{1}{2}u^2\Big]_0^{\pi/6} = \frac{1}{2}(\pi/6)^2 = \pi^2/72$

**105.** Let $u = \dfrac{x}{a}$. Then $du = \dfrac{1}{a}\, dx$, so $\displaystyle\int \frac{dx}{\sqrt{a^2-x^2}} = \int \frac{dx}{a\sqrt{1-(x/a)^2}} = \int \frac{du}{\sqrt{1-u^2}}$

$= \sin^{-1}u + C = \sin^{-1}(x/a) + C$

**107.** $f(x) = \tan^{-1}x + \cot^{-1}x \Rightarrow f'(x) = \dfrac{1}{1+x^2} - \dfrac{1}{1+x^2} = 0$, so $f(x) = C$, a

constant. Now $C = f(1) = \tan^{-1}1 + \cot^{-1}1 = \frac{\pi}{4} + \frac{\pi}{4} = \frac{\pi}{2}$ and so

$\tan^{-1}x + \cot^{-1}x = \pi/2$

## Exercises 6.9

1. $\sinh(-x) = \frac{1}{2}(e^{-x} - e^{-(-x)}) = \frac{1}{2}(e^{-x} - e^{x}) = -\frac{1}{2}(e^{x} - e^{-x}) = -\sinh x$

3. $\cosh x + \sinh x = \frac{1}{2}(e^{x} + e^{-x}) + \frac{1}{2}(e^{x} - e^{-x}) = \frac{1}{2}(2e^{x}) = e^{x}$

5. $\sinh x \cosh y + \cosh x \sinh y$
   $= (1/2)(e^{x} - e^{-x})(1/2)(e^{y} + e^{-y}) + (1/2)(e^{x} + e^{-x})(1/2)(e^{y} - e^{-y})$
   $= (1/4)[(e^{x+y} + e^{x-y} - e^{-x+y} - e^{-x-y}) + (e^{x+y} - e^{x-y} + e^{-x+y} - e^{-x-y})]$
   $= (1/4)(2e^{x+y} - 2e^{-x-y}) = (1/2)(e^{x+y} - e^{-(x+y)}) = \sinh(x+y)$

7. Divide both sides of the identity $\cosh^2 x - \sinh^2 x = 1$ by $\sinh^2 x$:
   $$\frac{\cosh^2 x}{\sinh^2 x} - 1 = \frac{1}{\sinh^2 x} \qquad \text{or} \qquad \coth^2 x - 1 = \operatorname{csch}^2 x$$

9. By Exercise 5, $\sinh 2x = \sinh(x+x) = \sinh x \cosh x + \cosh x \sinh x$
   $= 2 \sinh x \cosh x$.

11. By Exercise 10, $\cosh 2y = \cosh^2 y + \sinh^2 y = 1 + 2 \sinh^2 y \Rightarrow$
    $\sinh^2 y = (\cosh 2y - 1)/2$. Put $x = 2y$. Then
    $\sinh(x/2) = \pm\sqrt{(\cosh x - 1)/2}$.

13. $\tanh(\ln x) = \dfrac{\sinh(\ln x)}{\cosh(\ln x)} = \dfrac{(e^{\ln x} - e^{-\ln x})/2}{(e^{\ln x} + e^{-\ln x})/2} = \dfrac{x - 1/x}{x + 1/x} = \dfrac{x^2 - 1}{x^2 + 1}$

15. By Exercise 3, $(\cosh x + \sinh x)^n = (e^{x})^n = e^{nx} = \cosh nx + \sinh nx$

17. $\tanh x = 4/5 > 0$, so $x > 0$. $\coth x = 1/\tanh x = 5/4$, $\operatorname{sech}^2 x = 1 - \tanh^2 x = 1 - (4/5)^2 = 9/25 \Rightarrow \operatorname{sech} x = 3/5$ (since $\operatorname{sech} x > 0$),
    $\cosh x = 1/\operatorname{sech} x = 5/3$, $\sinh x = \tanh x \cosh x = (4/5)(5/3) = 4/3$,
    and $\operatorname{csch} x = 1/\sinh x = 3/4$.

19. (a) $\displaystyle\lim_{x\to\infty} \tanh x = \lim_{x\to\infty} \frac{e^{x} - e^{-x}}{e^{x} + e^{-x}} = \lim_{x\to\infty} \frac{1 - e^{-2x}}{1 + e^{-2x}} = \frac{1-0}{1+0} = 1$

    (b) $\displaystyle\lim_{x\to-\infty} \tanh x = \lim_{x\to-\infty} \frac{e^{x} - e^{-x}}{e^{x} + e^{-x}} = \lim_{x\to-\infty} \frac{e^{2x} - 1}{e^{2x} + 1} = \frac{0-1}{0+1} = -1$

    (c) $\displaystyle\lim_{x\to\infty} \sinh x = \lim_{x\to\infty} \frac{e^{x} - e^{-x}}{2} = \infty$

    (d) $\displaystyle\lim_{x\to-\infty} \sinh x = \lim_{x\to-\infty} \frac{e^{x} - e^{-x}}{2} = -\infty$

    (e) $\displaystyle\lim_{x\to\infty} \operatorname{sech} x = \lim_{x\to\infty} \frac{2}{e^{x} + e^{-x}} = 0$

(f) $\lim\limits_{x\to\infty} \coth x = \lim\limits_{x\to\infty} \dfrac{e^x+e^{-x}}{e^x-e^{-x}} = \lim\limits_{x\to\infty} \dfrac{1+e^{-2x}}{1-e^{-2x}} = \dfrac{1+0}{1-0} = 1$  [or use (a)]

(g) $\lim\limits_{x\to 0^+} \coth x = \lim\limits_{x\to 0^+} \dfrac{\cosh x}{\sinh x} = \infty$, since $\sinh x \to 0$ and $\coth x > 0$

(h) $\lim\limits_{x\to 0^-} \coth x = \lim\limits_{x\to 0^-} \dfrac{\cosh x}{\sinh x} = -\infty$, since $\sinh x \to 0$ and $\coth x < 0$

(i) $\lim\limits_{x\to -\infty} \operatorname{csch} x = \lim\limits_{x\to -\infty} \dfrac{2}{e^x-e^{-x}} = 0$

21.  Let $y = \sinh^{-1} x$. Then $\sinh y = x$ and, by Example 1(a), $\cosh y = \sqrt{1 + \sinh^2 y} = \sqrt{1+x^2}$. So, by Exercise 3, $e^y = \sinh y + \cosh y = x+\sqrt{1+x^2} \;\Rightarrow\; y = \ln(x+\sqrt{1+x^2})$

23.  (a) Let $y = \tanh^{-1} x$. Then $x = \tanh y = \dfrac{e^y-e^{-y}}{e^y+e^{-y}} = \dfrac{e^{2y}-1}{e^{2y}+1} \;\Rightarrow$
$xe^{2y}+x = e^{2y}-1 \;\Rightarrow\; e^{2y} = \dfrac{1+x}{1-x} \;\Rightarrow\; 2y = \ln\left[\dfrac{1+x}{1-x}\right] \;\Rightarrow\; y = \tfrac{1}{2}\ln\left[\dfrac{1+x}{1-x}\right]$

(b) Let $y = \tanh^{-1} x$. Then $x = \tanh y$, so from Exercise 14 we have
$e^{2y} = \dfrac{1+\tanh y}{1-\tanh y} = \dfrac{1+x}{1-x} \;\Rightarrow\; 2y = \ln\left[\dfrac{1+x}{1-x}\right] \;\Rightarrow\; y = \tfrac{1}{2}\ln\left[\dfrac{1+x}{1-x}\right]$

25.  (a) Let $y = \cosh^{-1} x$. Then $\cosh y = x$ and $y \geq 0 \;\Rightarrow\; \sinh y \dfrac{dy}{dx} = 1$
$\Rightarrow \dfrac{dy}{dx} = \dfrac{1}{\sinh y} = \dfrac{1}{\sqrt{\cosh^2 y - 1}} = \dfrac{1}{\sqrt{x^2-1}}$  (since $\sinh y \geq 0$ for $y \geq 0$)

[OR: Use Formula 6.81.]

(b) Let $y = \tanh^{-1} x$. Then $\tanh y = x \;\Rightarrow\; \operatorname{sech}^2 y \dfrac{dy}{dx} = 1 \;\Rightarrow$
$\dfrac{dy}{dx} = \dfrac{1}{\operatorname{sech}^2 y} = \dfrac{1}{1 - \tanh^2 y} = \dfrac{1}{1-x^2}$  [OR: Use Formula 6.82.]

(c) Let $y = \operatorname{csch}^{-1} x$. Then $\operatorname{csch} y = x \;\Rightarrow\; -\operatorname{csch} y \coth y \, dy/dx = 1$
$\Rightarrow \dfrac{dy}{dx} = -\dfrac{1}{\operatorname{csch} y \coth y}$  By Exercise 7, $\coth y = \pm\sqrt{\operatorname{csch}^2 y + 1} = \pm\sqrt{x^2+1}$. If $x > 0$, then $\coth y > 0$, so $\coth y = \sqrt{x^2+1}$. If $x < 0$, then $\coth y < 0$, so $\coth y = -\sqrt{x^2+1}$. In either case we have
$\dfrac{dy}{dx} = -\dfrac{1}{\operatorname{csch} y \coth y} = -1/|x|\sqrt{x^2+1}$.

(d) Let $y = \operatorname{sech}^{-1} x$. Then $\operatorname{sech} y = x \;\Rightarrow\; -\operatorname{sech} y \tanh y \, dy/dx = 1$
$\Rightarrow \dfrac{dy}{dx} = -\dfrac{1}{\operatorname{sech} y \tanh y} = -\dfrac{1}{\operatorname{sech} y \sqrt{1 - \operatorname{sech}^2 y}} = -\dfrac{1}{x\sqrt{1-x^2}}$

(Note that $y > 0$ and so $\tanh y > 0$.)

(e) Let $y = \coth^{-1} x$. Then $\coth y = x \Rightarrow -\operatorname{csch}^2 y \, dy/dx = 1 \Rightarrow$
$$\frac{dy}{dx} = -\frac{1}{\operatorname{csch}^2 y} = \frac{1}{1 - \coth^2 y} = \frac{1}{1-x^2} \quad \text{by Exercise 7.}$$

27. $f(x) = \tanh 3x \Rightarrow f'(x) = 3 \operatorname{sech}^2 3x$

29. $h(x) = \cosh(x^4) \Rightarrow h'(x) = \sinh(x^4) \cdot 4x^3$

31. $G(x) = x^2 \operatorname{sech} x \Rightarrow G'(x) = 2x \operatorname{sech} x - x^2 \operatorname{sech} x \tanh x$

33. $H(t) = \tanh(e^t) \Rightarrow H'(t) = \operatorname{sech}^2(e^t) \cdot e^t$

35. $y = x^{\cosh x} \Rightarrow \ln y = \cosh x \ln x \Rightarrow \frac{y'}{y} = \sinh x \ln x + \frac{\cosh x}{x} \Rightarrow$
$y' = x^{\cosh x}(\sinh x \ln x + \cosh x / x)$

37. $y = \cosh^{-1}(x^2) \Rightarrow y' = (1/\sqrt{(x^2)^2 - 1})(2x) = 2x/\sqrt{x^4 - 1}$

39. $y = x \ln(\operatorname{sech} 4x) \Rightarrow y' = \ln(\operatorname{sech} 4x) + x \frac{-\operatorname{sech} 4x \tanh 4x}{\operatorname{sech} 4x}(4)$
$= \ln(\operatorname{sech} 4x) - 4x \tanh 4x$

41. $y = \tanh^{-1}(x/a) \Rightarrow y' = \frac{1}{1-(x/a)^2}\left[\frac{1}{a}\right] = \frac{a}{a^2 - x^2}$

43. $y = \operatorname{csch}^{-1}(x^4) \Rightarrow y' = -4x^3/x^4\sqrt{(x^4)^2 + 1} = -4/x\sqrt{x^8 + 1}$

45. $y = \coth^{-1}\sqrt{x^2 + 1} \Rightarrow y' = \frac{1}{1-(x^2+1)} \frac{2x}{2\sqrt{x^2+1}} = -\frac{1}{x\sqrt{x^2+1}}$

47. $u = 2x \Rightarrow du = 2\,dx$, so $\int \sinh 2x \, dx = \frac{1}{2}\int \sinh u \, du$
$= (1/2)\cosh u + C = (1/2)\cosh 2x + C$

49. Let $u = \sinh x$. Then $du = \cosh x \, dx$, so $\int \coth x \, dx = \int \frac{\cosh x}{\sinh x} dx$
$= \int \frac{1}{u} du = \ln|u| + C = \ln|\sinh x| + C$

51. $u = \frac{x}{2} \Rightarrow du = \frac{1}{2}dx \Rightarrow \int \frac{1}{\sqrt{4+x^2}} dx = \frac{1}{2}\int \frac{1}{\sqrt{1+(x/2)^2}} dx = \int \frac{1}{\sqrt{1+u^2}} du$
$= \sinh^{-1} u + C = \sinh^{-1}(x/2) + C$

53. $\int_0^{1/2} \frac{1}{1-x^2} dx = \tanh^{-1} x \Big]_0^{1/2} = \tanh^{-1}(1/2) = \frac{1}{2}\ln\left[\frac{1+1/2}{1-1/2}\right]$ (Ex. 23)
$= \frac{1}{2}\ln\left[\frac{3/2}{1/2}\right] = \frac{1}{2}\ln 3$

55. The tangent to $y = \cosh x$ has slope 1 when $y' = \sinh x = 1 \Rightarrow$
$x = \sinh^{-1} 1 = \ln(1+\sqrt{2})$, by (6.80). Since $\sinh x = 1$ and $\cosh x = \sqrt{1 + \sinh^2 x}$, we have $\cosh x = \sqrt{2}$. The point is $(\ln(1+\sqrt{2}), \sqrt{2})$.

**Exercises 6.10**

NOTE: The use of l'Hospital's Rule is indicated by H (above the equal sign).

1. $\lim\limits_{x\to 2} \dfrac{x-2}{x^2-4} = \lim\limits_{x\to 2} \dfrac{x-2}{(x-2)(x+2)} = \lim\limits_{x\to 2} \dfrac{1}{x+2} = \dfrac{1}{4}$

3. $\lim\limits_{x\to -1} \dfrac{x^6-1}{x^4-1} \overset{H}{=} \lim\limits_{x\to -1} \dfrac{6x^5}{4x^3} = \dfrac{-6}{-4} = \dfrac{3}{2}$

5. $\lim\limits_{x\to 0} \dfrac{e^x-1}{\sin x} \overset{H}{=} \lim\limits_{x\to 0} \dfrac{e^x}{\cos x} = \dfrac{1}{1} = 1$

7. $\lim\limits_{x\to 0} \dfrac{\sin x}{x^3} \overset{H}{=} \lim\limits_{x\to 0} \dfrac{\cos x}{3x^2} = \infty$

9. $\lim\limits_{x\to 0} \dfrac{\tan x}{x + \sin x} \overset{H}{=} \lim\limits_{x\to 0} \dfrac{\sec^2 x}{1 + \cos x} = \dfrac{1}{1+1} = \dfrac{1}{2}$

11. $\lim\limits_{x\to \infty} \dfrac{\ln x}{x} \overset{H}{=} \lim\limits_{x\to \infty} \dfrac{1/x}{1} = 0$

13. $\lim\limits_{x\to \infty} \dfrac{e^x}{x^3} \overset{H}{=} \lim\limits_{x\to \infty} \dfrac{e^x}{3x^2} \overset{H}{=} \lim\limits_{x\to \infty} \dfrac{e^x}{6x} \overset{H}{=} \lim\limits_{x\to \infty} \dfrac{e^x}{6} = \infty$

15. $\lim\limits_{x\to a} \dfrac{\sqrt[3]{x} - \sqrt[3]{a}}{x - a} \overset{H}{=} \lim\limits_{x\to a} \dfrac{(1/3)x^{-2/3}}{1} = \dfrac{1}{3a^{2/3}}$

17. $\lim\limits_{x\to 0} \dfrac{e^x-1-x}{x^2} \overset{H}{=} \lim\limits_{x\to 0} \dfrac{e^x-1}{2x} \overset{H}{=} \lim\limits_{x\to 0} \dfrac{e^x}{2} = \dfrac{1}{2}$

19. $\lim\limits_{x\to 0} \dfrac{\sin x}{e^x} = \dfrac{0}{1} = 0$

21. $\lim\limits_{x\to 0} \dfrac{1 - \cos x}{x^2} \overset{H}{=} \lim\limits_{x\to 0} \dfrac{\sin x}{2x} \overset{H}{=} \lim\limits_{x\to 0} \dfrac{\cos x}{2} = \dfrac{1}{2}$

23. $\lim\limits_{x\to 2^-} \dfrac{\ln x}{\sqrt{2-x}} = \infty$ since $\sqrt{2-x} \to 0$ but $\ln x \to \ln 2$.

25. $\lim\limits_{x\to \infty} \dfrac{\ln \ln x}{\sqrt{x}} \overset{H}{=} \lim\limits_{x\to \infty} \dfrac{1/x \ln x}{1/2\sqrt{x}} = \lim\limits_{x\to \infty} \dfrac{2}{\sqrt{x} \ln x} = 0$

27. $\lim\limits_{x\to 0} \dfrac{\tan^{-1}(2x)}{3x} \overset{H}{=} \lim\limits_{x\to 0} \dfrac{2/(1+4x^2)}{3} = \dfrac{2}{3}$

29. $\lim\limits_{x\to 0} \dfrac{\tan \alpha x}{x} \overset{H}{=} \lim\limits_{x\to 0} \dfrac{\alpha \sec^2 \alpha x}{1} = \alpha$

31. $\displaystyle\lim_{x\to 0}\frac{\tan 2x}{\tanh 3x}\overset{H}{=}\lim_{x\to 0}\frac{2\,\sec^2 2x}{3\,\text{sech}^2 3x}=\frac{2}{3}$

33. $\displaystyle\lim_{x\to 0}\frac{x+\sin 3x}{x-\sin 3x}\overset{H}{=}\lim_{x\to 0}\frac{1+3\cos 3x}{1-3\cos 3x}=\frac{1+3}{1-3}=-2$

35. $\displaystyle\lim_{x\to 0}\frac{e^{4x}-1}{\cos x}=\frac{0}{1}=0$

37. $\displaystyle\lim_{x\to 0}\frac{\tan x-\sin x}{x^3}\overset{H}{=}\lim_{x\to 0}\frac{\sec^2 x-\cos x}{3x^2}\overset{H}{=}\lim_{x\to 0}\frac{2\sec^2 x\tan x+\sin x}{6x}\overset{H}{=}$

$\displaystyle\lim_{x\to 0}\frac{4\sec^2 x\tan^2 x+2\sec^4 x+\cos x}{6}=\frac{0+2+1}{6}=\frac{1}{2}$

39. $\displaystyle\lim_{x\to 0^+}\sqrt{x}\ln x=\lim_{x\to 0^+}\frac{\ln x}{x^{-1/2}}\overset{H}{=}\lim_{x\to 0^+}\frac{1/x}{-(1/2)x^{-3/2}}=\lim_{x\to 0^+}(-2\sqrt{x})=0$

41. $\displaystyle\lim_{x\to\infty}e^{-x}\ln x=\lim_{x\to\infty}\frac{\ln x}{e^x}\overset{H}{=}\lim_{x\to\infty}\frac{1/x}{e^x}=\lim_{x\to\infty}\frac{1}{xe^x}=0$

43. $\displaystyle\lim_{x\to\infty}x^3 e^{-x^2}=\lim_{x\to\infty}\frac{x^3}{e^{x^2}}\overset{H}{=}\lim_{x\to\infty}\frac{3x^2}{2xe^{x^2}}=\lim_{x\to\infty}\frac{3x}{2e^{x^2}}\overset{H}{=}\lim_{x\to\infty}\frac{3}{4xe^{x^2}}=0$

45. $\displaystyle\lim_{x\to\pi}(x-\pi)\cot x=\lim_{x\to\pi}\frac{x-\pi}{\tan x}\overset{H}{=}\lim_{x\to\pi}\frac{1}{\sec^2 x}=\frac{1}{(-1)^2}=1$

47. $\displaystyle\lim_{x\to 0}\left[\frac{1}{x^4}-\frac{1}{x^2}\right]=\lim_{x\to 0}\frac{1-x^2}{x^4}=\infty$

49. $\displaystyle\lim_{x\to 0}\left[\frac{1}{x}-\csc x\right]=\lim_{x\to 0}\left[\frac{1}{x}-\frac{1}{\sin x}\right]=\lim_{x\to 0}\frac{\sin x-x}{x\sin x}$

$\displaystyle\overset{H}{=}\lim_{x\to 0}\frac{\cos x-1}{\sin x+x\cos x}\overset{H}{=}\lim_{x\to 0}\frac{-\sin x}{2\cos x-x\sin x}=\frac{0}{2}=0$

51. $\displaystyle\lim_{x\to\infty}(x-\sqrt{x^2-1})=\lim_{x\to\infty}(x-\sqrt{x^2-1})\frac{x+\sqrt{x^2-1}}{x+\sqrt{x^2-1}}=\lim_{x\to\infty}\frac{x^2-(x^2-1)}{x+\sqrt{x^2-1}}=\lim_{x\to\infty}\frac{1}{x+\sqrt{x^2-1}}$

$=0$

53. $\displaystyle\lim_{x\to\infty}\left[\frac{x^3}{x^2-1}-\frac{x^3}{x^2+1}\right]=\lim_{x\to\infty}\frac{x^3(x^2+1)-x^3(x^2-1)}{(x^2-1)(x^2+1)}=\lim_{x\to\infty}\frac{2x^3}{x^4-1}=\lim_{x\to\infty}\frac{2/x}{1-1/x^4}$

$=0$

55. $y=x^{\sin x}\Rightarrow\ln y=\sin x\ln x$, so $\displaystyle\lim_{x\to 0^+}\ln y=\lim_{x\to 0^+}\sin x\ln x$

$\displaystyle=\lim_{x\to 0^+}\frac{\ln x}{\csc x}\overset{H}{=}\lim_{x\to 0^+}\frac{1/x}{-\csc x\cot x}=-\lim_{x\to 0^+}\frac{\sin x}{x}\lim_{x\to 0^+}\tan x=-1\cdot 0$

$=0\Rightarrow\displaystyle\lim_{x\to 0^+}x^{\sin x}=\lim_{x\to 0^+}e^{\ln y}=e^0=1$

57. $x^{1/\ln x} = e^{(\ln x)(1/\ln x)} = e^1 = e$, so $\lim\limits_{x \to 0^+} x^{1/\ln x} = \lim\limits_{x \to 0^+} e = e$

59. $y = (1-2x)^{1/x} \Rightarrow \ln y = (1/x)\ln(1-2x) \Rightarrow \lim\limits_{x \to 0} \ln y = \lim\limits_{x \to 0} \dfrac{\ln(1-2x)}{x}$

$\overset{H}{=} \lim\limits_{x \to 0} \dfrac{-2/(1-2x)}{1} = -2 \Rightarrow \lim\limits_{x \to 0} (1-2x)^{1/x} = \lim\limits_{x \to 0} e^{\ln y} = e^{-2}$

61. $y = (1+3/x+5/x^2)^x \Rightarrow \ln y = x \ln(1+3/x+5/x^2) \Rightarrow$

$\lim\limits_{x \to \infty} \ln y = \lim\limits_{x \to \infty} \dfrac{\ln(1+3/x+5/x^2)}{1/x} \overset{H}{=} \lim\limits_{x \to \infty} \dfrac{(-3/x^2-10/x^3)/(1+3/x+5/x^2)}{-1/x^2}$

$= \lim\limits_{x \to \infty} \dfrac{3+10/x}{1+3/x+5/x^2} = 3$, so $\lim\limits_{x \to \infty} (1+3/x+5/x^2)^x = \lim\limits_{x \to \infty} e^{\ln y} = e^3$

63. $y = x^{1/x} \Rightarrow \ln y = (1/x) \ln x \Rightarrow \lim\limits_{x \to \infty} \ln y = \lim\limits_{x \to \infty} \dfrac{\ln x}{x} \overset{H}{=} \lim\limits_{x \to \infty} \dfrac{1/x}{1} = 0$

$\Rightarrow \lim\limits_{x \to \infty} x^{1/x} = \lim\limits_{x \to \infty} e^{\ln y} = e^0 = 1$

65. $y = (\cot x)^{\sin x} \Rightarrow \ln y = \sin x \ln(\cot x) \Rightarrow \lim\limits_{x \to 0^+} \ln y =$

$\lim\limits_{x \to 0^+} \dfrac{\ln(\cot x)}{\csc x} \overset{H}{=} \lim\limits_{x \to 0^+} \dfrac{-\csc^2 x / \cot x}{-\csc x \cot x} = \lim\limits_{x \to 0^+} \dfrac{\csc x}{\cot^2 x} = \lim\limits_{x \to 0^+} \dfrac{\sin x}{\cos^2 x} = 0$

so $\lim\limits_{x \to 0^+} (\cot x)^{\sin x} = \lim\limits_{x \to 0^+} e^{\ln y} = e^0 = 1$

67. $y = \left[\dfrac{x}{x+1}\right]^x \Rightarrow \ln y = x \ln\left[\dfrac{x}{x+1}\right] \Rightarrow \lim\limits_{x \to \infty} \ln y = \lim\limits_{x \to \infty} x \ln\left[\dfrac{x}{x+1}\right]$

$= \lim\limits_{x \to \infty} \dfrac{\ln x - \ln(x+1)}{1/x} \overset{H}{=} \lim\limits_{x \to \infty} \dfrac{1/x - 1/(x+1)}{-1/x^2} = \lim\limits_{x \to \infty} \left[-x + \dfrac{x^2}{x+1}\right]$

$= \lim\limits_{x \to \infty} \dfrac{-x}{x+1} = -1$, so $\lim\limits_{x \to \infty} \left[\dfrac{x}{x+1}\right]^x = \lim\limits_{x \to \infty} e^{\ln y} = e^{-1} = \dfrac{1}{e}$

[OR: $\lim\limits_{x \to \infty} \left[\dfrac{x}{x+1}\right]^x = \lim\limits_{x \to \infty} \left[\left(\dfrac{x+1}{x}\right)^{-1}\right]^x = \left[\lim\limits_{x \to \infty} \left(1+\dfrac{1}{x}\right)^x\right]^{-1} = e^{-1}$]

69. $y = x^{x^x} \Rightarrow \ln y = x^x \ln x \Rightarrow \lim\limits_{x \to 0^+} \ln y = \lim\limits_{x \to 0^+} x^x \ln x = -\infty$ since

$\ln x \to -\infty$ and $x^x \to 1$ (from Example 9). So $\lim\limits_{x \to 0^+} x^{x^x} = \lim\limits_{x \to 0^+} e^{\ln y} = 0$

71. $\lim\limits_{x \to \infty} \dfrac{x}{2x + 3 \sin x} = \lim\limits_{x \to \infty} \dfrac{1}{2 + 3(\sin x / x)} = \dfrac{1}{2+3(0)} = \dfrac{1}{2}$

73. $\lim\limits_{x \to 0^+} \dfrac{x+1-e^x}{x^3} \overset{H}{=} \lim\limits_{x \to 0^+} \dfrac{1-e^x}{3x^2} \overset{H}{=} \lim\limits_{x \to 0^+} \dfrac{-e^x}{6x} = -\infty$ since $6x \to 0^+$ and $-e^x \to -1$

75. $\displaystyle\lim_{x\to 0} \frac{2x \sin x}{\sec x - 1} \overset{H}{=} \lim_{x\to 0} \frac{2 \sin x + 2x \cos x}{\sec x \tan x} \overset{H}{=} \lim_{x\to 0} \frac{4 \cos x - 2x \sin x}{\sec x \tan^2 x + \sec^3 x}$

$= 4$

77. $\displaystyle\lim_{x\to 0} \frac{\cos x - 1 + x^2/2}{x^4} \overset{H}{=} \lim_{x\to 0} \frac{-\sin x + x}{4x^3} \overset{H}{=} \lim_{x\to 0} \frac{-\cos x + 1}{12x^2}$

$\overset{H}{=} \displaystyle\lim_{x\to 0} \frac{\sin x}{24x} = \frac{1}{24}$

79. $y = f(x) = xe^{-x}$    A. $D = \mathbb{R}$   B. Intercepts are 0   C. No symmetry

D. $\displaystyle\lim_{x\to\infty} xe^{-x} = \lim_{x\to\infty} \frac{x}{e^x} \overset{H}{=} \lim_{x\to\infty} \frac{1}{e^x} = 0$, so $y = 0$ is a HA. $\displaystyle\lim_{x\to -\infty} xe^{-x} = -\infty$

E. $f'(x) = e^{-x} - xe^{-x} = e^{-x}(1-x) > 0 \Longleftrightarrow x < 1$, so $f$ is increasing on

$(-\infty, 1]$ and decreasing on $[1, \infty)$.        H.

F. Absolute maximum $f(1) = 1/e$

G. $f''(x) = e^{-x}(x-2) > 0 \Longleftrightarrow x > 2$

so $f$ is CU on $(2, \infty)$ and CD on $(-\infty, 2)$.

IP is $(2, 2/e^2)$

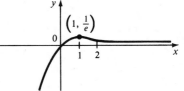

81. $y = f(x) = x \ln x$   A. $D = (0, \infty)$   B. x-intercept when $\ln x = 0 \Longleftrightarrow$

$x = 1$, no y-intercept   C. No symmetry   D. $\displaystyle\lim_{x\to\infty} x \ln x = \infty$,

$\displaystyle\lim_{x\to 0^+} x \ln x = \lim_{x\to 0^+} \frac{\ln x}{1/x} \overset{H}{=} \lim_{x\to 0^+} \frac{1/x}{-1/x^2} = \lim_{x\to 0^+} (-x) = 0$, no asymptotes

E. $f'(x) = \ln x + 1 = 0$ when $\ln x = -1 \Longleftrightarrow x = e^{-1}$.   $f'(x) > 0 \Longleftrightarrow$

$\ln x > -1 \Longleftrightarrow x > e^{-1}$, so $f$ is        H.

increasing on $[1/e, \infty)$ and

decreasing on $(0, 1/e]$.

F. $f(1/e) = -1/e$ is an absolute

minimum.   G. $f''(x) = 1/x > 0$, so

$f$ is CU on $(0, \infty)$. No IP.

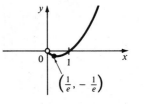

83. $y = f(x) = x^2 \ln x$   A. $D = (0, \infty)$   B. x-intercept when $\ln x = 0 \Longleftrightarrow$

$x = 1$, no y-intercept   C. No symmetry   D. $\displaystyle\lim_{x\to\infty} x^2 \ln x = \infty$,

$\displaystyle\lim_{x\to 0^+} x^2 \ln x = \lim_{x\to 0^+} \frac{\ln x}{1/x^2} \overset{H}{=} \lim_{x\to 0^+} \frac{1/x}{-2/x^3} = \lim_{x\to 0^+} \left(-\frac{x^2}{2}\right) = 0$, no asymptote

E. $f'(x) = 2x \ln x + x = x(2 \ln x + 1) > 0 \Longleftrightarrow \ln x > -1/2 \Longleftrightarrow$

$x > e^{-1/2}$, so $f$ is increasing on $[1/\sqrt{e}, \infty)$, decreasing on $(0, 1/\sqrt{e}]$.

F. $f(1/\sqrt{e}) = -1/2e$ is an absolute
minimum  G. $f''(x) = 2 \ln x + 3 > 0$
$\Leftrightarrow \ln x > -3/2 \Leftrightarrow x > e^{-3/2}$, so f
is CU on $(e^{-3/2}, \infty)$ and CD on
$(0, e^{-3/2})$.  IP is $(e^{-3/2}, -3/2e^3)$

H.

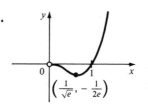

85.  $y = f(x) = xe^{-x^2}$  A. D = R  B. Intercepts are 0  C. $f(-x) = -f(x)$,
so the curve is symmetric about the origin.

D. $\lim\limits_{x \to \pm\infty} xe^{-x^2} = \overset{H}{\lim\limits_{x \to \pm\infty}} \dfrac{x}{e^{x^2}} = \lim\limits_{x \to \pm\infty} \dfrac{1}{2xe^{x^2}} = 0$, so y = 0 is a HA.

E. $f'(x) = e^{-x^2} - 2x^2 e^{-x^2} = e^{-x^2}(1-2x^2) > 0 \Leftrightarrow x^2 < 1/2 \Leftrightarrow |x| < 1/\sqrt{2}$
so f is increasing on $[-1/\sqrt{2}, 1/\sqrt{2}]$ and decreasing on $(-\infty, -1/\sqrt{2}]$ and
$[1/\sqrt{2}, \infty)$.  F. $f(1/\sqrt{2}) = 1/\sqrt{2e}$ is a local maximum, $f(-1/\sqrt{2}) = -1/\sqrt{2e}$
is a local minimum.

G. $f''(x) = -2xe^{-x^2}(1-2x^2) - 4xe^{-x^2}$

$= 2xe^{-x^2}(2x^2-3) > 0 \Leftrightarrow x > \sqrt{3/2}$ or
$-\sqrt{3/2} < x < 0$, so f is CU on $(\sqrt{3/2}, \infty)$
and $(-\sqrt{3/2}, 0)$ and CD on $(-\infty, -\sqrt{3/2})$
and $(0, \sqrt{3/2})$.  IP are $(0,0)$ and
$(\pm\sqrt{3/2}, \pm\sqrt{3/2}e^{-3/2})$

H.

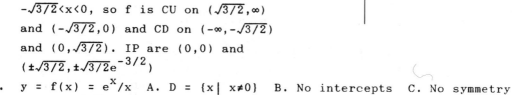

87.  $y = f(x) = e^x/x$  A. D = $\{x \mid x \neq 0\}$  B. No intercepts  C. No symmetry

D. $\lim\limits_{x \to \infty} \dfrac{e^x}{x} \overset{H}{=} \lim\limits_{x \to \infty} \dfrac{e^x}{1} = \infty$, $\lim\limits_{x \to -\infty} \dfrac{e^x}{x} = 0$, so y = 0 is a HA. $\lim\limits_{x \to 0^+} \dfrac{e^x}{x} = \infty$,

$\lim\limits_{x \to 0^-} \dfrac{e^x}{x} = -\infty$, so x = 0 is a VA.  E. $f'(x) = \dfrac{xe^x - e^x}{x^2} > 0 \Leftrightarrow (x-1)e^x > 0$

$\Leftrightarrow x > 1$, so f is increasing on $[1, \infty)$, decreasing on $(-\infty, 0)$ and $(0, 1]$

F. $f(1) = e$ is a local minimum

G. $f''(x) = \dfrac{x^2(xe^x) - 2x(xe^x - e^x)}{x^4}$

$= \dfrac{e^x(x^2 - 2x + 2)}{x^3} > 0 \Leftrightarrow x > 0$ since

$x^2 - 2x + 2 > 0$ for all x. So f is CU
on $(0, \infty)$ and CD on $(-\infty, 0)$. No IP.

H.

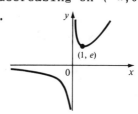

**89.** $y = f(x) = e^x/x^3$  A. $D = \{x \mid x \neq 0\}$  B. No intercepts  C. No symmetry

D. $\lim\limits_{x \to \infty} \dfrac{e^x}{x^3} \overset{H}{=} \lim\limits_{x \to \infty} \dfrac{e^x}{3x^2} \overset{H}{=} \lim\limits_{x \to \infty} \dfrac{e^x}{6x} \overset{H}{=} \lim\limits_{x \to \infty} \dfrac{e^x}{6} = \infty$,  $\lim\limits_{x \to -\infty} \dfrac{e^x}{x^3} = 0$, so $y = 0$

is a HA. $\lim\limits_{x \to 0^+} \dfrac{e^x}{x^3} = \infty$, $\lim\limits_{x \to 0^-} \dfrac{e^x}{x^3} = -\infty$, so $x = 0$ is a VA.  E. $f'(x) =$

$\dfrac{x^3 e^x - 3x^2 e^x}{x^6} = \dfrac{e^x(x-3)}{x^4} > 0 \Longleftrightarrow x > 3$, so $f$ is increasing on $[3,\infty)$ and

decreasing on $(-\infty,0)$ and $(0,3]$.　　H.

F. $f(3) = e^3/27$ is a local minimum

G. $f''(x) = \dfrac{e^x(x^2-6x+12)}{x^5} > 0 \Longleftrightarrow x > 0$

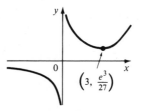

$\left(3, \dfrac{e^3}{27}\right)$

since $x^2-6x+12 > 0$ for all $x$.  So $f$
is CU on $(0,\infty)$ and CD on $(-\infty,0)$.

**91.** $y = f(x) = xe^{1/x}$  A. $D = \{x \mid x \neq 0\}$  B. No intercepts  C. No symmetry

D. $\lim\limits_{x \to \infty} xe^{1/x} = \infty$, $\lim\limits_{x \to -\infty} xe^{1/x} = -\infty$, no HA. $\lim\limits_{x \to 0^+} xe^{1/x} = \lim\limits_{x \to 0^+} \dfrac{e^{1/x}}{1/x}$

$\overset{H}{=} \lim\limits_{x \to 0^+} \dfrac{e^{1/x}(-1/x^2)}{-1/x^2} = \lim\limits_{x \to 0^+} e^{1/x} = \infty$, so $x = 0$ is a VA.  Also

$\lim\limits_{x \to 0^-} xe^{1/x} = 0$ since $\dfrac{1}{x} \to -\infty \Rightarrow e^{1/x} \to 0$.

E. $f'(x) = e^{1/x} + xe^{1/x}(-1/x^2) = e^{1/x}(1-1/x) > 0 \Longleftrightarrow 1/x < 1 \Longleftrightarrow x < 0$

or $x > 1$, so $f$ is increasing on $(-\infty,0)$ and $[1,\infty)$, decreasing on $(0,1]$

F. $f(1) = e$ is a local　　　　　　　H.

minimum

G. $f''(x) = e^{1/x}(-1/x^2)(1-1/x)$

$+ e^{1/x}(1/x^2) = e^{1/x}/x^3 > 0 \Longleftrightarrow$

$x > 0$, so $f$ is CU on $(0,\infty)$ and

CD on $(-\infty,0)$.  No IP.

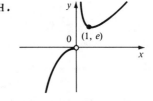

$(1, e)$

**93.** $y = f(x) = e^x + x$  A. $D = R$  B. x-intercept between $-1$ and $0$,

y-intercept $= f(0) = 1$  C. No symmetry  D. $\lim\limits_{x \to \infty} (e^x + x) = \infty$,

$\lim\limits_{x \to -\infty} (e^x + x) = -\infty$, so no HA.  But $f(x) - x = e^x \to 0$ as $x \to -\infty$, so

$y = x$ is a slant asymptote.

E. $f'(x) = e^x + 1 > 0$ for all x,
so f is increasing on R.

F. No extrema

G. $f''(x) = e^x > 0$ for all x, so
f is CU on R.

H.

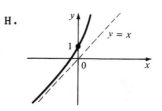

95. $y = f(x) = x - \ln(1+x)$   A. $D = \{x \mid x > -1\} = (-1, \infty)$   B. Intercepts
are 0.   C. No symmetry   D. $\lim_{x \to -1^+} (x - \ln(1+x)) = \infty$, so $x = -1$ is a

VA.   $\lim_{x \to \infty} (x - \ln(1+x)) = \lim_{x \to \infty} x \left[ 1 - \dfrac{\ln(1+x)}{x} \right] = \infty$, since $\lim_{x \to \infty} \dfrac{\ln(1+x)}{x}$

$\overset{H}{=} \lim_{x \to \infty} \dfrac{1/(1+x)}{1} = 0$   E. $f'(x) = 1 - \dfrac{1}{1+x} = \dfrac{x}{1+x} > 0 \Leftrightarrow x > 0$ since

$x+1 > 0$.   So f is increasing on
$[0, \infty)$ and decreasing on $(-1, 0]$.

F. $f(0) = 0$ is an absolute minimum.

G. $f''(x) = 1/(1+x)^2 > 0$, so f is
CU on $(-1, \infty)$.

H.

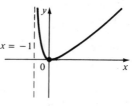

97. $y = f(x) = x^{1/\ln x}$

A. $D = \{x \mid x > 0, \ln x \neq 0\}$
$= (0,1) \cup (1, \infty)$

Note that $x^{1/\ln x} = (e^{\ln x})^{(1/\ln x)}$
$= e^1 = e$ for all $x \in D$.

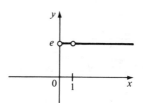

99. $y = f(x) = x^{\sqrt{x}}$   A. $D = (0, \infty)$   B. No intercepts   C. No symmetry

D. $\lim_{x \to \infty} x^{\sqrt{x}} = \infty$.   $\ln y = \sqrt{x} \ln x \Rightarrow \lim_{x \to 0^+} \ln y = \lim_{x \to 0^+} \sqrt{x} \ln x = $

$\lim_{x \to 0^+} \dfrac{\ln x}{1/\sqrt{x}} \overset{H}{=} \lim_{x \to 0^+} \dfrac{1/x}{(-1/2)x^{-3/2}} = \lim_{x \to 0^+} (-2\sqrt{x}) = 0 \Rightarrow \lim_{x \to 0^+} x^{\sqrt{x}} = e^0 = 1$

E. $\ln y = \sqrt{x} \ln x \Rightarrow \dfrac{y'}{y} = \dfrac{1}{2\sqrt{x}} \ln x + \dfrac{1}{\sqrt{x}} \Rightarrow f'(x) = x^{\sqrt{x}} \left[ \dfrac{\ln x + 2}{2\sqrt{x}} \right]$

$f'(x) > 0 \Leftrightarrow \ln x > -2 \Leftrightarrow x > e^{-2}$,   H.

so f is increasing on $[e^{-2}, \infty)$
and decreasing on $(0, e^{-2}]$.

F. $f(e^{-2}) = e^{-2/e}$ is an absolute
minimum.

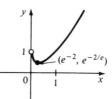

**101.** $y = f(x) = x^{x^2}$    A. $D = (0, \infty)$   B. No intercepts   C. No symmetry

D. $\lim\limits_{x \to \infty} x^{x^2} = \infty$, $\ln y = x^2 \ln x$  $\Rightarrow$  $\lim\limits_{x \to 0^+} \ln y = \lim\limits_{x \to 0^+} x^2 \ln x$

$= \lim\limits_{x \to 0^+} \dfrac{\ln x}{x^{-2}} \overset{H}{=} \lim\limits_{x \to 0^+} \dfrac{1/x}{-2x^{-3}} = \lim\limits_{x \to 0^+} (-\tfrac{1}{2}x^2) = 0$, so $\lim\limits_{x \to 0^+} x^{x^2} = e^0 = 1$

E. $\ln y = x^2 \ln x$  $\Rightarrow$  $y'/y = 2x \ln x + x$  $\Rightarrow$  $f'(x) = x^{x^2}(2x \ln x + x)$

$= x^{x^2+1}(2 \ln x + 1) > 0 \iff$

$\ln x > -1/2 \iff x > e^{-1/2}$, so

f is increasing on $[1/\sqrt{e}, \infty)$

and decreasing on $(0, 1/\sqrt{e}]$

F. $f(1/\sqrt{e}) = e^{-1/2e}$ is an

absolute minimum.

H.

**103.** $y = f(x) = (1+x)^{1/x}$    A. $D = \{x \mid 1+x > 0,\ x \neq 0\} = (-1, 0) \cup (0, \infty)$

B. No intercept   C. No symmetry   D. $\lim\limits_{x \to \infty} \ln y = \lim\limits_{x \to \infty} \dfrac{\ln(1+x)}{x} \overset{H}{=}$

$\lim\limits_{x \to \infty} \dfrac{1/(1+x)}{1} = 0$  $\Rightarrow$  $\lim\limits_{x \to \infty} (1+x)^{1/x} = e^0 = 1$, so $y = 1$ is a HA.

Also $\lim\limits_{x \to 0} (1+x)^{1/x} = e$, by definition of e, and $\lim\limits_{x \to -1^+} \ln y =$

$\lim\limits_{x \to -1^+} \dfrac{\ln(1+x)}{x} \overset{H}{=} \lim\limits_{x \to -1^+} \dfrac{1/(1+x)}{1} = \infty$  $\Rightarrow$  $\lim\limits_{x \to -1^+} (1+x)^{1/x} = \infty$, so $x = -1$

is a VA. E. $\dfrac{y'}{y} = \dfrac{x/(1+x) - \ln(1+x)}{x^2}$  $\Rightarrow$  $f'(x) = (1+x)^{1/x} \dfrac{x/(1+x) - \ln(1+x)}{x^2}$

$\Rightarrow$  $f'(x) < 0 \iff g(x) = \dfrac{x}{1+x} - \ln(1+x) < 0$.  But $g'(x) = -x/(1+x)^2$  $\Rightarrow$

$g'(0) = 0$ and $g'(x)$ changes from   H.

$+$ to $-$ at 0, so $g(0) = 0$ is the

maximum value of g.  Thus $g(x)$

$= \dfrac{x}{1+x} - \ln(1+x) \leq 0$ for all x

$\Rightarrow$  f is decreasing on D.

F. No extrema

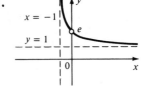

**105.** $\lim\limits_{x \to \infty} \dfrac{e^x}{x^n} \overset{H}{=} \lim\limits_{x \to \infty} \dfrac{e^x}{nx^{n-1}} \overset{H}{=} \lim\limits_{x \to \infty} \dfrac{e^x}{n(n-1)x^{n-2}} \overset{H}{=} \cdots \overset{H}{=} \lim\limits_{x \to \infty} \dfrac{e^x}{n!} = \infty$

**107.** $\lim\limits_{x \to 0^+} x^{\alpha} \ln x = \lim\limits_{x \to 0^+} \dfrac{\ln x}{x^{-\alpha}} \overset{H}{=} \lim\limits_{x \to 0^+} \dfrac{1/x}{-\alpha x^{-\alpha-1}} = \lim\limits_{x \to 0^+} \dfrac{x^{\alpha}}{-\alpha} = 0$ since $\alpha > 0$

**109.** (a) We show that $\lim_{x\to0} \dfrac{f(x)}{x^n} = 0$ for every integer $n \geq 0$. Let $y = \dfrac{1}{x^2}$.

Then $\lim_{x\to0} \dfrac{f(x)}{x^{2n}} = \lim_{x\to0} \dfrac{e^{-1/x^2}}{(x^2)^n} = \lim_{y\to\infty} \dfrac{y^n}{e^y} \overset{H}{=} \lim_{y\to\infty} \dfrac{ny^{n-1}}{e^y} \overset{H}{=} \cdots \overset{H}{=} \lim_{y\to\infty} \dfrac{n!}{e^y} = 0$

$\Rightarrow \lim_{x\to0} \dfrac{f(x)}{x^n} = \lim_{x\to0} x^n \dfrac{f(x)}{x^{2n}} = \lim_{x\to0} x^n \lim_{x\to0} \dfrac{f(x)}{x^{2n}} = 0$    Thus

$f'(0) = \lim_{x\to0} \dfrac{f(x)-f(0)}{x-0} = \lim_{x\to0} \dfrac{f(x)}{x} = 0$

(b) Using the Chain Rule and Quotient Rule we see that $f^{(n)}(x)$ exists for $x \neq 0$. In fact, we prove by induction that for each $n \geq 0$, there is a polynomial $p_n$ and nonnegative integer $k_n$ with $f^{(n)}(x) = p_n(x)f(x)/x^{k_n}$ for $x \neq 0$. This is true for $n = 0$; suppose it is true for the nth derivative. Then

$f^{(n+1)}(x) = [x^{k_n}(p_n'(x)f(x)+p_n(x)f'(x)) - k_n x^{k_n-1} p_n(x)f(x)]x^{-2k_n}$

$= [x^{k_n}p_n'(x) + p_n(x)(2/x^3) - k_n x^{k_n-1} p_n(x)]f(x)x^{-2k_n}$

$= [x^{k_n+3}p_n'(x) + 2p_n(x) - k_n x^{k_n+2} p_n(x)]f(x)x^{-(2k_n+3)}$

which has the desired form. Now we show by induction that $f^{(n)}(0) = 0$ for all $n$. By (a), $f'(0) = 0$. Suppose that $f^{(n)}(0) = 0$. Then

$f^{(n+1)}(0) = \lim_{x\to0} \dfrac{f^{(n)}(x)-f^{(n)}(0)}{x-0} = \lim_{x\to0} \dfrac{f^{(n)}(x)}{x} = \lim_{x\to0} \dfrac{p_n(x)f(x)/x^{k_n}}{x}$

$= \lim_{x\to0} \dfrac{p_n(x)f(x)}{x^{k_n+1}} = \lim_{x\to0} p_n(x) \lim_{x\to0} \dfrac{f(x)}{x^{k_n+1}} = p(0) \cdot 0 = 0$

**Chapter 6 Review**

**Review Exercises for Chapter 6**

**1.** $y = 7^x$

**3.** $y = e^{-x}$     $y = -e^{-x}$

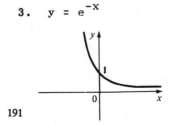

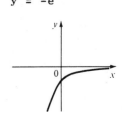

**5.** $y = \log_5 x$

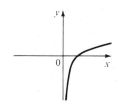

**7.** $y = \log_{1/5} x$

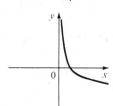

**9.** $e^x = 5 \;\Rightarrow\; x = \ln(e^x) = \ln 5$

**11.** $\log_{10}(e^x) = 1 \;\Rightarrow\; e^x = 10 \;\Rightarrow\; x = \ln(e^x) = \ln 10$

[OR: $1 = \log_{10}(e^x) = x \log_{10} e \;\Rightarrow\; x = 1/\log_{10} e$]

**13.** $2 = \ln(x^\pi) = \pi \ln x \;\Rightarrow\; \ln x = 2/\pi \;\Rightarrow\; x = e^{2/\pi}$

**15.** $\tan x = 4 \;\Rightarrow\; x = \tan^{-1}4 + n\pi = \arctan 4 + n\pi$, n an integer

**17.** $y = \log_{10}(x^2-x) \;\Rightarrow\; y' = \dfrac{1}{x^2-x}(\log_{10} e)(2x-1) = \dfrac{2x-1}{(\ln 10)(x^2-x)}$

**19.** $y = \dfrac{\sqrt{x+1}(2-x)^5}{(x+3)^7} \;\Rightarrow\; \ln|y| = \frac{1}{2}\ln(x+1) + 5\ln|2-x| - 7\ln(x+3) \;\Rightarrow\;$

$\dfrac{y'}{y} = \dfrac{1}{2(x+1)} + \dfrac{-5}{2-x} - \dfrac{7}{x+3} \;\Rightarrow\; y' = \dfrac{\sqrt{x+1}(2-x)^5}{(x+3)^7}\left[\dfrac{1}{2(x+1)} - \dfrac{5}{2-x} - \dfrac{7}{x+3}\right]$

**21.** $y = e^{cx}(c \sin x - \cos x) \;\Rightarrow\;$

$y' = ce^{cx}(c \sin x - \cos x) + e^{cx}(c \cos x + \sin x) = (c^2+1)e^{cx}\sin x$

**23.** $y = \ln(\sec^2 x) = 2\ln|\sec x| \;\Rightarrow\; y' = (2/\sec x)(\sec x \tan x) = 2 \tan x$

**25.** $y = xe^{-1/x} \;\Rightarrow\; y' = e^{-1/x} + xe^{-1/x}(1/x^2) = e^{-1/x}(1+1/x)$

**27.** $y = (\cos^{-1}x)^{\sin^{-1}x} \;\Rightarrow\; \ln y = \sin^{-1}x \ln(\cos^{-1}x) \;\Rightarrow\;$

$\dfrac{y'}{y} = \dfrac{1}{\sqrt{1-x^2}}\ln(\cos^{-1}x) + \sin^{-1}x \dfrac{1}{\cos^{-1}x}\left[-\dfrac{1}{\sqrt{1-x^2}}\right] \;\Rightarrow\;$

$y' = (\cos^{-1}x)^{\sin^{-1}x}\left[\dfrac{\ln(\cos^{-1}x) - \sin^{-1}x/\cos^{-1}x}{\sqrt{1-x^2}}\right]$

**29.** $y = e^{e^x} \;\Rightarrow\; y' = e^{e^x}e^x = e^{x+e^x}$

**31.** $y = \ln(1/x) + 1/\ln x = -\ln x + (\ln x)^{-1} \;\Rightarrow\; y' = -1/x - 1/x(\ln x)^2$

**33.** $y = 7^{\sqrt{2x}} \;\Rightarrow\; y' = 7^{\sqrt{2x}}(\ln 7)(1/2\sqrt{2x})(2) = 7^{\sqrt{2x}}(\ln 7)/\sqrt{2x}$

**35.** $y = xe^{\sec^{-1}x} \;\Rightarrow\; y' = e^{\sec^{-1}x}+xe^{\sec^{-1}x}(1/x\sqrt{x^2-1}) = e^{\sec^{-1}x}(1+1/\sqrt{x^2-1})$

**37.** $y = \ln(\cosh 3x) \;\Rightarrow\; y' = (1/\cosh 3x)(\sinh 3x)(3) = 3\tanh 3x$

**39.** $y = \cosh^{-1}(\sinh x) \;\Rightarrow\; y' = \cosh x/\sqrt{\sinh^2 x - 1}$

**41.** $y = \ln \sin x - \dfrac{1}{2} \sin^2 x$ $\Rightarrow$ $y' = \dfrac{\cos x}{\sin x} - \sin x \cos x$

$= \cot x - \sin x \cos x$

**43.** $y = \sin^{-1}\left[\dfrac{x-1}{x+1}\right]$ $\Rightarrow$ $y' = \dfrac{1}{\sqrt{1-[(x-1)/(x+1)]^2}} \dfrac{(x+1)-(x-1)}{(x+1)^2}$

$= \dfrac{1}{\sqrt{(x+1)^2-(x-1)^2}}\dfrac{2}{x+1} = \dfrac{2}{\sqrt{4x}(x+1)} = \dfrac{1}{\sqrt{x}(x+1)}$  [Note that the domain of $y'$ is $x > 0$.]

**45.** $y = \dfrac{1}{4}\left[\ln(x^2+x+1) - \ln(x^2-x+1)\right] + \dfrac{1}{2\sqrt{3}}\left[\tan^{-1}\left[\dfrac{2x+1}{\sqrt{3}}\right] + \tan^{-1}\left[\dfrac{2x-1}{\sqrt{3}}\right]\right]$ $\Rightarrow$

$y' = \dfrac{1}{4}\left[\dfrac{2x+1}{x^2+x+1} - \dfrac{2x-1}{x^2-x+1}\right] + \dfrac{1}{2\sqrt{3}}\left[\dfrac{2/\sqrt{3}}{1+[(2x+1)/\sqrt{3}]^2} + \dfrac{2/\sqrt{3}}{1+[(2x-1)/\sqrt{3}]^2}\right]$

$= \dfrac{1}{4}\left[\dfrac{2x+1}{x^2+x+1} - \dfrac{2x-1}{x^2-x+1}\right] + \dfrac{1}{4(x^2+x+1)} + \dfrac{1}{4(x^2-x+1)}$

$= \dfrac{1}{2}\left[\dfrac{x+1}{x^2+x+1} - \dfrac{x-1}{x^2-x+1}\right] = \dfrac{1}{x^4+x^2+1}$

**47.** $f(x) = 2^x$ $\Rightarrow$ $f'(x) = 2^x \ln 2$ $\Rightarrow$ $f''(x) = 2^x(\ln 2)^2$ $\Rightarrow$ $\cdots$ $\Rightarrow$
$f^{(n)}(x) = 2^x(\ln 2)^n$

**49.** $y = f(x) = \ln(e^x+e^{2x})$ $\Rightarrow$ $f'(x) = (e^x+2e^{2x})/(e^x+e^{2x})$ $\Rightarrow$ $f'(0) = 3/2$,
so the tangent line at $(0,\ln 2)$ is $y - \ln 2 = \dfrac{3}{2}x$ or $3x-2y+ \ln 4 = 0$

**51.** $y = [\ln(x+4)]^2$ $\Rightarrow$ $y' = 2 \ln(x+4)/(x+4) = 0$ $\Longleftrightarrow$ $\ln(x+4) = 0$ $\Longleftrightarrow$
$x+4 = 1$ $\Longleftrightarrow$ $x = -3$, so the tangent is horizontal at $(-3,0)$.

**53.** $\lim\limits_{x\to-\infty} 10^{-x} = \infty$ since $-x \to \infty$ as $x \to -\infty$

**55.** $\lim\limits_{x\to0^+} \ln(\tan x) = -\infty$ since $\tan x \to 0^+$ as $x \to 0^+$

**57.** $\lim\limits_{x\to-4^+} e^{1/(x+4)} = \infty$ since $\dfrac{1}{x+4} \to \infty$ as $x \to -4^+$

**59.** $\lim\limits_{x\to1} \cos^{-1}\left[\dfrac{x}{x+1}\right] = \cos^{-1}\left[\dfrac{1}{2}\right] = \dfrac{\pi}{3}$

**61.** $\lim\limits_{x\to\pi} \dfrac{\sin x}{x^2-\pi^2} \overset{H}{=} \lim\limits_{x\to\pi} \dfrac{\cos x}{2x} = \dfrac{-1}{2\pi}$

**63.** $\lim\limits_{x\to\infty} \dfrac{\ln(\ln x)}{\ln x} \overset{H}{=} \lim\limits_{x\to\infty} \dfrac{1/x \ln x}{1/x} = \lim\limits_{x\to\infty} \dfrac{1}{\ln x} = 0$

**65.** $\lim\limits_{x\to0} \dfrac{\ln(1-x)+x+x^2/2}{x^3} \overset{H}{=} \lim\limits_{x\to0} \dfrac{-1/(1-x)+1+x}{3x^2} \overset{H}{=} \lim\limits_{x\to0} \dfrac{-1/(1-x)^2+1}{6x} \overset{H}{=}$

$\lim\limits_{x\to0} \dfrac{-2/(1-x)^3}{6} = -\dfrac{2}{6} = -\dfrac{1}{3}$

**67.** $\displaystyle\lim_{x\to0^+} \sin x \, (\ln x)^2 = \lim_{x\to0^+} \frac{(\ln x)^2}{\csc x} \overset{H}{=} \lim_{x\to0^+} \frac{2 \ln x \, / \, x}{- \csc x \cot x}$

$\displaystyle = -2 \lim_{x\to0} \frac{\sin x}{x} \lim_{x\to0} \frac{\ln x}{\cot x} = -2 \lim_{x\to0} \frac{\ln x}{\cot x} \overset{H}{=} -2 \lim_{x\to0} \frac{1/x}{- \csc^2 x}$

$\displaystyle = 2 \lim_{x\to0} \frac{\sin^2 x}{x} = 2 \lim_{x\to0} \frac{\sin x}{x} \lim_{x\to0} \sin x = 2 \cdot 1 \cdot 0 = 0$

**69.** $\displaystyle\lim_{x\to1} (\ln x)^{\sin x} = (\ln 1)^{\sin 1} = 0^{\sin 1} = 0$

**71.** $\displaystyle\lim_{x\to0^+} \frac{\sqrt[3]{x}-1}{\sqrt[4]{x}-1} = \frac{0-1}{0-1} = 1$

**73.** $y = f(x) = \tan^{-1}(1/x)$  A. $D = \{x \mid x \neq 0\}$  B. No intercepts

C. $f(-x) = -f(x)$, so the curve is symmetric about the origin.

D. $\displaystyle\lim_{x\to\pm\infty} \tan^{-1}(1/x) = \tan^{-1}0 = 0$, so $y = 0$ is a HA.

$\displaystyle\lim_{x\to0^+} \tan^{-1}(1/x) = \frac{\pi}{2}$ and $\displaystyle\lim_{x\to0^-} \tan^{-1}(1/x) = -\frac{\pi}{2}$ since $\frac{1}{x} \to \pm\infty$ as $x \to 0^{\pm}$

E. $f'(x) = \dfrac{1}{1+(1/x)^2}(-1/x^2) = \dfrac{-1}{x^2+1}$  H.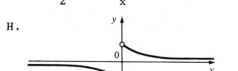

$\Rightarrow f'(x) < 0$, so $f$ is decreasing
on $(-\infty,0)$ and $(0,\infty)$. F. No extrema

G. $f''(x) = 2x/(x^2+1)^2 > 0 \iff x > 0$,
so $f$ is CU on $(0,\infty)$ and CD on $(-\infty,0)$

**75.** $y = f(x) = 2^{1/(x-1)}$  A. $D = \{x \mid x \neq 1\}$  B. No x-intercept,

y-intercept $= f(0) = 1/2$  C. No symmetry  D. $\displaystyle\lim_{x\to\pm\infty} 2^{1/(x-1)} = 2^0 = 1$

so $y = 1$ is a HA.  $\displaystyle\lim_{x\to1^+} 2^{1/(x-1)} = \infty$, so $x = 1$ is a VA  Also

$\displaystyle\lim_{x\to1^-} 2^{1/(x-1)} = 0$  E. $f'(x) = 2^{1/(x-1)}(-\ln 2)/(x-1)^2 < 0$ so $f$ is

decreasing on $(-\infty,1)$ and $(1,\infty)$.  H.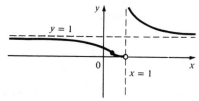

F. No extrema

G. $y'' = \dfrac{2^{1/(x-1)}(\ln 2)(2x-2 + \ln 2)}{(x-1)^4}$

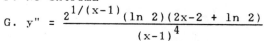

$> 0 \iff 2x-2 + \ln 2 > 0 \iff x > 1 - \frac{1}{2}\ln 2$

so $f$ is CU on $(1-\ln\sqrt{2},1)$ and $(1,\infty)$ and
CD on $(-\infty,1-\ln\sqrt{2})$.  IP when $x = 1-\ln\sqrt{2}$

**77.** $y = f(x) = e^x + e^{-3x}$   A. $D = R$   B. No x-intercept, y-intercept = $f(0)$ = 2   C. No symmetry   D. $\lim\limits_{x\to\pm\infty} (e^x + e^{-3x}) = \infty$, no asymptote

E. $f'(x) = e^x - 3e^{-3x} = e^{-3x}(e^{4x} - 3) > 0 \Leftrightarrow e^{4x} > 3 \Leftrightarrow 4x > \ln 3 \Leftrightarrow$ $x > \frac{1}{4} \ln 3$, so f is increasing   H.

on $[(\ln 3)/4, \infty)$ and decreasing on $(-\infty, (\ln 3)/4]$.   F. Absolute minimum $f((\ln 3)/4) = 3^{1/4} + 3^{-3/4}$
G. $f''(x) = e^x + 9e^{-3x} > 0$, so f is CU on $(-\infty, \infty)$

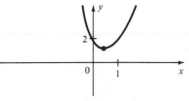

**79.** $y = f(x) = 2x^2 - \ln x$   A. $D = (0, \infty)$   B. No intercepts   C. No symmetry   D. $\lim\limits_{x\to 0^+} (2x^2 - \ln x) = \infty$, so $x = 0$ is a VA.

$\lim\limits_{x\to\infty} (2x^2 - \ln x) = \lim\limits_{x\to\infty} x^2 \left[2 - \frac{\ln x}{x^2}\right] = \infty$, since $\lim\limits_{x\to\infty} \frac{\ln x}{x^2} \overset{H}{=} \lim\limits_{x\to\infty} \frac{1/x}{2x}$

$= \lim\limits_{x\to\infty} \frac{1}{2x^2} = 0$   E. $f'(x) = 4x - \frac{1}{x} = \frac{4x^2 - 1}{x} > 0 \Leftrightarrow x^2 > 1/4$

$\Leftrightarrow x > 1/2$ (since $x > 0$), so f is   H. increasing on $[1/2, \infty)$ and decreasing on $(0, 1/2]$.   F. Absolute minimum $f(1/2) = \ln 2 + 1/2$
G. $f''(x) = 4 + 1/x^2 > 0$, so f is CU on $(0, \infty)$.

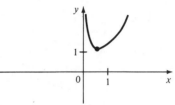

**81.** (a) $y(t) = y(0)e^{kt} = 1000e^{kt} \Rightarrow y(2) = 1000e^{2k} = 9000 \Rightarrow e^{2k} = 9$

$\Rightarrow 2k = \ln 9 \Rightarrow k = (1/2)\ln 9 = \ln 3 \Rightarrow y(t) = 1000e^{(\ln 3)t}$ $= 1000 \cdot 3^t$   (b) $y(3) = 1000 \cdot 3^3 = 27000$   (c) $1000 \cdot 3^t = 2000 \Rightarrow$ $3^t = 2 \Rightarrow t \ln 3 = \ln 2 \Rightarrow t = (\ln 2)/(\ln 3) \approx 0.63$ h

**83.** $s(t) = Ae^{-ct}\cos(\omega t + \delta) \Rightarrow v(t) = s'(t) = -cAe^{-ct}\cos(\omega t + \delta)$ $+ Ae^{-ct}[-\omega \sin(\omega t + \delta)] = -Ae^{-ct}[c \cos(\omega t + \delta) + \omega \sin(\omega t + \delta)]$
$a(t) = v'(t) = cAe^{-ct}[c \cos(\omega t + \delta) + \omega \sin(\omega t + \delta)] -$ $Ae^{-ct}[-\omega c \sin(\omega t + \delta) + \omega^2 \cos(\omega t + \delta)]$
$= Ae^{-ct}[(c^2 - \omega^2) \cos(\omega t + \delta) + 2c\omega \sin(\omega t + \delta)]$

**85.** $f(x) = e^{g(x)} \Rightarrow f'(x) = e^{g(x)} g'(x)$

**87.** $f(x) = \ln|g(x)| \Rightarrow f'(x) = g'(x)/g(x)$

**89.** $\int_0^{2\sqrt{3}} \frac{1}{x^2 + 4} dx = \frac{1}{2} \tan^{-1}\left[\frac{x}{2}\right]\Big]_0^{2\sqrt{3}} = \frac{1}{2}\left[\tan^{-1}\sqrt{3} - \tan^{-1} 0\right] = \frac{1}{2} \cdot \frac{\pi}{3} = \frac{\pi}{6}$

**91.** $\displaystyle\int_2^4 \frac{1+x-x^2}{x^2}\,dx = \int_2^4 (x^{-2} + \frac{1}{x} - 1)dx = \left[-\frac{1}{x} + \ln x - x\right]_2^4$

$\displaystyle = (-\frac{1}{4} + \ln 4 - 4) - (-\frac{1}{2} + \ln 2 - 2) = \ln 2 - \frac{7}{4}$

**93.** Let $u = \sqrt{x}$. Then $du = (1/2\sqrt{x})dx \Rightarrow \displaystyle\int \frac{e^{\sqrt{x}}}{\sqrt{x}}\,dx = 2\int e^u du = 2e^u + C$

$= 2e^{\sqrt{x}} + C$

**95.** Let $u = \ln(\cos x)$. Then $du = \dfrac{-\sin x}{\cos x}\,dx = -\tan x\,dx \Rightarrow$

$\displaystyle\int \tan x \ln(\cos x)dx = -\int u\,du = -\frac{1}{2}u^2 + C = -\frac{1}{2}[\ln(\cos x)]^2 + C$

**97.** Let $u = 1+x^4$. Then $du = 4x^3 dx \Rightarrow \displaystyle\int \frac{x^3}{1+x^4}\,dx = \frac{1}{4}\int \frac{1}{u}\,du$

$= \frac{1}{4}\ln|u| + C = \frac{1}{4}\ln(1+x^4) + C$

**99.** Let $u = 1 + \sec\theta$. Then $du = \sec\theta\tan\theta\,d\theta \Rightarrow$

$\displaystyle\int \frac{\sec\theta\tan\theta}{1 + \sec\theta}\,d\theta = \int \frac{1}{u}\,du = \ln|u| + C = \ln|1 + \sec\theta| + C$

**101.** $u = 3t \Rightarrow \displaystyle\int \cosh 3t\,dt = \frac{1}{3}\int \cosh u\,du = \frac{1}{3}\sinh u + C = \frac{1}{3}\sinh 3t + C$

**103.** $\cos x \le 1 \Rightarrow e^x\cos x \le e^x \Rightarrow \displaystyle\int_0^1 e^x\cos x\,dx \le \int_0^1 e^x dx = e^x\Big]_0^1 = e-1$

**105.** $f'(x) = \dfrac{d}{dx}\displaystyle\int_1^{\sqrt{x}} \frac{e^s}{s}\,ds = \frac{e^{\sqrt{x}}}{\sqrt{x}}\frac{d}{dx}\sqrt{x} = \frac{e^{\sqrt{x}}}{\sqrt{x}}\frac{1}{2\sqrt{x}} = \frac{e^{\sqrt{x}}}{2x}$

**107.** $f_{ave} = \dfrac{1}{4-1}\displaystyle\int_1^4 \frac{1}{x}\,dx = \frac{1}{3}\ln x\Big]_1^4 = \frac{1}{3}\ln 4$

**109.** $f(x) = \ln x + \tan^{-1}x \Rightarrow f(1) = \ln 1 + \tan^{-1}1 = \pi/4 \Rightarrow g(\pi/4) = 1$

$f'(x) = 1/x + 1/(1+x^2)$, so $g'(\pi/4) = 1/f'(1) = 1/(3/2) = 2/3$

**111.** $\displaystyle\lim_{x\to -1} F(x) = \lim_{x\to -1} \frac{b^{x+1}-a^{x+1}}{x+1} \overset{H}{=} \lim_{x\to -1} \frac{b^{x+1}\ln b - a^{x+1}\ln a}{1} = \ln b - \ln a$

$= F(-1)$, so $F$ is continuous at $-1$.

## CHAPTER SEVEN

### Exercises 7.1

1.  Let $u = x$, $dv = e^{2x}dx$ $\Rightarrow$ $du = dx$, $v = \frac{1}{2}e^{2x}$. Then by (7.2),
    $$\int xe^{2x}dx = \frac{1}{2}xe^{2x} - \int \frac{1}{2}e^{2x}dx = \frac{1}{2}xe^{2x} - \frac{1}{4}e^{2x} + C$$

3.  Let $u = x$, $dv = \sin 4x \, dx$ $\Rightarrow$ $du = dx$, $v = -\frac{1}{4}\cos 4x$.
    Then $\int x \sin 4x \, dx = -\frac{1}{4}x \cos 4x - \int -\frac{1}{4}\cos 4x \, dx$
    $= -\frac{1}{4}x \cos 4x + \frac{1}{16}\sin 4x + C$

5.  Let $u = x^2$, $dv = \cos 3x \, dx$ $\Rightarrow$ $du = 2x \, dx$, $v = \frac{1}{3}\sin 3x$. Then
    $I = \int x^2 \cos 3x \, dx = \frac{1}{3}x^2 \sin 3x - \frac{2}{3}\int x \sin 3x \, dx$ by (7.2). Next
    let $U = x$, $dV = \sin 3x \, dx$ $\Rightarrow$ $dU = dx$, $V = -\frac{1}{3}\cos 3x$ to get
    $\int x \sin 3x \, dx = -\frac{1}{3}x \cos 3x + \frac{1}{3}\int \cos 3x \, dx = -\frac{1}{3}x \cos 3x$
    $+ \frac{1}{9}\sin 3x + C_1$. Substituting for $\int x \sin 3x \, dx$, we get
    $I = \frac{1}{3}x^2 \sin 3x - \frac{2}{3}\left[-\frac{1}{3}x \cos 3x + \frac{1}{9}\sin 3x + C_1\right]$
    $= \frac{1}{3}x^2 \sin 3x + \frac{2}{9}x \cos 3x - \frac{2}{27}\sin 3x + C$, where $C = -\frac{2}{3}C_1$

7.  Let $u = (\ln x)^2$, $dv = dx$ $\Rightarrow$ $du = 2 \ln x \cdot \frac{1}{x}dx$, $v = x$. Then
    $I = \int (\ln x)^2 dx = x(\ln x)^2 - 2\int \ln x \, dx$. Taking $U = \ln x$, $dV = dx$
    $\Rightarrow$ $dU = \frac{1}{x}dx$, $V = x$, we find that
    $\int \ln x \, dx = x \ln x - \int x \cdot \frac{1}{x}dx = x \ln x - x + C_1$. Thus
    $I = x(\ln x)^2 - 2x \ln x + 2x + C$, where $C = -2C_1$

9.  $I = \int \theta \sin \theta \cos \theta \, d\theta = \frac{1}{4}\int 2\theta \sin 2\theta \, d\theta = \frac{1}{8}\int t \sin t \, dt$
    $[t = 2\theta \Rightarrow dt = d\theta/2]$. Let $u = t$, $dv = \sin t \, dt$ $\Rightarrow$ $du = dt$,
    $v = -\cos t$. Then $I = (1/8)(-t \cos t + \int \cos t \, dt)$
    $= (1/8)(-t \cos t + \sin t) + C = (1/8)(\sin 2\theta - 2\theta \cos 2\theta) + C$

11. Let $u = \ln t$, $dv = t^2$ $\Rightarrow$ $du = \frac{1}{t}dt$, $v = \frac{1}{3}t^3$. Then $\int t^2 \ln t \, dt$
    $= \frac{1}{3}t^3 \ln t - \int \frac{1}{3}t^3 \cdot \frac{1}{t}dt = \frac{1}{3}t^3 \ln t - \frac{1}{9}t^3 + C = \frac{1}{9}t^3(3 \ln t - 1) + C$

13. First let $u = \sin 3\theta$, $dv = e^{2\theta}d\theta$ $\Rightarrow$ $du = 3 \cos 3\theta \, d\theta$, $v = \frac{1}{2}e^{2\theta}$.
    Then $I = \int e^{2\theta} \sin 3\theta \, d\theta = \frac{1}{2}e^{2\theta} \sin 3\theta - \frac{3}{2}\int e^{2\theta}\cos 3\theta \, d\theta$. Next
    let $U = \cos 3\theta$, $dU = -3 \sin 3\theta \, d\theta$, $dV = e^{2\theta}d\theta$, $V = \frac{1}{2}e^{2\theta}$ to get

$\int e^{2\theta} \cos 3\theta \, d\theta = \frac{1}{2}e^{2\theta} \cos 3\theta + \frac{3}{2} \int e^{2\theta} \sin 3\theta \, d\theta$. Substituting in

the previous formula gives $I = \frac{1}{2}e^{2\theta} \sin 3\theta - \frac{3}{4}e^{2\theta} \cos 3\theta$

$- \frac{9}{4} \int e^{2\theta} \sin 3\theta \, d\theta$ or $\frac{13}{4} \int e^{2\theta} \sin 3\theta \, d\theta$

$= \frac{1}{2}e^{2\theta} \sin 3\theta - \frac{3}{4}e^{2\theta} \cos 3\theta + C_1$. Hence $\int e^{2\theta} \sin 3\theta \, d\theta$

$= \frac{1}{13}e^{2\theta}(2 \sin 3\theta - 3 \cos 3\theta) + C$, where $C = \frac{4}{13}C_1$

15. Let $u = y$, $dv = \sinh y \, dy$ $\Rightarrow$ $du = dy$, $v = \cosh y$. Then

$\int y \sinh y \, dy = y \cosh y - \int \cosh y \, dy = y \cosh y - \sinh y + C$

17. Let $u = t$, $dv = e^{-t} dt$ $\Rightarrow$ $du = dt$, $v = -e^{-t}$. Then (7.6) says

$\int_0^1 te^{-t}dt = -\left. te^{-t}\right]_0^1 + \int_0^1 e^{-t}dt = -\frac{1}{e} + \left[-e^{-t}\right]_0^1 = -\frac{1}{e}-\frac{1}{e}+1 = 1-\frac{2}{e}$

19. Let $u = x$, $dv = \cos 2x \, dx$ $\Rightarrow$ $du = dx$, $v = \frac{1}{2} \sin 2x \, dx$. Then

$\int_0^{\pi/2} x \cos 2x \, dx = \frac{1}{2}x \sin 2x\Big]_0^{\pi/2} - \frac{1}{2} \int_0^{\pi/2} \sin 2x \, dx$

$= 0 + \frac{1}{4} \cos 2x\Big]_0^{\pi/2} = \frac{1}{4}(-1-1) = -\frac{1}{2}$

21. Let $u = \cos^{-1}x$, $dv = dx$ $\Rightarrow$ $du = -dx/\sqrt{1-x^2}$, $v = x$. Then

$I = \int_0^1 \cos^{-1}x \, dx = x \cos^{-1}x\Big]_0^1 + \int_0^1 \frac{x \, dx}{\sqrt{1-x^2}} = 0 + \int_1^0 t^{-1/2}\left[-\frac{1}{2} \, dt\right]$

where $t = 1-x^2$ $\Rightarrow$ $dt = -2x \, dx$. Thus $I = \frac{1}{2} \int_0^1 t^{-1/2}dt = \sqrt{t}\Big]_0^1 = 1$

23. Let $u = \sin 3x$, $dv = \cos 5x \, dx$ $\Rightarrow$ $du = 3 \cos 3x \, dx$, $v = \frac{1}{5} \sin 5x$.

Then $I = \int \sin 3x \cos 5x \, dx = \frac{1}{5} \sin 3x \sin 5x - \frac{3}{5}\int \sin 5x \cos 3x \, dx$

Now use parts with $U = \cos 3x$, $dV = \sin 5x \, dx$ $\Rightarrow$ $dU = -3 \sin 3x \, dx$,

$V = -\frac{1}{5} \cos 5x$. Thus $\int \sin 5x \cos 3x \, dx = -\frac{1}{5} \cos 3x \cos 5x$

$- \frac{3}{5} \int \sin 3x \cos 5x \, dx$ $\Rightarrow$ $I = \frac{1}{5} \sin 3x \sin 5x$

$- \frac{3}{5}\left[-\frac{1}{5} \cos 3x \cos 5x - \frac{3}{5} I\right]$. Solving for $I$, we get

$I = \frac{1}{16}(5 \sin 3x \sin 5x + 3 \cos 3x \cos 5x) + C$

[Another method: Write $\sin 3x \cos 5x = (1/2)(\sin 8x - \sin 2x)$ as

in Section 7.2. This gives $I = (-1/16)\cos 8x + (1/4)\cos 2x + C$.]

25. Let $u = \ln(\sin x)$, $dv = \cos x \, dx$ $\Rightarrow$ $du = \frac{\cos x}{\sin x} \, dx$, $v = \sin x$.

Then $I = \int \cos x \ln(\sin x)dx = \sin x \ln(\sin x) - \int \cos x \, dx$

$= \sin x \ln(\sin x) - \sin x + C$ [Another method: Substitute

$t = \sin x$, so $dt = \cos x \, dx$. Then $I = \int \ln t \, dt = t \ln t - t + C$

[see Example 2] and so $I = \sin x (\ln \sin x - 1) + C$.]

**27.** Let $u = 2x+3$, $dv = e^x dx$ $\Rightarrow$ $du = 2\,dx$, $v = e^x$. Then $\int (2x+3)e^x\,dx$

$= (2x+3)e^x - \int e^x \cdot 2\,dx = (2x+3)e^x - 2e^x + C = (2x+1)e^x + C$

**29.** Let $w = \ln x$ $\Rightarrow$ $dw = dx/x$. Then $x = e^w$ and $dx = e^w\,dw$, so

$\int \cos(\ln x)dx = \int e^w \cos w\,dw = (1/2)e^w(\sin w + \cos w) + C$ [*by the*

*method of Example 4*] $= (1/2)x[\sin(\ln x) + \cos(\ln x)] + C$

**31.** $I = \int_1^4 \ln\sqrt{x}\,dx = \frac{1}{2}\int_1^4 \ln x\,dx = \frac{1}{2}[x \ln x - x]_1^4$ as in Example 2. So

$I = (1/2)[(4 \ln 4 - 4) - (0-1)] = 4 \ln 2 - 3/2$

**33.** Let $w = \sqrt{x}$, so that $x = w^2$ and $dx = 2w\,dw$. Then use $u = 2w$,

$dv = \sin w\,dw$. Thus $\int \sin \sqrt{x}\,dx = \int 2w \sin w\,dw$

$= - 2w \cos w + \int 2 \cos w\,dw = - 2w \cos w + 2 \sin w + C$

$= - 2\sqrt{x} \cos \sqrt{x} + 2 \sin\sqrt{x} + C$

**35.** $\int x^5 e^{x^2}\,dx = \int (x^2)^2 e^{x^2} x\,dx = \int t^2 e^t \frac{1}{2}\,dt$ [*where* $t = x^2$]

$= \frac{1}{2}(t^2 - 2t + 2)e^t + C$ [*by Example 3*] $= \frac{1}{2}(x^4 - 2x^2 + 2)e^{x^2} + C$

**37.** (a) Take $n = 2$ in Example 6 to get

$\int \sin^2 x\,dx = - \frac{1}{2} \cos x \sin x + \frac{1}{2} \int 1\,dx = \frac{x}{2} - \frac{\sin 2x}{4} + C$

(b) $\int \sin^4 x\,dx = - \frac{1}{4} \cos x \sin^3 x + \frac{3}{4} \int \sin^2 x\,dx$

$= - \frac{1}{4} \cos x \sin^3 x + \frac{3}{8}x - \frac{3}{16} \sin 2x + C$

**39.** (a) $\int_0^{\pi/2} \sin^n x\,dx = - \frac{1}{n} \cos x \sin^{n-1} x\Big]_0^{\pi/2} + \frac{n-1}{n} \int_0^{\pi/2} \sin^{n-2} x\,dx$

$= \frac{n-1}{n} \int_0^{\pi/2} \sin^{n-2} x\,dx$

(b) $\int_0^{\pi/2} \sin^3 x\,dx = \frac{2}{3} \int_0^{\pi/2} \sin x\,dx = - \frac{2}{3} \cos x\Big]_0^{\pi/2} = \frac{2}{3}$

$\int_0^{\pi/2} \sin^5 x\,dx = \frac{4}{5} \int_0^{\pi/2} \sin^3 x\,dx = \frac{4}{5} \cdot \frac{2}{3} = \frac{8}{15}$

(c) Let $m = (n-1)/2$, so that $n = 2m+1$, $m \geq 1$. The formula holds

for $m = 1$ (i.e., $n = 3$) by (b). Assume it holds for some value

of $m \geq 1$. Then $\int_0^{\pi/2} \sin^{2m+1} x\,dx = \frac{2 \cdot 4 \cdot 6 \cdots (2m)}{3 \cdot 5 \cdot 7 \cdots (2m+1)}$. By the

reduction forumula in Ex. 6, $\int_0^{\pi/2} \sin^{2m+3} x\,dx$

$= \frac{2m+2}{2m+3} \int_0^{\pi/2} \sin^{2m+1} x\,dx = \frac{2 \cdot 4 \cdot 6 \cdots [2(m+1)]}{2 \cdot 4 \cdot 6 \cdots [2(m+1)+1]}$ as desired. By

induction, the formula holds for all $m \geq 1$.

**41.** Let $u = (\ln x)^n$, $dv = dx$ $\Rightarrow$ $du = n(\ln x)^{n-1} \frac{1}{x}\,dx$, $v = x$. Then

$\int (\ln x)^n dx = x(\ln x)^n - n \int (\ln x)^{n-1} dx$ by (7.2)

43. Let $u = (x^2+a^2)^n$, $dv = dx$ $\Rightarrow$ $du = n(x^2+a^2)^{n-1}2x\,dx$, $v = x$.

    Then $\int (x^2+a^2)^n dx = x(x^2+a^2)^n - 2n \int x^2(x^2+a^2)^{n-1}dx$

    $= x(x^2+a^2)^n - 2n\left[\int (x^2+a^2)^n dx - a^2 \int (x^2+a^2)^{n-1}dx\right]$

    [since $x^2 = (x^2+a^2) - a^2$] $\Rightarrow$ $(2n+1) \int (x^2+a^2)^n dx$

    $= x(x^2+a^2)^n + 2na^2 \int (x^2+a^2)^{n-1}dx$ and $\int (x^2+a^2)^n dx$

    $= \dfrac{x(x^2+a^2)^n}{2n+1} + \dfrac{2na^2}{2n+1} \int (x^2+a^2)^{n-1}dx$ (provided $2n+1 \neq 0$)

45. Take $n = 3$ in #41 to get $\int (\ln x)^3 dx = x(\ln x)^3 - 3 \int (\ln x)^2 dx$

    $= x(\ln x)^3 - 3x(\ln x)^2 + 6x \ln x - 6x + C$ [by #7]

47. Let $u = \sin^{-1}x$, $dv = dx$ $\Rightarrow$ $du = dx/\sqrt{1-x^2}$, $v = x$.  Then area $=$

    $\int_0^1 \sin^{-1}x\,dx = x \sin^{-1}x\big]_0^1 - \int_0^1 \dfrac{x}{\sqrt{1-x^2}}dx = \dfrac{\pi}{2} + \left[\sqrt{1-x^2}\right]_0^1 = \dfrac{\pi}{2} - 1$

49. area $= \int_1^e \ln x\,dx = [x \ln x - x]_1^e$ [by Example 2]

    $= (e \ln e - e) - (1 \ln 1 - 1) = 0 - (-1) = 1$

51. Since $v(t) > 0$ for all $t$, the desired distance $s(t) = \int_0^t v(u)du$

    $= \int_0^t u^2 e^{-u}\,du = \int_0^t u^2\,d(-e^{-u}) = -u^2 e^{-u}\big]_0^t + \int_0^t 2u e^{-u}\,du$

    $= -t^2 e^{-t} + \int_0^t 2u\,d(-e^{-u}) = -t^2 e^{-t} - 2u e^{-u}\big]_0^t + \int_0^t 2e^{-u}\,du$

    $= -t^2 e^{-t}-2te^{-t} - 2e^{-u}\big]_0^t = -t^2 e^{-t}-2te^{-t}-2e^{-t}+2 = 2 - e^{-t}(t^2+2t+2)$ m

53. Take $g(x) = x$ in (7.1).

55. By #54, $\int_1^e \ln x\,dx = e \ln e - 1 \ln 1 - \int_{\ln 1}^{\ln e} e^y\,dy$

    $= e - \int_0^1 e^y\,dy = e - \left[e^y\right]_0^1 = e - (e-1) = 1$

Exercises 7.2

1. $\int_0^{\pi/2} \sin^2 3x\,dx = \int_0^{\pi/2} \frac{1}{2}(1 - \cos 6x)dx = \left[\dfrac{x}{2} - \dfrac{\sin 6x}{12}\right]_0^{\pi/2} = \dfrac{\pi}{4}$

3. $\int \cos^4 x\,dx = \int \left[\frac{1}{2}(1 + \cos 2x)\right]^2 dx = \frac{1}{4} \int (1 + 2\cos 2x + \cos^2 2x)dx$

   $= \dfrac{x}{4} + \dfrac{\sin 2x}{4} + \frac{1}{4} \int \frac{1}{2}(1 + \cos 4x)dx = \frac{1}{4}\left[x + \sin 2x + \dfrac{x}{2} + \dfrac{\sin 4x}{8}\right] + C$

   $= \dfrac{3}{8}x + \frac{1}{4} \sin 2x + \dfrac{1}{32} \sin 4x + C$

5.  $\int \sin^3x \cos^4x \, dx = \int \cos^4x(1 - \cos^2x)\sin x \, dx = \int u^4(1-u^2)(- du)$

   $[u = \cos x, \, du = - \sin x \, dx] = \int (u^6-u^4)du = \frac{1}{7} u^7 - \frac{1}{5} u^5 + C$

   $= \frac{1}{7} \cos^7x - \frac{1}{5} \cos^5x + C$

7.  $\int_0^{\pi/4} \sin^4x \cos^2x \, dx = \int_0^{\pi/4} \sin^2x (\sin x \cos x)^2dx$

   $= \int_0^{\pi/4} \frac{1}{2}(1 - \cos 2x)(\frac{1}{2} \sin 2x)^2dx = \frac{1}{8} \int_0^{\pi/4} (1 - \cos 2x)\sin^22x \, dx$

   $= \frac{1}{8} \int_0^{\pi/4} \sin^22x \, dx - \frac{1}{8} \int_0^{\pi/4} \sin^22x \cos 2x \, dx$

   $= \frac{1}{16}\int_0^{\pi/4}(1 - \cos 4x)dx - \frac{1}{16}\left[\frac{\sin^32x}{3}\right]_0^{\pi/4} = \frac{1}{16}\left[x - \frac{\sin 4x}{4} - \frac{\sin^32x}{3}\right]_0^{\pi/4}$

   $= \frac{1}{16}\left[\frac{\pi}{4} - 0 - \frac{1}{3}\right] = \frac{3\pi-4}{192}$

9.  $\int (1 - \sin 2x)^2dx = \int (1 - 2 \sin 2x + \sin^22x)dx$

   $= \int[1 - 2 \sin 2x + \frac{1}{2}(1 - \cos 4x)] \, dx = \int\left[\frac{3}{2} - 2 \sin 2x - \frac{1}{2} \cos 4x\right]dx$

   $= \frac{3}{2}x + \cos 2x - \frac{1}{8} \sin 4x + C$

11. $\int \cos^5x \sin^5x \, dx = \int u^5(1-u^2)^2du$ [where $u = \sin x$, $du = \cos x \, dx$]

   $= \int u^5(1-2u^2+u^4)du = \int(u^5-2u^7+u^9)du = \frac{u^{10}}{10} - \frac{u^8}{4} + \frac{u^6}{6} + C$

   $= \frac{1}{10} \sin^{10}x - \frac{1}{4} \sin^8x + \frac{1}{6} \sin^6x + C$

   OR: $\int \cos^5x \sin^5x \, dx = \int v^5(1-v^2)^2(- dv)$ [where $v = \cos x$ and $dv =$

   $- \sin x \, dx] = \int(-v^5+2v^7-v^9)dv = - \frac{v^{10}}{10} + \frac{v^8}{4} - \frac{v^6}{6} + C$

   $= - \frac{1}{10} \cos^{10}x + \frac{1}{4} \cos^8x - \frac{1}{6} \cos^6x + C$

13. $\int \cos^6x \, dx = \int\left[\frac{1}{2}(1 + \cos 2x)\right]^3dx$

   $= \frac{1}{8} \int\left[1 + 3 \cos 2x + 3 \cos^22x + \cos^32x\right]dx$

   $= \frac{1}{8} \int[1 + 3 \cos 2x + \frac{3}{2}(1 + \cos 4x) + (1 - \sin^22x)\cos 2x]dx$

   $= \frac{1}{8} \int\left\{\frac{5}{2} + 4 \cos 2x + \frac{3}{2} \cos 4x - \sin^22x \cos 2x\right\}dx$

   $= \frac{1}{8}\left[\frac{5}{2}x + 2 \sin 2x + \frac{3}{8} \sin 4x - \frac{1}{6} \sin^32x\right] + C$

15. $\int \sin^5x \, dx = \int(1 - \cos^2x)^2\sin x \, dx = \int(1-u^2)^2(-du)$ [where

   $u = \cos x, \, du = - \sin x \, dx] = \int(-1+2u^2-u^4)du$

   $= - \frac{u^5}{5} + \frac{2u^3}{3} - u + C = - \frac{\cos^5x}{5} + \frac{2 \cos^3x}{3} - \cos x + C$

17. $\int \sin^3x \sqrt{\cos x} \, dx = \int(1 - \cos^2x)\sqrt{\cos x} \sin x \, dx$

   $= \int(1-u^2)u^{1/2}(-du)$ [where $u = \cos x, \, du = - \sin x \, dx$]

   $= \int(u^{5/2} - u^{1/2})du = \frac{2}{7} u^{7/2} - \frac{2}{3} u^{3/2} + C$

   $= \frac{2}{7}(\cos x)^{7/2} - \frac{2}{3}(\cos x)^{3/2} + C = \left[\frac{2}{7} \cos^3x - \frac{2}{3} \cos x\right]\sqrt{\cos x} + C$

19. Let $u = \sqrt{x}$, so that $x = u^2$ and $dx = 2u\,du$. Then

$$\int \frac{\cos^2 \sqrt{x}}{\sqrt{x}}\,dx = \int \frac{\cos^2 u}{u}\,2u\,du = \int 2\cos^2 u\,du = \int(1 + \cos 2u)du$$

$$= u + \frac{1}{2}\sin 2u + C = \sqrt{x} + \frac{1}{2}\sin(2\sqrt{x}) + C$$

21. $\int \cos^2 x \tan^3 x\,dx = \int \frac{\sin^3 x}{\cos x}\,dx = \int \frac{(1-u^2)(-du)}{u}$   [ where $u = \cos x$,

$du = -\sin x\,dx$] $= \int\left[\frac{-1}{u} + u\right]du = -\ln|u| + \frac{1}{2}u^2 + C$

$= \frac{1}{2}\cos^2 x - \ln|\cos x| + C$

23. $\int \frac{1-\sin x}{\cos x}\,dx = \int(\sec x - \tan x)dx$

$= \ln|\sec x + \tan x| - \ln|\sec x| + C$ [by Example 8]

$= \ln|(\sec x + \tan x)\cos x| + C = \ln|1 + \sin x| + C$

$= \ln(1 + \sin x) + C$ since $1 + \sin x \geq 0$

OR: $\int \frac{1-\sin x}{\cos x}\,dx = \int \frac{1-\sin x}{\cos x} \cdot \frac{1+\sin x}{1+\sin x}\,dx = \int \frac{(1-\sin^2 x)dx}{\cos x(1+\sin x)}$

$= \int \frac{\cos x\,dx}{1+\sin x} = \int \frac{dw}{w}$   [where $w = 1 + \sin x$, $dw = \cos x\,dx$]

$= \ln|w| + C = \ln|1 + \sin x| + C = \ln(1 + \sin x) + C$

25. $\int \tan^2 x\,dx = \int(\sec^2 x - 1)dx = \tan x - x + C$

27. $\int \sec^4 x\,dx = \int(\tan^2 x + 1)\sec^2 x\,dx = \int \tan^2 x \sec^2 x\,dx + \int \sec^2 x\,dx$

$= \frac{1}{3}\tan^3 x + \tan x + C$

29. $\int_0^{\pi/4} \tan^4 x \sec^2 x\,dx = \int_0^1 u^4\,du$   [where $u = \tan x$, $du = \sec^2 x\,dx$]

$= \frac{1}{5}u^5\Big]_0^1 = \frac{1}{5}$

31. $\int \tan x \sec^3 x\,dx = \int \sec^2 x \sec x \tan x\,dx = \int u^2 du$   [where

$u = \sec x$, $du = \sec x \tan x\,dx$] $= \frac{1}{3}u^3 + C = \frac{1}{3}\sec^3 x + C$

33. $\int \tan^5 x\,dx = \int (\sec^2 x - 1)^2 \tan x\,dx$

$= \int \sec^4 x \tan x\,dx - 2\int \sec^2 x \tan x\,dx + \int \tan x\,dx$

$= \int \sec^3 x \sec x \tan x\,dx - 2\int \tan x \sec^2 x\,dx + \int \tan x\,dx$

$= (1/4)\sec^4 x - \tan^2 x + \ln|\sec x| + C$

[OR: $(1/4)\sec^4 x - \sec^2 x + \ln|\sec x| + C$]

35. $\int_0^{\pi/3} \tan^5 x \sec x\,dx = \int_0^{\pi/3} (\sec^2 x - 1)^2 \sec x \tan x\,dx$

$= \int_1^2 (u^2-1)^2 du$   [where $u = \sec x$, $du = \sec x \tan x\,dx$]

$= \int_1^2 (u^4 - 2u^2 + 1)du = \left[\frac{1}{5}u^5 - \frac{2}{3}u^3 + u\right]_1^2 = \left[\frac{32}{5} - \frac{16}{3} + 2\right] - \left[\frac{1}{5} - \frac{2}{3} + 1\right] = \frac{38}{15}$

37. $\int \tan x \sec^6 x\,dx = \int \sec^5 x \sec x \tan x\,dx = (1/6)\sec^6 x + C$

39. $\int \frac{\sec^2 x}{\cot x}\,dx = \int \tan x \sec^2 x\,dx = \frac{1}{2}\tan^2 x + C$

41. $\int_{\pi/6}^{\pi/2} \cot^2 x \, dx = \int_{\pi/6}^{\pi/2} (\csc^2 x - 1) dx = (-\cot x - x) \Big]_{\pi/6}^{\pi/2}$

$= (0 - \pi/2) - (-\sqrt{3} - \pi/6) = \sqrt{3} - \pi/3$

43. $\int \cot^4 x \csc^4 x \, dx = \int u^4 (u^2 + 1)(-du)$ [where $u = \cot x$, $du = -\csc^2 x \, dx$]

$= -\int (u^6 + u^4) du = -\frac{1}{7} u^7 - \frac{1}{5} u^5 + C = -\frac{1}{7} \cot^7 x - \frac{1}{5} \cot^5 x + C$

45. $\int \csc x \, dx = \int \frac{\csc x (\csc x - \cot x)}{\csc x - \cot x} dx = \int \frac{-\csc x \cot x + \csc^2 x}{\csc x - \cot x} dx$

$= \ln|\csc x - \cot x| + C$

47. $\int \frac{\cos^2 x}{\sin x} dx = \int \frac{1 - \sin^2 x}{\sin x} dx = \int (\csc x - \sin x) dx$

$= \ln|\csc x - \cot x| + \cos x + C$ by #45.

49. $\int \sin 5x \sin 2x \, dx = \int \frac{1}{2}[\cos(5x-2x) - \cos(5x+2x)] dx$

$= \frac{1}{2} \int (\cos 3x - \cos 7x) dx = \frac{1}{6} \sin 3x - \frac{1}{14} \sin 7x + C$

51. $\int \cos 3x \cos 4x \, dx = \int (1/2)[\cos(3x-4x) + \cos(3x+4x)] dx$

$= \frac{1}{2} \int (\cos x + \cos 7x) dx = \frac{1}{2} \sin x + \frac{1}{14} \sin 7x + C$

53. $\int \sin x \cos 5x \, dx = \int (1/2)[\sin(x-5x) + \sin(x+5x)] dx$

$= \frac{1}{2} \int (-\sin 4x + \sin 6x) dx = \frac{1}{8} \cos 4x - \frac{1}{12} \cos 6x + C$

55. $\int \frac{1 - \tan^2 x}{\sec^2 x} dx = \int (\cos^2 x - \sin^2 x) dx = \int \cos 2x \, dx = \frac{1}{2} \sin 2x + C$

57. $f_{ave} = \frac{1}{2\pi} \int_{-\pi}^{\pi} \sin^2 x \cos^3 x \, dx = \frac{1}{2\pi} \int_{-\pi}^{\pi} \sin^2 x (1 - \sin^2 x) \cos x \, dx$

$= \frac{1}{2\pi} \int_0^0 u^2 (1 - u^2) du$ [$u = \sin x$] $= 0$

59. For $0 < x < \pi/2$, we have $0 < \sin x < 1$, so $\sin^3 x < \sin x$. Hence the area is $\int_0^{\pi/2} (\sin x - \sin^3 x) dx = \int_0^{\pi/2} \sin x (1 - \sin^2 x) dx$

$= \int_0^{\pi/2} \cos^2 x \sin x \, dx$ [$u = \cos x$, $du = -\sin x \, dx$]

$= \int_1^0 u^2 (-du) = \int_0^1 u^2 \, du = u^3/3 \Big]_0^1 = \frac{1}{3}$

61. $s = f(t) = \int_0^t \sin \omega u \cos^2 \omega u \, du = \frac{-1}{\omega} \int_1^{\cos \omega t} y^2 \, dy$ [$y = \cos \omega u \Rightarrow$

$dy = -\omega \sin \omega u \, du] = \frac{-1}{\omega} \frac{1}{3} y^3 \Big]_1^{\cos \omega t} = \frac{1}{3\omega}(1 - \cos^3 \omega t)$

63. $\int_{-\pi}^{\pi} \sin mx \sin nx \, dx = \int_{-\pi}^{\pi} \frac{1}{2}[\cos(m-n)x - \cos(m+n)x] dx$. If $m \neq n$,

this $= \frac{1}{2}\left[\frac{1}{m-n} \sin(m-n)x - \frac{1}{m+n} \sin(m+n)x\right]_{-\pi}^{\pi} = 0$. If $m = n$, we get

$\int_{-\pi}^{\pi} \frac{1}{2}[1 - \cos(m+n)x] dx = \frac{x}{2}\Big]_{-\pi}^{\pi} - \frac{1}{2(m+n)} \sin(m+n)x \Big]_{-\pi}^{\pi} = \pi - 0 = \pi$

65. $\frac{1}{\pi} \int_{-\pi}^{\pi} f(x) \sin mx \, dx = \sum_{n=1}^{N} \frac{a_n}{\pi} \int_{-\pi}^{\pi} \sin mx \sin nx \, dx$. By #63, every

term is zero except the $m^{th}$ one, and that term is $\frac{a_m}{\pi} \cdot \pi = a_m$.

**Exercises 7.3**

1.  Let $x = \sin \theta$, where $-\pi/2 \le \theta \le \pi/2$. Then $dx = \cos \theta\, d\theta$ and
    $\sqrt{1-x^2} = |\cos \theta| = \cos \theta$ (since $\cos \theta > 0$ for $\theta$ in $[-\pi/2, \pi/2]$).
    Thus $\displaystyle\int_{1/2}^{\sqrt{3}/2} \frac{dx}{x^2\sqrt{1-x^2}} = \int_{\pi/6}^{\pi/3} \frac{\cos \theta\, d\theta}{\sin^2\theta \cos \theta} = \int_{\pi/6}^{\pi/3} \csc^2\theta\, d\theta$

    $= -\cot \theta\Big]_{\pi/6}^{\pi/3} = -\dfrac{1}{\sqrt{3}} - (-\sqrt{3}) = \dfrac{3}{\sqrt{3}} - \dfrac{1}{\sqrt{3}} = \dfrac{2}{\sqrt{3}}$

3.  Let $u = 1-x^2$. Then $du = -2x\, dx$, so $\displaystyle\int \frac{x}{\sqrt{1-x^2}}\, dx = -\frac{1}{2}\int \frac{du}{\sqrt{u}} = -\sqrt{u} + C$

    $= -\sqrt{1-x^2} + C$

5.  Let $2x = \sin \theta$, where $-\pi/2 \le \theta \le \pi/2$. Then $x = (1/2)\sin \theta$,
    $dx = (1/2)\cos \theta\, d\theta$, and $\sqrt{1-4x^2} = \sqrt{1-(2x)^2} = \cos \theta$.

    $\displaystyle\int \sqrt{1-4x^2}\, dx = \int \cos \theta\,(1/2)\cos \theta\, d\theta = (1/4)\int(1 + \cos 2\theta)d\theta$

    $= \dfrac{1}{4}(\theta + \dfrac{1}{2}\sin 2\theta) + C = \dfrac{1}{4}(\theta + \sin \theta \cos \theta) + C$

    $= \dfrac{1}{4}\left[\sin^{-1}(2x) + 2x\sqrt{1-4x^2}\right] + C$

7.  Let $x = 3\tan \theta$, where $-\pi/2 < \theta < \pi/2$. Then $dx = 3\sec^2\theta\, d\theta$ and
    $\sqrt{9+x^2} = 3\sec \theta$. $\displaystyle\int_0^3 \frac{dx}{\sqrt{9+x^2}} = \int_0^{\pi/4} \frac{3\sec^2\theta\, d\theta}{3\sec \theta} = \int_0^{\pi/4} \sec \theta\, d\theta$

    $= \Big[\ln|\sec \theta + \tan \theta|\Big]_0^{\pi/4} = \ln(\sqrt{2}+1) - \ln 1 = \ln(\sqrt{2}+1)$

9.  Let $x = 4\sec \theta$, where $0 \le \theta < \pi/2$ or $\pi \le \theta < 3\pi/2$. Then
    $dx = 4\sec \theta \tan \theta\, d\theta$ and $\sqrt{x^2-16} = 4|\tan \theta| = 4\tan \theta$. Thus
    $\displaystyle\int \frac{dx}{x^3\sqrt{x^2-16}} = \int \frac{4\sec \theta \tan \theta\, d\theta}{64\sec^3\theta\, 4\tan \theta} = \frac{1}{64}\int \cos^2\theta\, d\theta = \frac{1}{128}\int(1+\cos 2\theta)d\theta$

    $= \dfrac{1}{128}(\theta + \dfrac{1}{2}\sin 2\theta) + C = \dfrac{1}{128}(\theta + \sin \theta \cos \theta) + C$

    $= \dfrac{1}{128}\left[\sec^{-1}\dfrac{x}{4} + \dfrac{4\sqrt{x^2-16}}{x^2}\right] + C$ by the diagrams for $0 \le \theta < \dfrac{\pi}{2}$ and

    $\pi \le \theta < 3\pi/2$ (where the labels of the legs in the second diagram
    indicate the x- and y-coordinates of P rather than the lengths of

those sides). Henceforth we omit the second diagram from our solutions.

11. $9x^2 - 4 = (3x)^2 - 4$, so let $3x = 2 \sec \theta$, where $0 \le \theta < \pi/2$ or

$\pi \le \theta < 3\pi/2$. Then $dx = (2/3) \sec \theta \tan \theta \, d\theta$ and $\sqrt{9x^2 - 4} = 2 \tan \theta$.

$\int \dfrac{\sqrt{9x^2 - 4}}{x} \, dx = \int \dfrac{2 \tan \theta}{(2/3) \sec \theta} \, (2/3) \sec \theta \tan \theta \, d\theta = 2 \int \tan^2 \theta \, d\theta$

$= 2 \int (\sec^2 \theta - 1) d\theta = 2(\tan \theta - \theta) + C$

$= \sqrt{9x^2 - 4} - 2 \sec^{-1}(3x/2) + C$

13. Let $x = a \sin \theta$, where $-\pi/2 \le \theta \le \pi/2$. Then $dx = a \cos \theta \, d\theta$ and

$\int \dfrac{x^2 \, dx}{(a^2 - x^2)^{3/2}} = \int \dfrac{a^2 \sin^2 \theta \, a \cos \theta \, d\theta}{a^3 \cos^3 \theta} = \int \tan^2 \theta \, d\theta = \int (\sec^2 \theta - 1) d\theta$

$= \tan \theta - \theta + C = x/\sqrt{a^2 - x^2} - \sin^{-1}(x/a) + C$

15. Let $x = \sqrt{3} \tan \theta$, where $-\pi/2 < \theta < \pi/2$. Then

$\int \dfrac{dx}{x\sqrt{x^2 + 3}} = \int \dfrac{\sqrt{3} \sec^2 \theta \, d\theta}{\sqrt{3} \tan \theta \, \sqrt{3} \sec \theta} = \dfrac{1}{\sqrt{3}} \int \csc \theta \, d\theta$

$= \dfrac{1}{\sqrt{3}} \ln |\csc \theta - \cot \theta| + C = \dfrac{1}{\sqrt{3}} \ln \left| \dfrac{\sqrt{x^2 + 3} - \sqrt{3}}{x} \right| + C$

17. Let $u = 4 - 9x^2$, $du = -18x \, dx$. Then $x^2 = (1/9)(4 - u)$ and

$\int_0^{2/3} x^3 \sqrt{4 - 9x^2} \, dx = \int_4^0 (1/9)(4 - u) u^{1/2} (-1/18) du$

$= \dfrac{1}{162} \int_0^4 (4u^{1/2} - u^{3/2}) du = \dfrac{1}{162} \left[ \dfrac{8}{3} u^{3/2} - \dfrac{2}{5} u^{5/2} \right]_0^4 = \dfrac{1}{162} \left[ \dfrac{64}{3} - \dfrac{64}{5} \right] = \dfrac{64}{1215}$

OR: Let $3x = 2 \sin \theta$, where $-\pi/2 \le \theta \le \pi/2$.

19. Let $u = 1 + x^2$, $du = 2x \, dx$. Then $\int 5x\sqrt{1 + x^2} \, dx = \dfrac{5}{2} \int u^{1/2} \, du$

$= \dfrac{5}{3} u^{3/2} + C = \dfrac{5}{3}(1 + x^2)^{3/2} + C$

21. Let $x = \sqrt{2} \sec \theta$, where $0 \le \theta < \pi/2$ or $\pi \le \theta < 3\pi/2$. Then

$\int \dfrac{dx}{x^4 \sqrt{x^2 - 2}} = \int \dfrac{\sqrt{2} \sec \theta \tan \theta \, d\theta}{4 \sec^4 \theta \, \sqrt{2} \tan \theta} = \dfrac{1}{4} \int \cos^3 \theta \, d\theta$

$$= \tfrac{1}{4} \int (1 - \sin^2\theta) \cos\theta \, d\theta = \tfrac{1}{4}\left[\sin\theta - \tfrac{1}{3}\sin^3\theta\right] + C$$

$$= \tfrac{1}{4}\left[\sqrt{x^2-2}/x - (x^2-2)^{3/2}/(3x^3)\right] + C$$

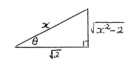

23. $2x-x^2 = -(x^2-2x+1)+1 = 1-(x-1)^2$. Let $u = x-1$. Then $du = dx$ and

$\int \sqrt{2x-x^2} \, dx = \int \sqrt{1-u^2} \, du = \int \cos^2\theta \, d\theta$  [*where* $u = \sin\theta$,

$-\pi/2 \le \theta \le \pi/2$] $= \tfrac{1}{2}\int(1 + \cos 2\theta)d\theta = \tfrac{1}{2}(\theta + \tfrac{1}{2}\sin 2\theta) + C$

$= (1/2)(\sin^{-1}u + u\sqrt{1-u^2}) + C = (1/2)(\sin^{-1}(x-1) + (x-1)\sqrt{2x-x^2}) + C$

25. $9x^2+6x-8 = (3x+1)^2-9$, so let $u = 3x+1$, $du = 3 \, dx$. Then

$\displaystyle\int \frac{dx}{\sqrt{9x^2+6x-8}} = \int \frac{(1/3) \, du}{\sqrt{u^2-9}}$. Now let $u = 3\sec\theta$, where $0 \le \theta < \pi/2$

or $\pi \le \theta < 3\pi/2$. Then $du = 3\sec\theta\tan\theta \, d\theta$ and $\sqrt{u^2-9}$

$= 3\tan\theta$, so $\displaystyle\int \frac{(1/3) \, du}{\sqrt{u^2-9}} = \int \frac{\sec\theta\tan\theta \, d\theta}{3\tan\theta} = \tfrac{1}{3}\int \sec\theta \, d\theta$

$= \tfrac{1}{3}\ln|\sec\theta + \tan\theta| + C_1 = \tfrac{1}{3}\ln\left|\dfrac{u + \sqrt{u^2-9}}{3}\right| + C_1$

$= \tfrac{1}{3}\ln\left|u + \sqrt{u^2-9}\right| + C = \tfrac{1}{3}\ln\left|3x + 1 + \sqrt{9x^2+6x-8}\right| + C$

27. $x^2+2x+2 = (x+1)^2+1$. Let $u = x+1$, $du = dx$. Then $\displaystyle\int \frac{dx}{(x^2+2x+2)^2}$

$= \displaystyle\int \frac{du}{(u^2+1)^2} = \int \frac{\sec^2\theta \, d\theta}{\sec^4\theta}$  [$u = \tan\theta$] $= \int \cos^2\theta \, d\theta$

$= (1/2)(\theta + \sin\theta\cos\theta) + C$ [*as in* #23]

$= \tfrac{1}{2}\left[\tan^{-1}u + \dfrac{u}{1+u^2}\right] + C = \tfrac{1}{2}\left[\tan^{-1}(x+1) + \dfrac{x+1}{x^2+2x+2}\right] + C$

29. $\int e^t\sqrt{9-e^{2t}} \, dt = \int \sqrt{9-u^2} \, du$  [*where* $u = e^t$, $du = e^t \, dt$]

$= \int 3\cos\theta \, 3\cos\theta \, d\theta$  [*where* $u = 3\sin\theta$, $-\pi/2 \le \theta \le \pi/2$]

$= 9\int \cos^2\theta \, d\theta = (9/2)(\theta + \sin\theta\cos\theta) + C$  [*as in* #23]

$= \tfrac{9}{2}\left[\sin^{-1}\left[\dfrac{u}{3}\right] + \dfrac{u}{3}\cdot\dfrac{\sqrt{9-u^2}}{3}\right] + C = \tfrac{9}{2}\sin^{-1}(e^t/3) + \tfrac{1}{2}e^t\sqrt{9-e^{2t}} + C$

31. (a)  Let $x = a\tan\theta$, where $-\pi/2 < \theta < \pi/2$. Then $\sqrt{x^2+a^2} = a\sec\theta$

and $\displaystyle\int \frac{dx}{\sqrt{x^2+a^2}} = \int \frac{a\sec^2\theta \, d\theta}{a\sec\theta} = \int \sec\theta \, d\theta = \ln|\sec\theta + \tan\theta| + C_1$

$= \ln\left|\dfrac{\sqrt{x^2+a^2}}{a} + \dfrac{x}{a}\right| + C_1 = \ln\left[x+\sqrt{x^2+a^2}\right] + C$, where $C = C_1 - \ln|a|$

(b)  Let $x = a \sinh t$, so that $dx = a \cosh t \, dt$ and $\sqrt{x^2 + a^2}$

$= a \cosh t$. Then $\displaystyle\int \frac{dx}{\sqrt{x^2 + a^2}} = \int \frac{a \cosh t \, dt}{a \cosh t} = t + C = \sinh^{-1}(x/a) + C$

33.  area of $\triangle POQ = \frac{1}{2}(r \cos \theta)(r \sin \theta) = \frac{1}{2}r^2 \sin \theta \cos \theta$. Area of region

PQR $= \displaystyle\int_{r \cos \theta}^{r} \sqrt{r^2 - x^2} \, dx$.  Let $x = r \cos u \Rightarrow dx = -r \sin u \, du$ for

$\theta \leq u \leq \pi/2$.  Then we obtain $\int \sqrt{r^2 - x^2} \, dx = \int r \sin u \, (-r \sin u) du$

$= -r^2 \int \sin^2 u \, du = -\frac{1}{2}r^2 (u - \sin u \cos u) + C = -\frac{1}{2}r^2 \cos^{-1}(x/r) +$

$\frac{1}{2}x\sqrt{r^2 - x^2} + C$.  Area of region PQR $= \frac{1}{2}\left[-r^2 \cos^{-1}(x/r) + x\sqrt{r^2 - x^2}\right]_{r\cos\theta}^{r}$

$= \frac{1}{2}\left[0 - [-r^2\theta + r \cos \theta \, r \sin\theta]\right] = \frac{1}{2}r^2\theta - \frac{1}{2}r^2 \sin \theta \cos \theta$, so area of

sector POR = area of $\triangle POQ$ + area of region PQR = $(1/2)r^2\theta$

**Exercises 7.4**

1.  $\dfrac{1}{(x-1)(x+2)} = \dfrac{A}{x-1} + \dfrac{B}{x+2}$          3.  $\dfrac{x+1}{x(x+2)} = \dfrac{A}{x} + \dfrac{B}{x+2}$

5.  $\dfrac{x^2 + 3x - 4}{(2x-1)^2(2x+3)} = \dfrac{A}{2x-1} + \dfrac{B}{(2x-1)^2} + \dfrac{C}{2x+3}$

7.  $\dfrac{1}{x^4 - x^3} = \dfrac{1}{x^3(x-1)} = \dfrac{A}{x} + \dfrac{B}{x^2} + \dfrac{C}{x^3} + \dfrac{D}{x-1}$

9.  $\dfrac{x^2 + 1}{x^2 - 1} = 1 + \dfrac{2}{(x-1)(x+1)} = 1 + \dfrac{A}{x-1} + \dfrac{B}{x+1}$

11.  $\dfrac{x^2 - 2}{x(x^2 + 2)} = \dfrac{A}{x} + \dfrac{Bx + C}{x^2 + 2}$

13.  $\dfrac{x^4 + x^2 + 1}{(x^2 + 1)(x^2 + 4)^2} = \dfrac{Ax + B}{x^2 + 1} + \dfrac{Cx + D}{x^2 + 4} + \dfrac{Ex + F}{(x^2 + 4)^2}$

15.  $\dfrac{x^4}{(x^2 + 9)^3} = \dfrac{Ax + B}{x^2 + 9} + \dfrac{Cx + D}{(x^2 + 9)^2} + \dfrac{Ex + F}{(x^2 + 9)^3}$

17.  $\dfrac{x^3 + x^2 + 1}{x^4 + x^3 + 2x^2} = \dfrac{x^3 + x^2 + 1}{x^2(x^2 + x + 2)} = \dfrac{A}{x} + \dfrac{B}{x^2} + \dfrac{Cx + D}{x^2 + x + 2}$

19. $\int \frac{x^2}{x+1} \, dx = \int (x - 1 + \frac{1}{x+1}) dx = \frac{1}{2}x^2 - x + \ln|x+1| + C$

21. $\frac{4x-1}{(x-1)(x+2)} = \frac{A}{x-1} + \frac{B}{x+2} \Rightarrow 4x-1 = A(x+2) + B(x-1)$. Take $x = 1$ to

get $3 = 3A$, then $x = -2$ to get $-9 = -3B \Rightarrow A = 1$, $B = 3$. Now

$\int_2^4 \frac{4x-1}{(x-1)(x+2)} \, dx = \int_2^4 \left[ \frac{1}{x-1} + \frac{3}{x+2} \right] dx = \ln(x-1) + 3 \ln(x+2) \Big]_2^4$

$= \ln 3 + 3 \ln 6 - \ln 1 - 3 \ln 4 = 4 \ln 3 - 3 \ln 2 = \ln(81/8)$

23. $\int \frac{6x-5}{2x+3} \, dx = \int \left[ 3 - \frac{14}{2x+3} \right] dx = 3x - 7 \ln|2x+3| + C$

25. $\frac{x^2+1}{x^2-x} = 1 + \frac{x+1}{x(x-1)} = 1 - \frac{1}{x} + \frac{2}{x-1}$, so $\int \frac{x^2+1}{x^2-x} \, dx$

$= x - \ln|x| + 2 \ln|x-1| + C = x + \ln[(x-1)^2/|x|] + C$

27. $\frac{2x+3}{(x+1)^2} = \frac{A}{x+1} + \frac{B}{(x+1)^2} \Rightarrow 2x+3 = A(x+1) + B$. Take $x = -1$ to get

$B = 1$, and equate coefficients of $x$ to get $A = 2$. Now

$\int_0^1 \frac{2x+3}{(x+1)^2} \, dx = \int_0^1 \left[ \frac{2}{x+1} + \frac{1}{(x+1)^2} \right] dx = \left[ 2 \ln(x+1) - \frac{1}{x+1} \right]_0^1$

$= 2 \ln 2 - \frac{1}{2} - (2 \ln 1 - 1) = 2 \ln 2 + \frac{1}{2}$

29. $\frac{1}{x(x+1)(2x+3)} = \frac{A}{x} + \frac{B}{x+1} + \frac{C}{2x+3}$. Multiply by $x$ to get

$\frac{1}{(x+1)(2x+3)} = A + x \left[ \frac{B}{x-1} + \frac{C}{2x+3} \right]$; set $x = 0$ to get $A = 1/3$.

Similarly, multiply by $x+1$ and set $x = -1$ to get $B = -1$; then

multiply by $2x+3$ and set $x = -3/2$ to get $C = 4/3$. Now

$\int \frac{dx}{x(x+1)(2x+3)} = \int \left[ \frac{1/3}{x} - \frac{1}{x+1} + \frac{4/3}{2x+3} \right] dx$

$= \frac{1}{3} \ln|x| - \ln|x+1| + \frac{2}{3} \ln|2x+3| + C$

31. $\frac{6x^2+5x-3}{x^3+2x^2-3x} = \frac{A}{x} + \frac{B}{x+3} + \frac{C}{x-1}$. Multiplying by $x$ and setting $x = 0$

gives $A = \frac{6x^2+5x-3}{(x+3)(x-1)} \Big|_{x=0} = \frac{-3}{-3} = 1$. Similarly, $B = \frac{6x^2+5x-3}{x(x-1)} \Big|_{x=-3}$

$= \frac{36}{12} = 3$ and $C = \frac{6x^2+5x-3}{x(x+3)} \Big|_{x=1} = \frac{8}{4} = 2 \Rightarrow \int_2^3 \frac{6x^2+5x-3}{x^3+2x^2-3x} \, dx$

$= \int_2^3 \left[ \frac{1}{x} + \frac{3}{x+3} + \frac{2}{x-1} \right] dx = \left[ \ln x + 3 \ln(x+3) + 2 \ln(x-1) \right]_2^3$

$= (\ln 3 + 3 \ln 6 + 2 \ln 2) - (\ln 2 + 3 \ln 5) = 4 \ln 6 - 3 \ln 5$

33. $\frac{1}{(x-1)^2(x+4)} = \frac{A}{x-1} + \frac{B}{(x-1)^2} + \frac{C}{x+4}$. Multiplying by $(x-1)^2$ and

setting $x = 1$ gives $B = \frac{1}{x+4} \Big|_{x=1} = \frac{1}{5}$. Multiplying by $x+4$ and

setting $x = -4$ gives $C = \frac{1}{(x-1)^2} \Big|_{x=-4} = \frac{1}{25}$. Multiply by $x-1$ to get

$\dfrac{1}{(x-1)(x+4)} = A + \dfrac{B}{x-1} + C\,\dfrac{x-1}{x+4}$. Now let $x \to \infty$ in this identity to

get $0 = A + 0 + C$, or $A = -C = -1/25$ $\Rightarrow$ $\displaystyle\int \dfrac{dx}{(x-1)^2(x+4)}$

$= \displaystyle\int \left[\dfrac{-1/25}{x-1} + \dfrac{1/5}{(x-1)^2} + \dfrac{1/25}{x+4}\right] dx = \dfrac{-1}{25}\ln|x-1| - \dfrac{1}{5}\dfrac{1}{x-1} + \dfrac{1}{25}\ln|x+4| + C$

$= 1/25[\ln|(x+4)/(x-1)| - 5/(x-1)] + C$

35. $\displaystyle\int \dfrac{5x^2+3x-2}{x^3+2x^2}\,dx = \int \dfrac{5x^2+3x-2}{x^2(x+2)}\,dx = \int \left[\dfrac{2}{x} - \dfrac{1}{x^2} + \dfrac{3}{x+2}\right] dx$

$= 2\ln|x| + 1/x + 3\ln|x+2| + C$

37. $\displaystyle\int \dfrac{2x^3+4x^2-12x+3}{x^3-3x+2}\,dx = \int \left[2 + \dfrac{4x^2-6x-1}{(x-1)^2(x+2)}\right] dx$

$= \displaystyle\int \left[2 + \dfrac{1}{x-1} - \dfrac{1}{(x-1)^2} + \dfrac{3}{x+2}\right] dx = 2x + \ln|x-1| + \dfrac{1}{x-1} + 3\ln|x+2| + C$

39. Let $u = x^3+3x^2+4$. Then $du = 3(x^2+2x)dx$ $\Rightarrow$

$\displaystyle\int \dfrac{x^2+2x}{x^3+3x^2+4}\,dx = \dfrac{1}{3}\int \dfrac{du}{u} = \dfrac{1}{3}\ln|x^3+3x^2+4| + C$

41. $\dfrac{1}{x(x-1)^3} = \dfrac{A}{x} + \dfrac{B}{x-1} + \dfrac{C}{(x-1)^2} + \dfrac{D}{(x-1)^3}$. Multiply by $x$ and set $x = 0$

to get $A = -1$. Multiply by $(x-1)^3$ and set $x = 1$ to get $D = 1$.

Multiply by $x$ and let $x \to \infty$ to get $0 = A+B$, so that $B = 1$. Finally

take $x = 2$ to get $1/2 = -1/2 + 1/1 + C/1 + 1/1 = 3/2 + C$, so that

$C = -1$. Now $\displaystyle\int \dfrac{dx}{x(x-1)^3} = \int \left[\dfrac{-1}{x} + \dfrac{1}{x-1} - \dfrac{1}{(x-1)^2} + \dfrac{1}{(x-1)^3}\right] dx$

$= -\ln|x| + \ln|x-1| + 1/(x-1) - 1/[2(x-1)^2] + C$

$= \ln|1-1/x| + 1/(x-1) - 1/[2(x-1)^2] + C$

43. $\dfrac{x^2}{(x+1)^3} = \dfrac{A}{x+1} + \dfrac{B}{(x+1)^2} + \dfrac{C}{(x+1)^3}$. Multiply by $(x+1)^3$ and set $x = -1$

to get $C = 1$. Multiply by $x+1$ and let $x \to \infty$ to get $A = 1$. Take

$x = 0$ to get $0 = A+B+C = 1+B+1$, so that $B = -2$. Now

$\displaystyle\int \dfrac{x^2 dx}{(x+1)^3} = \int \left[\dfrac{1}{x+1} - \dfrac{2}{(x+1)^2} + \dfrac{1}{(x+1)^3}\right] dx = \ln|x+1| + \dfrac{2}{x+1} - \dfrac{1}{2(x+1)^2} + C$

45. $\dfrac{1}{(x+1)(x^2-1)^2} = \dfrac{1}{(x+1)^3(x-1)^2} = \dfrac{A}{x+1} + \dfrac{B}{(x+1)^2} + \dfrac{C}{(x+1)^3} + \dfrac{D}{x-1} + \dfrac{E}{(x-1)^2}$.

Multiply by $(x+1)^3$ and set $x = -1$ to get $C = 1/4$. Multiply by

$(x-1)^2$ and set $x = 1$ to get $E = 1/8$. Multiply by $x+1$ and let $x \to \infty$

to get $0 = A+D$. Now take $x = 0$ to get $1 = A+B+1/4+A+1/8$, or $2A+B = $

$5/8$. Finally take $x = 2$ to get $1/27 = A/3+B/9+1/108-A/8$ or

$(-2/3)A+B/9 = -7/72$. This last equation says $-6A+B = -7/8$. From the two equations for A and B, we get A = 3/16 and B = 1/4.  Thus

$$\frac{1}{(x+1)(x^2-1)^2} = \frac{3/16}{x+1} + \frac{1/4}{(x+1)^2} + \frac{1/4}{(x+1)^3} - \frac{3/16}{x-1} + \frac{1/8}{(x+1)^2} \text{ and so}$$

$$\int \frac{dx}{(x+1)(x^2-1)^2} = \frac{3}{16} \ln\left|\frac{x+1}{x-1}\right| - \frac{1}{4(x+1)} - \frac{1}{8(x-1)} - \frac{1}{8(x+1)^2} + C$$

47.  $\dfrac{1}{x^4-x^2} = \dfrac{1}{x^2(x-1)(x+1)} = \dfrac{A}{x} + \dfrac{B}{x^2} + \dfrac{C}{x-1} + \dfrac{D}{x+1}$.  Multiply by $x^2$ and set

x = 0 to get B = -1.  Multiply by x-1 and set x = 1 to get C = 1/2.

Multiply by x+1 and set x = -1 to get D = - 1/2.  Multiply by x and

let x → ∞ to get 0 = A+C+D = A.  Now

$$\int \frac{dx}{x^4-x^2} = \int \left[\frac{-1}{x^2} + \frac{1/2}{x-1} - \frac{1/2}{x+1}\right]dx = \frac{1}{x} + \frac{1}{2}\ln\left|\frac{x-1}{x+1}\right| + C$$

49.  $\dfrac{x^3}{x^2+1} = \dfrac{(x^3+x)-x}{x^2+1} = x - \dfrac{x}{x^2+1}$,  so  $\displaystyle\int_0^1 \frac{x^3}{x^2+1} dx = \int_0^1 x\,dx - \frac{1}{2}\int_0^1 \frac{2x\,dx}{x^2+1}$

$$= \frac{x^2}{2}\Big|_0^1 - \frac{1}{2}\ln\left[x^2+1\right]\Big|_0^1 = \frac{1}{2} - \frac{1}{2}\ln 2 = \frac{1 - \ln 2}{2}$$

51.  Complete the square: $x^2+x+1 = (x+1/2)^2+3/4$ and let u = x+1/2.  Then

$$\int_0^1 \frac{x}{x^2+x+1} dx = \int_{1/2}^{3/2} \frac{u-1/2}{u^2+3/4} du = \int_{1/2}^{3/2} \frac{u}{u^2+3/4} du - \frac{1}{2}\int_{1/2}^{3/2} \frac{1}{u^2+3/4} du$$

$$= \frac{1}{2}\ln(u^2+3/4) - \frac{1}{2}\cdot\frac{1}{\sqrt{3}/2}\tan^{-1}\left[\frac{2u}{\sqrt{3}}\right]\Big]_{1/2}^{3/2} = \frac{1}{2}\ln 3 - \frac{1}{\sqrt{3}}\left[\frac{\pi}{3} - \frac{\pi}{6}\right]$$

$$= \ln\sqrt{3} - \pi/6\sqrt{3}$$

53.  $\dfrac{3x^2-4x+5}{(x-1)(x^2+1)} = \dfrac{A}{x-1} + \dfrac{Bx+C}{x^2+1}$  ⟹  $3x^2-4x+5 = A(x^2+1) + (Bx+C)(x-1)$

Take x = 1 to get 4 = 2A or A = 2.  Now (Bx+C)(x-1)

= $3x^2-4x+5-2(x^2+1) = x^2-4x+3$.  Equating coefficients of $x^2$ and then

comparing the constant terms, we get B = 1 and C = -3.  Hence

$$\int \frac{3x^2-4x+5}{(x-1)(x^2+1)} dx = \int \left[\frac{2}{x-1} + \frac{x-3}{x^2+1}\right]dx = 2\ln|x-1| + \int \frac{x\,dx}{x^2+1} - 3\int \frac{dx}{x^2+1}$$

$$= 2\ln|x-1| + (1/2)\ln(x^2+1) - 3\tan^{-1}x + C$$

$$= \ln(x-1)^2 + \ln\sqrt{x^2+1} - 3\tan^{-1}x + C$$

55.  $\dfrac{x^2+7x-6}{(x+1)(x^2-4x+7)} = \dfrac{A}{x+1} + \dfrac{Bx+C}{x^2-4x+7}$  ⟹  $x^2+7x-6$

= $A(x^2-4x+7) + (Bx+C)(x+1)$.  Take x = -1 to get -12 = 12A or

A = -1.  Then $2x^2+3x+1 = (Bx+C)(x+1)$, so B = 2 and C = 1.  Now

$$\int \frac{x^2+7x-6}{(x+1)(x^2-4x+7)} dx = \int \left[\frac{-1}{x+1} + \frac{2x+1}{x^2-4x+7}\right]dx$$

$$= - \ln|x+1| + \int \frac{2x-4}{x^2-4x+7} dx + 5\int \frac{dx}{(x-2)^2+3}$$

$$= - \ln|x+1| + \ln(x^2-4x+7) + \frac{5}{\sqrt{3}}\tan^{-1}\left[\frac{x-2}{\sqrt{3}}\right] + C$$

**57.** $\dfrac{1}{x^3-1} = \dfrac{1}{(x-1)(x^2+x+1)} = \dfrac{A}{x-1} + \dfrac{Bx+C}{x^2+x+1} \Rightarrow 1 = A(x^2+x+1)+(Bx+C)(x-1).$

Take $x = 1$ to get $A = 1/3$. Then equate coefficients of $x^2$ and 1 to

get $0 = 1/3+B$ and $1 = 1/3-C$; i.e., $B = -1/3$ and $C = -2/3$. Then

$\displaystyle\int \dfrac{dx}{x^3-1} = \int \dfrac{1/3}{x-1}\,dx + \int \dfrac{(-1/3)x-2/3}{x^2+x+1}\,dx = \dfrac{1}{3}\ln|x-1| - \dfrac{1}{3}\int \dfrac{x+2}{x^2+x+1}\,dx$

$= \dfrac{1}{3}\ln|x-1| - \dfrac{1}{3}\int \dfrac{x+1/2}{x^2+x+1}\,dx - \dfrac{1}{3}\int \dfrac{(3/2)dx}{(x+1/2)^2+3/4}$

$= \dfrac{1}{3}\ln|x-1| - \dfrac{1}{6}\ln(x^2+x+1) - \dfrac{1}{2}(2/\sqrt{3})\tan^{-1}[(x+1/2)/(\sqrt{3}/2)] + C$

$= \dfrac{1}{3}\ln|x-1| - \dfrac{1}{6}\ln(x^2+x+1) - (1/\sqrt{3})\tan^{-1}[(2x+1)/\sqrt{3}] + C$

**59.** $\dfrac{x^2-2x-1}{(x-1)^2(x^2+1)} = \dfrac{A}{x-1} + \dfrac{B}{(x-1)^2} + \dfrac{Cx+D}{x^2+1}.$ Multiply by $(x-1)^2$ and set

$x = 1$ to get $B = -1$. Multiply by $x$ and let $x \to \infty$ to get $C = -A$.

Now $\dfrac{x^2-2x-1}{(x-1)^2(x^2+1)} = \dfrac{A}{x-1} - \dfrac{1}{(x-1)^2} + \dfrac{-Ax+D}{x^2+1}$, so $x^2-2x-1$

$= A(x-1)(x^2+1) - (x^2+1) + (-Ax+D)(x-1)^2$. Equating constant terms

gives $-1 = -A-1+D$, so $D = A$. Now take $x = 2$ and get $-1 = 5A-5-2A+A$

or $A = 1$. We have $\displaystyle\int \dfrac{x^2-2x-1}{(x-1)^2(x^2+1)}\,dx = \int\left[\dfrac{1}{x-1} - \dfrac{1}{(x-1)^2} - \dfrac{x-1}{x^2+1}\right]dx$

$= \ln|x-1| + 1/(x-1) - (1/2)\ln(x^2+1) + \tan^{-1}x + C$

**61.** $\dfrac{3x^3-x^2+6x-4}{(x^2+1)(x^2+2)} = \dfrac{Ax+B}{x^2+1} + \dfrac{Cx+D}{x^2+2}$

$\Rightarrow 3x^3-x^2+6x-4 = (Ax+B)(x^2+2) + (Cx+D)(x^2+1) \Rightarrow$

$A+C = 3,\ B+D = -1,\ 2A+C = 6,\ 2B+D = -4 \Rightarrow A = 3,\ C = 0,\ B = -3,$

and $D = 2$. Now $\displaystyle\int \dfrac{3x^3-x^2+6x-4}{(x^2+1)(x^2+2)}\,dx = 3\int \dfrac{x-1}{x^2+1}\,dx + 2\int \dfrac{dx}{x^2+2}$

$= (3/2)\ln(x^2+1) - 3\tan^{-1}x + \sqrt{2}\tan^{-1}(x/\sqrt{2}) + C$

**63.** $\dfrac{1}{(x^2+3x+2)(x^2+2x+2)} = \dfrac{1}{(x+1)(x+2)(x^2+2x+2)} = \dfrac{A}{x+1} + \dfrac{B}{x+2} + \dfrac{Cx+D}{x^2+2x+2}$

$A = \dfrac{1}{(x+2)(x^2+2x+2)}\Big|_{x=-1} = 1$ and $B = \dfrac{1}{(x+1)(x^2+2x+2)}\Big|_{x=-2} = -\dfrac{1}{2}$

Multiply by $x$ and let $x \to \infty$ to get $0 = A+B+C = 1/2+C$ or $C = -1/2$.

Finally take $x = 0$ to get $1/4 = 1-1/4+D/2$ or $D = -1$. This gives

$\displaystyle\int \dfrac{dx}{(x^2+3x+2)(x^2+2x+2)} = \int\left[\dfrac{1}{x+1} - \dfrac{1/2}{x+2} - \dfrac{(1/2)x+1}{x^2+2x+2}\right]dx$

$= \ln|x+1| - (1/2)\ln|x+2| - \int \dfrac{(1/2)x+1/2}{x^2+2x+2}\,dx - \int \dfrac{(1/2)dx}{(x+1)^2+1}$

$= \ln|x+1| - (1/2)\ln|x+2| - (1/4)\ln(x^2+2x+2) - (1/2)\tan^{-1}(x+1) + C$

65. $\int \frac{x-3}{(x^2+2x+4)^2} dx = \int \frac{x-3}{[(x+1)^2+3]^2} dx = \int \frac{u-4}{(u^2+3)^2} du \quad [u = x+1]$

$= \frac{1}{2} \int \frac{2u\,du}{(u^2+3)^2} - 4 \int \frac{du}{(u^2+3)^2} = \frac{1}{2} \int \frac{dv}{v^2} - 4 \int \frac{\sqrt{3}\,\sec^2\theta\,d\theta}{9\,\sec^4\theta}$

$[v = u^2+3$ in the $1^{st}$ integral; $u = \sqrt{3}\,\tan\theta$ in the $2^{nd}]$

$= -1/(2v) - (4\sqrt{3}/9) \int \cos^2\theta\,d\theta = -1/[2(u^2+3)] - (2\sqrt{3}/9)(\theta + \sin\theta\,\cos\theta) + C$

$= -1/[2(x^2+2x+4)] - (2\sqrt{3}/9)\left[\tan^{-1}[(x+1)/\sqrt{3}] + \sqrt{3}(x+1)/(x^2+2x+4)\right] + C$

$= -1/[2(x^2+2x+4)] - (2\sqrt{3}/9)\{\tan^{-1}[(x+1)/\sqrt{3}]\} - (2/3)(x+1)/(x^2+2x+4) + C$

67. $\frac{x^4+1}{x(x^2+1)^2} = \frac{A}{x} + \frac{Bx+C}{x^2+1} + \frac{Dx+E}{(x^2+1)^2}$. Multiply by x and set x = 0 to get

A = 1. Multiply by x and let $x \to \infty$ to get 1 = A+B, so B = 0. Now

$\frac{C}{x^2+1} + \frac{Dx+E}{(x^2+1)^2} = \frac{x^4+1}{x(x^2+1)^2} - \frac{1}{x} = \frac{1}{x}\left[\frac{x^4+1-(x^4+2x^2+1)}{(x^2+1)^2}\right] = \frac{-2x}{(x^2+1)^2}$, so we

can take C = 0, D = -2, and E = 0. Hence

$\int \frac{x^4+1}{x(x^2+1)^2} dx = \int \left[\frac{1}{x} - \frac{2x}{(x^2+1)^2}\right]dx = \ln|x| + \frac{1}{x^2+1} + C$

69. $\frac{x^3+x^2+2x-3}{(x^2+2x+2)^2} = \frac{Ax+B}{x^2+2x+2} + \frac{Cx+D}{(x^2+2x+2)^2} \Rightarrow$

$x^3+x^2+2x-3 = (Ax+B)(x^2+2x+2)+Cx+D \Rightarrow A = 1, 2A+B = 1, 2A+2B+C = 2,$

$2B+D = -3 \Rightarrow A = 1, B = -1, C = 2, D = -1.$ Hence

$\int \frac{x^3+x^2+2x-3}{(x^2+2x+2)^2} dx = \int \frac{x-1}{x^2+2x+2} dx + \int \frac{2x-1}{(x^2+2x+2)^2} dx$

$= \frac{1}{2} \int \frac{(2x+2)dx}{x^2+2x+2} - 2 \int \frac{dx}{(x+1)^2+1} + \int \frac{(2x+2)dx}{(x^2+2x+2)^2} - 3 \int \frac{dx}{[(x+1)^2+1]^2}$

(*)

$= (1/2)\ln(x^2+2x+2) - 2\tan^{-1}(x+1) - 1/(x^2+2x+2)$

$\quad - (3/2)\tan^{-1}(x+1) - 3(x+1)/2(x^2+2x+2) + C$

$= (1/2)\ln(x^2+2x+2) - (7/2)\tan^{-1}(x+1) - (3x+5)/2(x^2+2x+2) + C$

NOTE: At the equality marked (*), we have used the fact that

$\int \frac{du}{(u^2+a^2)^2} = \int \frac{a\,\sec^2\theta\,d\theta}{a^4\,\sec^4\theta} \quad [u = a\,\tan\theta] = \frac{1}{a^3} \int \cos^2\theta\,d\theta$

$= \frac{1}{2a^3}(\theta + \sin\theta\,\cos\theta) + C = \frac{1}{2a^3}\left[\tan^{-1}\left[\frac{u}{a}\right] + \frac{au}{u^2+a^2}\right] + C$

$= (1/2a^3)\tan^{-1}(u/a) + u/2a^2(u^2+a^2) + C.$

This fact could also have been used in #65.

**71.** Let $u = x^2+4$. Then $\int \dfrac{8x\ dx}{(x^2+4)^3} = 4\int \dfrac{2x\ dx}{(x^2+4)^3} = 4\int u^{-3}du$

$= -2u^{-2} + C = -2/(x^2+4)^2 + C$

**73.** Let $u = \sin^2 x - 3\sin x + 2$. Then $du = (2\sin x \cos x - 3\cos x)dx$

so $\int \dfrac{(2\sin x - 3)\cos x}{\sin^2 x - 3\sin x + 2}\ dx = \int \dfrac{du}{u} = \ln|u| + C$

$= \ln|\sin^2 x - 3\sin x + 2| + C$

**75.** If $|x| < a$, then $\int \dfrac{dx}{a^2-x^2} = \int \dfrac{a\ \text{sech}^2 u\ du}{a^2\ \text{sech}^2 u}$    *[x = a tanh u]*

$= \dfrac{1}{a}u + C = \dfrac{1}{a}\tanh^{-1}(x/a) + C$.    If $|x| > a$, then

$\int \dfrac{dx}{a^2-x^2} = \int \dfrac{-a\ \text{csch}^2 u\ du}{-a^2\ \text{csch}^2 u}$ *[x = a coth u]* $= \dfrac{1}{a}u + C = \dfrac{1}{a}\coth^{-1}(x/a)+C$

**77.** $\int \dfrac{dx}{x^2-2x} = \int \dfrac{dx}{(x-1)^2-1} = \int \dfrac{du}{u^2-1}$    *[u = x-1]*    $= \dfrac{1}{2}\ln\left|\dfrac{u-1}{u+1}\right| + C$

*[by (7.20)]*    $= (1/2)\ln|(x-2)/x| + C$

**79.** $\int \dfrac{x\ dx}{x^2+x-1} = \dfrac{1}{2}\int \dfrac{(2x+1)dx}{x^2+x-1} - \dfrac{1}{2}\int \dfrac{dx}{(x+1/2)^2-5/4}$

$= (1/2)\ln|x^2+x-1| - \dfrac{1}{2}\int \dfrac{du}{u^2-(\sqrt{5}/2)^2}$    *[u = x+1/2]*

$= (1/2)\ln|x^2+x-1| - (1/2\sqrt{5})\ln|(u-\sqrt{5}/2)/(u+\sqrt{5}/2)| + C$

$= (1/2)\ln|x^2+x-1| - (1/2\sqrt{5})\ln|(2x+1-\sqrt{5})/(2x+1+\sqrt{5})| + C$

**81.** $\dfrac{x+1}{x-1} = 1 + \dfrac{2}{x-1} > 0$ for $2 \le x \le 3$, so the area $= \int_2^3 \left[1 + \dfrac{2}{x-1}\right]dx$

$= \left[x + 2\ln|x-1|\right]_2^3 = (3 + 2\ln 2) - (2 + 2\ln 1) = 1 + 2\ln 2$

**83.** There are only finitely many values of x where Q(x) = 0 (assuming
that Q is not the zero polynomial). At all other values of x,
F(x)/Q(x) = G(x)/Q(x), so F(x) = G(x). In other words, the values
of F and G agree at all except perhaps finitely many values of x.
By continuity of F and G, the polynomials F and G must agree at
those values of x too. (If a is a value of x such that Q(a) = 0,
then Q(x) ≠ 0 for all x sufficiently near a. Thus F(a) = lim F(x)
$\qquad\qquad$ x→a
[by continuity of F] = lim G(x) [since F(x) = G(x) wherever
$\qquad\qquad$ x→a
Q(x) ≠ 0] = G(a) [by continuity of G].)

## Exercises 7.5

1. Let $u = \sqrt{x}$. Then $x = u^2$, $dx = 2u\, du$ $\Rightarrow$ $\displaystyle\int_0^1 \frac{dx}{1+\sqrt{x}} = \int_0^1 \frac{2u\, du}{1+u}$

$$= 2 \int_0^1 \left[1 - \frac{1}{1+u}\right] du = 2\left[u - \ln(1+u)\right]_0^1 = 2(1 - \ln 2)$$

3. Let $u = \sqrt{x}$. Then $x = u^2$, $dx = 2u\, du$ $\Rightarrow$ $\displaystyle\int \frac{\sqrt{x}\, dx}{x+1} = \int \frac{u \cdot 2u\, du}{u^2+1}$

$$= 2 \int \left[1 - \frac{1}{u^2+1}\right] du = 2(u - \tan^{-1} u) + C = 2(\sqrt{x} - \tan^{-1}\sqrt{x}) + C$$

5. Let $u = \sqrt[3]{x}$. Then $x = u^3$, $dx = 3u^2\, du$ $\Rightarrow$ $\displaystyle\int \frac{dx}{x - \sqrt[3]{x}} = \int \frac{3u^2\, du}{u^3 - u}$

$$= 3 \int \frac{u\, du}{u^2-1} = \frac{3}{2} \ln|u^2-1| + C = \frac{3}{2} \ln|x^{2/3}-1| + C$$

7. Let $u = \sqrt{x-1}$. Then $x = u^2+1$, $dx = 2u\, du$ $\Rightarrow$ $\displaystyle\int_5^{10} \frac{x^2\, dx}{\sqrt{x-1}}$

$$= \int_2^3 \frac{(u^2+1)^2 2u\, du}{u} = 2 \int_2^3 (u^4+2u^2+1) du = 2\left[\frac{1}{5}u^5 + \frac{2}{3}u^3 + u\right]_2^3$$

$$= 2\left(\frac{243}{5} + 18 + 3\right) - 2\left(\frac{32}{5} + \frac{16}{3} + 2\right) = \frac{1676}{15}$$

9. Let $u = \sqrt{x}$. Then $x = u^2$, $dx = 2u\, du$ $\Rightarrow$ $\displaystyle\int \frac{dx}{\sqrt{1+\sqrt{x}}} = \int \frac{2u\, du}{\sqrt{1+u}}$

$$= 2 \int \frac{(v^2-1)2v\, dv}{v} \qquad [v = \sqrt{1+u},\ u = v^2-1,\ du = 2v\, dv]$$

$$= 4 \int (v^2-1) dv = \frac{4}{3}v^3 - 4v + C = \frac{4}{3}(1+\sqrt{x})^{3/2} - 4\sqrt{1+\sqrt{x}} + C$$

11. Let $u = \sqrt{x}$. Then $x = u^2$, $dx = 2u\, du$ $\Rightarrow$ $\displaystyle\int \frac{\sqrt{x}+1}{\sqrt{x}-1}\, dx = \int \frac{u+1}{u-1} 2u\, du$

$$= 2 \int \frac{u^2+u}{u-1}\, du = 2 \int \left[u + 2 + \frac{2}{u-1}\right] du = u^2 + 4u + 4\ln|u-1| + C$$

$$= x + 4\sqrt{x} + 4\ln|\sqrt{x} - 1| + C$$

13. Let $u = \sqrt[3]{x^2+1}$. Then $x^2 = u^3-1$, $2x\, dx = 3u^2\, du$ $\Rightarrow$

$$\int \frac{x^3\, dx}{\sqrt[3]{x^2+1}} = \int \frac{(u^3-1)(3/2)u^2\, du}{u} = (3/2) \int (u^4-u) du$$

$$= (3/10)u^5 - (3/4)u^2 + C = (3/10)(x^2+1)^{5/3} - (3/4)(x^2+1)^{2/3} + C$$

15. Let $u = \sqrt[6]{x}$. Then $x = u^6$, $dx = 6u^5\, du$ $\Rightarrow$ $\displaystyle\int \frac{dx}{\sqrt{x} + \sqrt[3]{x}} = \int \frac{6u^5\, du}{u^3+u^2}$

$$= 6 \int \frac{u^3\, du}{u+1} = 6 \int \left[u^2 - u + 1 - \frac{1}{u+1}\right] du$$

$$= 2u^3 - 3u^2 + 6u - 6\ln|u+1| + C = 2\sqrt{x} - 3\sqrt[3]{x} + 6\sqrt[6]{x} - 6\ln(\sqrt[6]{x}+1) + C$$

17. Let $u = \sqrt[4]{x}$. Then $x = u^4$, $dx = 4u^3\,du$ ⇒ $\displaystyle\int \frac{dx}{\sqrt{x} + \sqrt[4]{x}} = \int \frac{4u^3\,du}{u^2 + u}$

$= 4\displaystyle\int \frac{u^2\,du}{u+1} = 4\int\left[u - 1 + \frac{1}{u+1}\right]du = 2u^2 - 4u + 4\,\ln|u+1| + C$

$= 2\sqrt{x} - 4\sqrt[4]{x} + 4\,\ln(\sqrt[4]{x}+1) + C$

19. Let $u = \sqrt{x}$. Then $x = u^2$, $dx = 2u\,du$ ⇒ $\displaystyle\int \frac{dx}{(bx+c)\sqrt{x}} = \int \frac{2u\,du}{(bu^2+c)u}$

$= \dfrac{2}{b}\displaystyle\int \frac{du}{u^2 + (c/b)} = \frac{2}{b}\sqrt{\frac{b}{c}}\,\tan^{-1}\left[\sqrt{\frac{b}{c}}\,u\right] + C = (2/\sqrt{bc})\tan^{-1}(\sqrt{bx/c}) + C$

21. Let $u = \sqrt[3]{x-1}$. Then $x = u^3 + 1$, $dx = 3u^2\,du$ ⇒ $\displaystyle\int \frac{\sqrt[3]{x-1}}{x}\,dx$

$= \displaystyle\int \frac{u\cdot 3u^2\,du}{u^3+1} = 3\int\left[1 - \frac{1}{u^3+1}\right]du = 3\int\left[1 - \frac{1/3}{u+1} + \frac{(1/3)u - 2/3}{u^2 - u + 1}\right]du$

$= 3u - \ln|u+1| + \displaystyle\int \frac{u-2}{u^2-u+1}\,du = 3u - \ln|u+1| + \frac{1}{2}\int \frac{2u-1}{u^2-u+1}\,du$

$- \dfrac{3}{2}\displaystyle\int \frac{du}{(u-1/2)^2 + 3/4} = 3u - \ln|u+1| + \frac{1}{2}\ln(u^2 - u + 1)$

$- \dfrac{3}{2}\dfrac{2}{\sqrt{3}}\tan^{-1}[(2/\sqrt{3})(u-1/2)] + C = 3\sqrt[3]{x-1} - \ln|\sqrt[3]{x-1}+1|$

$+ (1/2)\ln[(x-1)^{2/3} - (x-1)^{1/3} + 1] - \sqrt{3}\tan^{-1}[(2\sqrt[3]{x-1}-1)/\sqrt{3}] + C$

23. Let $u = \sqrt{x}$. Then $x = u^2$, $dx = 2u\,du$ ⇒ $\displaystyle\int \sqrt{\frac{1-x}{x}}\,dx = \int \frac{\sqrt{1-u^2}}{u}\,2u\,du$

$= 2\displaystyle\int \sqrt{1-u^2}\,du = 2\int \cos^2\theta\,d\theta \quad [u = \sin\theta] = \theta + \sin\theta\cos\theta + C$

$= \sin^{-1}\sqrt{x} + \sqrt{x(1-x)} + C$

[OR: Let $u = \sqrt{(1-x)/x}$. This gives $I = \sqrt{x(1-x)} - \tan^{-1}\sqrt{(1-x)/x} + C$.]

25. Let $u = \sin x$. Then $du = \cos x\,dx$ ⇒ $\displaystyle\int \frac{\cos x\,dx}{\sin^2 x + \sin x} = \int \frac{du}{u^2 + u}$

$= \displaystyle\int \frac{du}{u(u+1)} = \int\left[\frac{1}{u} - \frac{1}{u+1}\right]du = \ln\left|\frac{u}{u+1}\right| + C = \ln\left|\frac{\sin x}{1 + \sin x}\right| + C$

27. Let $u = e^x$. Then $x = \ln u$, $dx = du/u$ ⇒ $\displaystyle\int \frac{e^{2x}\,dx}{e^{2x} + 3e^x + 2} = \int \frac{u^2(du/u)}{u^2 + 3u + 2}$

$= \displaystyle\int \frac{u\,du}{(u+1)(u+2)} = \int\left[\frac{-1}{u+1} + \frac{2}{u+2}\right]du = 2\,\ln|u+2| - \ln|u+1| + C$

$= \ln[(e^x+2)^2/(e^x+1)] + C$

29. Let $u = e^x$. Then $x = \ln u$, $dx = du/u$ ⇒ $\displaystyle\int \sqrt{1-e^x}\,dx = \int \sqrt{1-u}\,(du/u)$

$= \displaystyle\int \frac{\sqrt{1-u}\,du}{u} = \int \frac{v\cdot 2v\,dv}{v^2 - 1} \quad [v = \sqrt{1-u},\ u = v^2 - 1,\ du = 2v\,dv]$

$= 2\displaystyle\int\left[1 + \frac{1}{v^2-1}\right]dv = 2\left[v + \frac{1}{2}\ln\left|\frac{v-1}{v+1}\right|\right] + C$

$= 2\sqrt{1-e^x} + \ln\left[(1-\sqrt{1-e^x})/(1+\sqrt{1-e^x})\right] + C$

31. Let $t = \tan(x/2)$. Then $dx = 2dt/(1+t^2)$, $\cos x = (1-t^2)/(1+t^2)$ →

$$\int \frac{dx}{1 - \cos x} = \int \frac{2dt/(1+t^2)}{2t^2/(1+t^2)} = \int t^{-2} \, dt = -1/t + C = -\cot(x/2) + C$$

OR: $\int \frac{dx}{1 - \cos x} = \int \frac{1 + \cos x}{1 - \cos^2 x} \, dx = \int \frac{1 + \cos x}{\sin^2 x} \, dx$

$= \int (\csc^2 x + \cot x \csc x) \, dx = -\cot x - \csc x + C$

33. Let $t = \tan(x/2)$. Then, by (7.25), $\int_0^{\pi/2} \frac{dx}{\sin x + \cos x} = \int_0^1 \frac{2dt}{2t+1-t^2}$

$= -2 \int_0^1 \frac{dt}{t^2-2t-1} = -2 \int_0^1 \frac{dt}{(t-1)^2-2} = -\frac{1}{\sqrt{2}} \ln \left| \frac{t-1-\sqrt{2}}{t-1+\sqrt{2}} \right| \Big|_0^1$

$= -(1/\sqrt{2}) \left[ \ln 1 - \ln[(\sqrt{2}+1)/(\sqrt{2}-1)] \right] = (1/\sqrt{2}) \ln(\sqrt{2}+1)^2 = \sqrt{2} \ln(\sqrt{2}+1)$

or $-\sqrt{2} \ln(\sqrt{2}-1)$ [since $\sqrt{2}+1 = 1/(\sqrt{2}-1)$]

or $(1/\sqrt{2}) \ln(3+2\sqrt{2})$ [since $(\sqrt{2}+1)^2 = 3+2\sqrt{2}$]

35. Let $t = \tan(x/2)$. Then, by (7.25), $\int \frac{dx}{3 \sin x + 4 \cos x}$

$= \int \frac{2dt}{6t+4(1-t^2)} = \int \frac{-dt}{2t^2-3t-2} = -\int \left[ \frac{-2/5}{2t+1} + \frac{1/5}{t-2} \right] dt = \frac{1}{5} \ln \frac{(2t+1)}{|t-2|} + C$

$= (1/5) \ln\{[2 \tan(x/2)+1]/|\tan(x/2)-2|\} + C$

37. Let $t = \tan(x/2)$. Then $\int \frac{dx}{2 \sin x + \sin 2x} = \frac{1}{2} \int \frac{dx}{\sin x + \sin x \cos x}$

$= \frac{1}{2} \int \frac{2dt/(1+t^2)}{2t/(1+t^2)+2t(1-t^2)/(1+t^2)^2} = \frac{1}{2} \int \frac{(1+t^2)dt}{t(1+t^2)+t(1-t^2)}$

$= \frac{1}{4} \int \frac{(1+t^2)dt}{t} = \frac{1}{4} \int \left[ \frac{1}{t} + t \right] dt = \frac{1}{4} \ln|t| + \frac{1}{8} t^2 + C$

$= (1/4) \ln|\tan(x/2)| + (1/8) \tan^2(x/2) + C$

39. Let $t = \tan(x/2)$. Then $\int \frac{dx}{a \sin x + b \cos x} = \int \frac{2dt}{a(2t)+b(1-t^2)}$

$= \frac{-2}{b} \int \frac{dt}{t^2-2(a/b)t-1} = \frac{-2}{b} \int \frac{dt}{(t-a/b)^2-(1+a^2/b^2)}$

$= \frac{-1}{b} \frac{b}{\sqrt{a^2+b^2}} \ln \left| \frac{t-a/b-\sqrt{a^2+b^2}/b}{t-a/b+\sqrt{a^2+b^2}/b} \right| + C$

$= (1/\sqrt{a^2+b^2}) \ln |[b \tan(x/2)-a+\sqrt{a^2+b^2}]/[b \tan(x/2)-a-\sqrt{a^2+b^2}]| + C$

41. (a) Let $t = \tan(x/2)$. Then $\int \sec x \, dx = \int \frac{dx}{\cos x} = \int \frac{2dt}{1-t^2}$

$= \int \left[ \frac{1}{1-t} + \frac{1}{1+t} \right] dt = \ln|1+t| - \ln|1-t| + C = \ln|(1+t)/(1-t)| + C$

$= \ln |[1+\tan(x/2)]/[1-\tan(x/2)]| + C$

(b) $\tan(\pi/4+x/2) = \frac{\tan(\pi/4) + \tan(x/2)}{1 - \tan(\pi/4)\tan(x/2)} = \frac{1 + \tan(x/2)}{1 - \tan(x/2)}$.

Substituting in the formula from part (a), we get

$\int \sec x \, dx = \ln|\tan(\pi/4+x/2)| + C$

216

## Exercises 7.6

1. $\int \sin^2 x \cos^3 x \, dx = \int \sin^2 x \, (1 - \sin^2 x) \cos x \, dx$

    $= \int u^2 (1-u^2) du \quad [u = \sin x] \quad = \int (u^2 - u^4) du = u^3/3 - u^5/5 + C$

    $= (1/3) \sin^3 x - (1/5) \sin^5 x + C$

3. Let $u = 1-x^2$. Then $du = -2x \, dx \Rightarrow \int_0^{1/2} \dfrac{x \, dx}{\sqrt{1-x^2}} = -\int_1^{3/4} \dfrac{du}{2\sqrt{u}}$

    $= \int_{3/4}^{1} \dfrac{du}{2\sqrt{u}} = \sqrt{u}\Big]_{3/4}^{1} = 1 - \sqrt{3}/2$

5. Let $u = \sqrt{x-2}$. Then $x = u^2+2$, $dx = 2u \, du \Rightarrow \int \dfrac{\sqrt{x-2}}{x+2} dx = \int \dfrac{u \cdot 2u \, du}{u^2+4}$

    $= 2\int \left[1 - \dfrac{4}{u^2+4}\right] du = 2u - \dfrac{8}{2}\tan^{-1}\left(\dfrac{u}{2}\right) + C = 2\sqrt{x-2} - 4 \tan^{-1}(\sqrt{x-2}/2) + C$

7. Use integration by parts: $u = \ln(1+x^2)$, $dv = dx \Rightarrow$

    $du = \dfrac{2x}{1+x^2} dx$, $v = x$, so $\int \ln(1+x^2) dx = x \ln(1+x^2) - \int x \cdot \dfrac{2x \, dx}{1+x^2}$

    $= x \ln(1+x^2) - 2 \int \left[1 - \dfrac{1}{1+x^2}\right] dx = x \ln(1+x^2) - 2x + 2 \tan^{-1} x + C$

9. Let $u = 1+\sqrt{x}$. Then $x = (u-1)^2$, $dx = 2(u-1) du \Rightarrow \int_0^1 (1+\sqrt{x})^8 dx$

    $= \int_1^2 u^8 \cdot 2(u-1) du = 2 \int_1^2 (u^9 - u^8) du = \left[u^{10}/5 - 2u^9/9\right]_1^2$

    $= \dfrac{1024}{5} - \dfrac{1024}{9} - \dfrac{1}{5} + \dfrac{2}{9} = \dfrac{4097}{45}$

11. $\int \dfrac{x \, dx}{x^2 - 2x + 2} = \dfrac{1}{2}\int \dfrac{(2x-2) dx}{x^2 - 2x + 2} + \int \dfrac{dx}{(x-1)^2 + 1} = \dfrac{1}{2}\ln(x^2 - 2x + 2) + \tan^{-1}(x-1) + C$

13. Let $u = \sqrt{9-x^2}$. Then $u^2 = 9-x^2$, $u \, du = -x \, dx \Rightarrow \int \dfrac{\sqrt{9-x^2}}{x} dx$

    $= \int \dfrac{\sqrt{9-x^2}}{x^2} x \, dx = \int \dfrac{u}{9-u^2} (-u) du = \int \left[1 - \dfrac{9}{9-u^2}\right] du = u + 9 \int \dfrac{du}{u^2-9}$

    $= u + \dfrac{9}{2 \cdot 3} \ln\left|\dfrac{u-3}{u+3}\right| + C = \sqrt{9-x^2} + (3/2) \ln\left|\dfrac{\sqrt{9-x^2} - 3}{\sqrt{9-x^2} + 3}\right| + C$

    $= \sqrt{9-x^2} + (3/2) \ln\dfrac{(\sqrt{9-x^2}-3)^2}{x^2} + C = \sqrt{9-x^2} + 3 \ln[(3-\sqrt{9-x^2})/|x|] + C$

    OR: Put $x = 3 \sin \theta$.

15. Integrate by parts: $u = x^2$, $dv = \cosh x \, dx \Rightarrow du = 2x \, dx$,

    $v = \sinh x$, so $\int x^2 \cosh x \, dx = x^2 \sinh x - \int 2x \sinh x \, dx$

    $= x^2 \sinh x - \int (2x) d(\cosh x) = x^2 \sinh x - 2x \cosh x + \int 2 \cosh x \, dx$

    $= (x^2+2) \sinh x - 2x \cosh x + C$

217

17. Let $u = \sin x$. Then $\int \dfrac{\cos x\, dx}{1+\sin^2 x} = \int \dfrac{du}{1+u^2} = \tan^{-1} u + C$

$= \tan^{-1}(\sin x) + C$

19. $\int_0^1 \cos \pi x \tan \pi x\, dx = \int_0^1 \sin \pi x\, dx = \dfrac{-1}{\pi} \int_0^1 -\pi \sin \pi x\, dx$

$= \dfrac{-1}{\pi} [\cos \pi x]_0^1 = \dfrac{-1}{\pi}(-1-1) = \dfrac{2}{\pi}$

21. Integrate by parts twice, first with $u = e^{3x}$, $dv = \cos 5x\, dx$:

$\int e^{3x} \cos 5x\, dx = \frac{1}{5} e^{3x} \sin 5x - \int \frac{3}{5} e^{3x} \sin 5x\, dx = \frac{1}{5} e^{3x} \sin 5x$

$+ \frac{3}{25} e^{3x} \cos 5x - \frac{9}{25} \int e^{3x} \cos 5x\, dx$, so

$\frac{34}{25} \int e^{3x} \cos 5x\, dx = \frac{1}{25} e^{3x}(5 \sin 5x + 3 \cos 5x) + C_1$ and

$\int e^{3x} \cos 5x\, dx = \frac{1}{34} e^{3x}(5 \sin 5x + 3 \cos 5x) + C$

23. $\int \dfrac{dx}{x^3+x^2+x+1} = \int \dfrac{dx}{(x+1)(x^2+1)} = \int \left[\dfrac{1/2}{x+1} - \dfrac{(1/2)x-1/2}{x^2+1}\right] dx$

$= \frac{1}{2} \int \left[\dfrac{1}{x+1} - \dfrac{x}{x^2+1} + \dfrac{1}{x^2+1}\right] dx = \frac{1}{2} \ln|x+1| - \frac{1}{4} \ln(x^2+1) + \frac{1}{2} \tan^{-1} x + C$

25. Let $t = x^3$. Then $dt = 3x^2\, dx \Rightarrow I = \int x^5 e^{-x^3}\, dx = \frac{1}{3} \int t e^{-t}\, dt$

Now integrate by parts with $u = t$, $dv = e^{-t} dt$: $I = \dfrac{-1}{3} t e^{-t}$

$+ \frac{1}{3} \int e^{-t}\, dt = \dfrac{-1}{3} t e^{-t} - \frac{1}{3} e^{-t} + C = -\frac{1}{3} e^{-x^3}(x^3+1) + C$

27. Let $u = 3x+2$. Then $\int \dfrac{dx}{\sqrt{9x^2+12x-5}} = \int \dfrac{dx}{\sqrt{(3x+2)^2-9}} = \frac{1}{3} \int \dfrac{du}{\sqrt{u^2-9}}$

$= (1/3) \cosh^{-1}(u/3) + C_1 = (1/3) \cosh^{-1}[(3x+2)/3] + C_1$

$= (1/3) \ln|3x+2+\sqrt{9x^2+12x-5}| + C$  [by (6.81)]

[OR: Substitute $u = 3 \sec \theta$.]

29. $\int x^{1/3}(1-x^{1/2})dx = \int (x^{1/3}-x^{5/6})dx = \frac{3}{4}x^{4/3} - \frac{6}{11}x^{11/6} + C$

31. $\int \dfrac{2x+5}{x-3}\, dx = \int \dfrac{(2x-6)+11}{x-3}\, dx = \int \left[2 + \dfrac{11}{x-3}\right] dx = 2x + 11 \ln|x-3| + C$

33. $\int \sin^2 x \cos^4 x\, dx = \int (\sin x \cos x)^2 \cos^2 x\, dx$

$= \int \frac{1}{4} \sin^2 2x \cdot \frac{1}{2}(1 + \cos 2x)dx = \frac{1}{8} \int \sin^2 2x\, dx + \frac{1}{8} \int \sin^2 2x \cos 2x\, dx$

$= \frac{1}{16} \int (1 - \cos 4x)dx + \frac{1}{16} \int \sin^2 2x\, (2 \cos 2x)dx$

$= x/16 - (1/64) \sin 4x + (1/48) \sin^3 2x + C$

[OR: Write $\int \sin^2 x \cos^4 x\, dx = (1/8) \int (1 - \cos 2x)(1 + \cos 2x)^2 dx$.]

35. Let $u = 1-x^2$. Then $du = -2x\,dx \Rightarrow \displaystyle\int \frac{x\,dx}{1-x^2+\sqrt{1-x^2}} = \frac{-1}{2}\int \frac{du}{u+\sqrt{u}}$

$= -\displaystyle\int \frac{v\,dv}{v^2+v}$  $[v = \sqrt{u},\ u = v^2,\ du = 2v\,dv]$  $= -\displaystyle\int \frac{dv}{v+1}$

$= -\ln|v+1| + C = -\ln(\sqrt{1-x^2} + 1) + C$

37. Let $u = e^x$. Then $x = \ln u$, $dx = du/u \Rightarrow \displaystyle\int \frac{e^x\,dx}{e^{2x}-1} = \int \frac{u(du/u)}{u^2-1}$

$= \displaystyle\int \frac{du}{u^2-1} = \frac{1}{2}\ln\left|\frac{u-1}{u+1}\right| + C = \frac{1}{2}\ln|(e^x-1)/(e^x+1)| + C$

39. $\displaystyle\int_{-1}^{1} x^5 \cosh x\,dx = 0$ by Theorem 5.33 since $x^5 \cosh x$ is odd.

41. $\displaystyle\int_{-3}^{3} |x^3+x^2-2x|\,dx = \int_{-3}^{3} |(x+2)x(x-1)|\,dx = -\int_{-3}^{-2} (x^3+x^2-2x)\,dx$

$+ \displaystyle\int_{-2}^{0} (x^3+x^2-2x)\,dx - \int_{0}^{1} (x^3+x^2-2x)\,dx + \int_{1}^{3} (x^3+x^2-2x)\,dx.$

Let $f(x) = x^4/4 + x^3/3 - x^2$. Then $f'(x) = x^3+x^2-2x$, so

$\displaystyle\int_{-3}^{3} |x^3+x^2-2x|\,dx = -f(-2)+f(-3)+f(0)-f(-2)-f(1)+f(0)+f(3)-f(1) =$

$f(-3)-2f(-2)+2f(0)-2f(1)+f(3) = \dfrac{9}{4} - 2(-\dfrac{8}{3}) + 2\cdot 0 - 2(\dfrac{-5}{12}) + \dfrac{81}{4} = \dfrac{86}{3}$

43. Let $u = \ln(\sin x)$. Then $du = \cot x\,dx \Rightarrow \displaystyle\int \cos x \ln(\sin x)\,dx$

$= \displaystyle\int u\,du = u^2/2 + C = (1/2)[\ln(\sin x)]^2 + C$

45. $\dfrac{x}{(x^2+1)(x^2+4)} = \dfrac{Ax+B}{x^2+1} + \dfrac{Cx+D}{x^2+4} \Rightarrow x = (Ax+B)(x^2+4) + (Cx+D)(x^2+1) \Rightarrow$

$0 = A+C,\ 0 = B+D,\ 1 = 4A+C,$ and $0 = 4B+D \Rightarrow A = -C = 1/3,$

$B = D = 0.$  $\displaystyle\int \frac{x\,dx}{(x^2+1)(x^2+4)} = \frac{1}{3}\int \left[\frac{x}{x^2+1} - \frac{x}{x^2+4}\right]dx$

$= \dfrac{1}{6}\ln(x^2+1) - \dfrac{1}{6}\ln(x^2+4) + C = \dfrac{1}{6}\ln[(x^2+1)/(x^2+4)] + C$

47. Let $u = \sqrt[3]{x+c}$. Then $x = u^3-c \Rightarrow \displaystyle\int x \sqrt[3]{x+c}\,dx = \int (u^3-c)u\cdot 3u^2\,du$

$= 3\displaystyle\int (u^6-cu^3)\,du = 3u^7/7 - 3cu^4/4 + C$

$= 3[(x+c)^{7/3}/7 - c(x+c)^{4/3}/4] + C$

49. Let $u = \sqrt{x+1}$. Then $x = u^2-1 \Rightarrow \displaystyle\int \frac{dx}{x+4+4\sqrt{x+1}} = \int \frac{2u\,du}{u^2+3+4u}$

$= \displaystyle\int \left[\frac{-1}{u+1} + \frac{3}{u+3}\right]du = 3\ln|u+3| - \ln|u+1| + C$

$= 3\ln(\sqrt{x+1} + 3) - \ln(\sqrt{x+1} + 1) + C$

51. Use parts twice: $\displaystyle\int (x^2+4x-3) \sin 2x\,dx = \int (x^2+4x-3)\,d(-\tfrac{1}{2}\cos 2x)$

$= -\dfrac{1}{2}(x^2+4x-3)\cos 2x + \dfrac{1}{2}\displaystyle\int (2x+4)\cos 2x\,dx = -\dfrac{1}{2}(x^2+4x-3)\cos 2x$

$+ \dfrac{1}{2}\displaystyle\int (x+2)\,d(\sin 2x) = -\dfrac{1}{2}(x^2+4x-3)\cos 2x + \dfrac{1}{2}(x+2)\sin 2x$

$- \dfrac{1}{2}\displaystyle\int \sin 2x\,dx = -\dfrac{1}{2}(x^2+4x-3)\cos 2x + \dfrac{1}{2}(x+2)\sin 2x + \dfrac{1}{4}\cos 2x + C$

$= (1/2)(x+2)\sin 2x - (1/4)(2x^2+8x-7)\cos 2x + C$

53. Let $u = x^2$. Then $du = 2x\ dx$ $\Rightarrow$ $\displaystyle\int \frac{x\ dx}{\sqrt{16-x^4}} = \frac{1}{2} \int \frac{du}{\sqrt{16-u^2}}$

$= (1/2)\sin^{-1}(u/4) + C = (1/2)\sin^{-1}(x^2/4) + C$

55. Let $u = \csc 2x$. Then $du = -2\cot 2x \csc 2x\ dx$ $\Rightarrow$

$\displaystyle\int \cot^3 2x \csc^3 2x\ dx = \int \csc^2 2x(\csc^2 2x - 1)\cot 2x \csc 2x\ dx$

$= \displaystyle\int u^2(u^2-1)(-\tfrac{1}{2}\ du) = -\tfrac{1}{2}\int (u^4 - u^2)du = -\tfrac{1}{2}\left[\tfrac{1}{5}u^5 - \tfrac{1}{3}u^3\right] + C$

$= (1/6)\csc^3 2x - (1/10)\csc^5 2x + C$

57. Let $u = \arctan x$. Then $du = \dfrac{dx}{1+x^2}$ $\Rightarrow$ $\displaystyle\int \frac{e^{\arctan x}}{1+x^2}\ dx = \int e^u du$

$= e^u + C = e^{\arctan x} + C$

59. Integrate by parts 3 times, first with $u = t^3$, $dv = e^{-2t}\ dt$:

$\displaystyle\int t^3 e^{-2t}\ dt = -\tfrac{1}{2}t^3 e^{-2t} + \tfrac{1}{2}\int 3t^2 e^{-2t}\ dt = -\tfrac{1}{2}t^3 e^{-2t} - \tfrac{3}{4}t^2 e^{-2t}$

$\quad + \tfrac{1}{2}\displaystyle\int 3te^{-2t}\ dt = -e^{-2t}\left[\tfrac{1}{2}t^3 + \tfrac{3}{4}t^2\right] - \tfrac{3}{4}te^{-2t} + \tfrac{3}{4}\int e^{-2t}\ dt$

$= -e^{-2t}\left[\tfrac{1}{2}t^3 + \tfrac{3}{4}t^2 + \tfrac{3}{4}t + \tfrac{3}{8}\right] + C = -\tfrac{1}{8}e^{-2t}(4t^3 + 6t^2 + 6t + 3) + C$

61. $\displaystyle\int \sin x \sin 2x \sin 3x\ dx = \int \sin x \cdot (1/2)[\cos(2x-3x) - \cos(2x+3x)]dx$

$= (1/2)\displaystyle\int (\sin x \cos x - \sin x \cos 5x)dx$

$= (1/4)\displaystyle\int \sin 2x\ dx - (1/2)\int (1/2)[\sin(x+5x) + \sin(x-5x)]dx$

$= -(1/8)\cos 2x - (1/4)\displaystyle\int(\sin 6x - \sin 4x)dx$

$= -(1/8)\cos 2x + (1/24)\cos 6x - (1/16)\cos 4x + C$

63. As in Example 5, $\displaystyle\int \sqrt{\frac{1+x}{1-x}}\ dx = \int \frac{1+x}{\sqrt{1-x^2}}\ dx = \int \frac{dx}{\sqrt{1-x^2}} + \int \frac{x\ dx}{\sqrt{1-x^2}}$

$= \sin^{-1}x - \sqrt{1-x^2} + C$

[Another method: Substitute $u = \sqrt{(1+x)/(1-x)}$.]

65. $\displaystyle\int \frac{x+a}{x^2+a^2}\ dx = \frac{1}{2}\int \frac{2x\ dx}{x^2+a^2} + a\int \frac{dx}{x^2+a^2} = \frac{1}{2}\ln(x^2+a^2) + a\cdot\frac{1}{a}\tan^{-1}\left[\frac{x}{a}\right] + C$

$= \ln\sqrt{x^2+a^2} + \tan^{-1}(x/a) + C$

67. Let $u = x^5$. Then $du = 5x^4 dx$ $\Rightarrow$ $\displaystyle\int \frac{x^4\ dx}{x^{10}+16} = \int \frac{(1/5)du}{u^2+16}$

$= \tfrac{1}{5}\cdot\tfrac{1}{4}\tan^{-1}(\tfrac{u}{4}) + C = \tfrac{1}{20}\tan^{-1}(x^5/4) + C$

69. $\int \dfrac{\sin x \, dx}{1 + \sin x} = \int \dfrac{\sin x(1 - \sin x)}{\cos^2 x} dx = \int (\tan x \sec x - \tan^2 x) dx$

$= \int (\sec x \tan x - \sec^2 x + 1) dx = \sec x - \tan x + x + C$

OR: Let $t = \tan(x/2)$. Then $\int \dfrac{\sin x \, dx}{1 + \sin x} = \int \left[ 1 - \dfrac{1}{1 + \sin x} \right] dx$

$= x - \int \dfrac{2dt/(1+t^2)}{1+2t/(1+t^2)} = x - \int \dfrac{2 \, dt}{(t+1)^2} = x + \dfrac{2}{t+1} + C$

$= x + 2/[\tan(x/2) + 1] + C$

71. Integrate by parts with $u = x$, $dv = \sec x \tan x \, dx \Rightarrow du = dx$, $v = \sec x$: $\int x \sec x \tan x \, dx = x \sec x - \int \sec x \, dx$

$= x \sec x - \ln|\sec x + \tan x| + C$

73. $\int \dfrac{dx}{\sqrt{x+1} + \sqrt{x}} = \int (\sqrt{x+1} - \sqrt{x}) dx = \dfrac{2}{3}[(x+1)^{3/2} - x^{3/2}] + C$

75. Let $u = \sqrt{x}$. Then $du = \dfrac{dx}{2\sqrt{x}} \Rightarrow \int \dfrac{\arctan \sqrt{x}}{\sqrt{x}} dx = \int \tan^{-1} u \, 2 \, du$

$= 2u \tan^{-1} u - \int \dfrac{2u \, du}{1+u^2}$ [by parts] $= 2u \tan^{-1} u - \ln(1+u^2) + C$

$= 2\sqrt{x} \tan^{-1} \sqrt{x} - \ln(1+x) + C$

77. Let $u = 2x+1$. Then $du = 2 \, dx \Rightarrow \int \dfrac{dx}{\sqrt{4x^2+4x+5}} = \int \dfrac{(1/2) \, du}{\sqrt{u^2+4}}$

$= \dfrac{1}{2} \int \dfrac{2 \sec^2 \theta \, d\theta}{2 \sec \theta}$ [$u = 2 \tan \theta$, $du = 2 \sec^2 \theta \, d\theta$]

$= \dfrac{1}{2} \int \sec \theta \, d\theta = \dfrac{1}{2} \ln|\sec \theta + \tan \theta| + C_1 = \dfrac{1}{2} \ln \left| \dfrac{\sqrt{u^2+4}}{2} + \dfrac{u}{2} \right| + C_1$

$= \dfrac{1}{2} \ln(u + \sqrt{u^2+4}) + C = \dfrac{1}{2} \ln(2x + 1 + \sqrt{4x^2+4x+5}) + C$

79. Let $u = e^x$. Then $x = \ln u$, $dx = du/u \Rightarrow \int \dfrac{dx}{e^{3x}-e^x} = \int \dfrac{du/u}{u^3-u}$

$= \int \dfrac{du}{(u-1)u^2(u+1)} = \int \left[ \dfrac{1/2}{u-1} - \dfrac{1}{u^2} - \dfrac{1/2}{u+1} \right] du$

$= 1/u + (1/2)\ln|(u-1)/(u+1)| + C = e^{-x} + (1/2)\ln|(e^x-1)/(e^x+1)| + C$

Exercises 7.7

NOTATION:  T = Trapezoidal approximation, S = Simpson's approximation

1.  $\int_0^1 x^3 dx = x^4/4 \big]_0^1 = 0.25.$  $f(x) = x^3$, $\Delta x = \frac{1-0}{4} = 1/4$

   (a) $T = \frac{1/4}{2} [0^3 + 2(1/4)^3 + 2(1/2)^3 + 2(3/4)^3 + 1^3] = \frac{17}{64} = 0.265625$

   (b) $S = \frac{1/4}{3} [0^3 + 4(1/4)^3 + 2(1/2)^3 + 4(3/4)^3 + 1^3] = \frac{1}{4} = 0.250000$

3.  $\int_1^4 \sqrt{x}\, dx = \frac{2}{3} x^{3/2} \big]_1^4 = \frac{2}{3}(8-1) = \frac{14}{3} \approx 4.666667.$  $\Delta x = \frac{4-1}{6} = \frac{1}{2}$

   (a) $T = \frac{1/2}{2} [\sqrt{1} + 2\sqrt{1.5} + 2\sqrt{2} + 2\sqrt{2.5} + 2\sqrt{3} + 2\sqrt{3.5} + \sqrt{4}] \approx 4.661488$

   (b) $S = \frac{1/2}{3} [\sqrt{1} + 4\sqrt{1.5} + 2\sqrt{2} + 4\sqrt{2.5} + 2\sqrt{3} + 4\sqrt{3.5} + \sqrt{4}] \approx 4.666563$

5.  $\int_2^4 \frac{x\, dx}{x^2+1} = \frac{1}{2} \ln(x^2+1) \big]_2^4 = \frac{1}{2}(\ln 17 - \ln 5) \approx 0.611888.$  $\Delta x = .2$

   (a) $T = \frac{.2}{2} \Big[ \frac{2}{5} + 2\frac{2.2}{5.84} + 2\frac{2.4}{6.76} + 2\frac{2.6}{7.76} + 2\frac{2.8}{8.84} + 2\frac{3}{10} + 2\frac{3.2}{11.24}$
   $+ 2\frac{3.4}{12.56} + 2\frac{3.6}{13.96} + 2\frac{3.8}{15.44} + \frac{4}{17} \Big] \approx 0.612115$

   (b) $S = \frac{.2}{3} \Big[ \frac{2}{5} + 4\frac{2.2}{5.84} + 2\frac{2.4}{6.76} + 4\frac{2.6}{7.76} + 2\frac{2.8}{8.84} + 4\frac{3}{10} + 2\frac{3.2}{11.24}$
   $+ 4\frac{3.4}{12.56} + 2\frac{3.6}{13.96} + 4\frac{3.8}{15.44} + \frac{4}{17} \Big] \approx 0.611887$

7.  $f(x) = \sqrt{1+x^3}$, $\Delta x = \frac{1-(-1)}{8} = 1/4$

   (a) $T = (.25/2)[f(-1) + 2f(-3/4) + 2f(-1/2) + \cdots + 2f(1/2)$
   $+ 2f(3/4) + f(1)] \approx 1.913972$

   (b) $S = (.25/3)[f(-1) + 4f(-3/4) + 2f(-1/2) + 4f(-1/4) + 2f(0)$
   $+ 4f(1/4) + 2f(1/2) + 4f(3/4) + f(1)] \approx 1.934766$

9.  $f(x) = e^{-x^2}$, $\Delta x = (1-0)/10 = .1$

   (a) $T = (.1/2)[f(0) + 2f(.1) + 2f(.2) + \cdots + 2f(.8) + 2f(.9) + f(1)]$
   $\approx 0.746211$

   (b) $S = (.1/3)[f(0) + 4f(.1) + 2f(.2) + 4f(.3) + 2f(.4) + 4f(.5)$
   $+ 2f(.6) + 4f(.7) + 2f(.8) + 4f(.9) + f(1)] \approx 0.746825$

11.  $f(x) = \frac{\sin x}{x}$, $\Delta x = (\pi - \pi/2)/6 = \pi/12$

   (a) $T = (\pi/24) [f(\pi/2) + 2f(7\pi/12) + 2f(2\pi/3) + 2f(3\pi/4) + 2f(5\pi/6)$
   $+ 2f(11\pi/12) + f(\pi)] \approx 0.481672$

   (b) $S = (\pi/36) [f(\pi/2) + 4f(7\pi/12) + 2f(2\pi/3) + 4f(3\pi/4) + 2f(5\pi/6)$
   $+ 4f(11\pi/12) + f(\pi)] \approx 0.481172$

13. $f(x) = \cos(e^x)$, $\Delta x = (1/2-0)/8 = 1/16$

    (a) $T = (1/32) [f(0) + 2f(1/16) + 2f(1/8) + \cdots$
        $+ 2f(7/16) + f(1/2)] \approx 0.132465$

    (b) $S = (1/48) [f(0) + 4f(1/16) + 2f(1/8) + 4f(3/16) + 2f(1/4)$
        $+ 4f(5/16) + 2f(3/8) + 4f(7/16) + f(1/2)] \approx 0.132727$

15. $f(x) = x^5 e^x$, $\Delta x = (1-0)/10 = 1/10$

    (a) $T = (.1/2) [f(0) + 2f(.1) + 2f(.2) + \cdots + 2f(.9) + f(1)]$
        $\approx 0.409140$

    (b) $S = (.1/3) [f(0) + 4f(.1) + 2f(.2) + 4f(.3) + 2f(.4) + 4f(.5)$
        $+ 2f(.6) + 4f(.7) + 2f(.8) + 4f(.9) + f(1)] \approx 0.395802$

NOTATION FOR EXERCISES 17-23: $E_T$ AND $E_S$ DENOTE THE ERROR INVOLVED IN USING THE TRAPEZOIDAL RULE AND SIMPSON'S RULE.

17. (a) $f(x) = e^{-x^2}$, $f'(x) = -2xe^{-x^2}$, $f''(x) = (4x^2-2)e^{-x^2}$. For $0 \le x \le 1$, we have $e^{-x^2} \le 1$, $4x^2-2 \le 2$ (both factors are monotonic) so $|f''(x)| = |4x^2-2|e^{-x^2} \le 2 \cdot 1 = 2$. Taking $M = 2$, $a = 0$, $b = 1$, $n = 10$ in (7.30), we get $E_T \le 2 \cdot 1^3/12(10)^2 = 1/600 < 0.0017$.

    (b) $f'''(x) = (12x-8x^3)e^{-x^2}$, $f^{(4)}(x) = (16x^4-48x^2+12)e^{-x^2}$. For $0 \le x \le 1$, we have $16x^4-48x^2+12 \le 16x^4+12 \le 28$ and $e^{-x^2} \le 1$, so $|f^{(4)}(x)| = |16x^4-48x^2+12|e^{-x^2} \le 28$. Taking $M = 28$, $a = 0$, $b = 1$, $n = 10$ in (7.32), we get $E_S \le 28 \cdot 1^5/180(10)^4 < 0.000016$.

    [Note: A less crude estimate is obtained by computing the absolute maximum of $f^{(4)}(x)$ on $[0,1]$, as in Chapter 4, to be $M = 12$ and this gives $E_S \le 1/150,000 < 0.000007$.]

19. (a) $f(x) = x^5 e^x$, $f'(x) = (x^5+5x^4)e^x$, $f''(x) = (x^5+10x^4+20x^3)e^x$. Since $f''$ is positive and increasing on $[0,1]$, we have $|f''(x)| \le f''(1) = 31e$. Taking $M = 31e$, $a = 0$, $b = 1$, $n = 10$ in (7.30), we get $E_T \le 31e(1)^3/12(10)^2 = 31e/1200 < 0.0703$.

    (b) $f'''(x) = (x^5+15x^4+60x^3+60x^2)e^x$, $f^{(4)}(x) = (x^5+20x^4+120x^3+240x^2+120x)e^x$. Since $f^{(4)}$ is positive and increasing on $[0,1]$, we have $|f^{(4)}(x)| \le f^{(4)}(1) = 501e$. Taking $M = 501e$, $a = 0$, $b = 1$, $n = 10$ in (7.32), we get $E_S \le 501e(1)^5/180(10)^4 = 501e/1,800,000 < 0.00076$

21. From #17(a) we know that we can take M = 2. Then

$\dfrac{M(b-a)^3}{12n^2} < 10^{-5} \iff \dfrac{1}{6n^2} < 10^{-5} \iff 6n^2 > 10^5 \iff n > 129.099\ldots$

$\iff n \geq 130$. Take n = 130.

23. $f(x) = \sin(x^2)$, $f'(x) = 2x\cos(x^2)$, $f''(x) = 2\cos(x^2) - 4x^2\sin(x^2)$, so, by the Triangle Inequality, $|f''(x)| \leq 2|\cos(x^2)| + 4x^2|\sin(x^2)| \leq 2 + 4\sin 1 = M < 5.366$, since $0 \leq x \leq 1$. Thus $E_T$ will be less than 0.001 if $\dfrac{5.366}{12n^2} < \dfrac{1}{1000} \iff 12n^2 > 5366 \iff n \geq 22$. From the Trapezoidal Rule with n = 22, we obtain $\int_0^1 \sin(x^2)dx \approx 0.31045$ with an error less than 0.001. If Simpson's Rule is used, there is more work in finding derivatives: $f^{(4)}(x) = (16x^4-12)\sin(x^2)-48x^2\cos(x^2)$ but less work in computing the approximation since M = 12 sin 1 + 48 $\Rightarrow$ n = 6 suffices.

25. $\int_1^{3.2} y\, dx \approx \dfrac{.2}{2}[4.9 + 2(5.4) + 2(5.8) + 2(6.2) + 2(6.7) + 2(7.0)$

$\qquad + 2(7.3) + 2(7.5) + 2(8.0) + 2(8.2) + 2(8.3) + 8.3] = 15.4$

27. $\Delta t = 1 \text{ min} = \dfrac{1}{60}$ h, so distance $= \int_0^{1/6} v(t)dt \approx \dfrac{1/60}{3}[40 + 4(42) +$

$2(45) + 4(49) + 2(52) + 4(54) + 2(56) + 4(57) + 2(57) + 4(55) + 56]$

$\approx 8.6$ mi

29. Let f be a polynomial of degree $\leq 3$, say $f(x) = Ax^3+Bx^2+Cx+D$. It will suffice to show that Simpson's estimate is exact when there are two intervals (n = 2), because for a larger even number of intervals the sum of exact estimates is exact. As in the derivation of Simpson's Rule, we can assume that $x_0 = -h$, $x_1 = 0$, and $x_2 = h$. Then Simpson's approximation to $\int_{-h}^{h} f(x)dx$ is

$\dfrac{h}{3}[f(-h)+4f(0)+f(h)] = \dfrac{h}{3}[(-Ah^3+Bh^2-Ch+D) + 4D + (Ah^3+Bh^2+Ch+D)]$

$= \dfrac{h}{3}[2Bh^2+6D] = \dfrac{2}{3}Bh^3+2Dh$. The exact value of the integral is

$\int_{-h}^{h} (Ax^3+Bx^2+Cx+D)dx = \left[\dfrac{A}{4}x^4 + \dfrac{B}{3}x^3 + \dfrac{C}{2}x^2 + Dx\right]_{-h}^{h} = \dfrac{2}{3}Bh^3+2Dh$.

Thus Simpson's Rule is exact.

Exercises 7.8

1. $\int_2^\infty \frac{dx}{\sqrt{x+3}} = \lim_{t\to\infty} \int_2^t \frac{dx}{\sqrt{x+3}} = \lim_{t\to\infty} \left[2\sqrt{x+3}\right]_2^t = \lim_{t\to\infty} (2\sqrt{t+3} - 2\sqrt{5}) = \infty$.

Divergent.

3. $\int_{-\infty}^1 \frac{dx}{(2x-3)^2} = \lim_{t\to-\infty} \frac{1}{2} \int_t^1 \frac{2\ dx}{(2x-3)^2} = \lim_{t\to-\infty} \frac{1}{2}\left[-\frac{1}{2x-3}\right]_t^1$

$= \lim_{t\to-\infty} \left[\frac{1}{2} + \frac{1}{2(2t-3)}\right] = \frac{1}{2}$

5. $\int_{-\infty}^\infty x\ dx = \int_{-\infty}^0 x\ dx + \int_0^\infty x\ dx.$   $\int_{-\infty}^0 x\ dx = \lim_{t\to-\infty} \left[x^2/2\right]_t^0$

$= \lim_{t\to-\infty} (-t^2/2) = -\infty.$   Divergent.

7. $\int_0^\infty e^{-x}\ dx = \lim_{t\to\infty} \int_0^t e^{-x}\ dx = \lim_{t\to\infty} \left[-e^{-x}\right]_0^t = \lim_{t\to\infty} (-e^{-t}+1) = 1$

9. $\int_{-\infty}^\infty xe^{-x^2}dx = \int_{-\infty}^0 xe^{-x^2}dx + \int_0^\infty xe^{-x^2}\ dx.$   $\int_{-\infty}^0 xe^{-x^2}\ dx$

$= \lim_{t\to-\infty} -\frac{1}{2}\left[e^{-x^2}\right]_t^0 = \lim_{t\to-\infty} -\frac{1}{2}\left[1-e^{-t^2}\right] = -\frac{1}{2};\ \int_0^\infty xe^{-x^2}\ dx = \lim_{t\to\infty} -\frac{1}{2}\left[e^{-x^2}\right]_0^t$

$= \lim_{t\to\infty} -\frac{1}{2}\left[e^{-t^2}-1\right] = \frac{1}{2}.$   Therefore $\int_{-\infty}^\infty xe^{-x^2}\ dx = -\frac{1}{2} + \frac{1}{2} = 0.$

11. $\int_0^\infty \frac{dx}{(x+2)(x+3)} = \lim_{t\to\infty} \int_0^t \left[\frac{1}{x+2} - \frac{1}{x+3}\right]dx = \lim_{t\to\infty} \left[\ln\left[\frac{x+2}{x+3}\right]\right]_0^t$

$= \lim_{t\to\infty} \left[\ln\left[\frac{t+2}{t+3}\right] - \ln \frac{2}{3}\right] = \ln 1 - \ln \frac{2}{3} = \ln \frac{3}{2}$

13. $\int_0^\infty \cos x\ dx = \lim_{t\to\infty} [\sin x]_0^t = \lim_{t\to\infty} \sin t,$ which does not exist.

Divergent.

15. $\int_0^\infty \frac{5\ dx}{2x+3} = \frac{5}{2} \lim_{t\to\infty} \int_0^t \frac{2\ dx}{2x+3} = \frac{5}{2} \lim_{t\to\infty} \left[\ln(2x+3)\right]_0^t$

$= \frac{5}{2} \lim_{t\to\infty} [\ln(2t+3) - \ln 3] = \infty.$   Divergent.

17. $\int_{-\infty}^1 xe^{2x}\ dx = \lim_{t\to-\infty} \int_t^1 xe^{2x}dx = \lim_{t\to-\infty} \left[\frac{1}{2}xe^{2x} - \frac{1}{4}e^{2x}\right]_t^1$   [by parts]

$= \lim_{t\to-\infty} \left[\frac{1}{2}e^2 - \frac{1}{4}e^2 - \frac{1}{2}te^{2t} + \frac{1}{4}e^{2t}\right] = \frac{1}{4}e^2 - 0 + 0 = \frac{1}{4}e^2$   since

$\lim_{t\to-\infty} te^{2t} = \lim_{t\to-\infty} \frac{t}{e^{-2t}} \overset{H}{=} \lim_{t\to-\infty} \frac{1}{-2e^{-2t}} = \lim_{t\to-\infty} -\frac{1}{2}e^{2t} = 0.$

225

19. $\int_1^\infty \frac{\ln x}{x}\,dx = \lim_{t\to\infty}\left[\frac{(\ln x)^2}{2}\right]_1^t = \lim_{t\to\infty}\frac{(\ln t)^2}{2} = \infty.$ Divergent.

21. $\int_{-\infty}^\infty \frac{x\,dx}{1+x^2} = \int_{-\infty}^0 \frac{x\,dx}{1+x^2} + \int_0^\infty \frac{x\,dx}{1+x^2}$ and $\int_{-\infty}^0 \frac{x\,dx}{1+x^2} = \lim_{t\to-\infty}\left[\frac{1}{2}\ln(1+x^2)\right]_t^0$

$= \lim_{t\to-\infty}\left[0 - \frac{1}{2}\ln(1+t^2)\right] = -\infty.$ Divergent.

23. Integrate by parts with $u = \ln x$, $dv = dx/x^2 \Rightarrow du = dx/x$, $v = -1/x$.

$\int_1^\infty \frac{\ln x}{x^2}\,dx = \lim_{t\to\infty}\int_1^t \frac{\ln x}{x^2}\,dx = \lim_{t\to\infty}\left[-\frac{\ln x}{x} - \frac{1}{x}\right]_1^t$

$= \lim_{t\to\infty}\left[-\frac{\ln t}{t} - \frac{1}{t} + 0 + 1\right] = -0 - 0 + 0 + 1 = 1$ since $\lim_{t\to\infty}\frac{\ln t}{t} \overset{H}{=} \lim_{t\to\infty}\frac{1/t}{1} = 0$

25. $\int_0^\infty \frac{dx}{2^x} = \lim_{t\to\infty}\int_0^t 2^{-x}\,dx = \lim_{t\to\infty}\left[\frac{-1}{\ln 2}2^{-x}\right]_0^t = \frac{1}{\ln 2}\lim_{t\to\infty}(1-2^{-t}) = \frac{1}{\ln 2}$

27. $\int_0^3 \frac{dx}{\sqrt{x}} = \lim_{t\to0^+}\int_t^3 \frac{dx}{\sqrt{x}} = \lim_{t\to0^+}\left[2\sqrt{x}\right]_t^3 = \lim_{t\to0^+}(2\sqrt{3}-2\sqrt{t}) = 2\sqrt{3}$

29. $\int_{-1}^0 \frac{dx}{x^2} = \lim_{t\to0^-}\int_{-1}^t \frac{dx}{x^2} = \lim_{t\to0^-}\left[\frac{-1}{x}\right]_{-1}^t = \lim_{t\to0^-}\left[-\frac{1}{t} + \frac{1}{-1}\right] = \infty$ Divergent.

31. $\int_{-2}^3 \frac{dx}{x^4} = \int_{-2}^0 \frac{dx}{x^4} + \int_0^3 \frac{dx}{x^4}$ and $\int_{-2}^0 \frac{dx}{x^4} = \lim_{t\to0^-}\left[-\frac{1}{3}x^{-3}\right]_{-2}^t$

$= \lim_{t\to0^-}\left[-\frac{1}{3t^3} - \frac{1}{24}\right] = \infty.$ Divergent.

33. $\int_4^5 \frac{dx}{(5-x)^{2/5}} = \lim_{t\to5^-}\left[-\frac{5}{3}(5-x)^{3/5}\right]_4^t = \lim_{t\to5^-}\left[-\frac{5}{3}(5-t)^{3/5} + \frac{5}{3}\right] = 0 + \frac{5}{3} = \frac{5}{3}$

35. $\int_{\pi/4}^{\pi/2} \tan^2 x\,dx = \lim_{t\to\pi/2^-}\int_{\pi/4}^t (\sec^2 x - 1)\,dx = \lim_{t\to\pi/2^-}[\tan x - x]_{\pi/4}^t$

$= \frac{\pi}{4} - 1 + \lim_{t\to(\pi/2)^-}(\tan t - t) = \infty.$ Divergent.

37. $\int_0^2 \frac{x\,dx}{\sqrt{4-x^2}} = \lim_{t\to2^-}\int_0^t \frac{x\,dx}{\sqrt{4-x^2}} = \lim_{t\to2^-}\left[-\sqrt{4-x^2}\right]_0^t = \lim_{t\to2^-}\left[2 - \sqrt{4-t^2}\right] = 2$

39. $\int_0^\pi \sec x\,dx = \int_0^{\pi/2}\sec x\,dx + \int_{\pi/2}^\pi \sec x\,dx.$

$\int_0^{\pi/2}\sec x\,dx = \lim_{t\to\pi/2^-}\int_0^t \sec x\,dx = \lim_{t\to\pi/2^-}[\ln|\sec x + \tan x|]_0^t$

$= \lim_{t\to\pi/2^-}\ln|\sec t + \tan t| = \infty.$ Divergent.

41. $\int_{-2}^2 \frac{dx}{x^2-1} = \int_{-2}^{-1}\frac{dx}{x^2-1} + \int_{-1}^0 \frac{dx}{x^2-1} + \int_0^1 \frac{dx}{x^2-1} + \int_1^2 \frac{dx}{x^2-1}.$

$\int \frac{dx}{x^2-1} = \int \frac{dx}{(x-1)(x+1)} = \frac{1}{2}\ln\left|\frac{x-1}{x+1}\right| + C,$ so $\int_0^1 \frac{dx}{x^2-1}$

$= \lim_{t\to1^-}\left[\frac{1}{2}\ln\left|\frac{x-1}{x+1}\right|\right]_0^t = \lim_{t\to1^-}\frac{1}{2}\ln\left|\frac{t-1}{t+1}\right| = -\infty.$ Divergent

**43.** Let $u = \ln x$. Then $du = dx/x \Rightarrow \displaystyle\int_1^e \frac{dx}{x\sqrt[4]{\ln x}} = \lim_{t\to 1^+} \int_t^e \frac{dx}{x\sqrt[4]{\ln x}}$

$= \displaystyle\lim_{t\to 1^+} \int_{\ln t}^1 \frac{du}{\sqrt[4]{u}} = \lim_{t\to 1^+}\left[\frac{4}{3}u^{3/4}\right]_{\ln t}^1 = \lim_{t\to 1^+}\frac{4}{3}\left[1 - (\ln t)^{3/4}\right] = \frac{4}{3}$

**45.** Integrate by parts with $u = \ln x$, $dv = x\,dx$: $\displaystyle\int_0^1 x \ln x\,dx$

$= \displaystyle\lim_{t\to 0^+}\int_t^1 x \ln x\,dx = \lim_{t\to 0^+}\left[\frac{x^2}{2}\ln x - \frac{x^2}{4}\right]_t^1 = -\frac{1}{4} - \lim_{t\to 0^+}\frac{t^2}{2}\ln t$

$= -\frac{1}{4} - \frac{1}{2}\lim_{t\to 0^+}\frac{\ln t}{1/t^2} \overset{H}{=} -\frac{1}{4} - \frac{1}{2}\lim_{t\to 0^+}\frac{1/t}{-2/t^3} = -\frac{1}{4} + \frac{1}{4}\lim_{t\to 0^+}t^2 = -\frac{1}{4}$

**47.**

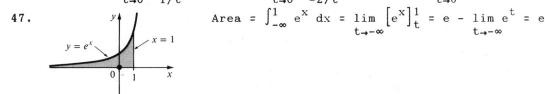

Area $= \displaystyle\int_{-\infty}^1 e^x\,dx = \lim_{t\to -\infty}\left[e^x\right]_t^1 = e - \lim_{t\to -\infty}e^t = e$

**49.**

Area $= \displaystyle\int_{-\infty}^\infty \frac{dx}{x^2-2x+5}$

$= \displaystyle\int_{-\infty}^0 \frac{dx}{(x-1)^2+4} + \int_0^\infty \frac{dx}{(x-1)^2+4}$

$= \displaystyle\lim_{t\to -\infty}\left[\frac{1}{2}\tan^{-1}\left[\frac{x-1}{2}\right]\right]_t^0 + \lim_{t\to \infty}\left[\frac{1}{2}\tan^{-1}\left[\frac{x-1}{2}\right]\right]_0^t$

$= \frac{1}{2}\tan^{-1}(-\frac{1}{2}) - \frac{1}{2}(-\frac{\pi}{2}) + \frac{1}{2}(\frac{\pi}{2}) - \frac{1}{2}\tan^{-1}(-\frac{1}{2}) = \frac{\pi}{2}$

**51.**

Area $= \displaystyle\int_0^{\pi/2}\tan x\,dx = \lim_{t\to\pi/2^-}\left[\ln(\sec x)\right]_0^t$

$= \infty$, so the area is infinite.

**53.** $\displaystyle\int_0^\infty \frac{dx}{\sqrt{x}(1+x)} = \int_0^1 \frac{dx}{\sqrt{x}(1+x)} + \int_1^\infty \frac{dx}{\sqrt{x}(1+x)} = \lim_{t\to 0^+}\int_t^1 \frac{dx}{\sqrt{x}(1+x)}$

$+ \displaystyle\lim_{t\to\infty}\int_1^t \frac{dx}{\sqrt{x}(1+x)} = \lim_{t\to 0^+}\int_{\sqrt{t}}^1 \frac{2\,du}{1+u^2} + \lim_{t\to\infty}\int_1^{\sqrt{t}}\frac{2\,du}{1+u^2} \qquad [u = \sqrt{x},\ x = u^2]$

$= \displaystyle\lim_{t\to 0^+}2\tan^{-1}u\Big]_{\sqrt{t}}^1 + \lim_{t\to\infty}2\tan^{-1}u\Big]_1^{\sqrt{t}} = \lim_{t\to 0^+}\left[2\left[\frac{\pi}{4}\right] - 2\tan^{-1}\sqrt{t}\right]$

$+ \displaystyle\lim_{t\to\infty}\left[2\tan^{-1}\sqrt{t} - 2\left[\frac{\pi}{4}\right]\right] = \frac{\pi}{2} - 0 + 2\left[\frac{\pi}{2}\right] - \frac{\pi}{2} = \pi$

55. If $p = 1$, then $\int_0^1 \frac{dx}{x^p} = \lim_{t \to 0^+} [\ln x]_t^1 = \infty$. Divergent. If $p \neq 1$,

then $\int_0^1 \frac{dx}{x^p} = \lim_{t \to 0^+} \int_t^1 \frac{dx}{x^p}$ [Note that the integral is not improper

if $p < 0$.] $= \lim_{t \to 0^+} \left[\frac{x^{-p+1}}{-p+1}\right]_t^1 = \lim_{t \to 0^+} \frac{1}{1-p}\left[1 - \frac{1}{t^{p-1}}\right]$. If $p > 1$, then

$p - 1 > 0$, so $1/t^{p-1} \to \infty$ as $t \to 0^+$ and the integral diverges.

Finally, if $p < 1$, then $\int_0^1 \frac{dx}{x^p} = \frac{1}{1-p} \lim_{t \to 0^+} (1-t^{1-p}) = \frac{1}{1-p}$. Thus the

integral converges $\iff p < 1$, and in that case its value is $1/(1-p)$.

57. First suppose $p = -1$. Then $\int_0^1 x^p \ln x \, dx = \int_0^1 \frac{\ln x}{x} \, dx$

$= \lim_{t \to 0^+} \int_t^1 \frac{\ln x}{x} \, dx = \lim_{t \to 0^+} \frac{1}{2}(\ln x)^2\Big]_t^1 = -\frac{1}{2} \lim_{t \to 0^+} (\ln t)^2 = -\infty$, so

the integral diverges. Now suppose $p \neq -1$. Then integration by

parts gives $\int x^p \ln x \, dx = \frac{x^{p+1}}{p+1} \ln x - \int \frac{x^p}{p+1} \, dx = \frac{x^{p+1}}{p+1} \ln x -$

$\frac{x^{p+1}}{(p+1)^2} + C$. If $p < -1$, then $p+1 < 0$, so $\int_0^1 x^p \ln x \, dx$

$= \lim_{t \to 0^+} \left[\frac{x^{p+1}}{p+1} \ln x - \frac{x^{p+1}}{(p+1)^2}\right]_t^1 = \frac{-1}{(p+1)^2} - \frac{1}{p+1} \lim_{t \to 0^+} \left[t^{p+1}\left(\ln t - \frac{1}{p+1}\right)\right]$

$= \infty$. If $p > -1$, then $p+1 > 0$ and $\int_0^1 x^p \ln x \, dx = \frac{-1}{(p+1)^2}$

$-\frac{1}{p+1} \lim_{t \to 0^+} \frac{\ln t - \frac{1}{p+1}}{t^{-(p+1)}} \overset{H}{=} \frac{-1}{(p+1)^2} - \frac{1}{p+1} \lim_{t \to 0^+} \frac{1/t}{-(p+1)t^{-(p+2)}}$

$= \frac{-1}{(p+1)^2} + \frac{1}{(p+1)^2} \lim_{t \to 0^+} t^{p+1} = \frac{-1}{(p+1)^2}$. Thus the integral converges

to $-1/(p+1)^2$ if $p > -1$ and diverges otherwise.

59. (a) $\int_{-\infty}^{\infty} x \, dx = \int_{-\infty}^0 x \, dx + \int_0^{\infty} x \, dx$. $\int_0^{\infty} x \, dx = \lim_{t \to \infty} \int_0^t x \, dx$

$= \lim_{t \to \infty} \left[\frac{t^2}{2} - \frac{0^2}{2}\right] = \infty$, so the integral is divergent.

(b) $\int_{-t}^t x \, dx = [x^2/2]_{-t}^t = t^2/2 - t^2/2 = 0$, so $\lim_{t \to \infty} \int_{-t}^t x \, dx = 0$.

Therefore $\int_{-\infty}^{\infty} x \, dx \neq \lim_{t \to \infty} \int_{-t}^t x \, dx$.

61. $\frac{\sin^2 x}{x^2} \leq \frac{1}{x^2}$ on $[1, \infty)$. $\int_1^{\infty} \frac{dx}{x^2}$ is convergent by Ex. 4, so $\int_1^{\infty} \frac{\sin^2 x}{x^2} \, dx$

is convergent by the Comparison Theorem.

63. $\int_{-\infty}^{\infty} e^{-x^2} dx = 2 \int_0^{\infty} e^{-x^2} dx$ [since $e^{-x^2}$ is an even function]

$= 2\left[\int_0^1 e^{-x^2} dx + \int_1^{\infty} e^{-x^2} dx\right]$ For $x \geq 1$, $x^2 \geq x \Rightarrow e^{-x^2} \leq e^{-x}$.

$\int_1^{\infty} e^{-x} dx$ converges. (In fact, $\int_0^{\infty} e^{-x} dx$ converges by #7.) Thus

$\int_1^{\infty} e^{-x^2} dx$ converges by the Comparison Theorem, and $\int_0^{\infty} e^{-x^2} dx$ also

converges. Hence $\int_{-\infty}^{\infty} e^{-x^2} dx$ converges.

Section 7.9

Exercises 7.9

1. By Formula 99, $\int e^{-3x} \cos 4x \, dx = \dfrac{e^{-3x}}{(-3)^2 + 4^2}(-3 \cos 4x + 4 \sin 4x) + C$

$= \dfrac{e^{-3x}}{25}(-3 \cos 4x + 4 \sin 4x) + C$

3. Let $u = 3x$. Then $du = 3 \, dx$, so $\int \dfrac{\sqrt{9x^2-1}}{x^2} dx = \int \dfrac{\sqrt{u^2-1}}{u^2/9} \dfrac{du}{3}$

$= 3 \int \dfrac{\sqrt{u^2-1}}{u^2} du = -\dfrac{3\sqrt{u^2-1}}{u} + 3 \ln|u + \sqrt{u^2-1}| + C$ [by Formula 42]

$= -\dfrac{\sqrt{9x^2-1}}{x} + 3 \ln|3x + \sqrt{9x^2-1}| + C$

5. $\int x^2 e^{3x} dx = \dfrac{1}{3}x^2 e^{3x} - \dfrac{2}{3} \int x e^{3x} dx$ [Formula 97]

$= \dfrac{1}{3}x^2 e^{3x} - \dfrac{2}{3}\left[\dfrac{1}{9}(3x-1)e^{3x}\right] + C$ [Formula 96]

$= (1/27)(9x^2 - 6x + 2)e^{3x} + C$

7. Let $u = x^2$. Then $du = 2x \, dx$, so $\int x \sin^{-1}(x^2) dx$

$= \dfrac{1}{2}\int \sin^{-1} u \, du = \dfrac{1}{2}\left[u \sin^{-1} u + \sqrt{1-u^2}\right] + C$ [Formula 87]

$= \dfrac{1}{2}\left[x^2 \sin^{-1}(x^2) + \sqrt{1-x^4}\right] + C$

9. Let $u = e^x$. Then $du = e^x dx$, so $\int e^x \operatorname{sech}(e^x) dx$

$= \int \operatorname{sech} u \, du = \tan^{-1}|\sinh u| + C$ [Formula 107]

$= \tan^{-1}(\sinh(e^x)) + C$

11. Let $u = x+2$. Then $\int \sqrt{5-4x-x^2}\, dx = \int \sqrt{9-(x+2)^2}\, dx$

$= \int \sqrt{9-u^2}\, du = \frac{u}{2}\sqrt{9-u^2} + \frac{9}{2}\sin^{-1}\frac{u}{3} + C$    [Formula 30]

$= \frac{x+2}{2}\sqrt{5-4x-x^2} + \frac{9}{2}\sin^{-1}\left[\frac{x+2}{3}\right]+C$

13. $\int \sec^5 x\, dx = \frac{1}{4}\tan x \sec^3 x + \frac{3}{4}\int \sec^3 x\, dx$    [Formula 77]

$= \frac{1}{4}\tan x \sec^3 x + \frac{3}{4}(\frac{1}{2}\tan x \sec x + \frac{1}{2}\int \sec x\, dx)$    [Formula 77 again]

$= \frac{1}{4}\tan x \sec^3 x + \frac{3}{8}\tan x \sec x + \frac{3}{8}\ln|\sec x + \tan x| + C$   [Formula 14]

15. Let $u = \sin x$. Then $du = \cos x\, dx$, so $\int \sin^2 x \cos x \ln(\sin x)dx$

$= \int u^2 \ln u\, du = \frac{u^3}{9}(3\ln u - 1) + C$    [Formula 101]

$= (1/9)\sin^3 x [3\ln(\sin x) - 1] + C$

17. $\int \sqrt{2 + 3\cos x}\, \tan x\, dx = -\int \frac{\sqrt{2 + 3\cos x}}{\cos x}(-\sin x\, dx)$

$= -\int \frac{\sqrt{2+3u}}{u}\, du$   $[u = \cos x]$   $= -2\sqrt{2+3u} - 2\int \frac{du}{u\sqrt{2+3u}}$    [Formula 58]

$= -2\sqrt{2+3u} - 2\cdot\frac{1}{\sqrt{2}}\ln\left|\frac{\sqrt{2+3u} - \sqrt{2}}{\sqrt{2+3u} + \sqrt{2}}\right| + C$    [Formula 57]

$= -2\sqrt{2 + 3\cos x} - \sqrt{2}\ln\left|\frac{\sqrt{2 + 3\cos x} - \sqrt{2}}{\sqrt{2 + 3\cos x} + \sqrt{2}}\right| + C$

19. $\int_0^{\pi/2} \cos^5 x\, dx = \frac{1}{5}\left[\cos^4 x \sin x\right]_0^{\pi/2} + \frac{4}{5}\int_0^{\pi/2} \cos^3 x\, dx$    [Formula 74]

$= 0 + \frac{4}{5}\left[\frac{1}{3}(2 + \cos^2 x)\sin x\right]_0^{\pi/2}$    [Formula 68]   $= \frac{4}{15}(2-0) = \frac{8}{15}$

## Review Exercises for Chapter 7

1. $\int \frac{x-1}{x+1}\, dx = \int \left[1 - \frac{2}{x+1}\right]dx = x - 2\ln|x+1| + C$

3. Let $u = \arctan x$. Then $du = dx/(1+x^2)$, so $\int \frac{(\arctan x)^5}{1+x^2}\, dx$

$= \int u^5\, du = \frac{1}{6}u^6 + C = \frac{1}{6}(\arctan x)^6 + C$

5. Let $u = \sin x$. Then $\int \frac{\cos x\, dx}{e^{\sin x}} = \int e^{-u}\, du = -e^{-u} + C = \frac{-1}{e^{\sin x}} + C$

7. Use integration by parts with $u = \ln x$, $dv = x^4 dx$ $\Rightarrow$ $du = dx/x$,

$v = x^5/5$: $\int x^4 \ln x\, dx = \dfrac{x^5}{5} \ln x - \dfrac{1}{5}\int x^4\, dx = \dfrac{x^5}{5} \ln x - \dfrac{x^5}{25} + C$

9. Let $u = x^2$. Then $du = 2x\, dx$, so $\int x \sin(x^2)dx = (1/2) \int \sin u\, du$

$= -(1/2)\cos u + C = -(1/2)\cos(x^2) + C$

11. $\int \dfrac{dx}{2x^2 - 5x + 2} = \int \left[\dfrac{-2/3}{2x-1} + \dfrac{1/3}{x-2}\right]dx = -\dfrac{1}{3} \ln|2x-1| + \dfrac{1}{3} \ln|x-2| + C$

$= (1/3)\ln|(x-2)/(2x-1)| + C$

13. Let $u = \sec x$. Then $du = \sec x \tan x\, dx$, so $\int \tan^7 x \sec^3 x\, dx$

$= \int \tan^6 x \sec^2 x \sec x \tan x\, dx = \int (u^2-1)^3 u^2\, du$

$= \int (u^8 - 3u^6 + 3u^4 - u^2)du = u^9/9 - 3u^7/7 + 3u^5/5 - u^3/3 + C$

$= \dfrac{1}{9} \sec^9 x - \dfrac{3}{7} \sec^7 x + \dfrac{3}{5} \sec^5 x - \dfrac{1}{3} \sec^3 x + C$

15. Let $u = \sqrt{1+2x}$. Then $x = \dfrac{1}{2}(u^2-1)$, $dx = u\, du$, so $\int \dfrac{dx}{\sqrt{1+2x} + 3}$

$= \int \dfrac{u\, du}{u+3} = \int \left[1 - \dfrac{3}{u+3}\right]du = u - 3 \ln|u+3| + C$

$= \sqrt{1+2x} - 3 \ln(\sqrt{1+2x} + 3) + C$

17. $u = \sqrt{x}$ $\Rightarrow$ $du = \dfrac{dx}{2\sqrt{x}}$ $\Rightarrow$ $\int \dfrac{e^{\sqrt{x}}\, dx}{\sqrt{x}} = 2 \int e^u\, du = 2e^u + C = 2e^{\sqrt{x}} + C$

19. Let $u = x+1$. Then $\int \ln(x^2 + 2x + 2)dx = \int \ln(u^2 + 1)du$

$= u \ln(u^2 + 1) - \int u \cdot \dfrac{2u\, du}{u^2 + 1}$    [integration by parts]

$= u \ln(u^2 + 1) - 2 \int \left[1 - \dfrac{1}{u^2 + 1}\right]du = u \ln(u^2 + 1) - 2u + 2 \tan^{-1} u + C$

$= (x+1)\ln(x^2 + 2x + 2) - 2(x+1) + 2 \tan^{-1}(x+1) + C$

[OR: Integrate by parts with $u = \ln(x^2 + 2x + 2)$, $dv = dx$. This gives $(x+1)\ln(x^2 + 2x + 2) - 2x + 2 \tan^{-1}(x+1) + C_1$]

21. $\int \dfrac{dx}{x^3 + x} = \int \left[\dfrac{1}{x} - \dfrac{x}{x^2 + 1}\right]dx = \ln|x| - \dfrac{1}{2} \ln(x^2 + 1) + C$

23. $\int \cot^2 x\, dx = \int (\csc^2 x - 1)dx = -\cot x - x + C$

25. Let $x = \sec \theta$. Then $\int \dfrac{dx}{(x^2-1)^{3/2}} = \int \dfrac{\sec \theta \tan \theta}{\tan^3 \theta} d\theta = \int \dfrac{\sec \theta}{\tan^2 \theta} d\theta$

$= \int \dfrac{\cos \theta\, d\theta}{\sin^2 \theta} = -\dfrac{1}{\sin \theta} + C = -\dfrac{x}{\sqrt{x^2-1}} + C$

27. $\int \dfrac{2x^2 + 3x + 11}{x^3 + x^2 + 3x - 5} dx = \int \left[\dfrac{2}{x-1} - \dfrac{1}{x^2 + 2x + 5}\right]dx = 2 \ln|x-1| - \int \dfrac{dx}{(x+1)^2 + 4}$

$= 2 \ln|x-1| - \dfrac{1}{2} \tan^{-1}\left(\dfrac{x+1}{2}\right) + C$

29. Let $u = e^x$. Then $\int e^{x+e^x} dx = \int e^{e^x} e^x\, dx = \int e^u\, du = e^u + C = e^{e^x} + C$

31. Let $u = \ln x$. Then $\int \frac{\ln(\ln x)}{x}\,dx = \int \ln u\,du = u \ln u - u + C$

[by parts] $= (\ln x)[\ln(\ln x) - 1] + C$

33. $\int_0^{\pi/2} \cos^3 x \sin 2x\,dx = \int_0^{\pi/2} 2 \cos^4 x \sin x\,dx = -\frac{2}{5} \cos^5 x\Big]_0^{\pi/2} = \frac{2}{5}$

35. $\int_0^3 \frac{dx}{x^2-x-2} = \int_0^3 \frac{dx}{(x+1)(x-2)} = \int_0^2 \frac{dx}{(x+1)(x-2)} + \int_2^3 \frac{dx}{(x+1)(x-2)}$

$\int_2^3 \frac{dx}{x^2-x-2} = \lim_{t \to 2^+} \int_t^3 \left[\frac{-1/3}{x+1} + \frac{1/3}{x-2}\right]dx = \lim_{t \to 2^+} \left[\frac{1}{3}\ln\left|\frac{x-2}{x+1}\right|\right]_t^3$

$= \lim_{t \to 2^+} \left[\frac{1}{3} \ln \frac{1}{4} - \frac{1}{3} \ln\left|\frac{t-2}{t+1}\right|\right] = \infty$   Divergent

37. $\int_0^1 \frac{t^2-1}{t^2+1}\,dt = \int_0^1 \left[1 - \frac{2}{t^2+1}\right]dt = \left[t - 2\tan^{-1}t\right]_0^1 = (1-2\cdot\frac{\pi}{4})-0 = 1 - \frac{\pi}{2}$

39. $\int_0^\infty \frac{dx}{(x+2)^4} = \lim_{t \to \infty} \left[\frac{-1}{3(x+2)^3}\right]_0^t = \lim_{t \to \infty} \left[\frac{1}{3\cdot 2^3} - \frac{1}{3(t+2)^3}\right] = \frac{1}{24}$

41. Let $u = \ln x$. Then $\int_1^e \frac{dx}{x\sqrt{\ln x}} = \lim_{t \to 1^+} \int_t^e \frac{dx}{x\sqrt{\ln x}} = \lim_{t \to 1^+} \int_{\ln t}^1 \frac{du}{\sqrt{u}}$

$= \lim_{t \to 1^+} \left[2\sqrt{u}\right]_{\ln t}^1 = \lim_{t \to 1^+}(2 - 2\sqrt{\ln t}) = 2$

43. Let $u = \sqrt{x}+2$. Then $x = (u-2)^2$, $dx = 2(u-2)du$, so $\int_1^4 \frac{\sqrt{x}\,dx}{\sqrt{x} + 2}$

$= \int_3^4 \frac{2(u-2)^2 du}{u} = \int_3^4 \left[2u - 8 + \frac{8}{u}\right]du = \left[u^2 - 8u + 8 \ln u\right]_3^4$

$= (16 - 32 + 8 \ln 4) - (9 - 24 + 8 \ln 3) = -1 + 8 \ln 4 - 8 \ln 3$

$= 8 \ln (4/3) - 1$

45. Let $u = 2x+1$. Then $\int_{-\infty}^\infty \frac{dx}{4x^2+4x+5} = \int_{-\infty}^\infty \frac{(1/2)du}{u^2+4}$

$= \frac{1}{2} \int_{-\infty}^0 \frac{du}{u^2+4} + \frac{1}{2} \int_0^\infty \frac{du}{u^2+4} = \frac{1}{2} \lim_{t \to -\infty} \left[\frac{1}{2}\tan^{-1}(\frac{u}{2})\right]_t^0 + \frac{1}{2} \lim_{t \to \infty} \left[\frac{1}{2}\tan^{-1}(\frac{u}{2})\right]_0^t$

$= \frac{1}{4}[0 - (-\pi/2)] + \frac{1}{4}[\pi/2 - 0] = \frac{\pi}{4}$

47. Let $x = \sec\theta$. Then $\int_1^2 \frac{\sqrt{x^2-1}}{x}\,dx = \int_0^{\pi/3} \frac{\tan\theta}{\sec\theta} \sec\theta \tan\theta\,d\theta$

$= \int_0^{\pi/3} \tan^2\theta\,d\theta = \int_0^{\pi/3} (\sec^2\theta - 1)d\theta = \tan\theta - \theta\Big]_0^{\pi/3} = \sqrt{3} - \frac{\pi}{3}$

49. $\int_0^\infty e^{ax} \cos bx\,dx = \lim_{t \to \infty} \int_0^t e^{ax} \cos bx\,dx$.   Integrate by parts

twice: $\int e^{ax} \cos bx\,dx = \frac{1}{b} e^{ax} \sin bx - \frac{a}{b} \int e^{ax} \sin bx\,dx$

$= \frac{1}{b} e^{ax}\sin bx + \frac{a}{b^2} e^{ax}\cos bx - \frac{a^2}{b^2} \int e^{ax} \cos bx\,dx$, so

$\left[1 + \dfrac{a^2}{b^2}\right] \int e^{ax} \cos bx \, dx = \dfrac{1}{b} e^{ax} \sin bx + \dfrac{a}{b^2} e^{ax} \cos bx + C_1$   Thus

$\int e^{ax} \cos bx \, dx = \dfrac{e^{ax}}{a^2+b^2}(b \sin bx + a \cos bx) + C.$   Now

$\displaystyle\int_0^\infty e^{ax} \cos bx \, dx = \lim_{t\to\infty} \left[ \dfrac{e^{ax}}{a^2+b^2}(a \cos bx + b \sin bx) \right]_0^t$

$= \displaystyle\lim_{t\to\infty} \dfrac{e^{at}}{a^2+b^2}(a \cos bt + b \sin bt) - \dfrac{a}{a^2+b^2}.$   If $a \geq 0$, the limit does

not exist and the integral is divergent. If $a < 0$, the limit is 0 (since $|e^{at} \cos bt| \leq e^{at}$ and $|e^{at} \sin bt| \leq e^{at}$), so the integral converges to $- a/(a^2+b^2)$.

51.  Let $u = e^x$.  Then $du = e^x \, dx$, so $\int e^x \sqrt{1-e^{2x}} \, dx = \int \sqrt{1-u^2} \, du$

$= \dfrac{u}{2} \sqrt{1-u^2} + \dfrac{1}{2} \sin^{-1}u + C$ [Formula 30] $= \dfrac{1}{2}\left[ e^x \sqrt{1-e^{2x}} + \sin^{-1}(e^x) \right] + C$

53.  Let $u = x+\dfrac{1}{2}$.  Then $du = dx$, so $\int \sqrt{x^2+x+1} \, dx = \int \sqrt{(x+1/2)^2+3/4} \, dx$

$= \int \sqrt{u^2+(\sqrt{3}/2)^2} \, du = \dfrac{u}{2}\sqrt{u^2+3/4} + \dfrac{3}{8} \ln|u+\sqrt{u^2+3/4}| + C$  [Formula 21]

$= \dfrac{2x+1}{4}\sqrt{x^2+x+1} + \dfrac{3}{8} \ln|x+\dfrac{1}{2}+\sqrt{x^2+x+1}| + C$

55.  $f(x) = \sqrt{1+x^4}$, $\Delta x = (b-a)/n = (1-0)/10 = 1/10$.

(a) $T = \dfrac{.1}{2}[f(0) + 2f(.1) + 2f(.2) + \cdots + 2f(.8) + 2f(.9) + f(1)]$

$\approx 1.090608$

(b) $S = \dfrac{.1}{3}[f(0) + 4f(.1) + 2f(.2) + 4f(.3) + 2f(.4) + 4f(.5)$

$+ 2f(.6) + 4f(.7) + 2f(.8) + 4f(.9) + f(1)] \approx 1.089429$

57.  $f(x) = (1+x^4)^{1/2}$, $f'(x) = (1/2)(1+x^4)^{-1/2}(4x^3) = 2x^3(1+x^4)^{-1/2}$,

$f''(x) = (2x^6+6x^2)(1+x^4)^{-3/2}$.  Thus $|f''(x)| \leq 8\cdot 1^{-3/2} = 8$ on $[0,1]$.

By taking $M = 8$, we find that the error in #55(a) is bounded by

$M(b-a)^3/(12n^2) = 8/1200 = 1/150 < .0067$

59.  $\int_1^4 \dfrac{e^x}{x} \, dx \approx \dfrac{(4-1)/6}{3}[f(1) + 4f(1.5) + 2f(2) + 4f(2.5) + 2f(3)$

$+ 4f(3.5) + f(4)] \approx 17.74$

61. For x in $[0, \pi/2]$, $0 \leq \cos^2 x \leq \cos x$. For x in $[\pi/2, \pi]$,

$\cos x \leq 0 \leq \cos^2 x$. Thus area $= \int_0^{\pi/2} (\cos x - \cos^2 x) dx$

$+ \int_{\pi/2}^{\pi} (\cos^2 x - \cos x) dx = \left[ \sin x - \frac{x}{2} - \frac{1}{4} \sin 2x \right]_0^{\pi/2}$

$+ \left[ \frac{x}{2} + \frac{1}{4} \sin 2x - \sin x \right]_{\pi/2}^{\pi} = \left[ \left[ 1 - \frac{\pi}{4} \right] - 0 \right] + \left[ \frac{\pi}{2} - \left[ \frac{\pi}{4} - 1 \right] \right] = 2$

63. For $n \geq 0$, $\int_0^{\infty} x^n dx = \lim_{t \to \infty} \left[ x^{n+1}/(n+1) \right]_0^t = \infty$. For $n < 0$,

$\int_0^{\infty} x^n dx = \int_0^1 x^n dx + \int_1^{\infty} x^n dx$. Both integrals are improper. By

(7.34) the second integral diverges if $-1 \leq n < 0$. By Ex. 55 of

Section 7.8, the first integral diverges if $n \leq -1$. Thus

$\int_0^{\infty} x^n dx$ is divergent for every $n < 0$ as well as every $n \geq 0$.

65. By the Fundamental Theorem of Calculus, $\int_0^{\infty} f'(x) dx = \lim_{t \to \infty} \int_0^t f'(x) dx$

$= \lim_{t \to \infty} [f(t) - f(0)] = \lim_{t \to \infty} f(t) - f(0) = 0 - f(0) = -f(0)$.

# CHAPTER 8

**Exercises 8.1    ABBREVIATION: V = Volume**

1.  $V = \int_0^1 \pi(x^2)^2 dx = \pi\left[\dfrac{x^5}{5}\right]_0^1 = \dfrac{\pi}{5}$

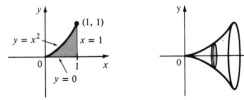

3.  $V = \int_0^1 \pi(-x+1)^2 dx = \pi\int_0^1(x^2-2x+1)dx = \pi\left[\dfrac{x^3}{3}-x^2+x\right]_0^1 = \pi(\dfrac{1}{3}-1+1) = \dfrac{\pi}{3}$

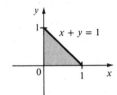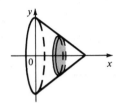

5.  $V = \int_0^4 \pi(\sqrt{y})^2 dy = \pi\left[\dfrac{y^2}{2}\right]_0^4 = 8\pi$

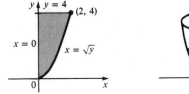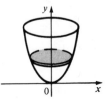

7.  $V = \pi\int_0^1\left[(\sqrt{x})^2-(x^2)^2\right]dx = \pi\int_0^1(x-x^4)dx = \pi\left[\dfrac{x^2}{2}-\dfrac{x^5}{5}\right]_0^1 = \pi\left[\dfrac{1}{2}-\dfrac{1}{5}\right] = \dfrac{3\pi}{10}$

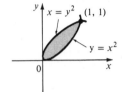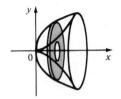

9. $V = \pi \int_0^2 \left[(2y)^2 - (y^2)^2\right]dy = \pi \int_0^2 (4y^2 - y^4)dy = \pi\left[\frac{4}{3}y^3 - \frac{y^5}{5}\right]_0^2$

$= \pi(\frac{32}{3} - \frac{32}{5}) = \frac{64\pi}{15}$

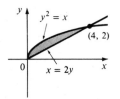 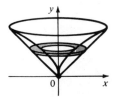

11. $V = \pi \int_{-1}^1 \left[(2-x^4)^2 - 1^2\right]dx = 2\pi \int_0^1 (3 - 4x^4 + x^8)dx = 2\pi\left[3x - \frac{4}{5}x^5 + \frac{x^9}{9}\right]_0^1$

$= 2\pi(3 - \frac{4}{5} + \frac{1}{9}) = \frac{208\pi}{45}$

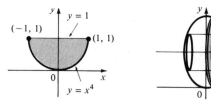

13. $V = \pi \int_0^8 (\frac{x}{4})^2 dx = \frac{\pi}{16}\left[\frac{x^3}{3}\right]_0^8 = \frac{32\pi}{3}$

15. $V = \pi \int_0^2 (8-4y)^2 dy = \pi\left[64y - 32y^2 + \frac{16}{3}y^3\right]_0^2 = \pi(128 - 128 + \frac{128}{3}) = 128\pi/3$

17. $V = \pi \int_0^8 (x^{2/3} - \frac{x^2}{16})dx = \pi\left[\frac{3}{5}x^{5/3} - \frac{x^3}{48}\right]_0^8 = \pi(\frac{96}{5} - \frac{32}{3}) = \frac{128\pi}{15}$

19. $V = \pi \int_0^8 \left[(2 - \frac{x}{4})^2 - (2 - \sqrt[3]{x})^2\right]dx = \pi \int_0^8 (-x + \frac{x^2}{16} + 4x^{1/3} - x^{2/3})dx$

$= \pi\left[-\frac{x^2}{2} + \frac{x^3}{48} + 3x^{4/3} - \frac{3}{5}x^{5/3}\right]_0^8 = \pi(-32 + \frac{32}{3} + 48 - \frac{96}{5}) = \frac{112\pi}{15}$

21. $V = \pi \int_0^8 (2^2 - x^{2/3})dx = \pi\left[4x - \frac{3}{5}x^{5/3}\right]_0^8 = \pi(32 - \frac{96}{5}) = \frac{64\pi}{5}$

23. $V = \pi \int_0^8 (2 - \sqrt[3]{x})^2 dx = \pi \int_0^8 (4 - 4x^{1/3} + x^{2/3})dx = \pi\left[4x - 3x^{4/3} + \frac{3}{5}x^{5/3}\right]_0^8$

$= \pi(32 - 48 + \frac{96}{5}) = \frac{16\pi}{5}$

25. $V = \pi \int_0^2 (x^2 - 1)^2 dx = \pi\left[\frac{x^5}{5} - \frac{2}{3}x^3 + x\right]_0^2 = \pi\left[\frac{32}{5} - \frac{16}{3} + 2\right] = \frac{46\pi}{15}$

27. $V = \pi \int_0^1 (e^x)^2 dx = \frac{\pi}{2} \int_0^1 e^{2x}2dx = \frac{\pi}{2}\left[e^{2x}\right]_0^1 = \pi(e^2 - 1)/2$

29. $V = \pi \int_{-1}^{1} (\sec^2 x - 1^2)dx = \pi\left[\tan x - x\right]_{-1}^{1}$

$= \pi[(\tan 1 - 1) - (-\tan 1 + 1)] = 2\pi(\tan 1 - 1)$

31. $V = \pi \int_{0}^{\pi/2} \sin^2 y \, dy = \frac{\pi}{2}\left[y - \frac{1}{2}\sin 2y\right]_{0}^{\pi/2} = \frac{\pi^2}{4}$

33. $V = \pi \int_{0}^{\pi/4}\left[1^2 - \tan^2 x\right]dx = \pi \int_{0}^{\pi/4} (2 - \sec^2 x)dx = \pi\left[2x - \tan x\right]_{0}^{\pi/4} = \frac{\pi^2}{2} - \pi$

35. $x - 1 = (x-4)^2 + 1 \iff x^2 - 9x + 18 = 0 \iff x = 3$ or $6$, so

$V = \pi \int_{3}^{6}\left[\{6 - (x-4)^2\}^2 - (8-x)^2\right]dx = \pi \int_{3}^{6} (x^4 - 16x^3 + 83x^2 - 144x + 36)dx$

$= \pi\left[\frac{x^5}{5} - 4x^4 + \frac{83x^3}{3} - 72x^2 + 36x\right]_{3}^{6}$

$= \pi\left[(\frac{7776}{5} - 5184 + 5976 - 2592 + 216) - (\frac{243}{5} - 324 + 747 - 648 + 108)\right] = \frac{198\pi}{5}$

37. $V = \pi \int_{0}^{\pi/2}\left[(1 + \cos x)^2 - 1^2\right]dx = \pi \int_{0}^{\pi/2} (2\cos x + \cos^2 x)dx$

$= \pi\left[2\sin x + \frac{x}{2} + \frac{\sin 2x}{4}\right]_{0}^{\pi/2} = \pi(2 + \frac{\pi}{4}) = 2\pi + \frac{\pi^2}{4}$

39. $V = \pi \int_{-3}^{-2}(-x-2)^2 dx + \pi \int_{-2}^{0}(x+2)^2 dx = \pi \int_{-3}^{0}(x+2)^2 dx = \frac{\pi}{3}(x+2)^3\Big]_{-3}^{0}$

$= \frac{\pi}{3}[8 - (-1)] = 3\pi$

41. $V = \pi \int_{0}^{1} 3^2 dx + \pi \int_{1}^{4} 1^2 dx + \pi \int_{4}^{5} 3^2 dx$

$= 9\pi + 3\pi + 9\pi = 21\pi$

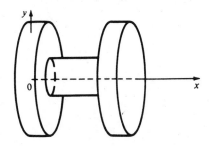

43. $V = \pi \int_{0}^{h}(-\frac{r}{h}y + r)^2 dy$

$= \pi \int_{0}^{h}\left[\frac{r^2}{h^2} y^2 - \frac{2r^2}{h} y + r^2\right]dy$

$= \pi\left[\frac{r^2}{3h^2}y^3 - \frac{r^2}{h}y^2 + r^2 y\right]_{0}^{h} = \frac{1}{3}\pi r^2 h$

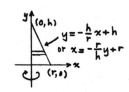

**45.** $V = \pi \int_{r-h}^{r} (r^2-y^2)dy = \pi\left[r^2 y - \frac{y^3}{3}\right]_{r-h}^{r}$

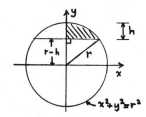

$$= \pi\left[\left[r^3 - \frac{r^3}{3}\right] - \left[r^2(r-h) - \frac{(r-h)^3}{3}\right]\right]$$

$$= \pi h^2(r - h/3)$$

**47.** A typical cross section at height y above the base has dimensions $b(1 - \frac{y}{h})$ and $2b(1 - \frac{y}{h})$, so

$$V = \int_0^h A(y)dy = \int_0^h 2b^2(1 - \frac{y}{h})^2 dy = 2b^2\int_0^h (1 - \frac{2y}{h} + \frac{y^2}{h^2})dy$$

$$= 2b^2\left[y - \frac{y^2}{h} + \frac{y^3}{3h^2}\right]_0^h = 2b^2\left[h-h+\frac{h}{3}\right] = \frac{2}{3}b^2 h$$

$(= \frac{1}{3}Bh$, where B is the area of the base).

**49.** $V = \int_{-r}^{r} \pi\left[(R + \sqrt{r^2-y^2})^2 - (R - \sqrt{r^2-y^2})^2\right]dy$

$$= 2\pi\int_0^r 4R\sqrt{r^2-y^2}\, dy = 8\pi R\int_0^r \sqrt{r^2-y^2}\, dy$$

$$= 8\pi R\left[\frac{y}{2}\sqrt{r^2-y^2} + \frac{r^2}{2}\sin^{-1}\frac{y}{r}\right]_0^r \quad \text{[integration formula 30]}$$

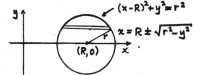

$$= 8\pi R\left[(0 + \frac{r^2}{2}\cdot\frac{\pi}{2}) - 0\right] = 2\pi^2 R r^2$$

$\left[\text{Other ways to evaluate } \int_0^r \sqrt{r^2-y^2}\, dy:\right.$

1. Make the substitution $y = r\sin\theta$.     OR

2.  Observe that the integral represents a quarter of the area of a

circle with radius r, so $\int_0^r \sqrt{r^2-y^2}\, dy = \frac{1}{4}(\pi r^2).\Big]$

**51.** $V = \int_{-r}^{r}(2\sqrt{r^2-x^2})^2 dx = 2\int_0^r 4(r^2-x^2)dx = 8\left[r^2 x - \frac{x^3}{3}\right]_0^r = \frac{16}{3}r^3$

**53.** $V = \int_{-2}^{2} A(x)dx = 2\int_0^2 A(x)dx$

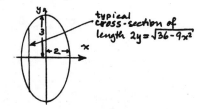

typical cross-section of length $2y = \sqrt{36-9x^2}$

$$= 2\int_0^2 \frac{1}{2}(\sqrt{2}y)^2 dx = 2\int_0^2 y^2 dx$$

$$= \frac{1}{2}\int_0^2 (36-9x^2)dx = \frac{9}{2}\int_0^2 (4-x^2)dx$$

$$= \frac{9}{2}\left[4x - \frac{x^3}{3}\right]_0^2 = \frac{9}{2}(8 - \frac{8}{3}) = 24$$

**55.** The cross section of the base corresponding
to the coordinate y has length $2x = 2\sqrt{y}$,
so $V = \int_0^1 A(y)dy = \int_0^1 (2x)^2 dy = \int_0^1 4x^2 dy$

$$= \int_0^1 4y \, dy = 2y^2 \Big]_0^1 = 2$$

57. Assume that the base of each isosceles triangle lies in the base of S. Then $V = \int_0^2 A(x)dx = \int_0^2 \frac{1}{2}y^2 dx = \frac{1}{2} \int_0^2 (1 - \frac{x}{2})^2 dx$

$$= \frac{4}{\pi} \left[ \frac{\pi}{8} \int_0^2 (1 - \frac{x}{2})^2 dx \right] = \frac{1}{2} \left[ \frac{2}{3}(\frac{x}{2} - 1)^3 \right]_0^2 = \frac{1}{3}$$

59. The cross-sections perpendicular to the y-axis in Fig. 8.15 are rectangles. The rectangle corresponding to the coordinate y has a base of length $2\sqrt{16-y^2}$ in the xy-plane and a height of $y/\sqrt{3}$. (Since $\angle BAC = 30°$, $|BC| = |AB|/\sqrt{3}$.) Thus $A(y) = (2/\sqrt{3})y\sqrt{16-y^2}$ and

$$V = \int_0^4 A(y)dy = \int_0^4 A(y)dy = \frac{2}{\sqrt{3}} \int_0^4 \sqrt{16-y^2} \, y \, dy$$

$$= \frac{2}{\sqrt{3}} \int_{16}^0 u^{1/2}(-\frac{1}{2}du) \left[ u=16-y^2, \ du = -2y \, dy \right] = \frac{1}{\sqrt{3}} \int_0^{16} u^{1/2}du$$

$$= \frac{1}{\sqrt{3}} \frac{2}{3} \left[ u^{3/2} \right]_0^{16} = \frac{2}{3\sqrt{3}}(64) = \frac{128}{3\sqrt{3}}$$

61. For each k satisfying $-c < k < c$, the cross-section of the ellipsoid in the plane $z=k$ is the ellipse $\frac{x^2}{a^2} + \frac{y^2}{b^2} = 1 - \frac{k^2}{c^2}$, whose interior has area $\pi ab(1 - k^2/c^2)$. See Example 7.3.2. Thus the volume enclosed by the ellipsoid is $V = \int_{-c}^c A(z)dz = 2 \int_0^c A(z)dz$

$$= 2\pi ab \int_0^c (1 - \frac{z^2}{c^2})dz = 2\pi ab \left[ z - \frac{z^3}{3c^2} \right]_0^c = \frac{4}{3}\pi abc$$

63. Take the cylinder to be $x^2+y^2 = R^2$ and drill out the interior of the cylinder $y^2+z^2 = r^2$. Taking cross-sections perpendicular to the y-axis, we compute $A(y) = (2\sqrt{R^2-y^2})(2\sqrt{r^2-y^2})$. (The cross-sections are rectangles with length 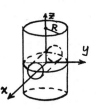 $2\sqrt{R^2-y^2}$ in the x-direction and height $2\sqrt{r^2-y^2}$ in the z-direction.) Thus the volume cut out is $V = 2\int_0^r A(y)dy = 8 \int_0^r \sqrt{R^2-y^2} \sqrt{r^2-y^2} \, dy$.

65. $V = \int_1^\infty \pi(\frac{1}{x})^2 dx = \pi \lim_{t\to\infty} \int_1^t \frac{dx}{x^2} = \pi \lim_{t\to\infty} \left[ \frac{-1}{x} \right]_1^t = \pi \lim_{t\to\infty} (1 - \frac{1}{t}) = \pi < \infty$

Exercises 8.2

1. $V = \int_1^2 2\pi x \cdot x^2 dx = 2\pi\left[\frac{x^4}{4}\right]_1^2 = 2\pi(\frac{15}{4}) = \frac{15\pi}{2}$

3. $V = \int_0^4 2\pi x\sqrt{4+x^2}\ dx = \pi \int_0^4 \sqrt{x^2+4}\ 2x\ dx = \pi\frac{2}{3}(x^2+4)^{3/2}\Big|_0^4$

   $= (2\pi/3)(20\sqrt{20}-8) = (16\pi/3)(5\sqrt{5}-1)$

5. $V = \int_0^2 2\pi x \cdot (4-x^2)dx = 2\pi \int_0^2(4x-x^3)dx = 2\pi\left[2x^2-x^4/4\right]_0^2 = 2\pi(8-4) = 8\pi$

   [Note: If we integrated from -2 to 2, we would be generating the volume twice.]

7. $V = \int_0^1 2\pi x(x^2-x^3)dx = 2\pi(\frac{x^4}{4} - \frac{x^5}{5})\Big|_0^1 = 2\pi(\frac{1}{4} - \frac{1}{5}) = \frac{\pi}{10}$

9. $V = \int_{2\pi}^{3\pi} 2\pi x \sin x\ dx = 2\pi[-x\cos x + \sin x]_{2\pi}^{3\pi}$ (by parts)

   $= 2\pi[(+3\pi+0)-(-2\pi+0)] = 2\pi(5\pi) = 10\pi^2$

11. The two curves intersect at $(2,2)$ and $(4,2)$. For $2 < x < 4$,
    $-x^2+6x-6 > x^2-6x+10$. [To see this, just notice that $(x^2-6x+10)$
    $-(-x^2+6x-6) = 2x^2-12x+16 = 2(x-2)(x-4) < 0$ for $2 < x < 4$.] Thus

    $V = \int_2^4 2\pi x[(-x^2+6x-6)-(x^2-6x+10)]dx = 2\pi \int_2^4 x(-2x^2+12x-16)dx$

    $= 4\pi \int_2^4 (-x^3+6x^2-8x)dx = 4\pi\left[-\frac{x^4}{4} + 2x^3-4x^2\right]_2^4$

    $= 4\pi[(-64+128-64)-(-4+16-16)] = 16\pi$

13. $V = \int_1^e 2\pi x \ln x\ dx = 2\pi\left[\frac{x^2}{4}(2\ln x - 1)\right]_1^e = 2\pi\left[\frac{e^2}{4} - (-\frac{1}{4})\right] = \frac{\pi}{2}(e^2+1)$

    [Integrate by parts or use Formula 101 with $n=1$.]

15. $V = \int_0^{16} 2\pi y \sqrt[4]{y}\ dy = 2\pi \int_0^{16} y^{5/4}\ dy$

    $= 2\pi \frac{4}{9} y^{9/4}\Big|_0^{16} = (\frac{8\pi}{9})(512-0) = \frac{4096\pi}{9}$

17. $V = \int_0^9 2\pi y \cdot 2\sqrt{y}\ dy = 4\pi \int_0^9 y^{3/2}dy = 4\pi \frac{2}{5} y^{5/2}\Big|_0^9 = \frac{8\pi}{5}(243) = \frac{1944\pi}{5}$

19. The two curves intersect at $(0,0)$ and $(0,6)$, so

    $V = \int_0^6 2\pi y(-y^2+6y)dy = 2\pi\left[\frac{-y^4}{4} + 2y^3\right]_0^6 = 2\pi(-324+432) = 216\pi$

21. $V = \int_0^1 2\pi y[(2-y)-y^2]dy = 2\pi\left[y^2 - \frac{y^3}{3} - \frac{y^4}{4}\right]_0^1 = 2\pi\left[1-\frac{1}{3}-\frac{1}{4}\right] = \frac{5\pi}{6}$

**23.** $V = \int_1^4 2\pi x\sqrt{x}\, dx = 2\pi\left[\frac{2}{5}x^{5/2}\right]_1^4 = \frac{4\pi}{5}(32-1) = \frac{124\pi}{5}$

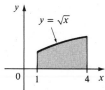
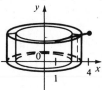

**25.** $V = \int_1^2 2\pi(x-1)x^2 dx = 2\pi\left[\frac{x^4}{4} - \frac{x^3}{3}\right]_1^2 = 2\pi\left[(4-\frac{8}{3}) - (\frac{1}{4}-\frac{1}{3})\right] = \frac{17\pi}{6}$

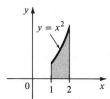
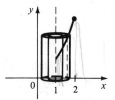

**27.** $V = \int_{-1}^0 2\pi(1-x)e^{-x} dx = 2\pi\left[xe^{-x}\right]_{-1}^0$ [by parts] $= 2\pi(0+e) = 2\pi e$

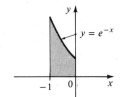
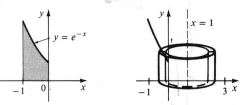

**29.** $V = \int_1^2 2\pi(y-1)\ln y\, dy = 2\pi\int_1^2 (y\ln y - \ln y)dy$

$= 2\pi\left[\frac{y^2}{2}\ln y - y\ln y - \frac{y^2}{4} + y\right]_1^2$ [by parts]

$= 2\pi[(2\ln 2 - 2\ln 2 - 1 + 2) - (-1/4 + 1)] = \pi/2$

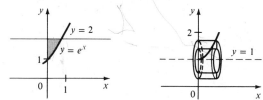
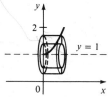

**31.** Use disks: $V = \int_{-2}^1 \pi(x^2+x-2)^2 dx = \pi\int_{-2}^1 (x^4+2x^3-3x^2-4x+4)dx$

$= \pi\left[\frac{x^5}{5} + \frac{x^4}{2} - x^3 - 2x^2 + 4x\right]_{-2}^1 = \pi\left[(\frac{1}{5}+\frac{1}{2}-1-2+4)-(-\frac{32}{5}+8+8-8-8)\right] = \pi(\frac{33}{5}+\frac{3}{2})$

$= 81\pi/10$

241

33. Use shells and integrate by parts: $V = 2\pi \int_{3\pi/4}^{5\pi/4} x \sec^2 x \, dx$

$= 2\pi x \tan x \Big]_{3\pi/4}^{5\pi/4} - \int_{3\pi/4}^{5\pi/4} \tan x \, dx =$

$2\pi \left[ (5\pi/4)\cdot 1 - (3\pi/4)(-1) - \ln|\sec x| \Big]_{3\pi/4}^{5\pi/4} \right] = 2\pi(2\pi - \ln\sqrt{2} + \ln\sqrt{2}) = 4\pi^2$

35. Use disks: $V = \pi \int_{-1}^{1} (1-y^2)^2 dy = 2\pi \int_0^1 (y^4 - 2y^2 + 1) dy$

$= 2\pi \left[ \frac{y^5}{5} - \frac{2y^3}{3} + y \right]_0^1 = 2\pi \left[ \frac{1}{5} - \frac{2}{3} + 1 \right] = \frac{16\pi}{15}$

37. $V = \pi \int_0^2 (\sqrt{1-(y-1)^2})^2 dy = \pi \int_0^2 (2y - y^2) dy = \pi \left[ y^2 - \frac{y^3}{3} \right]_0^2 = \pi(4 - \frac{8}{3}) = \frac{4\pi}{3}$

39. $V = 2 \int_0^r 2\pi x \sqrt{r^2 - x^2} \, dx = -2\pi \int_0^r (r^2 - x^2)^{1/2}(-2x) dx$

$= -2\pi \frac{2}{3}(r^2 - x^2)^{3/2} \Big]_0^r = -(\frac{4\pi}{3})(0 - r^3) = \frac{4}{3}\pi r^3$

41. $V = 2\pi \int_0^r x(-\frac{h}{r}x + h) dx = 2\pi h \int_0^r (-\frac{x^2}{r} + x) dx$

$= 2\pi h \left[ -\frac{x^3}{3r} + \frac{x^2}{2} \right]_0^r = 2\pi h(\frac{r^2}{6}) = \frac{\pi r^2 h}{3}$

43. If $a < b \leq 0$, then a typical cylindrical shell has radius $-x$ and height $f(x)$, so $V = \int_a^b 2\pi(-x)f(x)dx = - \int_a^b 2\pi x f(x)dx$.

Section 8.3

Exercises 8.3

1. $x^2 = 64y^3$, $y = (\frac{x}{8})^{2/3}$ $\Rightarrow$ $\frac{dy}{dx} = \frac{1}{12}(\frac{x}{8})^{-1/3}$ $\Rightarrow$ $1 + (\frac{dy}{dx})^2 = 1 + \frac{1}{144}(\frac{x}{8})^{-2/3}$

$= 1 + \frac{1}{36x^{2/3}}$ $\qquad L = \int_8^{64} \sqrt{1 + \frac{1}{(6\sqrt[3]{x})^2}} \, dx = \int_2^4 \sqrt{1 + \frac{1}{(6u)^2}} \, 3u^2 du$

[Let $u = \sqrt[3]{x}$. Then $x = u^3$, $dx = 3u^2 du$] $\qquad = \int_2^4 \frac{\sqrt{(6u)^2 + 1}}{6u} \, 3u^2 du$

$= \int_2^4 \frac{1}{2}\sqrt{(6u)^2 + 1} \, u \, du = \int_{12}^{24} \sqrt{v^2 + 1} \, \frac{v}{12} \frac{dv}{6}$ $\qquad$ [$v = 6u$ $\Rightarrow$ $dv = 6 \, du$]

$= \frac{1}{144} \int_{12}^{24} \sqrt{v^2 + 1} \, 2v \, dv = \frac{1}{144} \frac{2}{3}(v^2 + 1)^{3/2} \Big]_{12}^{24} = \frac{577^{3/2} - 145^{3/2}}{216}$

3.  $y^2 = (x-1)^3$, $y = (x-1)^{3/2}$, $\frac{dy}{dx} = \frac{3}{2}(x-1)^{1/2}$, $1 + (\frac{dy}{dx})^2 = 1 + \frac{9}{4}(x-1)$

$L = \int_1^2 \sqrt{1 + \frac{9}{4}(x-1)}\ dx = \int_1^2 \sqrt{\frac{9}{4}x - \frac{5}{4}}\ dx = \frac{4}{9} \frac{2}{3}(\frac{9}{4}x - \frac{5}{4})^{3/2}\Big]_1^2 = \frac{13\sqrt{13}-8}{27}$

5.  $12xy = 4y^4 + 3$, $x = \frac{y^3}{3} + \frac{y-1}{4}$, $\frac{dx}{dy} = y^2 - \frac{y^{-2}}{4}$, $(\frac{dx}{dy})^2 = y^4 - \frac{1}{2} + \frac{y^{-4}}{16}$,

$1 + (\frac{dx}{dy})^2 = y^4 + \frac{1}{2} + \frac{y^{-4}}{16} \Rightarrow \sqrt{1 + (\frac{dx}{dy})^2} = y^2 + \frac{y^{-2}}{4}$

$L = \int_1^2 (y^2 + \frac{y^{-2}}{4})dy = \Big[\frac{y^3}{3} - \frac{1}{4y}\Big]_1^2 = (\frac{8}{3} - \frac{1}{8}) - (\frac{1}{3} - \frac{1}{4}) = \frac{59}{24}$

7.  $y = \frac{1}{3}(x^2+2)^{3/2}$, $\frac{dy}{dx} = \frac{1}{2}(x^2+2)^{1/2}(2x) = x\sqrt{x^2+2}$, $1+(\frac{dy}{dx})^2 = 1+x^2(x^2+2)$

$= (x^2+1)^2$. $L = \int_0^1 (x^2+1)dx = \Big[\frac{x^3}{3} + x\Big]_0^1 = \frac{4}{3}$

9.  $y = \frac{x^4}{4} + \frac{1}{8x^2} \Rightarrow \frac{dy}{dx} = x^3 - \frac{1}{4x^3} \Rightarrow 1+(\frac{dy}{dx})^2 = x^6 + \frac{1}{2} + \frac{1}{16x^6}$

$L = \int_1^3 (x^3 + \frac{1}{4}x^{-3})dx = \Big[\frac{x^4}{4} - \frac{x^{-2}}{8}\Big]_1^3 = (\frac{81}{4} - \frac{1}{72}) - (\frac{1}{4} - \frac{1}{8}) = \frac{181}{9}$

11.  $y = \ln(\cos x)$, $y' = -\tan x$, $1+(y')^2 = 1 + \tan^2 x = \sec^2 x$
$L = \int_0^{\pi/4} \sec x\ dx = \ln(\sec x + \tan x)]_0^{\pi/4} = \ln(\sqrt{2}+1)$

13.  $y = \ln(1-x^2)$, $\frac{dy}{dx} = \frac{-2x}{1-x^2}$, $1+(\frac{dy}{dx})^2 = 1 + \frac{4x^2}{(1-x^2)^2} = \frac{(1+x^2)^2}{(1-x^2)^2}$,

$L = \int_0^{1/2} \frac{1+x^2}{1-x^2}\ dx = \int_0^{1/2}\Big[-1 + \frac{2}{(1-x)(1+x)}\Big]dx = \int_0^{1/2}\Big[-1 + \frac{1}{1+x} + \frac{1}{1-x}\Big]dx$

$= \Big[-x + \ln(1+x) - \ln(1-x)\Big]_0^{1/2} = -\frac{1}{2} + \ln\frac{3}{2} - \ln\frac{1}{2} - 0 = \ln 3 - \frac{1}{2}$

15.  $y = e^x$, $y' = e^x$, $1+(y')^2 = 1+e^{2x} \Rightarrow L = \int_0^1 \sqrt{1+e^{2x}}\ dx = \int_1^e \sqrt{1+u^2}\ \frac{du}{u}$

$[u = e^x \Rightarrow x = \ln u,\ dx = \frac{du}{u}] = \int_1^e \frac{\sqrt{1+u^2}}{u^2}\ u\ du = \int_{\sqrt{2}}^{\sqrt{1+e^2}} \frac{v}{v^2-1}\ v\ dv$

$[v = \sqrt{1+u^2} \Rightarrow v^2 = 1+u^2,\ v\ dv = u\ du]$

$= \int_{\sqrt{2}}^{\sqrt{1+e^2}} (1 + \frac{1/2}{v-1} - \frac{1/2}{v+1})dv = \Big[v + \frac{1}{2}\ln\frac{v-1}{v+1}\Big]_{\sqrt{2}}^{\sqrt{1+e^2}} = \sqrt{1+e^2} - \sqrt{2} +$

$\frac{1}{2}\ln\frac{\sqrt{1+e^2}-1}{\sqrt{1+e^2}+1} - \frac{1}{2}\ln\frac{\sqrt{2}-1}{\sqrt{2}+1} = \sqrt{1+e^2} - \sqrt{2} + \ln(\sqrt{1+e^2}-1) - 1 - \ln(\sqrt{2}-1)$

OR: Use Formula 23 for $\int \frac{\sqrt{1+u^2}}{u}du$, or substitute $u = \tan\theta$.

17. $y = \cosh x$, $y' = \sinh x$, $1+(y')^2 = 1 + \sinh^2 x = \cosh^2 x$,
$L = \int_0^1 \cosh x \, dx = \left[\sinh x\right]_0^1 = \sinh 1 = \frac{1}{2}(e-e^{-1})$

19. $y = x^3$, $y' = 3x^2$, $1+(y')^2 = 1+9x^4$ $\Rightarrow$ $L = \int_0^1 \sqrt{1+9x^4} \, dx$

21. $y = \sin x$, $1+(y')^2 = 1 + \cos^2 x$ $\Rightarrow$ $L = \int_0^\pi \sqrt{1 + \cos^2 x} \, dx$

23. $y = e^x \cos x$, $y' = e^x(\cos x - \sin x)$, $(y')^2 = e^{2x}(1 - \sin 2x)$ $\Rightarrow$
$L = \int_0^{\pi/2} \sqrt{1 + e^{2x}(1 - \sin 2x)} \, dx$

25. $y = 2x^{3/2}$, $y' = 3x^{1/2}$, $1+(y')^2 = 1+9x$. The arc length function
with starting point $P_0(1,2)$ is $s(x) = \int_1^x \sqrt{1+9t} \, dt$
$= \frac{2}{27}(1+9t)^{3/2} \Big]_1^x = \frac{2}{27}[(1+9x)^{3/2}-10\sqrt{10}]$

27. $y = 4500 - \frac{x^2}{8000}$, $\frac{dy}{dx} = -\frac{x}{4000}$, $(\frac{dy}{dx})^2 = \frac{x^2}{16,000,000}$
When $y = 4500$ m, $x = 0$ m. When $y = 0$ m, $x = 6000$ m. Therefore
$L = \int_0^{6000} \sqrt{1+(x/4000)^2} \, dx = \int_0^{3/2} \sqrt{1+u^2} \; 4000 \, du$ $\quad [u = \frac{x}{4000}]$
$= 4000 \left[\frac{u}{2}\sqrt{1+u^2} + \frac{1}{2}\ln(u+\sqrt{1+u^2})\right]_0^{3/2}$ $\quad$ [Formula 21 or $u = \tan \theta$]
$= 4000 \left[\frac{3}{4}\sqrt{\frac{13}{4}} + \frac{1}{2}\ln\left(\frac{3}{2} + \sqrt{\frac{13}{4}}\right)\right] = 1500\sqrt{13} + 2000 \ln\frac{3+\sqrt{13}}{2} \approx 7798$ m

29. $y^{2/3} = 1-x^{2/3}$ $\Rightarrow$ $y = (1-x^{2/3})^{3/2}$
$\Rightarrow$ $\frac{dy}{dx} = \frac{3}{2}(1-x^{2/3})^{1/2}(-\frac{2}{3}x^{-1/3})$
$= -x^{-1/3}(1-x^{2/3})^{1/2}$ $\Rightarrow$
$(\frac{dy}{dx})^2 = x^{-2/3}(1-x^{2/3}) = x^{-2/3}-1$

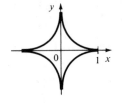

Thus $L = 4\int_0^1 \sqrt{1+(x^{-2/3}-1)} \, dx = 4\int_0^1 x^{-1/3} \, dx = 4 \lim_{t\to 0^+} \frac{3}{2}x^{2/3} \Big]_t^1 = 6$

31. $y = x^3$, $1+(y')^2 = 1+(3x^2)^2 = 1+9x^4$ $\Rightarrow$ $L = \int_0^1 \sqrt{1+9x^4} \, dx$. Let
$f(x) = \sqrt{1+9x^4}$. Then by Simpson's rule with $n = 10$,
$L \approx \frac{1/10}{3}[f(0)+4f(.1)+2f(.2)+4f(.3)+2f(.4)+4f(.5)+2f(.6)+4f(.7)$
$+2f(.8)+4f(.9)+f(1)] \approx 1.548$

33. $y = \sin x$, $1+(\frac{dy}{dx})^2 = 1 + \cos^2 x$, $L = \int_0^\pi \sqrt{1 + \cos^2 x} \, dx$. Let
$g(x) = \sqrt{1 + \cos^2 x}$. Then $L \approx \frac{\pi/10}{3}[g(0)+4g(\frac{\pi}{10})+2g(\frac{\pi}{5})+4g(\frac{3\pi}{10})+2g(\frac{2\pi}{5})$
$+4g(\frac{\pi}{2})+2g(\frac{3\pi}{5})+4g(\frac{7\pi}{10})+2g(\frac{4\pi}{5})+4g(\frac{9\pi}{10})+g(\pi)] \approx 3.820$

Exercises 8.4

1. $y = \sqrt{x}$, $1+(\frac{dy}{dx})^2 = 1+(1/2\sqrt{x})^2 = 1 + \frac{1}{4x}$ → $S = \int_4^9 2\pi y \sqrt{1+(\frac{dy}{dx})^2}\, dx$

$= \int_4^9 2\pi\sqrt{x}\sqrt{1+\frac{1}{4x}}\, dx = 2\pi\int_4^9 \sqrt{x+\frac{1}{4}}\, dx = 2\pi \cdot \frac{2}{3}(x+\frac{1}{4})^{3/2}\Big]_4^9$

$= \frac{4\pi}{3}\cdot\frac{1}{8}(4x+1)^{3/2}\Big]_4^9 = \frac{\pi}{6}(37\sqrt{37}-17\sqrt{17})$

3. $2y = x+4$, $y = \frac{1}{2}x+2$, $y' = \frac{1}{2}$, $\sqrt{1+(y')^2} = \frac{\sqrt{5}}{2}$ → $S = 2\pi\int_0^2 \frac{1}{2}(x+4)\frac{\sqrt{5}}{2}\, dx$

$= \frac{\pi\sqrt{5}}{2}(\frac{x^2}{2} + 4x)\Big]_0^2 = 5\sqrt{5}\pi$

5. $y = x^3+\frac{1}{12x}$, $\frac{dy}{dx} = 3x^2-\frac{1}{12x^2}$, $1+(\frac{dy}{dx})^2 = 9x^4+\frac{1}{2}+\frac{1}{144x^4}$

→ $S = 2\pi\int_1^2 (x^3+\frac{1}{12x})(3x^2+\frac{1}{12x^2})dx = 2\pi\int_1^2(3x^5+\frac{x}{3}+\frac{1}{144x^3})dx$

$= 2\pi(\frac{x^6}{2} + \frac{x^2}{6} - \frac{1}{288x^2})\Big]_1^2 = 2\pi[32+\frac{2}{3}-\frac{1}{288\cdot 4})-(\frac{1}{2}+\frac{1}{6}-\frac{1}{288})] = \frac{12289\pi}{192}$

7. $y = \sin x$, $\sqrt{1+(\frac{dy}{dx})^2} = \sqrt{1 + \cos^2 x}$ → $S = 2\pi\int_0^\pi \sin x \sqrt{1 + \cos^2 x}\, dx$

$= 2\pi\int_{-1}^1 \sqrt{1+u^2}\, du$   [$u = -\cos x$ → $du = \sin x\, dx$]

$= 4\pi\int_0^1 \sqrt{1+u^2}\, du = 4\pi\int_0^{\pi/4} \sec^3\theta\, d\theta$   [$u = \tan\theta$]

$= 2\pi(\sec\theta \tan\theta + \ln|\sec\theta + \tan\theta|)\Big]_0^{\pi/4} = 2\pi[\sqrt{2} + \ln(\sqrt{2}+1)]$

9. $y = \cosh x$, $1+(\frac{dy}{dx})^2 = 1+\sinh^2 x = \cosh^2 x$ → $S = 2\pi\int_0^1 \cosh x \cosh x\, dx$

$= 2\pi\int_0^1 \frac{1}{2}(1+\cosh 2x)dx = \pi(x+\frac{1}{2}\sinh 2x)\Big]_0^1 = \pi(1+\frac{1}{2}\sinh 2)$ or $\pi(1+\frac{e^2-e^{-2}}{4})$

11. $x = \frac{y^4}{2} + \frac{1}{16y^2}$, $\frac{dx}{dy} = 2y^3-\frac{1}{8y^3}$, $1+(\frac{dx}{dy})^2 = 4y^6+\frac{1}{2}+\frac{1}{64y^6}$ →

$S = 2\pi\int_1^3 y[2y^3+\frac{1}{8y^3}]dy = 2\pi\int_1^3(2y^4+\frac{1}{8}y^{-2})dy = 2\pi(\frac{2}{5}y^5-\frac{1}{8}y^{-1})\Big]_1^3$

$= 2\pi[\frac{2}{5}(243) - \frac{1}{24} - \frac{2}{5} + \frac{1}{8}] = \frac{5813\pi}{30}$

13. $x = 1+2y^2$, $1+(\frac{dx}{dy})^2 = 1+(4y)^2 = 1+16y^2$ → $S = 2\pi\int_1^2 y\sqrt{1+16y^2}dy$

$= \frac{\pi}{16}\int_1^2 (16y^2+1)^{1/2}32y\, dy = \frac{\pi}{16}\cdot\frac{2}{3}(16y^2+1)^{3/2}\Big]_1^2 = \frac{\pi}{24}(65\sqrt{65}-17\sqrt{17})$

15. $y = \sqrt{x^2-1}$, $\dfrac{dy}{dx} = x/\sqrt{x^2-1}$, $1+(\dfrac{dy}{dx})^2 = 1 + \dfrac{x^2}{x^2-1}$ $\Rightarrow$

$S = 2\pi\int_1^2 \sqrt{x^2-1}\ \sqrt{1 + x^2/(x^2-1)}\ dx = 2\pi\int_1^2\sqrt{x^2-1+x^2}\ dx = 2\pi\int_1^2\sqrt{2x^2-1}\ dx$

$= 2\sqrt{2}\pi\int_1^2\sqrt{x^2-1/2}\ dx = 2\sqrt{2}\pi\left[\dfrac{x}{2}\sqrt{x^2-1/2} - \dfrac{1}{4}\ln|x+\sqrt{x^2-1/2}|\right]_1^2$ [Formula 39

or $x = (1/\sqrt{2})\sec\theta] = 2\sqrt{2}\pi[\sqrt{7/2} - \dfrac{1}{4}\ln(2+\sqrt{7/2}) - \dfrac{1}{2}\sqrt{1/2} + \dfrac{1}{4}\ln(1+\sqrt{1/2})]$

$= 2\pi(\sqrt{7}-1/2) + (\sqrt{2}\pi/2)[\ln(1+1/\sqrt{2}) - \ln(2+\sqrt{7/2})]$

17. $y = \sqrt[3]{x}$, $x = y^3$, $1+(\dfrac{dx}{dy})^2 = 1+9y^4$ $\Rightarrow$ $S = 2\pi\int_1^2 x\sqrt{1+(\dfrac{dx}{dy})^2}\ dy$

$= 2\pi\int_1^2 y^3\sqrt{1+9y^4}\ dy = \dfrac{2\pi}{36}\int_1^2 \sqrt{1+9y^4}\ 36y^3 dy = \dfrac{\pi}{18}\cdot\dfrac{2}{3}(1+9y^4)^{3/2}\Big]_1^2$

$= \pi(145\sqrt{145}-10\sqrt{10})/27$

19. $y^2 = x^3$, $x = y^{2/3}$, $1+(\dfrac{dx}{dy})^2 = 1+\dfrac{4}{9}y^{-2/3}$ $\Rightarrow$ $S = 2\pi\int_1^8 y^{2/3}\sqrt{1+\dfrac{4}{9}y^{-2/3}}\ dy$

$= 2\pi\int_1^2 u^2\sqrt{1+\dfrac{4}{9u^2}}\ 3u^2 du$ $\quad [u = y^{1/3} \Rightarrow y = u^3,\ dy = 3u^2 du]$

$= 2\pi\int_1^2 u^3\sqrt{9u^2+4}\ du = 2\pi\int_{13}^{40}\dfrac{1}{9}(v-4)v^{1/2}\dfrac{1}{18}\ dv$ $\quad [v = 9u^2+4 \Rightarrow$

$dv = 18u\cdot du,\ u^2 = \dfrac{1}{9}(v-4)]$ $\quad = \dfrac{\pi}{81}\int_{13}^{40}(v^{3/2}-4v^{1/2})dv$

$= \dfrac{\pi}{81}\left[\dfrac{2}{5}v^{5/2} - \dfrac{8}{3}v^{3/2}\right]_{13}^{40} = \dfrac{\pi}{81}\left[\dfrac{3200}{5}\sqrt{40} - \dfrac{320}{3}\sqrt{40} - \dfrac{338}{5}\sqrt{13} + \dfrac{104}{3}\sqrt{13}\right]$

$= \dfrac{\pi}{243}(3200\sqrt{10} - \dfrac{494}{5}\sqrt{13})$

21. $x = e^{2y}$, $1+(\dfrac{dx}{dy})^2 = 1+4e^{4y}$ $\Rightarrow$ $S = 2\pi\int_0^{1/2} e^{2y}\sqrt{1+(2e^{2y})^2}\ dy$

$= 2\pi\int_2^{2e}\sqrt{1+u^2}\dfrac{1}{4}du$ $\quad [u = 2e^{2y},\ du = 4e^{2y}dy]$ $\quad = \dfrac{\pi}{2}\int_2^{2e}\sqrt{1+u^2}\ du$

$= \dfrac{\pi}{2}\left[\dfrac{u}{2}\sqrt{1+u^2} + \dfrac{1}{2}\ln|u+\sqrt{1+u^2}|\right]_2^{2e}$ $\quad [u = \tan\theta$ or Formula 21]

$= \dfrac{\pi}{2}\left[e\sqrt{1+4e^2} + \dfrac{1}{2}\ln(2e+\sqrt{1+4e^2}) - \sqrt{5} - \dfrac{1}{2}\ln(2+\sqrt{5})\right]$

$= \dfrac{\pi}{4}\left[2e\sqrt{1+4e^2} - 2\sqrt{5} + \ln\{(2e+\sqrt{1+4e^2})/(2+\sqrt{5})\}\right]$

23. $x = (1/2\sqrt{2})(y^2 - \ln y)$, $\dfrac{dx}{dy} = (1/2\sqrt{2})(2y-\dfrac{1}{y})$, $1+(\dfrac{dx}{dy})^2 = 1 + \dfrac{1}{8}(2y-\dfrac{1}{y})^2$

$= 1 + \dfrac{1}{8}(4y^2-4+\dfrac{1}{y^2}) = \dfrac{1}{8}(4y^2+4+\dfrac{1}{y^2}) = \left[\dfrac{1}{2\sqrt{2}}(2y+\dfrac{1}{y})\right]^2$

$\Rightarrow S = 2\pi\int_1^2 (1/2\sqrt{2})(y^2- \ln y)(1/2\sqrt{2})(2y + \dfrac{1}{y})dy$

$= \dfrac{\pi}{4}\int_1^2(2y^3 + y - 2y\ln y - \dfrac{\ln y}{y})dy = \dfrac{\pi}{4}\left[\dfrac{y^4}{2} + \dfrac{y^2}{2} - y^2\ln y + \dfrac{y^2}{2}\right.$

246

$$-\frac{1}{2}(\ln y)^2\Big]_1^2 = \frac{\pi}{8}\Big[y^4 + 2y^2 - 2y^2\ln y - (\ln y)^2\Big]_1^2$$

$$= \frac{\pi}{8}[16 + 8 - 8\ln 2 - (\ln 2)^2 - 1 - 2] = \frac{\pi}{8}[21 - 8\ln 2 - (\ln 2)^2]$$

**25.** $S = 2\pi\int_0^1 x^4\sqrt{1+(4x^3)^2}\,dx = 2\pi\int_0^1 x^4\sqrt{16x^6+1}\,dx \approx 2\pi\cdot\frac{1/10}{3}[f(0)+4f(.1)$

$+2f(.2)+4f(.3)+2f(.4)+4f(.5)+2f(.6)+4f(.7)+2f(.8)+4f(.9)+f(1)]$

(where $f(x) = x^4\sqrt{16x^6+1}$) $\approx 3.44$

**27.** The curve $8y^2 = x^2(1-x^2)$ actually consists of two loops in the region described by the inequalities $|x| \le 1$, $|y| \le \sqrt{2}/8$. (The maximum value of $|y|$ is attained when $|x| = 1/\sqrt{2}$.) If we consider the loop in the region $x \ge 0$, the surface area S it generates when rotated about the x-axis is calculated as follows:

$16y\frac{dy}{dx} = 2x-4x^3$, so $(\frac{dy}{dx})^2 = \Big[\frac{x-2x^3}{8y}\Big]^2 = \frac{x^2(1-2x^2)^2}{64y^2} = \frac{x^2(1-2x^2)^2}{8x^2(1-x^2)}$

$= \frac{(1-2x^2)^2}{8(1-x^2)}$ for $x \ne 0, \pm 1$. (The formula also holds for $x = 0$ by

continuity.) $1+(\frac{dy}{dx})^2 = 1+\frac{(1-2x^2)^2}{8(1-x^2)} = \frac{9-12x^2+4x^4}{8(1-x^2)} = \frac{(3-2x^2)^2}{8(1-x^2)} \Rightarrow S =$

$2\pi\int_0^1 \frac{\sqrt{x^2(1-x^2)}}{2\sqrt{2}}\cdot\frac{3-2x^2}{2\sqrt{2}\sqrt{1-x^2}}\,dx = \frac{\pi}{4}\int_0^1 x(3-2x^2)dx = \frac{\pi}{4}\Big[\frac{3x^2}{2} - \frac{x^4}{2}\Big]_0^1 = \frac{\pi}{4}(\frac{3}{2}-\frac{1}{2})$

$= \pi/4$

**29.** $S = 2\pi\int_0^{3a}\Big[\sqrt{a}x^{1/2} - \frac{x^{3/2}}{3\sqrt{a}}\Big]\Big(\frac{\sqrt{a}}{2}x^{-1/2} + \frac{x^{1/2}}{2\sqrt{a}}\Big)dx = 2\pi\int_0^{3a}(\frac{a}{2}+\frac{x}{2}-\frac{x}{6}-\frac{x^2}{6a})dx$

$= 2\pi\Big[\frac{a}{2}x + \frac{x^2}{6} - \frac{x^3}{18a}\Big]_0^{3a} = 2\pi(\frac{3a^2}{2} + \frac{9a^2}{6} - \frac{27a^3}{18a}) = 3\pi a^2$

**31.** $S = 2\pi\int_1^\infty y\sqrt{1+(\frac{dy}{dx})^2}\,dx = 2\pi\int_1^\infty \frac{1}{x}\sqrt{1+\frac{1}{x^4}}\,dx = 2\pi\int_1^\infty \frac{\sqrt{x^4+1}}{x^3}\,dx > 2\pi\int_1^\infty \frac{x^2}{x^3}\,dx$

$= 2\pi\int_1^\infty \frac{dx}{x} = 2\pi \lim_{t\to\infty} [\ln x]_1^t = 2\pi \lim_{t\to\infty} \ln t = \infty$

**33.** $\frac{x^2}{a^2} + \frac{y^2}{b^2} = 1 \Rightarrow \frac{y(dy/dx)}{b^2} = \frac{-x}{a^2} \Rightarrow \frac{dy}{dx} = \frac{-b^2x}{a^2y} \Rightarrow 1+(\frac{dy}{dx})^2 = 1 + \frac{b^4x^2}{a^4y^2}$

$= \frac{b^4x^2+a^4y^2}{a^4y^2} = \frac{b^4x^2+a^4b^2(1-x^2/a^2)}{a^4b^2(1-x^2/a^2)} = \frac{a^4b^2+b^4x^2-a^2b^2x^2}{a^4b^2-a^2b^2x^2} = \frac{a^4+b^2x^2-a^2x^2}{a^4-a^2x^2}$

$= \frac{a^4-(a^2-b^2)x^2}{a^2(a^2-x^2)}$ . The ellipsoid's surface area is twice the area

generated by rotating the first quadrant portion of the ellipse about the x-axis. Thus $S = 2\int_0^a 2\pi y \sqrt{1+(\frac{dy}{dx})^2}\ dx$

$= 4\pi\int_0^a \frac{b}{a}\sqrt{a^2-x^2}\ \frac{\sqrt{a^4-(a^2-b^2)x^2}}{a\sqrt{a^2-x^2}}\ dx = \frac{4\pi b}{a^2}\int_0^a \sqrt{a^4-(a^2-b^2)x^2}\ dx$

$= \frac{4\pi b}{a^2}\int_0^{a\sqrt{a^2-b^2}} \sqrt{a^4-u^2}\ \frac{du}{\sqrt{a^2-b^2}}$ $\qquad [u = \sqrt{a^2-b^2}\ x]$

$= \frac{4\pi b}{a^2\sqrt{a^2-b^2}} \left[\frac{u}{2}\sqrt{a^4-u^2} + \frac{a^4}{2}\sin^{-1}\frac{u}{a^2}\right]_0^{a\sqrt{a^2-b^2}}$ $\qquad [Formula\ 30]$

$= \frac{4\pi b}{a^2\sqrt{a^2-b^2}} \left[\frac{a\sqrt{a^2-b^2}}{2}\sqrt{a^4-a^2(a^2-b^2)} + \frac{a^4}{2}\sin^{-1}\frac{\sqrt{a^2-b^2}}{a}\right]$

$= 2\pi\left[b^2 + a^2 b\ \sin^{-1}(\sqrt{a^2-b^2}/a)/\sqrt{a^2-b^2}\right]$

35. In the derivation of (8.20), we computed a typical contribution to the surface area to be $2\pi\frac{y_{i-1}+y_i}{2}|P_{i-1}P_i|$, the area of a frustum of a cone. When $f(x)$ is not necessarily positive, the approximations $y_i = f(x_i) \approx f(x_i^*)$ and $y_{i-1} = f(x_{i-1}) \approx f(x_i^*)$ must be replaced by $y_i = |f(x_i)| \approx |f(x_i^*)|$ and $y_{i-1} = |f(x_{i-1})| \approx |f(x_i^*)|$. Thus $2\pi\frac{y_{i-1}+y_i}{2}|P_{i-1}P_i| \approx 2\pi|f(x_i^*)|\sqrt{1+[f'(x_i^*)]^2}\Delta x_i$. Continuing with the rest of the derivation as before, we obtain $S = \int_a^b 2\pi|f(x)|\sqrt{1+[f'(x)]^2}\ dx$.

37. For the upper semicircle, $f(x) = \sqrt{r^2-x^2}$, $f'(x) = -x/\sqrt{r^2-x^2}$. The surface area generated is $S_1 = \int_{-r}^r 2\pi\left[r-\sqrt{r^2-x^2}\right]\sqrt{1+x^2/(r^2-x^2)}\ dx$

$= 4\pi\int_0^r \left[r-\sqrt{r^2-x^2}\right]\left[r/\sqrt{r^2-x^2}\right]dx$. For the lower semicircle,

$f(x) = -\sqrt{r^2-x^2}$, $f'(x) = x/\sqrt{r^2-x^2}$ and the surface area generated is

$S_2 = \int_{-r}^r 2\pi\left[r+\sqrt{r^2-x^2}\right]\sqrt{1+x^2/(r^2-x^2)}\ dx = 4\pi\int_0^r\left[r+\sqrt{r^2-x^2}\right]\left[r/\sqrt{r^2-x^2}\right]dx$.

Thus the total area is $S = S_1+S_2 = 8\pi\int_0^r (r^2/\sqrt{r^2-x^2})dx$

$= 8\pi r^2\sin^{-1}(\frac{x}{r})\Big]_0^r = 8\pi r^2(\frac{\pi}{2}) = 4\pi^2 r^2$.

**Exercises 8.5**

1.  By (8.26), $W = Fd = (900)(8) = 7200$ J

3.  By (8.28), $W = \int_0^{10}(5x^2+1)dx = \left[\frac{5}{3}x^3+x\right]_0^{10} = \frac{5000}{3} + 10 = \frac{5030}{3}$ ft-lb

5.  $10 = f(x) = kx = \frac{k}{3}$ (4 inches $= \frac{1}{3}$ foot), so $k = 30$ (The units for $k$
    are pounds per foot.) Now 6 inches $= \frac{1}{2}$ foot, so $W = \int_0^{1/2} 30x\ dx$
    $= \left[15x^2\right]_0^{1/2} = \frac{15}{4}$ ft-lb.

7.  If $\int_0^{.12} kx\ dx = 2$ J, then $2 = \left[\frac{1}{2}kx^2\right]_0^{.12} = \frac{1}{2}k(.0144) = .0072k$ and
    $k = 2500/9 \approx 277.78$. Thus the work needed to stretch the spring
    from 35 cm to 40 cm is $\int_{.05}^{.10}\frac{2500}{9}x\ dx = \frac{1250}{9}x^2\Big]_{1/20}^{1/10}$
    $= \frac{1250}{9}(\frac{1}{100} - \frac{1}{400}) = \frac{25}{24} \approx 1.04$ J.

9.  $f(x) = kx$, so $30 = \frac{2500}{9}x$ and $x = \frac{270}{2500}$ m $= 10.8$ cm

11. First notice that the exact height of the building is immaterial.
    The portion of the rope from $x$ ft to $x+\Delta x$ ft below the top of the
    building weighs $\frac{1}{2}\Delta x$ lb and must be lifted $x$ ft (approximately), so
    its contribution to the total work is $\frac{1}{2}x\Delta x$ ft-lb. The total work is
    $W = \int_0^{50}\frac{1}{2}x\ dx = \frac{1}{4}x^2\Big]_0^{50} = \frac{2500}{4} = 625$ ft-lb.

13. The work needed to lift the cable is $\int_0^{500}2x\ dx = x^2\Big]_0^{500} = 250{,}000$
    ft-lb. The work needed to lift the coal is 800 lb $\cdot$ 500 ft
    $= 400{,}000$ ft-lb. Thus the total work required is $650{,}000$ ft-lb.

15. A "slice" of water $\Delta x$ m thick and lying at a depth of $x$ m ($0 \le x \le 1/2$)
    has volume $2\Delta x$ m$^3$, a mass of $2000\Delta x$ kg, weighs $(9.8)(2000\Delta x)$
    $= 19600\ \Delta x$ N, and requires $19600x\Delta x$ J of work for its removal.
    Thus $W = \int_0^{1/2}19600x\ dx = 9800x^2\Big]_0^{1/2} = 2450$J. (This answer is
    approximate because the value $g = 9.8$ m/s$^2$ is approximate. Thus we
    should really write $W \approx 2.45 \times 10^3$ J)

17. A "slice" of water $\Delta x$ m thick and lying $x$ ft above the bottom has
    volume $8x\Delta x$ m$^3$ and weighs $\approx (9.8\times10^3)(8x\Delta x)$ N. It must be lifted
    $5-x$ m by the pump, so the work needed is $\approx (9.8\times10^3)(5-x)(8x\Delta x)$ J.
    The total work required is $W \approx \int_0^3(9.8\times10^3)(5-x)8x\ dx$
    $= (9.8\times10^3)\int_0^3(40x-8x^2)dx = (9.8\times10^3)\left[20x^2-\frac{8}{3}x^3\right]_0^3 = (9.8\times10^3)(180-72)$
    $= (9.8\times10^3)(108) = 1058.4 \times 10^3 \approx 1.06 \times 10^6$ J.

19. Measure depth x downward from the flat top of the tank, so that

$0 \leq x \leq 2$ ft. Then $\Delta W = (62.5)(2\sqrt{4-x^2})(8\Delta x)(x+1)$ ft-lb, so

$W \approx (62.5)(16)\int_0^2 (x+1)\sqrt{4-x^2} \, dx = 1000\left[\int_0^2 x\sqrt{4-x^2} \, dx + \int_0^2 \sqrt{4-x^2} \, dx\right]$

$= 1000\left[\int_0^4 u^{1/2} \frac{1}{2} \, du + \frac{1}{4}\pi(2^2)\right]$    $[u = 4-x^2, \; du = -2x \, dx]$

$= 1000\left\{\left[\frac{1}{2}\cdot\frac{2}{3}u^{3/2}\right]_0^4 + \pi\right\} = 1000(\frac{8}{3} + \pi) \approx 5.8 \times 10^3$ ft-lb.

Note: The second integral was computed by noticing that it represents the area of a quarter-circle of radius 2.

21. The only change needed is to replace 1000 kg/m$^3$ by 680 kg/m$^3$, so we multiply the answer to #17 by .680 to get $\approx .720\times10^6 = 7.20\times10^5$ J

23. $V = \pi r^2 x$, so V is a function of x and P can also be regarded as a function of x. If $V_1 = \pi r^2 x_1$ and $V_2 = \pi r^2 x_2$, then $W = \int_{x_1}^{x_2} F(x) dx$

$= \int_{x_1}^{x_2} \pi r^2 P(V(x)) dx = \int_{x_1}^{x_2} P(V(x)) dV(x)$   $[V(x) = \pi r^2 x$, so $dV(x) = \pi r^2 dx]$

$= \int_{V_1}^{V_2} P(V) dV$   by the Substitution Rule.

25. $W = \int_a^b F(r) dr = \int_a^b G\frac{m_1 m_2}{r^2} \, dr = Gm_1 m_2\left[\frac{-1}{r}\right]_a^b = Gm_1 m_2\left[\frac{1}{a} - \frac{1}{b}\right]$

**Exercises 8.6**

1.  (a) $P = \rho g d = (1000 \text{ kg/m}^3)(9.8 \text{ m/s}^2)(1 \text{ m}) = 9800 \text{ Pa} = 9.8 \text{ kPa}$

    (b) $F = \rho g d A = PA = (9800 \text{ N/m}^2)(2\text{m}^2) = 1.96 \times 10^4 \text{ N}$

    (c) $F = \int_0^1 \rho g x \cdot 1 \, dx = 9800 \int_0^1 x \, dx = 4900 x^2 \Big]_0^1 = 4.90 \times 10^3 \text{ N}$

3.  $F = \int_0^{10} \rho g x \cdot 2\sqrt{100-x^2} \, dx$

    $= 9.8 \times 10^3 \int_0^{10} \sqrt{100-x^2} \; 2x \, dx$

    $= 9.8 \times 10^3 \int_{100}^0 u^{1/2} (-du) \quad [u = 100-x^2]$

    $= 9.8 \times 10^3 \int_0^{100} u^{1/2} du = 9.8 \times 10^3 \times \frac{2}{3} u^{3/2} \Big]_0^{100}$

    $= \frac{2}{3} \cdot 9.8 \times 10^6 \approx 6.5 \times 10^6 \text{ N}$

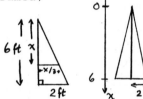

5.  $F = \int_{-r}^{r} \rho g(x+r) \cdot 2\sqrt{r^2-x^2} \, dx = \rho g \int_{-r}^{r} \sqrt{r^2-x^2} \; 2x \, dx + 2\rho g r \int_{-r}^{r} \sqrt{r^2-x^2} \, dx$

    $= \rho g \cdot 0 + 2\rho g r \cdot \frac{1}{2}\pi r^2 \quad \text{[by the same method used in Example 2]}$

    $= \rho g \cdot \pi r^3 = 1000 g \pi r^3 \text{ N (metric units assumed)}$

7.  $F = \int_0^6 \delta x \cdot \frac{2x}{3} \, dx = \frac{2}{9} \delta x^3 \Big]_0^6$

    $= 48\delta \approx 48 \times 62.5$

    $= 3000 \text{ lb}$

9.  $F = \int_2^6 \delta(x-2)\frac{2x}{3} \, dx = \frac{2}{3}\delta \int_2^6 (x^2-2x) dx = \frac{2}{3}\delta(\frac{x^3}{3} - x^2)\Big]_2^6 = \frac{2}{3}\delta \left[ 36 - (-\frac{4}{3}) \right]$

    $= \frac{224}{9}\delta \approx 1.56 \times 10^3 \text{ lb}$

11. $F = \int_0^8 \delta x \cdot (12+x) dx = \delta \int_0^8 (12x+x^2) dx$

    $= \delta(6x^2+\frac{x^3}{3})\Big]_0^8 = \delta(384+\frac{512}{3})$

    $= (62.5)\frac{1664}{3} \approx 3.47 \times 10^4 \text{ lb}$

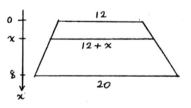

13. $F = \int_0^{4\sqrt{3}} \rho g(4\sqrt{3}-x)(2x/\sqrt{3}) dx = 8\rho g \int_0^{4\sqrt{3}} x \, dx$

    $- (2\rho g/\sqrt{3})\int_0^{4\sqrt{3}} x^2 dx = 4\rho g x^2 \Big]_0^{4\sqrt{3}} - (2\rho g/3\sqrt{3})x^3 \Big]_0^{4\sqrt{3}}$

    $= 192\rho g - (2\rho g/3\sqrt{3})64 \cdot 3\sqrt{3} = 192\rho g - 128\rho g = 64\rho g$

    $\approx 64(840)(9.8) \approx 5.27 \times 10^5 \text{ N}$

15. (a) $F = \rho g d A \approx (1000)(9.8)(.8)(.2)^2 \approx 314$ N

(b) $F = \int_{.8}^{1} \rho g x (.2) dx = .2\rho g \frac{x^2}{2}\Big]_{.8}^{1} = (.2\rho g)(.18) = .036\rho g$

$\approx (.036)(1000)(9.8) \approx 353$ N

17. $F = \int_{0}^{2} \rho g x \cdot 3 \cdot \sqrt{2}\, dx = 3\sqrt{2}\rho g \int_{0}^{2} x\, dx$

$= 3\sqrt{2}\rho g (\frac{x^2}{2})\Big]_{0}^{2} = 6\sqrt{2}\rho g \approx 8.32 \times 10^4$ N

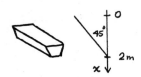

19. Assume that the pool is filled with water.

(a) $F = \int_{0}^{3} \delta x\, 20\, dx = 20\delta\, \frac{x^2}{2}\Big]_{0}^{3} = 20\delta \cdot \frac{9}{2} = 90\delta$

$\approx 5625$ lb $\approx 5.63 \times 10^3$ lb

(b) $F = \int_{0}^{9} \delta x\, 20\, dx = 20\delta\, \frac{x^2}{2}\Big]_{0}^{9} = 810\delta \approx 50625$ lb $\approx 5.06 \times 10^4$ lb

(c) $F = \int_{0}^{3} \delta x\, 40\, dx + \int_{3}^{9} \delta x (40)\frac{9-x}{6}\, dx = 40\delta\, \frac{x^2}{2}\Big]_{0}^{3} + \frac{20}{3}\delta \int_{3}^{9}(9x-x^2)dx$

$= 180\delta + \frac{20}{3}\delta \left[\frac{9}{2}x^2 - \frac{x^3}{3}\right]_{3}^{9} = 180\delta + \frac{20}{3}\delta\left[(\frac{729}{2} - 243) - (\frac{81}{2} - 9)\right] = 780\delta$

$\approx 4.88 \times 10^4$ lb

(d) $F = \int_{3}^{9} \delta x\, 20\frac{\sqrt{409}}{3}\, dx$

$= \frac{1}{3}(20\sqrt{409})\delta\, \frac{x^2}{2}\Big]_{3}^{9} = \frac{1}{3}\cdot 10\sqrt{409}\delta(81-9)$

$\approx 3.03 \times 10^5$ lb

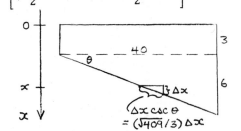

## Exercises 8.7

1.  $m_1 = 4$, $m_2 = 8$; $P_1(-1,2)$, $P_2(2,4)$.  $m = m_1+m_2 = 12$.

$M_x = 4\cdot 2 + 8\cdot 4 = 40$; $M_y = 4\cdot(-1) + 8\cdot 2 = 12$; $\bar{x} = \dfrac{M_y}{m} = 1$ and $\bar{y} = \dfrac{M_x}{m}$

$= \dfrac{10}{3}$, so the center of mass is $(\bar{x},\bar{y}) = (1,\dfrac{10}{3})$.

3.  $M_x = 4\cdot(-2) + 2\cdot 4 + 5\cdot(-3) = -15$;  $M_y = 4\cdot(-1) + 2\cdot(-2) + 5\cdot 5 = 17$,

$m = 4+2+5 = 11$, $(\bar{x},\bar{y}) = (\dfrac{17}{11},-\dfrac{15}{11})$

5.  $A = \int_0^2 x^2 dx = \dfrac{x^3}{3}\Big]_0^2 = \dfrac{8}{3}$, $\bar{x} = \dfrac{1}{A}\int_0^2 x\cdot x^2 dx = \dfrac{3}{8}\dfrac{x^4}{4}\Big]_0^2 = \dfrac{3}{8}\cdot 4 = \dfrac{3}{2}$,

$\bar{y} = \dfrac{1}{A}\int_0^2 \dfrac{1}{2}(x^2)^2 dx = \dfrac{3}{8}\cdot\dfrac{1}{2}\dfrac{x^5}{5}\Big]_0^2 = \dfrac{3}{16}\dfrac{32}{5} = \dfrac{6}{5}$.  Centroid $(\bar{x},\bar{y}) = (\dfrac{3}{2},\dfrac{6}{5})$

7.  $A = \int_{-1}^2 (3x+5)dx = \dfrac{3x^2}{2} + 5x\Big]_{-1}^2 = (6+10)-(\dfrac{3}{2}-5) = 16+\dfrac{7}{2} = \dfrac{39}{2}$,

$\bar{x} = \dfrac{1}{A}\int_{-1}^2 x(3x+5)dx = \dfrac{2}{39}\int_{-1}^2 (3x^2+5x)dx = \dfrac{2}{39}(x^3+\dfrac{5x^2}{2})\Big]_{-1}^2 = $

$\dfrac{2}{39}[(8+10)-(-1+\dfrac{5}{2})] = \dfrac{11}{13}$; $\bar{y} = \dfrac{1}{A}\int_{-1}^2 \dfrac{1}{2}(3x+5)^2 dx = \dfrac{1}{39}\int_{-1}^2 (9x^2+30x+25)dx$

$= \dfrac{1}{39}(3x^3+15x^2+25x)\Big]_{-1}^2 = \dfrac{1}{39}[(24+60+50)-(-3+15-25)] = \dfrac{49}{13}$

$(\bar{x},\bar{y}) = (\dfrac{11}{13},\dfrac{49}{13})$

9.  $A = \int_0^1 \sqrt{x^2+1}\ dx = \left[\dfrac{x}{2}\sqrt{1+x^2} + \dfrac{1}{2}\ln|x+\sqrt{1+x^2}|\right]_0^1$    [Formula 21]

$= (\dfrac{1}{2}\sqrt{2}+\dfrac{1}{2}\ln(1+\sqrt{2})) - (\dfrac{1}{2}\ln 1) = (\sqrt{2} + \ln(1+\sqrt{2}))/2$, $\bar{x} = \dfrac{1}{A}\int_0^1 x\sqrt{x^2+1}\ dx$

$= \dfrac{1}{A}\dfrac{1}{3}(x^2+1)^{3/2}\Big]_0^1 = \dfrac{1}{A}\dfrac{1}{3}[2\sqrt{2}-1] = 2(2\sqrt{2}-1)/3(\sqrt{2}+\ln(1+\sqrt{2}))$

$\bar{y} = \dfrac{1}{A}\int_0^1 \dfrac{1}{2}(\sqrt{x^2+1})^2 dx = \dfrac{1}{2A}\int_0^1(x^2+1)dx = \dfrac{1}{2A}\left[\dfrac{x^3}{3} + x\right]_0^1 = \dfrac{1}{2A}\cdot\dfrac{4}{3} = \dfrac{2}{3A}$

$= 4/3(\sqrt{2}+\ln(1+\sqrt{2}))$

$(\bar{x},\bar{y}) = (2(2\sqrt{2}-1)/3(\sqrt{2}+\ln(1+\sqrt{2})),\ 4/3(\sqrt{2}+\ln(1+\sqrt{2})))$

11.  By symmetry, $\bar{x} = 0$ and $A = 2\int_0^{\pi/4}\cos 2x\ dx = \sin 2x\Big]_0^{\pi/4} = 1$,

$\bar{y} = \dfrac{1}{A}\int_{-\pi/4}^{\pi/4}\dfrac{1}{2}\cos^2 2x\ dx = \int_0^{\pi/4}\cos^2 2x\ dx = \dfrac{1}{2}\int_0^{\pi/4}(1 + \cos 4x)dx$

$= \dfrac{1}{2}(x + \dfrac{1}{4}\sin 4x)]_0^{\pi/4} = \dfrac{1}{2}(\dfrac{\pi}{4} + \dfrac{1}{4}\cdot 0) = \dfrac{\pi}{8}$.   $(\bar{x},\bar{y}) = (0,\dfrac{\pi}{8})$

13.  $A = \int_0^1 e^x dx = e^x\Big]_0^1 = e-1$, $\bar{x} = \dfrac{1}{A}\int_0^1 xe^x dx = \dfrac{1}{e-1}(xe^x-e^x)\Big]_0^1$

$= \dfrac{1}{e-1}[0-(-1)] = \dfrac{1}{e-1}$, $\bar{y} = \dfrac{1}{A}\int_0^1 \dfrac{1}{2}(e^x)^2 dx = \dfrac{1}{e-1}\dfrac{1}{4}e^{2x}\Big]_0^1 = \dfrac{1}{4(e-1)}(e^2-1)$

$=(e+1)/4$.    $(\bar{x},\bar{y}) = (\dfrac{1}{e-1}, \dfrac{e+1}{4})$

15. $A = \int_0^1 (\sqrt{x}-x)dx = \frac{2}{3}x^{3/2} - \frac{x^2}{2}\Big]_0^1 = \frac{2}{3} - \frac{1}{2} = \frac{1}{6}$, $\quad \bar{x} = \frac{1}{A}\int_0^1 x(\sqrt{x}-x)dx$

$= 6\int_0^1 (x^{3/2}-x^2)dx = 6\left[\frac{2}{5}x^{5/2} - \frac{x^3}{3}\right]_0^1 = 6(\frac{2}{5}-\frac{1}{3}) = \frac{2}{5}$, $\quad \bar{y} = \frac{1}{A}\int_0^1 \frac{1}{2}\left[(\sqrt{x})^2-x^2\right]dx$

$= 3\int_0^1 (x-x^2)dx = 3(\frac{x^2}{2} - \frac{x^3}{3})\Big]_0^1 = 3(\frac{1}{2}-\frac{1}{3}) = \frac{1}{2}$. $\quad (\bar{x},\bar{y}) = (\frac{2}{5},\frac{1}{2})$

17. $A = \int_0^{\pi/4}(\cos x - \sin x)dx = \sin x + \cos x\Big]_0^{\pi/4} = \sqrt{2}-1$,

$\bar{x} = \frac{1}{A}\int_0^{\pi/4}x(\cos x - \sin x)dx = \frac{1}{A}[x(\sin x + \cos x) + \cos x - \sin x]_0^{\pi/4}$

$= \frac{1}{A}[\frac{\pi}{4}\sqrt{2}-1] = (\frac{\pi}{4}\sqrt{2}-1)/(\sqrt{2}-1)$

$\bar{y} = \frac{1}{A}\int_0^{\pi/4}\frac{1}{2}(\cos^2 x - \sin^2 x)dx = \frac{1}{2A}\int_0^{\pi/4}\cos 2x\, dx = \frac{1}{4A}\sin 2x\Big]_0^{\pi/4} = \frac{1}{4A}$

$= 1/4(\sqrt{2}-1)$. $\quad (\bar{x},\bar{y}) = ((\pi\sqrt{2}-4)/4(\sqrt{2}-1),\ 1/4(\sqrt{2}-1))$

19. This is an ellipse centered at the origin. By symmetry, $(\bar{x},\bar{y}) = (0,0)$.

21. By symmetry, $M_y = 0$ and $\bar{x} = 0$. $A = \frac{1}{2}bh = \frac{1}{2}(2)(2) = 2$.

$M_x = 2\rho\int_0^1 \frac{1}{2}(2-2x)^2 dx = 4\int_0^1 (1-x)^2 dx = 4\int_0^1 (1-2x+x^2)dx$

$= 4(x-x^2+\frac{x^3}{3})\Big]_0^1 = 4(1-1+\frac{1}{3}) = \frac{4}{3}$.

$\bar{y} = \frac{1}{\rho A}M_x = \frac{2}{3}$. $\quad (\bar{x},\bar{y}) = (0,\frac{2}{3})$

23. By symmetry, $M_y = 0 \Rightarrow \bar{x} = 0$. $A = $ area of triangle + area of square

$= 1+4 = 5$, so $m = \rho A = 4\cdot 5 = 20$. $M_x = \rho\cdot 2\int_0^1 \frac{1}{2}[(1-x)^2-(-2)^2]dx$

$= 4\int_0^1 (x^2-2x-3)dx = 4(\frac{x^3}{3} - x^2 - 3x)\Big]_0^1 = 4(\frac{1}{3}-1-3) = 4(-\frac{11}{3}) = -44/3$

$\bar{y} = \frac{1}{m}M_x = \frac{1}{20}(-44/3) = -11/15$. $\quad (\bar{x},\bar{y}) = (0,-11/15)$

25. Choose x- and y-axes so that the base (one side of the triangle) lies along the x-axis with the other vertex along the positive y-axis as shown. From geometry, we know the

medians intersect at a point 2/3 of the way from each vertex (along the median) to the opposite side. The median from B goes to the midpoint $((a+c)/2,0)$ of side AC, so the point of intersection of the medians is $(\frac{2}{3}\cdot\frac{a+c}{2},\frac{1}{3}\cdot b) = (\frac{(a+c)}{3},\frac{b}{3})$. This can also be verified by finding the equations of two medians, and solving them simultaneously to find their point of intersection. Now let us compute the location of the centroid of the triangle. The area is

$A = \frac{1}{2}(c-a)b$. $\bar{x} = \frac{1}{A}\left[\int_a^0 x\cdot\frac{b}{a}(a-x)dx + \int_0^c x\cdot\frac{b}{c}(c-x)dx\right] = \frac{1}{A}\left[\frac{b}{a}\int_a^0 (ax-x^2)dx\right.$

$\left. + \frac{b}{c}\int_0^c (cx-x^2)dx\right] = \frac{b}{Aa}\left[\frac{ax^2}{2} - \frac{x^3}{3}\right]\Big]_a^0 + \frac{b}{Ac}\left[\frac{cx^2}{2} - \frac{x^3}{3}\right]\Big]_0^c = \frac{b}{Aa}\left[-\frac{a^3}{2} + \frac{a^3}{3}\right] +$

$\frac{b}{Ac}\left[\frac{c^3}{2} - \frac{c^3}{3}\right] = \frac{2}{a(c-a)}\cdot\frac{-a^3}{6} + \frac{2}{c(c-a)}\cdot\frac{c^3}{6} = \frac{1}{3(c-a)}(c^2-a^2) = (a+c)/3$ and

$\bar{y} = \frac{1}{A}\left[\int_a^0 \frac{1}{2}[\frac{b}{a}(a-x)]^2 dx + \int_0^c \frac{1}{2}[\frac{b}{c}(c-x)]^2 dx\right] = \frac{1}{A}\left[\frac{b^2}{2a^2}\int_a^0 (a^2-2ax+x^2)dx +\right.$

$\left.\frac{b^2}{2c^2}\int_0^c (c^2-2cx+x^2)dx\right] = \frac{1}{A}\left\{\frac{b^2}{2a^2}(a^2x-ax^2+\frac{x^3}{3})\Big]_a^0 + \frac{b^2}{2c^2}(c^2x-cx^2+\frac{x^3}{3})\Big]_0^c\right\}$

$= \frac{1}{A}\left\{\frac{b^2}{2a^2}(-a^3+a^3-\frac{a^3}{3}) + \frac{b^2}{2c^2}(c^3-c^3+\frac{c^3}{3})\right\} = \frac{1}{A}\{\frac{b^2}{6}(-a+c)\} = \frac{2}{(c-a)b}\cdot\frac{(c-a)b^2}{6}$

$= b/3$. Thus $(\bar{x},\bar{y}) = ((a+c)/3, b/3)$ as claimed. Remarks. Actually the computation of $\bar{y}$ is all that is needed. By considering each side of the triangle in turn to be the base, we see that the centroid is 1/3 of the way from each side to the opposite vertex and must therefore be the intersection of the medians.

The computation of $\bar{y}$ in this problem (and many others) can be simplified by using horizontal rather than vertical approximating rectangles. (We use this approach again in the solution to Review Exercise 21.) If the length of a thin rectangle at coordinate $y$ is $l(y)$, then its area is $l(y)\Delta y$, its mass is $\rho l(y)\Delta y$, and its moment about the x-axis is $\Delta M_x = \rho y l(y)\Delta y$. Thus $M_x = \int \rho y l(y)dy$ and

$\bar{y} = \frac{\int\rho y l(y)dy}{\rho A} = \frac{1}{A}\int yl(y)dy$. In this problem $l(y) = \frac{c-a}{b}(b-y)$ by

similar triangles, so $\bar{y} = \frac{1}{A}\int_0^b \frac{c-a}{b}y(b-y)dy$

$= \frac{2}{b^2}\int_0^b (by-y^2)dy = \frac{2}{b^2}(\frac{by^2}{2} - \frac{y^3}{3})\Big]_0^b = \frac{2}{b^2}\cdot\frac{b^3}{6}$

$= \frac{b}{3}$. Notice that only one integral is needed when this method is used.

27. A cone of height h and radius r can be
    generated by rotating a right triangle about
    one of its legs as shown. By #25, $\bar{x} = \frac{r}{3}$, so
    the volume of the cone is $V = Ad = \frac{1}{2}rh \cdot 2\pi\frac{r}{3}$
    $= \frac{1}{3}\pi r^2 h$ by the Theorem of Pappus.

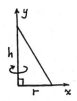

29. Suppose the region lies between two curves $y = f(x)$ and $y = g(x)$
    where $f(x) \geq g(x)$, as illustrated in Fig. 8.56. Take a partition P
    by points $x_i$ with $a = x_0 < x_1 < \cdots < x_n = b$ and choose $x_i^*$ to be the
    midpoint of the $i^{th}$ subinterval; that is, $x_i^* = (x_{i-1}+x_i)/2$. Then
    the centroid of the $i^{th}$ approximating rectangle $R_i$ is its center
    $C_i(x_i^*, \frac{1}{2}[f(x_i^*)+g(x_i^*)])$. Its area is $[f(x_i^*)-g(x_i^*)]\varDelta x_i$, so its mass
    is $\rho[f(x_i^*)-g(x_i^*)]\varDelta x_i$. Thus $M_y(R_i) = \rho[f(x_i^*)-g(x_i^*)]\varDelta x_i \cdot x_i^*$
    $= \rho x_i^*[f(x_i^*)-g(x_i^*)]\varDelta x_i$ and $M_x(R_i) = \rho[f(x_i^*)-g(x_i^*)]\varDelta x_i \cdot \frac{1}{2}[f(x_i^*)+g(x_i^*)]$
    $= \rho \cdot \frac{1}{2}[f(x_i^*)^2-g(x_i^*)^2]\varDelta x_i$. Summing over i and taking the limit as
    $\|P\| \to 0$, we get $M_y = \lim_{\|P\| \to 0} \sum_i \rho x_i^*[f(x_i^*)-g(x_i^*)\varDelta x_i = \rho\int_a^b x[f(x)-g(x)]dx$
    and $M_x = \lim_{\|P\| \to 0} \sum_i \rho \cdot \frac{1}{2}[f(x_i^*)^2-g(x_i^*)^2]\varDelta x_i = \rho\int_a^b \frac{1}{2}[f(x)^2-g(x)^2]dx$. Thus
    $\bar{x} = \frac{M_y}{m} = \frac{M_y}{\rho A} = \frac{1}{A}\int_a^b x[f(x)-g(x)]dx$ and
    $\bar{y} = \frac{M_x}{m} = \frac{M_x}{\rho A} = \frac{1}{A}\int_a^b \frac{1}{2}[f(x)^2-g(x)^2]dx$

## Exercises 8.8

1. $\dfrac{dy}{dx} = y^2 \Rightarrow \dfrac{dy}{y^2} = dx \ (y \neq 0\ ) \Rightarrow \displaystyle\int \dfrac{dy}{y^2} = \int dx \Rightarrow \dfrac{-1}{y} = x+C \Rightarrow -y = \dfrac{1}{x+C} \Rightarrow$

$y = \dfrac{-1}{x+C}$ ; $y = 0$ is also a solution.

3. $yy' = x \Rightarrow \displaystyle\int y\,dy = \int x\,dx \Rightarrow \dfrac{y^2}{2} = \dfrac{x^2}{2} + C_1 \Rightarrow y^2 = x^2 + 2C_1 \Rightarrow$

$x^2 - y^2 = C\ (= -2C_1)$. This represents a family of hyperbolas.

5. $x^2 y' + y = 0 \Rightarrow \dfrac{dy}{dx} = \dfrac{-y}{x^2} \Rightarrow \displaystyle\int \dfrac{dy}{y} = \int \dfrac{-dx}{x^2}\ (y \neq 0) \Rightarrow \ln|y| = \dfrac{1}{x} + K \Rightarrow$

$|y| = e^K e^{1/x} \Rightarrow y = Ce^{1/x}$, where now we allow C to be any constant.

7. $\dfrac{dy}{dx} = \dfrac{x\sqrt{x^2+1}}{ye^y} \Rightarrow \displaystyle\int ye^y\,dy = \int x\sqrt{x^2+1}\,dx \Rightarrow (y-1)e^y = \dfrac{1}{3}(x^2+1)^{3/2} + C$

9. $\dfrac{du}{dt} = e^{u+2t} = e^u e^{2t} \Rightarrow \displaystyle\int e^{-u}\,du = \int e^{2t}\,dt \Rightarrow -e^{-u} = \dfrac{1}{2}e^{2t} + C_1 \Rightarrow$

$e^{-u} = -\dfrac{1}{2}e^{2t} + C$ [where $C = -C_1$ and the right-hand side is positive,

since $e^{-u} > 0$] $\Rightarrow -u = \ln(C - \dfrac{1}{2}e^{2t}) \Rightarrow u = -\ln(C - \dfrac{1}{2}e^{2t})$

11. $\dfrac{dy}{dx} = y^2+1$, $y(1) = 0$. $\displaystyle\int \dfrac{dy}{y^2+1} = \int dx \Rightarrow \tan^{-1}y = x+C$. $y = 0$ when $x = 1$,

so $1+C = \tan^{-1}0 = 0$ and $C = -1$. Thus $\tan^{-1}y = x-1$ and $y = \tan(x-1)$

13. $e^y y' = \dfrac{3x^2}{1+y}$, $y(2) = 0$. $\displaystyle\int e^y(1+y)\,dy = \int 3x^2\,dx \Rightarrow ye^y = x^3 + C$.

$y(2) = 0 \Rightarrow 0 = 2^3 + C \Rightarrow C = -8$, so $ye^y = x^3 - 8$

15. $\dfrac{dy}{dx} = \dfrac{\sin x}{\sin y}$, $y(0) = \dfrac{\pi}{2}$. $\displaystyle\int \sin y\,dy = \int \sin x\,dx \Rightarrow -\cos y = -\cos x + C$

$y(0) = \dfrac{\pi}{2} \Rightarrow -\cos \dfrac{\pi}{2} = -\cos 0 + C \Rightarrow C = 1$, so $-\cos y = -\cos x + 1$

and $\cos y = \cos x - 1$

17. $xe^{-t}\dfrac{dx}{dt} = t$, $x(0) = 1$. $\displaystyle\int x\,dx = \int te^t\,dt \Rightarrow \dfrac{x^2}{2} = (t-1)e^t + C$. $x(0) = 1$

$\Rightarrow \dfrac{1}{2} = (0-1)e^0 + C \Rightarrow C = \dfrac{3}{2}$, so $x^2 = 2(t-1)e^t + 3 \Rightarrow x = \sqrt{2(t-1)e^t+3}$

19. $\dfrac{du}{dt} = \dfrac{2t+1}{2(u-1)}$, $u(0) = -1$. $\displaystyle\int 2(u-1)\,du = \int(2t+1)\,dt \Rightarrow u^2 - 2u = t^2 + t + C$.

$u(0) = -1 \Rightarrow (-1)^2 - 2(-1) = 0^2 + 0 + C \Rightarrow C = 3$, so $u^2 - 2u = t^2 + t + 3$; the

quadratic formula gives $u = 1 - \sqrt{t^2+t+4}$.

21. Let $y = f(x)$. Then $\frac{dy}{dx} = x^3 y$ and $y(0) = 1$. $\frac{dy}{y} = x^3 dx$ $(y \neq 0)$, so

$\int \frac{dy}{y} = \int x^3 dx$ and $\ln|y| = \frac{x^4}{4} + C$; $y(0) = 1 \Rightarrow C = 0$, so $\ln|y| = \frac{x^4}{4}$,

$|y| = e^{x^4/4}$ and $y = f(x) = e^{x^4/4}$ (since $y(0) = 1$).

23. (a) Let $y(t)$ be the amount of salt (in kg) after t minutes. Then $y(0) = 15$. The amount of liquid in the tank is 1000 L at all times, so the concentration at time t (in minutes) is $y(t)/1000$ kg/L and $\frac{dy}{dt} = -(\frac{y(t)}{1000} \frac{kg}{L})(10\frac{L}{min}) = -\frac{y(t)}{100} \frac{kg}{min}$. $\int \frac{dy}{y} = -\frac{1}{100} \int dt$

$\Rightarrow \ln y = -\frac{t}{100} + C$. $y(0) = 15 \Rightarrow \ln 15 = C$, so $\ln y = \ln 15 - \frac{t}{100}$. It follows that $\ln(\frac{y}{15}) = -\frac{t}{100}$ and $\frac{y}{15} = e^{-t/100}$, so $y = 15e^{-t/100}$ kg

(b) After 20 min., $y = 15e^{-20/100} = 15e^{-.2} \approx 12.3$ kg.

25. $\frac{dx}{dt} = k(a-x)(b-x)$, $a \neq b$. $\int \frac{dx}{(a-x)(b-x)} = \int k \, dt \Rightarrow \frac{1}{b-a} \int (\frac{1}{a-x} - \frac{1}{b-x}) dx = \int k \, dt \Rightarrow \frac{1}{b-a}(-\ln|a-x| + \ln|b-x|) = kt+C \Rightarrow \ln\left|\frac{b-x}{a-x}\right| = (b-a)(kt+C)$.

Here the concentrations $[A] = a-x$ and $[B] = b-x$ cannot be negative, so $(b-x)/(a-x) \geq 0$ and $|(b-x)/(a-x)| = (b-x)/(a-x)$. We now have $\ln\left[\frac{b-x}{a-x}\right] = (b-a)(kt+C)$. Since $x(0) = 0$, $\ln\left[\frac{b}{a}\right] = (b-a)C$. Hence $\ln\left[\frac{b-x}{a-x}\right] = (b-a)kt + \ln\left[\frac{b}{a}\right]$, $\frac{b-x}{a-x} = \frac{b}{a}e^{(b-a)kt}$, and

$x = \frac{b[e^{(b-a)kt}-1]}{\frac{b}{a}e^{(b-a)kt}-1} = \frac{ab[e^{(b-a)kt}-1]}{be^{(b-a)kt}-a} \frac{\text{moles}}{L}$

27. Let $P(t)$ be the world population in the year t. Then $\frac{dP}{dt} = .02P$, so $\int \frac{dP}{P} = \int .02 \, dt$ and $\ln P = .02t+C \Rightarrow P(t) = Ae^{.02t}$. $P(1986) = 5 \times 10^9$

$\Rightarrow P(t) = 5 \times 10^9 e^{.02(t-1986)}$ (a) The population in 2000 will be approximately $P(2000) = 5e^{.28} \times 10^9 \approx 6.6$ billion $\Rightarrow$ number of $ft^2/\text{person} = 3.7 \times 10^{11}/6.6 \times 10^9 \approx 56$ (b) The predicted population for the year 2100 is $P(2100) = 5e^{2.28} \times 10^9 \approx 49$ billion $\Rightarrow$ $3.7 \times 10^{11}/49 \times 10^9 \approx 7.6 \, ft^2/\text{person}$. (c) The prediction for 2500 is $P(2500) = 5e^{10.28} \times 10^9 \approx 146$ trillion $\Rightarrow 3.7 \times 10^{11}/146 \times 10^{12} \approx 0.0025$ $ft^2/\text{person}$.

29. (a) Our assumption is that $\frac{dy}{dt} = ky(1-y)$, where $y$ is the fraction of the population that has heard the rumor.

(b) Take $M = 1$ in the solution of (8.51) to get $y = \dfrac{y_0}{y_0 + (1-y_0)e^{-kt}}$

(c) Let $t$ be the number of hours since 8 A.M. Then $y_0 = y(0) = \frac{80}{1000}$ $= .08$ and $y(4) = \frac{1}{2}$, so $\frac{1}{2} = y(4) = \dfrac{.08}{.08 + .92e^{-4k}}$ . Thus $.08 + .92e^{-4k} =$ $.16$, $e^{-4k} = \frac{.08}{.92} = \frac{2}{23}$, and $e^{-k} = \left[\frac{2}{23}\right]^{1/4}$, so $y = \dfrac{.08}{.08 + .92(2/23)^{t/4}}$ $= \dfrac{2}{2 + 23(2/23)^{t/4}}$ and $\left[\frac{2}{23}\right]^{t/4} = \frac{2}{23}\frac{1-y}{y}$ or $\left[\frac{2}{23}\right]^{t/4-1} = \frac{1-y}{y}$ . It follows that $\frac{t}{4} - 1 = \ln\left[\frac{1-y}{y}\right]/\ln\left[\frac{2}{23}\right]$, so $t = 4\left[1 + \ln\left[\frac{1-y}{y}\right]/\ln\left[\frac{2}{23}\right]\right]$. When $y = .9$, $(1-y)/y = 1/9$, so $t = 4[1 - (\ln 9)/\ln(23)] \approx 7.6$ h or 7 hr 36 min. Thus 90% of the population will have heard the rumor by 3:36 P.M.

31. $y$ increases most rapidly when $y'$ is maximal, that is, $y'' = 0$. But $y' = ky(M-y) \Rightarrow y'' = ky'(M-y) + ky(-y') = ky'(M-2y)$ $= k^2 y(M-y)(M-2y)$. Since $0 < y < M$, we see that $y'' = 0 \iff y = M/2$.

# Chapter 8 Review

Review Exercises for Chapter 8

1.  $V = \int_1^3 \pi(\sqrt{x-1})^2 dx = \pi \int_1^3 (x-1)dx = \pi(\frac{x^2}{2} - x)\Big]_1^3 = \pi\left[(\frac{9}{2}-3)-(\frac{1}{2}-1)\right] = 2\pi$

3.  $V = \int_1^3 2\pi y \cdot (-y^2+4y-3)dy = 2\pi \int_1^3 (-y^3+4y^2-3y)dy$

$= 2\pi\left[-\frac{y^4}{4} + \frac{4}{3}y^3 - \frac{3}{2}y^2\right]_1^3 = 2\pi\left[(-\frac{81}{4} + 36 - \frac{27}{2}) - (-\frac{1}{4} + \frac{4}{3} - \frac{3}{2})\right] = \frac{16\pi}{3}$

5.  $V = \int_a^{a+h} 2\pi x \cdot 2\sqrt{x^2-a^2} \, dx = 2\pi \int_0^{2ah+h^2} u^{1/2} \, du \qquad [u = x^2-a^2, \ du = 2x \ dx]$

$= 2\pi \cdot \frac{2}{3} u^{3/2}\Big]_0^{2ah+h^2} = \frac{4\pi}{3}(2ah+h^2)^{3/2}$

7.  $V = \int_0^1 \pi[(1-x^3)^2 - (1-x^2)^2]dx = \pi\int_0^1 (x^6-x^4-2x^3+2x^2)dx$

$= \pi\left[\frac{x^7}{7} - \frac{x^5}{5} - \frac{x^4}{2} + \frac{2x^3}{3}\right]\Big]_0^1 = \pi(\frac{1}{7} - \frac{1}{5} - \frac{1}{2} + \frac{2}{3}) = \frac{23\pi}{210}$

9.  Take the base to be the disk $x^2+y^2 \le 9$.  Then $V = \int_{-3}^3 A(x)dx$, where

$A(x_0)$ is the area of the isosceles right triangle whose hypotenuse

lies along the line $x = x_0$ in the xy-plane.  $A(x) = \frac{1}{4}(2\sqrt{9-x^2})^2$

$= 9-x^2$, so $V = 2\int_0^3 A(x)dx = 2\int_0^3 (9-x^2)dx = 2\left[9x-\frac{x^3}{3}\right]\Big]_0^3 = 2(27-9) = 36$

11.  $3x = 2(y-1)^{3/2}$, $2 \le y \le 5$.  $x = \frac{2}{3}(y-1)^{3/2}$, so $\frac{dx}{dy} = (y-1)^{1/2}$ and $1+(\frac{dx}{dy})^2$

$= 1+(y-1) = y \Rightarrow L = \int_2^5 \sqrt{1+\left[\frac{dx}{dy}\right]^2} \, dy = \int_2^5 \sqrt{y} \, dy = \frac{2}{3}y^{3/2}\Big]_2^5 = \frac{2}{3}(5\sqrt{5}-2\sqrt{2})$

13.  $y = \frac{x^3}{6} + \frac{1}{2x}$, $1 \le x \le 2$.  $\frac{dy}{dx} = \frac{x^2}{2} - \frac{1}{2x^2} = \frac{1}{2}\left[x^2-\frac{1}{x^2}\right]$, so $1+(\frac{dy}{dx})^2$

$= 1+\frac{1}{4}\left[x^4-2+\frac{1}{x^4}\right] = \frac{1}{4}\left[x^4+2+\frac{1}{x^4}\right] = \left[\frac{1}{2}\left[x^2+\frac{1}{x^2}\right]\right]^2 \Rightarrow$

$S = \int_1^2 2\pi y \sqrt{1+(\frac{dy}{dx})^2} \, dx = 2\pi\int_1^2 \left[\frac{x^3}{6} + \frac{1}{2x}\right]\frac{1}{2}\left[x^2+\frac{1}{x^2}\right]dx$

$= \pi\int_1^2 \left[\frac{x^5}{6} + \frac{2x}{3} + \frac{x^{-3}}{2}\right]dx = \pi\left[\frac{x^6}{36} + \frac{x^2}{3} - \frac{x^{-2}}{4}\right]\Big]_1^2$

$= \pi\left[\left[\frac{64}{36} + \frac{4}{3} - \frac{1}{16}\right] - \left[\frac{1}{36} + \frac{1}{3} - \frac{1}{4}\right]\right] = \frac{47\pi}{16}$

15.  $y = \dfrac{1}{x^2}$, $1 \le x \le 2$.  $\dfrac{dy}{dx} = \dfrac{-2}{x^3}$, so $1 + (\dfrac{dy}{dx})^2 = 1 + \dfrac{4}{x^6}$.  $L = \displaystyle\int_1^2 \sqrt{1 + \dfrac{4}{x^6}}\, dx$.

By Simpson's Rule with n = 10, $L \approx \dfrac{1/10}{3}[f(1) + 4f(1.1) + 2f(1.2)$

$+ 4f(1.3) + 2f(1.4) + 4f(1.5) + 2f(1.6) + 4f(1.7) + 2f(1.8) + 4f(1.9) + f(2)]$

$\approx 1.297$   (where $f(x) = \sqrt{1 + 4/x^6}$ ).

17.  $30\ N = f(x) = kx = k(.03m)$, so $k = \dfrac{30}{.03} = 1000\ \dfrac{N}{m}$

$W = \displaystyle\int_0^{.08} kx\, dx = 1000 \int_0^{.08} x\, dx = 500x^2 \Big]_0^{.08} = 500(.08)^2 = 3.2\ J$

19.  $W = \displaystyle\int_0^4 \pi(2\sqrt{y})^2\, 62.5(4-y)\, dy = 250\pi \int_0^4 y(4-y)\, dy$

$= 250\pi(2y^2 - \dfrac{y^3}{3}) \Big]_0^4 = 250\pi(32 - \dfrac{64}{3})$

$= 8000\pi/3$ ft-lb

21.  By Exercise 8.6.20, if a plate is immersed vertically in a fluid of density $\rho$ and the width of the plate at a depth of x meters is $w(x)$, then $F = \displaystyle\int_a^b \rho g x w(x)\, dx$, where a and b are the depths of the top and bottom of the plate.  If $\bar{x}$ is the depth of the centroid of the plate, then $\bar{x} = \dfrac{1}{A} \displaystyle\int_a^b xw(x)\, dx$.  (To see this, note that a small element of A at depth $x_i^*$ has mass $\rho \cdot w(x_i^*)\Delta x_i$ and distance $x_i^*$ from the surface.  The resulting Riemann sum $\displaystyle\sum_i \rho x_i^* w(x_i^*)\Delta x_i$ is an approximation for the moment $\displaystyle\int_a^b \rho x w(x)\, dx$.  $\bar{x} = \displaystyle\int_a^b \rho xw(x)\, dx / \int_a^b \rho w(x)\, dx$

$= \displaystyle\int_a^b xw(x)\, dx / \int_a^b w(x)\, dx = \dfrac{1}{A}\int_a^b xw(x)\, dx.$)  Thus  $F = \rho g \displaystyle\int_a^b xw(x)\, dx = \rho g \bar{x} A$

$= \bar{P}A$, where $\bar{P}$ is the pressure at the centroid and A is the area of the plate.  A less general argument based more directly on formulas in 8.7 can be given for regions bounded by two curves given as the graphs of functions.  (See formulas (8.40) in 8.7.)

23.  $A = \displaystyle\int_{-2}^1 [(4-x^2) - (x+2)]\, dx = \int_{-2}^1 (2 - x - x^2)\, dx = [2x - \dfrac{1}{2}x^2 - \dfrac{1}{3}x^3]_{-2}^1$

$= (2 - \dfrac{1}{2} - \dfrac{1}{3}) - (-4 + 2 + \dfrac{8}{3}) = \dfrac{9}{2} \Rightarrow \bar{x} = \dfrac{1}{A}\displaystyle\int_{-2}^1 x(2 - x - x^2)\, dx = \dfrac{2}{9}\int_{-2}^1 (2x - x^2 - x^3)\, dx$

$= \dfrac{2}{9}[x^2 - \dfrac{1}{3}x^3 - \dfrac{1}{4}x^4]_{-2}^1 = \dfrac{2}{9}[(1 - \dfrac{1}{3} - \dfrac{1}{4}) - (4 + \dfrac{8}{3} - 4)] = -\dfrac{1}{2}$   and

$\bar{y} = \dfrac{1}{A}\displaystyle\int_{-2}^1 \dfrac{1}{2}[(4-x^2)^2 - (x+2)^2]\, dx = \dfrac{1}{9}\int_{-2}^1 (x^4 - 9x^2 - 4x + 12)\, dx$

$= \dfrac{1}{9}[\dfrac{1}{5}x^5 - 3x^3 - 2x^2 + 12x]_{-2}^1 = \dfrac{1}{9}[(\dfrac{1}{5} - 3 - 2 + 12) - (-\dfrac{32}{5} + 24 - 8 - 24)] = \dfrac{12}{5}$

25.   $y^2 \dfrac{dy}{dx} = x + \sin x \Rightarrow \int y^2 dy = \int (x + \sin x)dx \Rightarrow \dfrac{y^3}{3} = \dfrac{x^2}{2} - \cos x + C$

  $\Rightarrow y^3 = \dfrac{3}{2}x^2 - 3 \cos x + K \qquad$ [where K = 3C]

  $\Rightarrow y = \sqrt[3]{3x^2/2 - 3 \cos x + K}$

27.   $y' = \dfrac{1}{x^2 y - 2x^2 + y - 2} \Rightarrow \dfrac{dy}{dx} = \dfrac{1}{(x^2+1)(y-2)} \Rightarrow \int (y-2)dy = \int \dfrac{dx}{x^2+1} \Rightarrow$

  $\dfrac{y^2}{2} - 2y = \tan^{-1}x + K \Rightarrow y = 2 \pm \sqrt{2 \tan^{-1}x + C}$, where C = 4+2K.

29.   First note that, in this question, "weighs" is used in the informal
sense, so what we really require is Barbara's mass m in kg as a
function of t. Barbara's net intake of calories per day at time t
(measured in days) is c(t) = 1600 - 850 - 15m(t) = 750 - 15m(t)
where m(t) is her mass at time t. We are given that m(0) = 60 kg
and $\dfrac{dm}{dt} = \dfrac{c(t)}{10,000}$, so $\dfrac{dm}{dt} = \dfrac{750-15m}{10,000} = \dfrac{150-3m}{2,000} = \dfrac{-3(m-50)}{2,000}$ with
m(0) = 60. From $\int \dfrac{dm}{m-50} = \int \dfrac{-3 \, dt}{2,000}$, we get $\ln|m-50| = \dfrac{-3t}{2,000} + C$.
Since m(0) = 60, C = ln 10. Now ln(|m-50|/10) = -3t/2000, so
$|m-50| = 10e^{-3t/2000}$. The quantity m-50 is continuous and
initially positive; the right-hand side is never zero. Thus m-50
is positive for all t, and m(t) = 50+10e$^{-3t/2000}$ kg. As t→∞,
m(t)→50 kg. Thus Barbara's mass gradually settles down to 50 kg.

# CHAPTER 9

## Exercises 9.1

1. (a)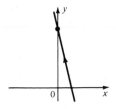

   (b) $x = 1-t$, $y = 2+3t$

   $y = 2+3(1-x) = 5-3x$, so $3x+y = 5$

3. (a)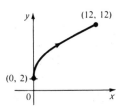

   (b) $x = 3t^2$, $y = 2+5t$, $0 \leq t \leq 2$

   $x = 3\left[\dfrac{y-2}{5}\right]^2 = \dfrac{3}{25}(y-2)^2$, $2 \leq y \leq 12$

5. (a)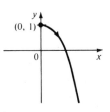

   (b) $x = \sqrt{t}$, $y = 1-t$

   $y = 1-t = 1-x^2$, $x \geq 0$

7. (a)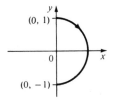

   (b) $x = \sin\theta$, $y = \cos\theta$, $0 \leq \theta \leq \pi$

   $x^2+y^2 = \sin^2\theta + \cos^2\theta = 1$, $0 \leq x \leq 1$

9. (a)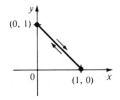

   (b) $x = \sin^2\theta$, $y = \cos^2\theta$

   $x+y = \sin^2\theta + \cos^2\theta = 1$, $0 \leq x \leq 1$

11. (a)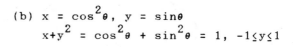

(b) $x = e^t$, $y = e^t$

$y = x$, $x > 0$

13. (a)

(1, 1)

(b) $x = \cos^2 t$, $y = \cos^4 t$

$y = x^2$, $0 \le x \le 1$

15. (a)

(0, 1)

(1, 0)

(0, −1)

(b) $x = \cos^2 \theta$, $y = \sin\theta$

$x + y^2 = \cos^2 \theta + \sin^2 \theta = 1$, $-1 \le y \le 1$

17. (a)

(e, 1)

0   (1, 0)

(b) $x = e^t$, $y = \sqrt{t}$

$x = e^{y^2}$, $0 \le y \le 1$

or: $y = \sqrt{\ln x}$, $1 \le x \le e$

19. (a)

0   1

(b) $x = \cosh t$, $y = \sinh t$

$x^2 - y^2 = \cosh^2 t - \sinh^2 t = 1$, $x \ge 1$

21. $x^2 + y^2 = \cos^2 \pi t + \sin^2 \pi t = 1$, $1 \le t \le 2$, so the particle moves counterclockwise along the circle $x^2 + y^2 = 1$ from $(-1,0)$ to $(1,0)$, along the lower half of the circle.

23. $x = 8t - 3$, $y = 2 - t$, $0 \le t \le 1$, $\Rightarrow$ $x = 8(2-y) - 3 = 13 - 8y$, so the particle moves along the line $x + 8y = 13$ from $(-3,2)$ to $(5,1)$.

25. $(x/2)^2+(y/3)^2 = \sin^2 t + \cos^2 t = 1$, so the particle moves once clockwise along the ellipse $\frac{x^2}{4} + \frac{y^2}{9} = 1$, starting and ending at $(0,3)$.

27. $x = 3(t^2-3)$ and $y = t(t^2-3)$

| t: | -4 | -3 | -2 | $-\sqrt{3}$ | -1.5 | -1 | 0 | 1 | 1.5 | $\sqrt{3}$ | 2 | 3 | 4 |
|---|---|---|---|---|---|---|---|---|---|---|---|---|---|
| x: | 39 | 18 | 3 | 0 | -2.25 | -6 | -9 | -6 | -2.25 | 0 | 3 | 18 | 39 |
| y: | -52 | -18 | -2 | 0 | 1.125 | 2 | 0 | -2 | -1.125 | 0 | 2 | 18 | 52 |

Note that replacing t by -t changes the sign of y but leaves x unchanged, so the curve is symmetric about the x-axis. Also $3y = tx$, so $9y^2 = t^2 x^2 = \left[\frac{x}{3} + 3\right]x^2$ and $27y^2 = x^3+9x^2$. For x near 0, $27y^2 \approx 9x^2$, so $y \approx \pm x/\sqrt{3}$. For large x, $27y^2 \approx x^3$, so $y \approx \pm(x/3)^{3/2}$. The graph has a loop in it.

**27.**

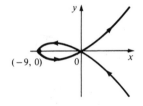

**29.**

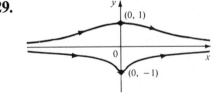

29. $x = \tan\theta + \sin\theta$, $y = \cos\theta$. Replacing $\theta$ by $-\theta$ changes the sign of x and leaves y unchanged, so the curve is symmetric about the y-axis. Clearly $|y| \leq 1$, but $x \to \pm\infty$ as $\theta \to \pm\pi/2$. The full curve is obtained from values of $\theta$ in $[-\pi, \pi]$ (but is not defined for $\theta = \pm\pi/2$). $y>0$ for $\theta$ in $(-\pi/2, \pi/2)$; $y<0$ for $\theta$ in $[-\pi, -\pi/2)$ and $(\pi/2, \pi]$. As $\theta \to (\pi/2)^-$, $x \to \infty$ and $y \to 0^+$. As $\theta \to (\pi/2)^+$, $x \to -\infty$ and $y \to 0^-$.

| $\theta$: | $-\pi$ | $-5\pi/6$ | $-3\pi/4$ | $-2\pi/3$ | $-5\pi/9$ | $-4\pi/9$ | $-\pi/3$ | $-\pi/4$ | $-\pi/6$ | 0 |
|---|---|---|---|---|---|---|---|---|---|---|
| x: | 0 | .077 | .293 | .866 | 4.69 | -6.66 | -2.60 | -1.71 | -1.08 | 0 |
| y: | -1 | -.866 | -.707 | -.5 | -.174 | .174 | .5 | .707 | .866 | 1 |

31. Clearly the curve passes through $(x_1, y_1)$ when t=0 and through $(x_2, y_2)$ when t=1. For $0<t<1$, x is strictly between $x_1$ and $x_2$ and y is strictly between $y_1$ and $y_2$. For every value of t, x and y

satisfy the relation $y-y_1 = \frac{y_2-y_1}{x_2-x_1} \cdot (x-x_1)$, which is the equation of

the straight line through $(x_1,y_1)$ and $(x_2,y_2)$. Finally, any point

$(x,y)$ on that line satisfies $\frac{y-y_1}{y_2-y_1} = \frac{x-x_1}{x_2-x_1}$; if we call that common

value t, then the given parametric equations yield the point $(x,y)$;

and any $(x,y)$ on the line between $(x_1,y_1)$ and $(x_2,y_2)$ yields a

value of t in $[0,1]$. This proves that the given parametric

equations exactly specify the line segment from $(x_1,y_1)$ to $(x_2,y_2)$.

33. The case $\pi/2 < \theta < \pi$ is illustrated to the

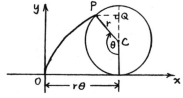

right. C has coordinates $(r\theta,r)$ as before,

and Q has coordinates $(r\theta,r+r(\cos(\pi-\theta))) =$

$(r\theta,r(1-\cos\theta))$, so P has coordinates

$(r\theta - r\sin(\pi-\theta)),r(1-\cos\theta)) =$

$(r(\theta-\sin\theta),r(1-\cos\theta))$. Again we have the

parametric equations $x = r(\theta-\sin\theta)$, $y = r(1-\cos\theta)$.

35. (a) The center Q of the smaller circle has

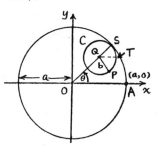

coordinates $((a-b)\cos\theta,(a-b)\sin\theta)$. Arc PS

on circle C has length $a\theta$ since it is equal

in length to arc AS (the smaller circle

rolls without slipping against the larger).

Thus $\angle PQS = a\theta/b$ and $\angle PQT = a\theta/b - \theta$, so P

has coordinates

$x = (a-b)\cos\theta + b\cos(\angle PQT) = (a-b)\cos\theta + b\cos\left[\frac{a-b}{b}\theta\right]$, and

$y = (a-b)\sin\theta - b\sin(\angle PQT) = (a-b)\sin\theta - b\sin\left[\frac{a-b}{b}\theta\right]$.

(b) If $b = a/4$, then $a-b = \frac{3a}{4}$ and $\frac{a-b}{b} = 3$,

so $x = \frac{3a}{4}\cos\theta + \frac{a}{4}\cos 3\theta =$

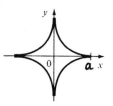

$\frac{3a}{4}\cos\theta + \frac{a}{4}(4\cos^3\theta - 3\cos\theta) = a\cos^3\theta$,

and $y = \frac{3a}{4}\sin\theta - \frac{a}{4}\sin 3\theta =$

$\frac{3a}{4}\sin\theta - \frac{a}{4}(3\sin\theta - 4\sin^3\theta) = a\sin^3\theta$

The curve is symmetric about the origin.

37. The coordinates of T are (r cos θ, r sin θ).
Since TP was unwound from arc TA, TP has
length rθ. Also ∠PTQ = ∠PTR - ∠QTR =
π/2 - θ, so P has coordinates

x = r cos θ + rθ cos(π/2-θ)

= r cos θ + rθ sin θ, and

y = r sin θ - rθ sin(π/2-θ)

= r sin θ - rθ cos θ.

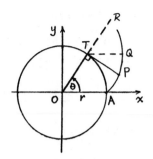

Section 9.2

Exercises 9.2

1. $x=t^2+t$, $y=t^2-t$, $t=0$. $\frac{dy}{dt}=2t-1$, $\frac{dx}{dt}=2t+1$, so $\frac{dy}{dx} = \frac{dy/dt}{dx/dt} = \frac{2t-1}{2t+1}$. When

t=0, x=y=0 and $\frac{dy}{dx}=-1$. The tangent is y-0 = (-1)(x-0) or y = -x.

3. $x=t^2+t$, $y=\sqrt{t}$; t=4. $\frac{dy}{dt} = \frac{1}{2\sqrt{t}}$, $\frac{dx}{dt}=2t+1$, so $\frac{dy}{dx} = \frac{dy/dt}{dx/dt} = \frac{1}{2\sqrt{t}(2t+1)}$.

When t=4, (x,y) = (20,2) and $\frac{dy}{dx} = \frac{1}{36}$, so the equation of the

tangent is y-2 = $\frac{1}{36}$(x-20) or x-36y+52=0.

5. $x=2\sin θ$, $y=3\cos θ$; θ=π/4. $\frac{dx}{dθ}=2\cos θ$, $\frac{dy}{dθ}=-3\sin θ$, $\frac{dy}{dx}=\frac{dy/dθ}{dx/dθ}=-\frac{3}{2}\tan θ$.

When θ=π/4, (x,y) = (√2, 3√2/2), and $\frac{dy}{dx}=-\frac{3}{2}$, so the equation of the

tangent is y-3√2/2 = (-3/2)(x-√2) or 3x+2y=6√2.

7. (a) x=1-t, $y=1-t^2$; (1,1). $\frac{dy}{dt}=-2t$, $\frac{dx}{dt}=-1$, $\frac{dy}{dx}=\frac{dy/dt}{dx/dt}=2t$. At (1,1),

t=0, so $\frac{dy}{dx}=0$, and the tangent is y-1 = 0(x-1) or y=1.

(b) y = $1-t^2$ = $1-(1-x)^2$ = $2x-x^2$, so $\left[\frac{dy}{dx}\right]_{x=1}$ = $(2-2x)_{x=1}$= 0, and the

tangent is y=1.

9. (a) x = 5 cos t, y = 5 sin t; (3,4). $\frac{dy}{dt}$ = 5 cos t, $\frac{dx}{dt}$ = -5 sin t,

$\frac{dy}{dx}$ = $\frac{dy/dt}{dx/dt}$ = -cot t. At (3,4), t=$\tan^{-1}$(y/x)=$\tan^{-1}$(4/3), so $\frac{dy}{dx}$

= -3/4, and the tangent is y-4 = -(3/4)(x-3), or 3x+4y=25.

(b) $x^2+y^2=25$, so $2x+2y\cdot\frac{dy}{dx}=0$, or $\frac{dy}{dx} = -x/y$. At $(3,4)$, $\frac{dy}{dx} = -3/4$, and as in (a) the tangent is $3x+4y=25$.

11. $x=t^2+t$, $y=t^2+1$. $\frac{dy}{dx} = \frac{dy/dt}{dx/dt} = \frac{2t}{2t+1} = 1-\frac{1}{2t+1}$. $\frac{d}{dt}\left[\frac{dy}{dx}\right] = \frac{2}{(2t+1)^2}$

$\frac{d^2y}{dx^2} = \frac{d}{dx}\left[\frac{dy}{dx}\right] = \frac{d(dy/dx)/dt}{dx/dt} = \frac{2}{(2t+1)^3}$

13. $x=\sqrt{t+1}$, $y=t^2-3t$. $\frac{dy}{dx} = \frac{dy/dt}{dx/dt} = \frac{2t-3}{1/2\sqrt{t+1}} = 2(2t-3)\sqrt{t+1}$

$\frac{d}{dt}\left[\frac{dy}{dx}\right] = 4\sqrt{t+1} + 2(2t-3)\cdot\frac{1}{2\sqrt{t+1}} = \frac{1}{\sqrt{t+1}}(4t+4+2t-3) = \frac{6t+1}{\sqrt{t+1}}$

$\frac{d^2y}{dx^2} = \frac{d}{dx}\left[\frac{dy}{dx}\right] = \frac{d(dy/dx)/dt}{dx/dt} = 2\sqrt{t+1}\cdot\frac{6t+1}{\sqrt{t+1}} = 2(6t+1)$

15. $x=\sin\pi t$, $y=\cos\pi t$. $\frac{dy}{dx} = \frac{dy/dt}{dx/dt} = \frac{-\pi\sin\pi t}{\pi\cos\pi t} = -\tan\pi t$

$\frac{d^2y}{dx^2} = \frac{d}{dx}\left[\frac{dy}{dx}\right] = \frac{d(dy/dx)/dt}{dx/dt} = \frac{-\pi\sec^2\pi t}{\pi\cos\pi t} = -\sec^3\pi t$

17. $x=e^{-t}$, $y=te^{2t}$. $\frac{dy}{dx} = \frac{dy/dt}{dx/dt} = \frac{(2t+1)e^{2t}}{-e^{-t}} = -(2t+1)e^{3t}$

$\frac{d}{dt}\left[\frac{dy}{dx}\right] = -3(2t+1)e^{3t}-2e^{3t} = -(6t+5)e^{3t}$

$\frac{d^2y}{dx^2} = \frac{d}{dx}\left[\frac{dy}{dx}\right] = \frac{d(dy/dx)/dt}{dx/dt} = \frac{-(6t+5)e^{3t}}{-e^{-t}} = (6t+5)e^{4t}$

19. $x=1-2\cos t$, $y=2+3\sin t$. $\frac{dx}{dt}=2\sin t$,

$\frac{dy}{dt}=3\cos t$. $\frac{dy}{dt}=0 \Longleftrightarrow \cos t=0$ (so that

$\sin t=\pm 1$) $\Longleftrightarrow y=5$ or $-1 \Longleftrightarrow (x,y)=(1,5)$ or

$(1,-1)$. $\frac{dx}{dt}=0 \Longleftrightarrow \sin t=0 \Longleftrightarrow y=2$ and $x=3$ or

$-1 \Longleftrightarrow (x,y)=(3,2)$ or $(-1,2)$. Thus the

tangent is horizontal at $(1,5)$ and $(1,-1)$, and is vertical at $(3,2)$

and $(-1,2)$. The curve is the ellipse $\frac{(x-1)^2}{4} + \frac{(y-2)^2}{9} = 1$.

21.  $x=t(t^2-3)=t^3-3t$, $y=3(t^2-3)$.  $\frac{dx}{dt}=3t^2-3=3(t-1)(t+1)$; $\frac{dy}{dt}=6t$.

$\frac{dy}{dt}=0 \iff t=0 \iff (x,y)=(0,-9)$.  $\frac{dx}{dt}=0 \iff t=\pm1 \iff (x,y)=(-2,-6)$ or

$(2,-6)$.  So there is a horizontal tangent at $(0,-9)$ and there are

vertical tangents at $(-2,-6)$ and $(2,-6)$.  The graph is the

reflection about the line $y=x$ of the graph in Exercise 9.1.27.

**21.**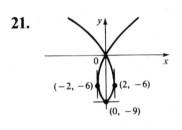

(−2, −6)  (2, −6)

(0, −9)

**23.**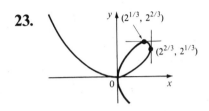

$y$ $(2^{1/3}, 2^{2/3})$

$(2^{2/3}, 2^{1/3})$

23.  $x = \frac{3t}{1+t^3}$, $y = \frac{3t^2}{1+t^3}$.  $\frac{dx}{dt} = \frac{(1+t^3)3-3t(3t^2)}{(1+t^3)^2} = \frac{3-6t^3}{(1+t^3)^2}$,

$\frac{dy}{dt} = \frac{(1+t^3)(6t)-3t^2(3t^2)}{(1 + t^3)^2} = \frac{6t-3t^4}{(1+t^3)^2} = \frac{3t(2-t^3)}{(1+t^3)^2}$  $\frac{dy}{dt}=0 \iff t=0$ or $\sqrt[3]{2} \iff$

$(x,y)=(0,0)$ or $(2^{1/3},2^{2/3})$.  $\frac{dx}{dt}=0 \iff t^3=1/2 \iff t=2^{-1/3} \iff$

$(x,y)=(2^{2/3},2^{1/3})$.  There are horizontal tangents at $(0,0)$ and

$(2^{1/3},2^{2/3})$, and there are vertical tangents at $(2^{2/3},2^{1/3})$ and

$(0,0)$.  (The vertical tangent at $(0,0)$ is undetectable by the methods

of this section because that tangent is the limiting position of

the curve as $t\to\pm\infty$.)

25.  $x=\cos t$, $y=\sin t \cos t$.  $\frac{dx}{dt}=-\sin t$,

$\frac{dy}{dt} = -\sin^2 t+\cos^2 t = \cos 2t$.  $(x,y)=(0,0) \iff$

$\cos t=0 \iff t$ is an odd multiple of $\pi/2$.

When $t=\pi/2$, $\frac{dx}{dt}=-1$ and $\frac{dy}{dt}=-1$, so $\frac{dy}{dx}=1$.  When

$t=3\pi/2$, $\frac{dx}{dt}=1$ and $\frac{dy}{dt}=-1$, so $\frac{dy}{dx}=-1$.  Thus $y=x$

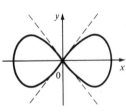

and $y=-x$ are both tangent to the curve at $(0,0)$.

27.  The curve crosses itself at $(0,0)$, when $t=\pm\sqrt{3}$, so $\frac{dx}{dt}=6$ and $\frac{dy}{dt}=\pm6\sqrt{3}$.

Thus $\frac{dy}{dx}=\pm\sqrt{3}$, and the tangents at $(0,0)$ are $y=\sqrt{3}x$ and $y=-\sqrt{3}x$.

29.  (a) $x=r\theta-d \sin\theta$, $y=r-d \cos\theta$; $\frac{dx}{d\theta}=r-d \cos\theta$, $\frac{dy}{d\theta}=d \sin\theta$.  So $\frac{dy}{dx}=\frac{d \sin\theta}{r-d \cos\theta}$.

(b) If $0<d<r$, then $|d \cos\theta|\leq d<r$, so $r-d \cos\theta \geq r-d > 0$.

This shows that $\frac{dx}{d\theta}$ never vanishes, so the trochoid can have no

vertical tangents if $d<r$.                269

31. The line with parametric equations $x=-7t$, $y=12t-5$ is $y=12(-x/7)-5$, which has slope $-12/7$. The curve $x=t^3+4t$, $y=6t^2$ has slope $\frac{dy}{dx} = \frac{dy/dt}{dx/dt} = \frac{12t}{3t^2+4}$. This equals $-12/7 \Longleftrightarrow 3t^2+4=-7t \Longleftrightarrow$

    $(3t+4)(t+1)=0 \Longleftrightarrow t=-1$ or $t=-4/3 \Longleftrightarrow (x,y)=(-5,6)$ or $(-208/27,32/3)$.

33. By symmetry of the ellipse about the x- and y-axes,

    $$A = 4\int_0^a y\ dx = 4\int_{\pi/2}^0 b\sin\theta\ (-a\sin\theta)\ d\theta = 4ab\int_0^{\pi/2}\sin^2\theta\ d\theta$$

    $$= 4ab\int_0^{\pi/2}\frac{1-\cos 2\theta}{2}\ d\theta = 2ab\left[\theta-\frac{\sin 2\theta}{2}\right]_0^{\pi/2} = 2ab(\pi/2) = \pi ab.$$

35. $$A = \int_0^1 (y-1)dx = \int_{\pi/2}^0 (e^t-1)(-\sin t)dt = \int_0^{\pi/2} (e^t\sin t - \sin t)dt$$

    $$= \frac{e^t}{2}(\sin t - \cos t) + \cos t\Big]_0^{\pi/2} = (e^{\pi/2}-1)/2 \quad \text{[Formula 98]}$$

37. $$A = \int_0^{2\pi r} y\ dx = \int_0^{2\pi} (r-d\cos\theta)(r-d\cos\theta)\ d\theta$$

    $$= \int_0^{2\pi} (r^2-2dr\cos\theta+d^2\cos^2\theta)d\theta = r^2\theta-2dr\sin\theta+\frac{d^2}{2}(\theta+\frac{\sin 2\theta}{2})\Big]_0^{2\pi} = 2\pi r^2+\pi d^2$$

Section 9.3

Section 9.3

1. $x=1+2\sin\pi t$, $y=3-2\cos\pi t$, $0\le t\le 1$. $\left[\frac{dx}{dt}\right]^2+\left[\frac{dy}{dt}\right]^2=(2\pi\cos\pi t)^2+(2\pi\sin\pi t)^2$

    $=4\pi^2 \Rightarrow L = \int_0^1 \sqrt{(dx/dt)^2+(dy/dt)^2}\ dt = \int_0^1 2\pi\ dt = 2\pi$

3. $x=5t^2+1$, $y=4-3t^2$, $0\le t\le 2$. $\left[\frac{dx}{dt}\right]^2+\left[\frac{dy}{dt}\right]^2=(10t)^2+(-6t)^2=136t^2 \Rightarrow$

    $L = \int_0^2 \sqrt{136t^2}\ dt = \int_0^2 \sqrt{136}t\ dt = 2\sqrt{34}t^2/2\Big]_0^2 = 4\sqrt{34}$

5. $x=e^t\cos t$, $y=e^t\sin t$, $0\le t\le\pi$.

    $\left[\frac{dx}{dt}\right]^2+\left[\frac{dy}{dt}\right]^2 = [e^t(\cos t - \sin t)]^2+[e^t(\sin t + \cos t)]^2$

    $= e^{2t}(2\cos^2 t+2\sin^2 t)^2=2e^{2t} \Rightarrow L = \int_0^\pi \sqrt{2}e^t\ dt = \sqrt{2}(e^\pi-1)$

7.  $x = \ln \sin t$, $y = t$, $\pi/4 \le t \le \pi/2$.

$$\left[\frac{dx}{dt}\right]^2 + \left[\frac{dy}{dt}\right]^2 = \left[\frac{\cos t}{\sin t}\right]^2 + 1^2 = \cot^2 t + 1 = \csc^2 t \quad \Rightarrow \quad L = \int_{\pi/4}^{\pi/2} \csc t \, dt$$

$$= \ln|\csc t - \cot t| \Big]_{\pi/4}^{\pi/2} = \ln|1-0| - \ln|\sqrt{2}-1| = \ln\left[\frac{1}{\sqrt{2}-1}\right] = \ln(\sqrt{2}+1)$$

9.  $x = 2 - 3\sin^2\theta$, $y = \cos 2\theta$, $0 \le \theta \le \pi/2$. $\left[\frac{dx}{d\theta}\right]^2 + \left[\frac{dy}{d\theta}\right]^2 = (-6 \sin \theta \cos \theta)^2$

$+ (-2 \sin 2\theta)^2 = (-3 \sin 2\theta)^2 + (-2 \sin 2\theta)^2 = 13 \sin^2 2\theta \quad \Rightarrow$

$$L = \int_0^{\pi/2} \sqrt{13} \sin 2\theta \, d\theta = -\frac{\sqrt{13}}{2} \cos 2\theta \Big]_0^{\pi/2} = -\frac{\sqrt{13}}{2}(-1-1) = \sqrt{13}$$

11. $x = \sin^2\theta$, $y = \cos^2\theta$, $0 \le \theta \le 3\pi$. $\left[\frac{dx}{d\theta}\right]^2 + \left[\frac{dy}{d\theta}\right]^2 =$

$(2 \sin \theta \cos \theta)^2 + (-2 \cos \theta \sin \theta)^2 = 8 \sin^2\theta \cos^2\theta = 2 \sin^2 2\theta \quad \Rightarrow$

$$\text{distance} = \int_0^{3\pi} \sqrt{2} \, |\sin 2\theta| \, d\theta = 6\sqrt{2} \int_0^{\pi/2} \sin 2\theta \, d\theta \quad \text{[by symmetry]}$$

$$= -3\sqrt{2} \cos 2\theta \Big]_0^{\pi/2} = -3\sqrt{2}(-1-1) = 6\sqrt{2}$$

The full curve is traversed as $\theta$ goes from 0 to $\pi/2$, because the
curve is the segment of $x+y=1$ that lies in the first quadrant
(since $x, y \ge 0$), and this segment is completely traversed as $\theta$ goes
from 0 to $\pi/2$. Thus $L = \int_0^{\pi/2} \sin 2\theta \, d\theta = \sqrt{2}$, as above.

13. $x = a \sin\theta$, $y = b \cos\theta$, $0 \le \theta \le 2\pi$. $\left[\frac{dx}{d\theta}\right]^2 + \left[\frac{dy}{d\theta}\right]^2 = (a \cos\theta)^2 + (-b \sin\theta)^2$

$= a^2\cos^2\theta + b^2\sin^2\theta = a^2(1-\sin^2\theta) + b^2\sin^2\theta = a^2 - (a^2-b^2)\sin^2\theta$

$= a^2 - c^2\sin^2\theta = a^2\left[1 - \frac{c^2}{a^2}\sin^2\theta\right] = a^2\left[1 - e^2\sin^2\theta\right]$ So $L =$

$$4\int_0^{\pi/2}\left[a^2(1 - e^2\sin^2\theta)\right]^{1/2} d\theta \quad \text{[by symmetry]} = 4a\int_0^{\pi/2}\sqrt{1 - e^2\sin^2\theta} \, d\theta$$

15. $\left[\frac{dx}{dt}\right]^2 + \left[\frac{dy}{dt}\right]^2 = \left[2t-2/t^2\right]^2 + (4/\sqrt{t})^2 = 4t^2 + \frac{8}{t} + \frac{4}{t^4} = 4\left[t + \frac{1}{t^2}\right]^2$

$$S = \int_1^9 2\pi y \sqrt{(dx/dt)^2 + (dy/dt)^2} \, dt = \int_1^9 2\pi \, (8\sqrt{t}) \, 2\left[t + \frac{1}{t^2}\right] dt$$

$$= 32\pi\int_1^9 (t^{3/2} + t^{-3/2}) \, dt = 32\pi \, (\frac{2}{5}t^{5/2} - 2t^{-1/2}) \Big]_1^9$$

$$= 32\pi[(2 \cdot 243/5 - 2/3) - (2/5 - 2)] = 47104\pi/15$$

17. $\left[\frac{dx}{dt}\right]^2 + \left[\frac{dy}{dt}\right]^2 = 2e^{2t}$ as in #5, so $S = \int_0^{\pi/2} 2\pi y \sqrt{(dx/dt)^2 + (dy/dt)^2} \, dt$

$$= \int_0^{\pi/2} 2\pi \, e^t \sin t \cdot \sqrt{2}e^t \, dt = 2\sqrt{2}\pi\int_0^{\pi/2} e^{2t}\sin t \, dt$$

$$= 2\sqrt{2}\pi \cdot \frac{e^{2t}}{5}(2 \sin t - \cos t)\Big]_0^{\pi/2} \quad \text{[Formula 98]}$$

$$= \frac{2\sqrt{2}\pi}{5}[2e^{\pi}-(-1)] = 2\sqrt{2}\pi(2e^{\pi}+1)/5$$

19. $\left[\dfrac{dx}{d\theta}\right]^2 + \left[\dfrac{dy}{d\theta}\right]^2 = (-2\sin\theta+2\sin2\theta)^2+(2\cos\theta-2\cos2\theta)^2$

$$= 4[(\sin^2\theta-2\sin\theta\sin2\theta+\sin^22\theta)+(\cos^2\theta-2\cos\theta\cos2\theta+\cos^22\theta)]$$

$$= 4[1+1-2(\cos2\theta\cos\theta+\sin2\theta\sin\theta)] = 8[1-\cos(2\theta-\theta)] = 8(1-\cos\theta)$$

Note that $x(2\pi-\theta)=x(\theta)$ and $y(2\pi-\theta)=-y(\theta)$, so the piece of the curve from $\theta=0$ to $\theta=\pi$ generates the same surface as the piece from $\theta=\pi$ to $\theta=2\pi$. Also note that $y=2\sin\theta-\sin2\theta=2\sin\theta(1-\cos\theta)$. So

$$S = \int_0^{\pi} 2\pi\cdot2\sin\theta(1-\cos\theta)\cdot2\sqrt{2}(1-\cos\theta)^{1/2}d\theta = 8\sqrt{2}\pi\int_0^{\pi}(1-\cos\theta)^{3/2}\sin\theta\ d\theta$$

$$= 8\sqrt{2}\pi\int_0^2 u^{3/2}du \quad [u=1-\cos\theta,\ du=\sin\theta\ d\theta] \quad = 8\sqrt{2}\pi(\tfrac{2}{5})u^{5/2}\Big]_0^2 = 128\pi/5$$

21. $\left[\dfrac{dx}{dt}\right]^2 + \left[\dfrac{dy}{dt}\right]^2 = (6t)^2+(6t^2)^2=36t^2(1+t^2) \quad \Rightarrow$

$$S = \int_0^5 2\pi x\ \sqrt{(dx/dt)^2+(dy/dt)^2}\ dt = \int_0^5 2\pi\cdot3t^2\cdot6t\sqrt{1+t^2}\ dt$$

$$= 18\pi\int_0^5 t^2(1+t^2)^{1/2}2t\ dt = 18\pi\int_1^{26}(u-1)\sqrt{u}\ du \quad [u=1+t^2]$$

$$= 18\pi\int_1^{26}(u^{3/2}-u^{1/2})\ du = 18\pi\left[(2/5)u^{5/2}-(2/3)u^{3/2}\right]_1^{26}$$

$$= 18\pi\left[(\tfrac{2}{5}\cdot676\sqrt{26}-\tfrac{2}{3}\cdot26\sqrt{26})-(\tfrac{2}{5}-\tfrac{2}{3})\right] = 24\pi(949\sqrt{26}+1)/5$$

23. As in #13, we can show that $\left[\dfrac{dx}{d\theta}\right]^2 + \left[\dfrac{dy}{d\theta}\right]^2 = a^2(1 - e^2\cos^2\theta)$.

(a) $S = \displaystyle\int_0^{\pi} 2\pi\cdot b \sin\theta\cdot a(1-e^2\cos^2\theta)^{1/2}d\theta = 2\pi ab\int_{-e}^{e}(1-u^2)^{1/2}(1/e)\ du$

$[u = -e\cos\theta,\ du = e\sin\theta\ d\theta] \qquad = \dfrac{4\pi ab}{e}\int_0^e(1-u^2)^{1/2}du$

$= \dfrac{4\pi ab}{e}\displaystyle\int_0^{\sin^{-1}e}\cos^2v\ dv \quad [u = \sin v,\ du = \cos v\ dv]$

$= \dfrac{2\pi ab}{e}\displaystyle\int_0^{\sin^{-1}e}(1+\cos 2v)\ dv = \dfrac{2\pi ab}{e}\left[v + \dfrac{\sin 2v}{2}\right]\Big]_0^{\sin^{-1}e}$

$= \dfrac{2\pi ab}{e}(v + \sin v \cos v)\Big]_0^{\sin^{-1}e} = \dfrac{2\pi ab}{e}\left[\sin^{-1}e+e\sqrt{1-e^2}\right] \qquad \text{But}$

$\sqrt{1-e^2} = \sqrt{1-c^2/a^2} = \sqrt{(a^2-c^2)/a^2} = \sqrt{b^2/a^2} = \dfrac{b}{a}$, so $S = \dfrac{2\pi ab}{e}\sin^{-1}e+2\pi b^2$

(b) $S = \displaystyle\int_{-\pi/2}^{\pi/2} 2\pi \cdot a \cos \theta \cdot a (1-e^2\cos^2\theta)^{1/2} d\theta$

$= 4\pi a^2 \displaystyle\int_0^{\pi/2} \cos \theta \; [(1-e^2)+e^2\sin^2\theta]^{1/2} d\theta$

$= \dfrac{4\pi a^2(1-e^2)}{e} \displaystyle\int_0^{\pi/2} \dfrac{e}{\sqrt{1-e^2}} \cdot \cos\theta \left[1+\left[\dfrac{e \sin\theta}{\sqrt{1-e^2}}\right]^2\right]^{1/2} d\theta$

$= \dfrac{4\pi a^2(1-e^2)}{e} \displaystyle\int_0^{e/(1-e^2)^{1/2}} (1+u^2)^{1/2} du \qquad \left[ u = \dfrac{e \sin\theta}{\sqrt{1-e^2}} \right]$

$= \dfrac{4\pi a^2(1-e^2)}{e} \displaystyle\int_0^{\sin^{-1}e} \sec^3 v \; dv \quad [u = \tan v, \; du = \sec^2 v]$

$= \dfrac{2\pi a^2(1-e^2)}{e} \left[ \sec v \tan v + \ln|\sec v + \tan v| \right]_0^{\sin^{-1}e}$

$= \dfrac{2\pi a^2(1-e^2)}{e} \left[ \dfrac{1}{\sqrt{1-e^2}} \cdot \dfrac{e}{\sqrt{1-e^2}} + \ln\left| \dfrac{1}{\sqrt{1-e^2}} + \dfrac{e}{\sqrt{1-e^2}} \right| \right]$

$= 2\pi a^2 + \dfrac{2\pi a^2(1-e^2)}{e} \cdot \ln \sqrt{\dfrac{1+e}{1-e}} = 2\pi a^2 + \dfrac{2\pi b^2}{e} \dfrac{1}{2} \ln \dfrac{1+e}{1-e}$ [since

$1-e^2 = b^2/a^2$] $= 2\pi [a^2 + (b^2/2e) \ln \{(1+e)/(1-e)\}]$

## Section 9.4

### Exercises 9.4

1. $(1, \pi/2)$                    3. $(-1, \pi/5)$

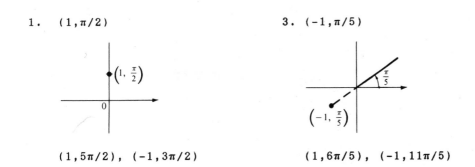

     $(1, 5\pi/2), \; (-1, 3\pi/2)$               $(1, 6\pi/5), \; (-1, 11\pi/5)$

**5.** (4,-2π/3)

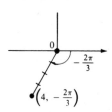

(4,4π/3), (-4,π/3)

**7.** (-1,π)

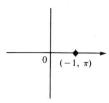

(-1,3π), (1,0)

**9.**

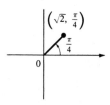

x=√2cos(π/4)=1,y=√2sin(π/4)=1

**11.**

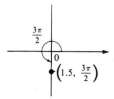

(0,-1.5)

**13.**

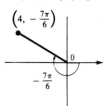

x=4cos(-7π/6)=4(-√3/2)=-2√3
y=4sin(-7π/6)=4(1/2)=2

**15.**

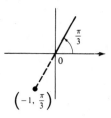

x=-cos(π/3)=-1/2
y=-sin(π/3)=-√3/2

**17.** (x,y)=(-1,1), r=√((-1)²+1²)=√2, tanθ=y/x=-1 and (x,y) is in quadrant II, so θ=3π/4. Coordinates (√2,3π/4)

**19.** (x,y)=(2√3,-2)
r=√(12+4)=4
tanθ = y/x = -1/√3
(x,y) is in quadrant IV,
so θ = 11π/6.
(4,11π/6)

**21.** r > 1

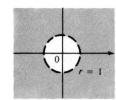

**23.** $0 \leq r \leq 2$, $\pi/2 \leq \theta \leq \pi$

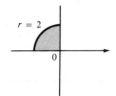

**25.** $3 < r < 4$, $-\pi/2 \leq \theta \leq \pi$

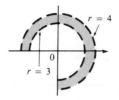

**27.** $(1, \pi/6)$ is $(\sqrt{3}/2, 1/2)$ Cartesian and $(3, 3\pi/4)$ is $(-3/\sqrt{2}, 3/\sqrt{2})$ Cartesian. The square of the distance between them is $(\sqrt{3}/2 + 3/\sqrt{2})^2 + (1/2 - 3/\sqrt{2})^2 = (40 + 6\sqrt{6} - 6\sqrt{2})/4$, so the distance is $(1/2)\sqrt{40 + 6\sqrt{6} - 6\sqrt{2}}$.

**29.** Since $y = r \sin \theta$, the equation $r \sin \theta = 2$ becomes $y = 2$.

**31.** $r = \dfrac{1}{1 - \cos\theta} \Leftrightarrow r - r\cos\theta = 1 \Leftrightarrow r = 1 + r\cos\theta \Leftrightarrow$
$r^2 = (1 + r\cos\theta)^2 \Leftrightarrow x^2 + y^2 = (1+x)^2 = 1 + 2x + x^2 \Leftrightarrow y^2 = 1 + 2x$

**33.** $r^2 = \sin 2\theta = 2\sin\theta\cos\theta \Leftrightarrow r^4 = 2r\sin\theta \, r\cos\theta \Leftrightarrow (x^2 + y^2)^2 = 2yx$

**35.** $y = 5 \Leftrightarrow r\sin\theta = 5$

**37.** $x^2 + y^2 = 25 \Leftrightarrow r^2 = 25 \Leftrightarrow r = 5$

**39.** $2xy = 1 \Leftrightarrow 2 \, r\cos\theta \, r\sin\theta = 1 \Leftrightarrow r^2 \sin 2\theta = 1 \Leftrightarrow r^2 = \csc 2\theta$

**41.** $r = 5$

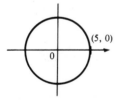

**43.** $\theta = 3\pi/4$

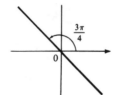

**45.** $r = 2\sin\theta \Leftrightarrow r^2 = 2r\sin\theta$
$\Leftrightarrow x^2 + y^2 = y \Leftrightarrow x^2 + (y-1)^2 = 1$

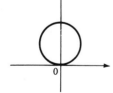

**47.** $r = -\cos\theta \Leftrightarrow r^2 = -r\cos\theta \Leftrightarrow$
$x^2 + y^2 = -x \Leftrightarrow (x + 1/2)^2 + y^2 = 1/4$

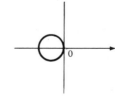

49.  $r = \cos\theta - \sin\theta \iff$
$r^2 = r\cos\theta - r\sin\theta \iff$
$x^2 + y^2 = x - y \iff$
$(x - 1/2)^2 + (y + 1/2)^2 = 1/2$

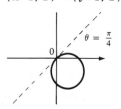

51.  $r = 3(1 - \cos\theta)$

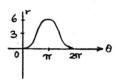

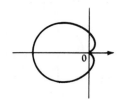

53.  $r = \theta$, $\theta \geq 0$

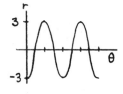

55.  $r = 1/\theta$

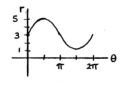

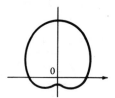

57.  $r = 1 - 2\cos\theta$

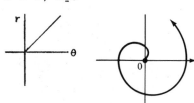

59.  $r = 3 + 2\sin\theta$

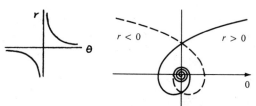

61.  $r = -3\cos2\theta$

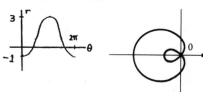

63.  $r = \sin3\theta$

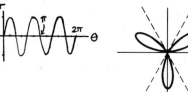

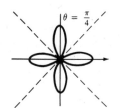

**65.** r = 2 cos4θ

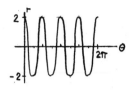

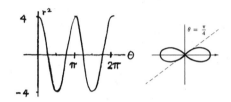

**67.** r = sin5θ

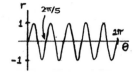

**69.** $r^2$ = 4 cos2θ

**71.** r = 4 + 2 secθ

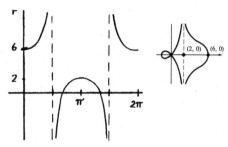

**73.** r = sinθ tanθ $\Longleftrightarrow$ $r^2$ = r sinθ tanθ $\Longleftrightarrow$
$x^2+y^2$ = y·(y/x) $\Longleftrightarrow$ $x^2$ = $y^2$(1/x − 1) $\Longleftrightarrow$
$y^2$ = $x^3$/(1−x). The right side $\geq$ 0 for
0≤x<1, so the curve is defined only for
0≤x<1.

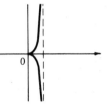

**75.** r = 3 cosθ $\Rightarrow$ x = r cosθ = 3 $\cos^2\theta$, y = r sinθ = 3 sinθ cosθ $\Rightarrow$
$\dfrac{dy}{dx} = \dfrac{dy/d\theta}{dx/d\theta} = \dfrac{-3\sin^2\theta + 3\cos^2\theta}{-6\cos\theta\sin\theta} = \dfrac{\cos2\theta}{-\sin2\theta} = -\cot2\theta = 1/\sqrt{3}$ when θ=π/3

**77.** $\dfrac{dy}{dx} = \dfrac{dy/d\theta}{dx/d\theta} = \dfrac{(dr/d\theta)\sin\theta + r\cos\theta}{(dr/d\theta)\cos\theta - r\sin\theta} = \dfrac{\sin\theta + \theta\cos\theta}{\cos\theta - \theta\sin\theta} = -2/\pi$ when θ=π/2

**79.** $\dfrac{dy}{dx} = \dfrac{(dr/d\theta)\sin\theta + r\cos\theta}{(dr/d\theta)\cos\theta - r\sin\theta} = \dfrac{-\sin^2\theta+(1+\cos\theta)\cos\theta}{-\sin\theta\cos\theta-(1+\cos\theta)\sin\theta} = -1$ when θ=π/6

**81.** $\dfrac{dy}{dx} = \dfrac{(dr/d\theta)\sin\theta + r\cos\theta}{(dr/d\theta)\cos\theta - r\sin\theta} = \dfrac{3\cos3\theta\sin\theta + \sin3\theta\cos\theta}{3\cos3\theta\cos\theta - \sin3\theta\sin\theta} = \sqrt{3}$ when θ=π/3

**83.** $\dfrac{dy}{d\theta}$ = (dr/dθ)sinθ + r cosθ = $-3\sin^2\theta+3\cos^2\theta$ = 3cos2θ = 0 $\Rightarrow$ 2θ = π/2

or 3π/2 $\Longleftrightarrow$ θ = π/4 or 3π/4. So the tangent is horizontal at
(3/$\sqrt{2}$,π/4) and (−3/$\sqrt{2}$,3π/4) [same as (3/$\sqrt{2}$,−π/4)]).
$\dfrac{dx}{d\theta}$ = (dr/dθ)cosθ − r sinθ = −6sinθcosθ = −3sin2θ = 0 $\Rightarrow$ 2θ = 0 or π
$\Longleftrightarrow$ θ = 0 or π/2. So the tangent is vertical at (3,0) and (0,π/2).

85. $\frac{dy}{d\theta}$ = (dr/d$\theta$)sin$\theta$ + r cos$\theta$ = -2sin2$\theta$sin$\theta$ + cos2$\theta$cos$\theta$ =

-4sin$^2\theta$cos$\theta$ + (cos$^3\theta$ - sin$^2\theta$cos$\theta$) = cos$\theta$(cos$^2\theta$ - 5sin$^2\theta$) =

cos$\theta$(1 - 6sin$^2\theta$) = 0 $\Rightarrow$ cos$\theta$ = 0 or sin$\theta$ = ±1/$\sqrt{6}$ $\Rightarrow$ $\theta$ = $\pi$/2, 3$\pi$/2, $\alpha$,

$\pi$-$\alpha$, $\pi$+$\alpha$, or 2$\pi$-$\alpha$ [where $\alpha$ = sin$^{-1}$(1/$\sqrt{6}$)]. So the tangent is

horizontal at (1,3$\pi$/2), (1,$\pi$/2), (2/3,$\alpha$), (2/3,$\pi$-$\alpha$), (2/3,$\pi$+$\alpha$), and

(2/3,2$\pi$-$\alpha$). $\frac{dx}{d\theta}$ = (dr/d$\theta$)cos$\theta$ - r sin$\theta$ = -2sin2$\theta$cos$\theta$ - cos2$\theta$sin$\theta$ =

-4sin$\theta$cos$^2\theta$ - (2cos$^2\theta$-1)sin$\theta$ = sin$\theta$(1-6cos$^2\theta$) = 0 $\Rightarrow$ sin$\theta$ = 0 or

cos$\theta$ = ±1/$\sqrt{6}$ $\Rightarrow$ $\theta$ = 0, $\pi$, $\pi$/2-$\alpha$, $\pi$/2+$\alpha$, 3$\pi$/2-$\alpha$, or 3$\pi$/2+$\alpha$. So the

tangent is vertical at (1,0), (1,$\pi$), (2/3,3$\pi$/2-$\alpha$), (2/3,3$\pi$/2+$\alpha$),

(2/3,$\pi$/2-$\alpha$), and (2/3,$\pi$/2+$\alpha$).

87. $\frac{dy}{d\theta}$ = -sin$^2\theta$ + (1+cos$\theta$)cos$\theta$ = 2cos$^2\theta$ + cos$\theta$ - 1 =

(2cos$\theta$-1)(cos$\theta$+1) = 0 $\Rightarrow$ cos$\theta$ = 1/2 or -1 $\Rightarrow$ $\theta$ = $\pi$/3, $\pi$, or 5$\pi$/3 $\Rightarrow$

horizontal tangent at (3/2,$\pi$/3), (0,$\pi$), and (3/2,5$\pi$/3).

$\frac{dx}{d\theta}$ = -sin$\theta$cos$\theta$ -(1+cos$\theta$)sin$\theta$ = -sin$\theta$(1+2cos$\theta$) = 0 $\Rightarrow$ sin$\theta$ = 0 or

cos$\theta$ = -1/2 $\Rightarrow$ $\theta$ = 0, $\pi$, 2$\pi$/3, or 4$\pi$/3 $\Rightarrow$ vertical tangent at (2,0),

(1/2,2$\pi$/3), and (1/2, 4$\pi$/3). [Note that the tangent is horizontal,

not vertical when $\theta$=$\pi$, since $\lim\limits_{\theta\to\pi} \frac{dy/d\theta}{dx/d\theta}$ = 0.]

89. r = a sin$\theta$ + b cos$\theta$ $\Rightarrow$ r$^2$ = ar sin$\theta$ + br cos$\theta$ $\Rightarrow$ x$^2$+y$^2$ = ay + bx $\Rightarrow$

(x-b/2)$^2$+(y-a/2)$^2$ = (a$^2$+b$^2$)/4 and this is a circle with center

(b/2,a/2) and radius (1/2)$\sqrt{a^2+b^2}$.

91. Following the hint, tan$\Psi$ = tan($\phi$-$\theta$) = $\frac{\tan\phi - \tan\theta}{1 + \tan\phi\,\tan\theta}$ =

$\frac{dy/dx - \tan\theta}{1 + (dy/dx)\tan\theta}$ = $\frac{(dy/d\theta)/(dx/d\theta) - \tan\theta}{1 + (\tan\theta)(dy/d\theta)/(dx/d\theta)}$ = $\frac{dy/d\theta - (\tan\theta)dx/d\theta}{dx/d\theta + (\tan\theta)dy/d\theta}$ =

$\frac{[(dr/d\theta)\sin\theta+r\cos\theta]-\tan\theta[(dr/d\theta)\cos\theta-r\sin\theta]}{[(dr/d\theta)\cos\theta-r\sin\theta]+\tan\theta[(dr/d\theta)\sin\theta+r\cos\theta]}$ =

$\frac{r \cos^2\theta + r \sin^2\theta}{(dr/d\theta)\cos^2\theta + (dr/d\theta)\sin^2\theta}$ = $\frac{r}{dr/d\theta}$.

Exercises 9.5

1. $A = \int_0^\pi \frac{1}{2}r^2 d\theta = \int_0^\pi \frac{1}{2}\theta^2 d\theta = \left.\frac{\theta^3}{6}\right]_0^\pi = \pi^3/6$

3. $A = \int_0^{\pi/6} \frac{1}{2}(2\cos\theta)^2 d\theta = \int_0^{\pi/6} (1+\cos2\theta)d\theta = \left.\theta + \frac{1}{2}\sin2\theta\right]_0^{\pi/6} = \frac{\pi}{6} + \frac{\sqrt{3}}{4}$

5. $A = \int_{\pi/2}^{3\pi/2} \frac{1}{2}(\theta^2)^2 d\theta = \left.\theta^5/10\right]_{\pi/2}^{3\pi/2} = 121\pi^5/160$

7. $A = \int_0^{\pi/6} \frac{1}{2}\sin^2 2\theta \, d\theta = \frac{1}{4}\int_0^{\pi/6} (1-\cos4\theta)d\theta = \left.\frac{\theta}{4} - \frac{\sin4\theta}{16}\right]_0^{\pi/6} = \frac{4\pi-3\sqrt{3}}{96}$

9. $A = \int_0^\pi \frac{1}{2}(5\sin\theta)^2 d\theta = \frac{25}{4}\int_0^\pi (1-\cos2\theta)d\theta$

$= \frac{25}{4}\left[\theta - \frac{1}{2}\sin2\theta\right]_0^\pi = 25\pi/4$

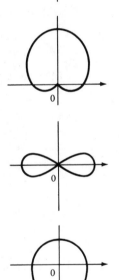

11. $A = 2\int_{-\pi/2}^{\pi/2} \frac{1}{2}(1+\sin\theta)^2 d\theta$

$= \int_{-\pi/2}^{\pi/2} (1+2\sin\theta+\sin^2\theta)d\theta$

$= \left.\theta\right]_{-\pi/2}^{\pi/2} + 0 + 2\int_0^{\pi/2} \sin^2\theta \, d\theta = \pi + 2\pi/4 = 3\pi/2$

13. $A = 4\int_0^{\pi/4} \frac{1}{2}r^2 d\theta = 8\int_0^{\pi/4} \cos2\theta \, d\theta$

$= \left.4\sin2\theta\right]_0^{\pi/4} = 4$

15. $A = 2\int_{-\pi/2}^{\pi/2} \frac{1}{2}(4-\sin\theta)^2 d\theta$

$= \int_{-\pi/2}^{\pi/2} (16-8\sin\theta+\sin^2\theta)d\theta$

$= 16\pi + 0 + \int_{-\pi/2}^{\pi/2} \sin^2\theta \, d\theta = 33\pi/2$

17. $A = 8\int_0^{\pi/4} \frac{1}{2}\sin^2 4\theta \, d\theta = 2\int_0^{\pi/4} (1-\cos8\theta) \, d\theta$

$= \left.2\theta - \frac{1}{4}\sin8\theta\right]_0^{\pi/4} = \pi/2$

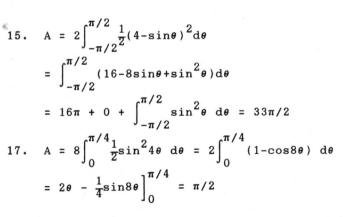

19. $A = 2\int_0^{\pi/6} \frac{1}{2}\cos^2 3\theta \, d\theta = \frac{1}{2}\int_0^{\pi/6}(1+\cos 6\theta) \, d\theta = \frac{1}{2}(\theta + \frac{1}{6}\sin 6\theta)\Big]_0^{\pi/6} = \pi/12$

21. $A = \int_0^{\pi/5} \frac{1}{2}\sin^2 5\theta \, d\theta = \frac{1}{4}\int_0^{\pi/5}(1-\cos 10\theta) \, d\theta = \frac{1}{4}(\theta - \frac{1}{10}\sin 10\theta)\Big]_0^{\pi/5} = \pi/20$

23. This is a limacon, with inner loop traced out between $\theta = 7\pi/6$ and $11\pi/6$.  $A = 2\int_{7\pi/6}^{3\pi/2} \frac{1}{2}(1 + 2\sin\theta)^2 d\theta = \int_{7\pi/6}^{3\pi/2}(1+4\sin\theta+4\sin^2\theta) \, d\theta$

    $= \theta - 4\cos\theta + 2\theta - \sin 2\theta\Big]_{7\pi/6}^{3\pi/2} = \pi - 3\sqrt{3}/2$

25. $1-\cos\theta = 3/2 \iff \cos\theta = -1/2 \Rightarrow \theta = 2\pi/3 \text{ or } 4\pi/3 \Rightarrow$

    $A = \int_{2\pi/3}^{4\pi/3} \frac{1}{2}[(1-\cos\theta)^2 - (3/2)^2] \, d\theta = \int_{2\pi/3}^{\pi}[-5/4 - 2\cos\theta + \cos^2\theta] \, d\theta$

    $= -(\frac{5}{4}\theta + 2\sin\theta)\Big]_{2\pi/3}^{\pi} + \int_{2\pi/3}^{\pi} \frac{1+\cos 2\theta}{2} \, d\theta = 9\sqrt{3}/8 - \pi/4$

27. $4\sin\theta = 2 \iff \sin\theta = 1/2 \Rightarrow \theta = \pi/6 \text{ or } 5\pi/6 \Rightarrow$

    $A = 2\int_{\pi/6}^{\pi/2} \frac{1}{2}[(4\sin\theta)^2 - 2^2] \, d\theta$

    $= \int_{\pi/6}^{\pi/2}[16\sin^2\theta - 4] \, d\theta = \int_{\pi/6}^{\pi/2}[8(1-\cos 2\theta) - 4] \, d\theta$

    $= 4\theta - 4\sin 2\theta\Big]_{\pi/6}^{\pi/2} = 4\pi/3 + 2\sqrt{3}$

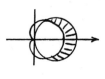

29. $3\cos\theta = 1+\cos\theta \iff \cos\theta = 1/2 \Rightarrow \theta = \pi/3 \text{ or } -\pi/3$

    $\Rightarrow A = 2\int_0^{\pi/3} \frac{1}{2}[(3\cos\theta)^2 - (1+\cos\theta)^2] \, d\theta$

    $= \int_0^{\pi/3}[8\cos^2\theta - 2\cos\theta - 1] \, d\theta = 2\sin 2\theta - 2\sin\theta + 3\theta\Big]_0^{\pi/3}$

    $= \pi$

31. $A = 2\int_0^{\pi/4} \frac{1}{2}\sin^2\theta \, d\theta = \int_0^{\pi/4} \frac{1-\cos 2\theta}{2} \, d\theta$

    $= \frac{1}{2}\theta - \frac{1}{4}\sin 2\theta\Big]_0^{\pi/4} = \pi/8 - 1/4$

$r=\sin\theta$

$r=\cos\theta$

33. $\sin 2\theta = \cos 2\theta \Rightarrow \tan 2\theta = 1 \Rightarrow 2\theta = \pi/4 \Rightarrow \theta = \pi/8$

    $\Rightarrow A = 16\int_0^{\pi/8} \frac{1}{2}\sin^2 2\theta \, d\theta = 4\int_0^{\pi/8}(1-\cos 4\theta) \, d\theta$

    $= 4\theta - \sin 4\theta\Big]_0^{\pi/8} = \frac{\pi}{2} - 1$

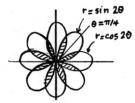

$r=\sin 2\theta$

$\theta=\pi/4$

$r=\cos 2\theta$

35. $A = 2\left\{\int_{-\pi/2}^{-\pi/6} \frac{1}{2}(3+2\sin\theta)^2 d\theta + \int_{-\pi/6}^{\pi/2} \frac{1}{2}\cdot 2^2 d\theta\right\}$

$= \int_{-\pi/2}^{-\pi/6} (9+12\sin\theta+4\sin^2\theta)d\theta + 4\theta\Big]_{-\pi/6}^{\pi/2}$

$= 9\theta-12\cos\theta+2\theta-\sin2\theta\Big]_{-\pi/2}^{-\pi/6} + 8\pi/3$

$= 19\pi/3 - 11\sqrt{3}/2$

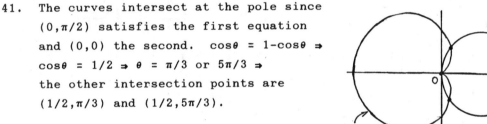

37. $A = 2\left\{\int_{0}^{2\pi/3} \frac{1}{2}\left[\frac{1}{2}+\cos\theta\right]^2 d\theta - \int_{2\pi/3}^{\pi} \frac{1}{2}\left[\frac{1}{2}+\cos\theta\right]^2 d\theta\right\}$

$= \frac{\theta}{4}+\sin\theta+\frac{\theta}{2}+\frac{1}{4}\sin2\theta\Big]_{0}^{2\pi/3} -(\frac{\theta}{4}+\sin\theta+\frac{\theta}{2}+\frac{1}{4}\sin2\theta)\Big]_{2\pi/3}^{\pi}$

$= (\pi + 3\sqrt{3})/4$

39. The two circles intersect at the pole since
(0,0) satisfies the first equation and $(0,\pi/2)$
the second. The other intersection point
$(1/\sqrt{2},\pi/4)$ is where $\sin\theta = \cos\theta$.

41. The curves intersect at the pole since
$(0,\pi/2)$ satisfies the first equation
and (0,0) the second. $\cos\theta = 1-\cos\theta \Rightarrow$
$\cos\theta = 1/2 \Rightarrow \theta = \pi/3$ or $5\pi/3 \Rightarrow$
the other intersection points are
$(1/2,\pi/3)$ and $(1/2,5\pi/3)$.

43. The pole is a point of intersection, (0,0) on
both curves. $\sin\theta = \sin2\theta = 2\sin\theta\cos\theta \iff$
$\sin\theta(1-2\cos\theta) = 0 \iff \sin\theta = 0$ or $\cos\theta = 1/2 \Rightarrow$
$\theta = 0, \pi, \pi/3, -\pi/3 \Rightarrow (\sqrt{3}/2,\pi/3)$ and
$(\sqrt{3}/2,2\pi/3)$ are the other intersection points.

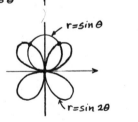

45. $L = \int_{0}^{3\pi/4} \sqrt{r^2 + (dr/d\theta)^2}\, d\theta = \int_{0}^{3\pi/4} \sqrt{(5\cos\theta)^2+(-5\sin\theta)^2}\, d\theta$

$= 5\int_{0}^{3\pi/4} \sqrt{\cos^2\theta+\sin^2\theta}\, d\theta = 5\int_{0}^{3\pi/4} d\theta = 15\pi/4$

47. $L = \int_{0}^{2\pi} \sqrt{(2^\theta)^2+((\ln2)2^\theta)^2}\, d\theta = \int_{0}^{2\pi} 2^\theta\sqrt{1+\ln^2 2}\, d\theta = \sqrt{1+\ln^2 2}\,(2^\theta/\ln2)\Big]_{0}^{2\pi}$

$= \sqrt{1+\ln^2 2}\,(2^{2\pi}-1)/\ln2$

49.  $L = \int_0^{2\pi} \sqrt{(\theta^2)^2 + (2\theta)^2} \, d\theta = \int_0^{2\pi} \theta \sqrt{\theta^2 + 4} \, d\theta = \frac{1}{2} \cdot \frac{2}{3} (\theta^2 + 4)^{3/2} \Big]_0^{2\pi}$

$= \frac{8}{3} [(\pi^2 + 1)^{3/2} - 1]$

51.  $L = 2\int_0^{2\pi} \sqrt{\cos^8(\theta/4) + \cos^6(\theta/4)\sin^2(\theta/4)} \, d\theta$

$= 2\int_0^{2\pi} |\cos^3(\theta/4)| \sqrt{\cos^2(\theta/4) + \sin^2(\theta/4)} \, d\theta = 2\int_0^{2\pi} |\cos^3(\theta/4)| \, d\theta$

$= 8\int_0^{\pi/2} \cos^3 u \, du \ [u = \theta/4] = 8(\sin u - \sin^3 u/3) \Big]_0^{\pi/2} = 16/3$

(Note that the curve is retraced after every interval of length $4\pi$)

53.  (a) From (9.9), $S = \int_a^b 2\pi y \sqrt{(dx/d\theta)^2 + (dy/d\theta)^2} \, d\theta$

$= \int_a^b 2\pi y \sqrt{r^2 + (dr/d\theta)^2} \, d\theta$ [see the derivation of (9.16)]

$= \int_a^b 2\pi r \sin\theta \sqrt{r^2 + (dr/d\theta)^2} \, d\theta$

(b) $r^2 = \cos 2\theta \Rightarrow 2r \frac{dr}{d\theta} = -2\sin 2\theta \Rightarrow \left[\frac{dr}{d\theta}\right]^2 = \frac{\sin^2 2\theta}{r^2} = \frac{\sin^2 2\theta}{\cos 2\theta}$

$S = 2\int_0^{\pi/4} 2\pi \sqrt{\cos 2\theta} \sin\theta \sqrt{\cos 2\theta + (\sin^2 2\theta)/\cos 2\theta} \, d\theta = 4\pi \int_0^{\pi/4} \sin\theta \, d\theta$

$= -4\pi \cos\theta \Big]_0^{\pi/4} = -4\pi[1/\sqrt{2} - 1] = 2\pi(2 - \sqrt{2})$

Section 9.6

Exercises 9.6

1.  $x^2 = -8y$.  $4p = -8$, so $p = -2$.  The vertex
is $(0,0)$, the focus is $(0,-2)$, and the
directrix is $y = 2$.

3.  $y^2=x$.  p=1/4 and the vertex is (0,0), so the focus is (1/4,0), and the directrix x = -1/4.

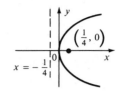

5.  $x+1 = 2(y-3)^2 \Rightarrow (y-3)^2 = \frac{1}{2}(x+1) \Rightarrow$ p=1/8 $\Rightarrow$ vertex (-1,3), focus (-7/8,3), directrix x = -9/8

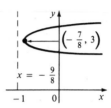

7.  $2x+y^2-8y+12=0 \Rightarrow (y-4)^2 = -2(x-2) \Rightarrow$ p = -1/2 $\Rightarrow$ vertex (2,4), focus (3/2,4), directrix x = 5/2

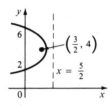

9.  $\frac{x^2}{16}+\frac{y^2}{4} = 1 \Rightarrow$ a=4, b=2, c=$\sqrt{16-4}$=2$\sqrt{3}$ $\Rightarrow$ center (0,0), vertices ($\pm4$,0), foci ($\pm2\sqrt{3}$,0)

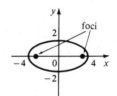

11.  $25x^2+9y^2 = 225 \Longleftrightarrow \frac{x^2}{9}+\frac{y^2}{25} = 1 \Rightarrow$ a=5, b=3, c=4 $\Rightarrow$ center (0,0), vertices (0,$\pm5$), foci (0,$\pm4$)

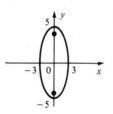

13.  $\frac{x^2}{144} - \frac{y^2}{25} = 1 \Rightarrow$ a=12, b=5, c=$\sqrt{144+25}$=13 $\Rightarrow$ center (0,0), vertices ($\pm12$,0), foci ($\pm13$,0), asymptotes y = $\pm5x/12$

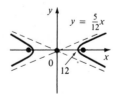

15. $9y^2 - x^2 = 9 \Rightarrow y^2 - \frac{x^2}{9} = 1 \Rightarrow a=1, \; b=3, \; c=\sqrt{10}$

    $\Rightarrow$ center $(0,0)$, vertices $(0,\pm 1)$, foci $(0,\pm\sqrt{10})$, asymptotes $y = \pm x/3$

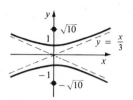

17. $9x^2 - 18x + 4y^2 = 27 \iff \frac{(x-1)^2}{4} + \frac{y^2}{9} = 1 \Rightarrow a=3,$

    $b=2, \; c=\sqrt{5} \Rightarrow$ center $(1,0)$, vertices $(1,\pm 3)$, foci $(1,\pm\sqrt{5})$

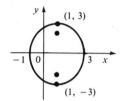

19. $2y^2 - 4y - 3x^2 + 12x = -8 \iff \frac{(x-2)^2}{6} - \frac{(y-1)^2}{9} = 1$

    $\Rightarrow a=\sqrt{6}, \; b=3, \; c=\sqrt{15} \Rightarrow$ center $(2,1)$, vertices $(2\pm\sqrt{6},1)$, foci $(2\pm\sqrt{15},1)$, asymptotes $y-1 = \pm(3/\sqrt{6})(x-2)$

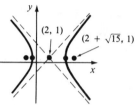

21. Vertex at $(0,0)$, $p=3$, opens upward $\Rightarrow x^2 = 4py = 12y$

23. Vertex at $(2,0)$, $p=1$, opens to right $\Rightarrow y^2 = 4p(x-2) = 4(x-2)$

25. Parabola must have equation $y^2 = 4px$, so $(-4)^2 = 4p\cdot 1 \Rightarrow p=4 \Rightarrow y^2 = 16x$

27. Center $(0,0)$, $c=1$, $a=2 \Rightarrow b = \sqrt{2^2 - 1^2} = \sqrt{3} \Rightarrow \frac{x^2}{4} + \frac{y^2}{3} = 1$

29. Center $(3,0)$, $c=1$, $a=3 \Rightarrow b=\sqrt{8}=2\sqrt{2} \Rightarrow \frac{(x-3)^2}{8} + \frac{y^2}{9} = 1$

31. Center $(2,2)$, $c=2$, $a=3 \Rightarrow b=\sqrt{5} \Rightarrow \frac{(x-2)^2}{9} + \frac{(y-2)^2}{5} = 1$

33. Center $(0,0)$, vertical axis, $c=3$, $a=1 \Rightarrow b=\sqrt{8}=2\sqrt{2} \Rightarrow y^2 - \frac{x^2}{8} = 1$

35. Center $(4,3)$, horizontal axis, $c=3$, $a=2 \Rightarrow b=\sqrt{5} \Rightarrow \frac{(x-4)^2}{4} - \frac{(y-3)^2}{5} = 1$

37. Center $(0,0)$, axis horizontal, $a=3$, $b/a=2 \Rightarrow b=6 \Rightarrow \frac{x^2}{9} - \frac{y^2}{36} = 1$

39. In Figure 9.45, we see that the point on the ellipse closest to a focus is the closer vertex (which is a distance a-c from it) while the furthest point is the other vertex (at a distance of a+c). So for this lunar orbit, $(a-c)+(a+c) = 2a = (1728+110) + (1728+314)$, or $a = 1940$; and $(a+c)-(a-c) = 2c = 314-110$, or $c = 102$. Thus $b^2 = a^2 - c^2 = 3,753,196$, and the equation is $x^2/3763600 + y^2/3753196 = 1$.

41. (a) Set up the coordinate system so that A is $(-200,0)$ and B is $(200,0)$. $|PA|-|PB| = 1200 \cdot 980 = 1,176,000$ ft $= 2450/11$ mi $=2a \Rightarrow$ $a=1225/11$, and $c=200$ so $b^2 = c^2-a^2 = 3339375/121 \Rightarrow$

$$\frac{121x^2}{1500625} + \frac{121y^2}{3339375} = 1$$

(b) Due north of B $\Rightarrow$ x=200 $\Rightarrow$ $\frac{121 \cdot 200^2}{1500625} + \frac{121y^2}{3339375} = 1 \Rightarrow$ y $= 133575/539$

$\approx 248$ mi.

43. The upper branch of this hyperbola is concave upward. The function is $y = f(x) = a\sqrt{1 + x^2/b^2} = (a/b)\sqrt{b^2+x^2}$, so $y' = (a/b)x(b^2+x^2)^{-1/2}$ and $y'' = (a/b)\left[(b^2+x^2)^{-1/2} - x^2(b^2+x^2)^{-3/2}\right] = ab(b^2+x^2)^{-3/2} > 0$ for all x, and so concave upward.

45. (a) ellipse  (b) hyperbola  (c) empty graph (no curve)
    (d) In case (a), $a^2=k$, $b^2=k-16$, and $c^2 = a^2-b^2 = 16$, so the foci are at $(\pm 4,0)$. In case (b), k-16 < 0, so $a^2=k$, $b^2=16-k$, and $c^2 = a^2+b^2 = 16$, and so again the foci are at $(\pm 4,0)$.

47. Use the parametrization x=2cost, y=sint, $0 \leq t \leq 2\pi$ to get

$$L = 4\int_0^{\pi/2} \sqrt{(dx/dt)^2+(dy/dt)^2}\ dt = 4\int_0^{\pi/2} \sqrt{4\sin^2 t+\cos^2 t}\ dt$$

$$= 4\int_0^{\pi/2} \sqrt{3\sin^2 t + 1}\ dt.\ \text{Using Simpson's Rule with n=10,}$$

$$L \approx 4 \cdot \frac{\pi/20}{3}\left\{f(0)+4f(\pi/20)+2f(\pi/10)+\ldots+2f(2\pi/5)+4f(9\pi/20)+f(\pi/2)\right\}$$

with $f(t) = \sqrt{3\sin^2 t + 1}$, so $L \approx 9.69$.

49. $2yy' = 4p \Rightarrow y' = 2p/y \Rightarrow$ slope of tangent at P is $2p/y_1$. Slope of FP is $y_1/(x_1-p)$, so by Formula 1.16, $\tan \alpha = \dfrac{-2p/y_1 + y_1/(x_1-p)}{1 + (2p/y_1)y_1/(x_1-p)}$

$$= \frac{-2p(x_1-p) + y_1^2}{y_1(x_1-p) + 2py_1} = \frac{-2px_1 + 2p^2 + 4px_1}{y_1(p+x_1)} = \frac{2p(p+x_1)}{y_1(p+x_1)} = 2p/y_1 = \text{slope of}$$

tangent at P = $\tan \beta$. Since $0 \leq \alpha, \beta \leq \pi/2$, this proves that $\alpha = \beta$.

## Exercises 9.7

1.  $r = \dfrac{ed}{1 + e\cos\theta} = \dfrac{(2/3)(3)}{1 + (2/3)\cos\theta} = \dfrac{6}{3+2\cos\theta}$

3.  $r = \dfrac{ed}{1 + e\sin\theta} = \dfrac{1\cdot 2}{1 + 1\cdot\sin\theta} = \dfrac{2}{1+\sin\theta}$

5.  $r = 5\sec\theta \iff x = r\cos\theta = 5$, so $r = \dfrac{ed}{1 + e\cos\theta} = \dfrac{4\cdot 5}{1+4\cos\theta} = \dfrac{20}{1+4\cos\theta}$

7.  Focus $(0,0)$, vertex $(5,\pi/2)$ ⇒ directrix $y=10$ ⇒
    $r = \dfrac{ed}{1 + e\sin\theta} = \dfrac{10}{1+\sin\theta}$

9.  $e=3$ ⇒ hyperbola; $ed=4$ ⇒ $d=4/3$ ⇒ directrix
    $x=4/3$; vertices $(1,0)$ and $(-2,\pi)=(2,0)$;
    center $(3/2,0)$; asymptotes parallel to
    $\theta = \pm\cos^{-1}(-1/3)$

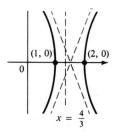

11. $e=1$ ⇒ parabola; $ed=2$ ⇒ $d=2$ ⇒ directrix
    $x=-2$; vertex $(-1,0)=(1,\pi)$

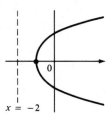

13. $r = \dfrac{3}{1 + (1/2)\sin\theta}$ ⇒ $e=1/2$ ⇒ ellipse; $ed=3$ ⇒
    $d=6$ ⇒ directrix $y=6$; vertices $(2,\pi/2)$ and
    $(6,3\pi/2)$; center $(2,3\pi/2)$

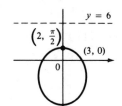

15. $r = \dfrac{1/4}{1 - (3/4)\cos\theta}$ ⇒ $e=3/4$ ⇒ ellipse; $ed=1/4$
    ⇒ $d=1/3$ ⇒ directrix $x=-1/3$; vertices $(1,0)$
    and $(1/7,\pi)$; center $(3/7,0)$; other focus
    $(6/7,0)$

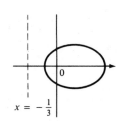

17.  $r = \dfrac{7/2}{1 - (5/2)\sin\theta} \Rightarrow e=5/2 \Rightarrow$ hyperbola;

ed=7/2 $\Rightarrow$ d=7/5 $\Rightarrow$ directrix y=-7/5; center

$(5/3,3\pi/2)$; vertices $(-7/3,\pi/2)=(7/3,3\pi/2)$

and $(1,3\pi/2)$

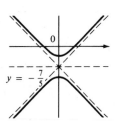

19.  $r = \dfrac{5/2}{1 + \sin\theta} \Rightarrow e=1 \Rightarrow$ parabola; ed=5/2 $\Rightarrow$

d=5/2 $\Rightarrow$ directrix y=5/2; vertex $(5/4,\pi/2)$

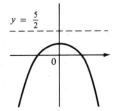

21.  $|PF| = e|P\ell| \Rightarrow r = e[d - r\cos(\pi-\theta)]$

$= e(d + r\cos\theta) \Rightarrow r(1 - e\cos\theta) = ed$

$\Rightarrow r = \dfrac{ed}{1 - e\cos\theta}$

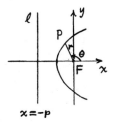

$x=-p$

23.  $|PF| = e|P\ell| \Rightarrow r = e[d - r\sin(\theta-\pi)]$

$= e(d + r\sin\theta) \Rightarrow r(1 - e\sin\theta) = ed$

$\Rightarrow r = \dfrac{ed}{1 - e\sin\theta}$

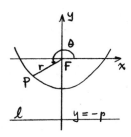

25.  The parabolas intersect at the two points where $\dfrac{c}{1+\cos\theta} = \dfrac{d}{1-\cos\theta} \Rightarrow$

$\cos\theta = \dfrac{c-d}{c+d} \Rightarrow r = (c+d)/2$. For the first parabola, $\dfrac{dr}{d\theta}$

$= c\sin\theta/(1+\cos\theta)^2$ so $\dfrac{dy}{dx} = \dfrac{(dr/d\theta)\sin\theta + r\cos\theta}{(dr/d\theta)\cos\theta - r\sin\theta}$

$= \dfrac{c\sin^2\theta + c\cos\theta(1+\cos\theta)}{c\sin\theta\cos\theta - c\sin\theta(1+\cos\theta)} = \dfrac{1+\cos\theta}{-\sin\theta}$, and similarly for the

second, $\dfrac{dy}{dx} = \dfrac{1-\cos\theta}{\sin\theta} = \dfrac{\sin\theta}{1+\cos\theta}$. Since the product of these slopes is

-1, the parabolas intersect at right angles.

Exercises 9.8

1. $X = 1 \cdot \cos 30° + 4 \sin 30° = 2 + \sqrt{3}/2$, $Y = -1 \cdot \sin 30° + 4 \cos 30° = 2\sqrt{3} - 1/2$

3. $X = -2 \cos 60° + 4 \sin 60° = -1 + 2\sqrt{3}$, $Y = 2 \sin 60° + 4 \cos 60° = \sqrt{3} + 2$

5. $\cot 2\theta = \dfrac{A-C}{B} = 0 \Rightarrow 2\theta = \pi/2 \Leftrightarrow \theta = \pi/4 \Rightarrow$

   [Equations 9.34] $x = (X-Y)/\sqrt{2}$ and
   $y = (X+Y)/\sqrt{2}$. Substituting these into the
   curve equation gives $0 = (x-y)^2 - (x+y) =$
   $2Y^2 - \sqrt{2}X$ or $Y^2 = X/\sqrt{2}$. (Parabola, vertex
   $(0,0)$, directrix $X = -1/4\sqrt{2}$, focus $(1/4\sqrt{2}, 0)$)

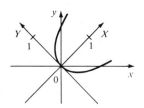

7. $\cot 2\theta = \dfrac{A-C}{B} = 0 \Rightarrow 2\theta = \pi/2 \Leftrightarrow \theta = \pi/4 \Rightarrow$

   [Equations 9.34] $x = (X-Y)/\sqrt{2}$ and
   $y = (X+Y)/\sqrt{2}$. Substituting these into the
   curve equation gives
   $1 = \dfrac{X^2 - 2XY + Y^2}{2} + \dfrac{X^2 - Y^2}{2} + \dfrac{X^2 + 2XY + Y^2}{2} \Rightarrow$
   $3X^2 + Y^2 = 2 \Rightarrow X^2/(2/3) + Y^2/2 = 1$. (An
   ellipse, center $(0,0)$, foci on $Y$-axis with
   $a = \sqrt{2}$, $b = \sqrt{6}/3$, $c = 2\sqrt{3}/3$)

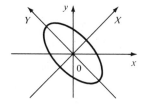

9. $\cot 2\theta = \dfrac{97 - 153}{192} = \dfrac{-7}{24} \Rightarrow \tan 2\theta = -24/7 \Rightarrow$

   $\dfrac{\pi}{2} < 2\theta < \pi$ and $\cos 2\theta = -7/25 \Rightarrow \dfrac{\pi}{4} < \theta < \dfrac{\pi}{2}$, $\cos\theta = 3/5$,

   $\sin\theta = 4/5 \Rightarrow x = X\cos\theta - Y\sin\theta = \dfrac{3X - 4Y}{5}$ and

   $y = X\sin\theta + Y\cos\theta = \dfrac{4X + 3Y}{5}$. Substituting,

   we get $\dfrac{97}{25}(3X-4Y)^2 + \dfrac{192}{25}(3X-4Y)(4X+3Y) + \dfrac{153}{25}(4X+3Y)^2 = 225$

   which simplifies to $X^2 + Y^2/9 = 1$ (an ellipse with foci on $Y$-axis,
   centered at origin, $a = 3$ and $b = 1$).

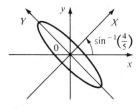

11. $\cot 2\theta = \dfrac{A-C}{B} = 1/\sqrt{3} \Rightarrow \theta = \pi/6 \Rightarrow x = \dfrac{\sqrt{3}X - Y}{2}$,

    $y = \dfrac{X + \sqrt{3}Y}{2}$. Substituting into curve

    equation and simplifying gives
    $4X^2 - 12Y^2 - 8X = 0 \Rightarrow (X-1)^2 - 3Y^2 = 1$ (a
    hyperbola with foci on $X$-axis, centered at
    $(1,0)$, $a = 1$, $b = 1/\sqrt{3}$, $c = 2/\sqrt{3}$)

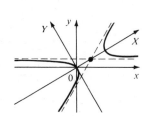

13. (a) $\cot 2\theta = \frac{A-C}{B} = \frac{-7}{24}$ so as in Exercise 9 $x = \frac{3X-4Y}{5}$ and $y = \frac{4X+3Y}{5}$.

Substituting and simplifying we get $100X^2 - 25Y + 25 = 0 \Rightarrow$
$4X^2 = Y-1$, which is a parabola.

(b) The vertex is $(0,1)$ and $p=1/16$, so the XY-coordinates of the
focus are $(0,17/16)$, and the xy-coordinates are
$x = 0\cdot 3/5 - (17/16)(4/5) = -17/20$ and $y = 0\cdot 4/5 + (17/16)(3/5)$
$=51/80$. (c) The directrix is $Y=15/16$, so $-x\cdot(4/5) + y\cdot(3/5) = 15/16$
$\Rightarrow 64x-48y+75 = 0$.

15. A rotation through $\theta$ changes (9.33) to $A(X\cos\theta-Y\sin\theta)^2 +$
$B(X\cos\theta-Y\sin\theta)(X\sin\theta+Y\cos\theta) + C(X\sin\theta+Y\cos\theta)^2 + D(X\cos\theta-Y\sin\theta) +$
$E(X\sin\theta+Y\cos\theta) + F = 0$. Comparing this to (9.36), we see that
$A' + C' = A(\cos^2\theta+\sin^2\theta) + C(\sin^2\theta+\cos^2\theta) = A + C$.

17. Choose $\theta$ so that $B'=0$. Then $B^2-4AC = (B')^2-4A'C' = -4A'C'$. But
$A'C'$ will be 0 for a parabola, negative for a hyperbola (where the
$X^2$ and $Y^2$ coefficients are of opposite sign), and positive for an
ellipse (same sign for $X^2$ and $Y^2$ coefficients). So $B^2-4AC$: $=0$ for
a parabola, $>0$ for a hyperbola, and $<0$ for an ellipse. Note that
the transformed equation takes the form $A'X^2+C'Y^2+D'X+E'Y+F = 0$, or
by completing the square (assuming $A'C'\neq0$), $A'(X')^2+C'(Y')^2 = F'$,
so that if $F'=0$, the graph is either a pair of intersecting lines
or a point depending on the signs of $A'$ and $C'$. If $F'\neq0$ and
$A'C'>0$, then the graph is either an ellipse, a point, or nothing,
and if $A'C'<0$, the graph is a hyperbola. If $A'$ or $C'$ is 0, we
cannot complete the square, so we get $A'(X')^2+E'Y+F = 0$ or
$C'(Y')^2+D'X+F' = 0$. This is a parabola, a straight line (if only
the second degree coefficient is nonzero), a pair of parallel lines
(if the first degree coefficient is zero and the other two have
opposite signs), or an empty graph (if the first degree coefficient
is zero and the other two have the same sign).

Review Exercises for Chapter 9

1. $x=1-t^2$, $y=1-t$ $(-1\leq t\leq 1)$
$x=1-(1-y)^2=2y-y^2$ $(0\leq y\leq 2)$

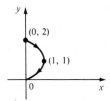

3. $x=1+\sin t$, $y=2+\cos t$ $\Rightarrow$
$(x-1)^2+(y-2)^2=\sin^2 t+\cos^2 t=1$

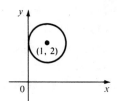

5. $r = 1+3\cos\theta$

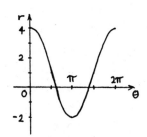

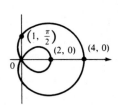

7. $r^2 = \sec 2\theta$ $\Rightarrow$ $r^2\cos 2\theta = 1$ $\Rightarrow$
$r^2(\cos^2\theta-\sin^2\theta) = 1$
$\Rightarrow x^2-y^2=1$, a hyperbola

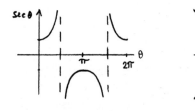

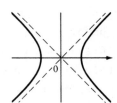

9. $r = 2\cos^2(\theta/2) = 1 + \cos\theta$

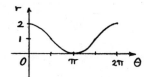

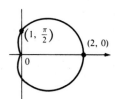

11.  $r = \dfrac{1}{1+\cos\theta}$  e=1 ⇒ parabola; d=1 ⇒ directrix

x=1 and vertex (1/2,0); y-intercepts are

(1,π/2) and (1,3π/2)

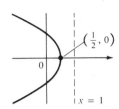

13.  $x^2+y^2=4x$ ⟺ $r^2 = 4r\cos\theta$ ⟺ $r = 4\cos\theta$

15.  $x=t^2+2t$, $y=t^3-t$.  $\dfrac{dy}{dx} = \dfrac{dy/dt}{dx/dt} = (3t^2-1)/(2t+2) = 1/2$ when t=1

17.  $\dfrac{dy}{dx} = \dfrac{(dr/d\theta)\sin\theta + r\cos\theta}{(dr/d\theta)\cos\theta - r\sin\theta} = \dfrac{\sin\theta + \theta\cos\theta}{\cos\theta - \theta\sin\theta} = \dfrac{1/\sqrt{2}+(\pi/4)(1/\sqrt{2})}{1/\sqrt{2}-(\pi/4)(1/\sqrt{2})} = \dfrac{4+\pi}{4-\pi}$

19.  $\dfrac{dy}{dx} = \dfrac{dy/dt}{dx/dt} = \dfrac{t\cos t + \sin t}{-t\sin t + \cos t}$   $\dfrac{d^2y}{dx^2} = \dfrac{\frac{d}{dt}\left[\frac{dy}{dx}\right]}{dx/dt}$

$\dfrac{d}{dt}\left[\dfrac{dy}{dx}\right] = [(-t\sin t + \cos t)(-t\sin t + 2\cos t) -$
$\dfrac{(t\cos t + \sin t)(-t\cos t - 2\sin t)}{(-t\sin t + \cos t)^2}]$

$= (t^2+2)/(-t\sin t + \cos t)^2$ ⇒
$d^2y/dx^2 = (t^2 + 2)/(-t\sin t + \cos t)^3$

21.  dx/dt = -2a sin t + 2a sin 2t = 2a sin t(2 cos t - 1) = 0 ⟺

sin t = 0 or cos t = 1/2 ⇒ t = 0, π/3, π, or 5π/3

dy/dt = 2a cos t - 2a cos 2t = 2a(1 + cos t - 2 cos²t)

$= 2a(1 - \cos t)(1 + 2\cos t) = 0$ ⇒ t = 0, 2π/3, or 4π/3

Thus the graph has vertical tangents where t = π/3, π, and 5π/3,
and horizontal tangents where t = 2π/3 and 4π/3.  To determine what
the slope is where t=0, we use l'Hospital's rule to evaluate

$\lim_{t\to0} \dfrac{dy/dt}{dx/dt} = 0$, so there is a horizontal tangent there.

| t | x | y |
|---|---|---|
| 0 | a | 0 |
| π/3 | 3a/2 | √3a/2 |
| 2π/3 | -a/2 | 3√3a/2 |
| π | -3a | 0 |
| 4π/3 | -a/2 | -3√3a/2 |
| 5π/3 | 3a/2 | -√3a/2 |

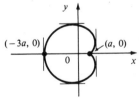

23.  This curve will have 5 "petals", each petal corresponding to a
segment of values of θ for which cos5θ is positive (since

$\cos5\theta = r^2/9 \geq 0$).  $A = 5\displaystyle\int_{-\pi/10}^{\pi/10} \frac{1}{2}r^2 d\theta = \frac{5}{2}\displaystyle\int_{-\pi/10}^{\pi/10} 9\cos5\theta\ d\theta$

$= 45\displaystyle\int_{0}^{\pi/10} \cos5\theta\ d\theta = 9\sin5\theta\Big]_{0}^{\pi/10} = 9$

25. The curves intersect where
4cosθ=2; that is, at
(2,π/3) and (2,-π/3).

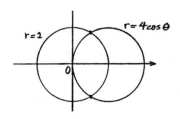

27. Curves intersect where $2\sin\theta = \sin\theta+\cos\theta \Rightarrow$
$\sin\theta = \cos\theta \Rightarrow \theta=\pi/4$, and also at the origin
(at which $\theta=3\pi/4$ on the second curve).

$$A = \int_0^{\pi/4} \frac{1}{2}(2\sin\theta)^2 d\theta + \int_{\pi/4}^{3\pi/4} \frac{1}{2}(\sin\theta+\cos\theta)^2 d\theta$$

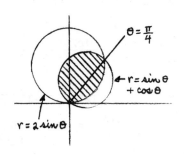

$$= \int_0^{\pi/4}(1-\cos2\theta)d\theta + \frac{1}{2}\int_{\pi/4}^{3\pi/4}(1+\sin2\theta)d\theta$$

$$= \theta-\frac{1}{2}\sin2\theta\Big]_0^{\pi/4} + \frac{\theta}{2} - \frac{\cos2\theta}{4}\Big]_{\pi/4}^{3\pi/4} = \pi/2 - 1/2$$

29. $x=3t^2$, $y=2t^3$. $L = \int_0^2 \sqrt{(dx/dt)^2+(dy/dt)^2}dt = \int_0^2 \sqrt{(6t)^2+(6t^2)^2}dt$

$$= 6\int_0^2 t\sqrt{1+t^2}dt = 2(1+t^2)^{3/2}\Big]_0^2 = 2(5\sqrt{5} - 1)$$

31. $L = \int_\pi^{2\pi} \sqrt{r^2+(dr/d\theta)^2}d\theta = \int_\pi^{2\pi} \sqrt{(1/\theta)^2+(-1/\theta^2)^2}d\theta = \int_\pi^{2\pi} \frac{\sqrt{\theta^2+1}}{\theta^2}d\theta$

$$= -\frac{\sqrt{\theta^2+1}}{\theta} + \ln|\theta+\sqrt{\theta^2+1}|\Big]_\pi^{2\pi} \quad \text{[Formula 24]}$$

$$= \frac{\sqrt{\pi^2+1}}{\pi} - \frac{\sqrt{4\pi^2+1}}{2\pi} + \ln\left[\frac{2\pi+\sqrt{4\pi^2+1}}{\pi+\sqrt{\pi^2+1}}\right]$$

33. $S = \int_1^4 2\pi y\sqrt{(dx/dt)^2+(dy/dt)^2}dt$

$$= \int_1^4 2\pi [t^3/3+1/2t^2] \sqrt{(2/\sqrt{t})^2+(t^2-t^{-3})^2}dt$$

$$= 2\pi\int_1^4 [t^3/3+1/2t^2] \sqrt{(t^2+t^{-3})^2}dt = 2\pi\int_1^4 (t^5/3 + 5/6 + t^{-5}/2)dt$$

$$= 2\pi\left[t^6/18 + 5t/6 - t^{-4}/8\right]_1^4 = 471295\pi/1024$$

**35.** Ellipse, center $(0,0)$, $a=3$, $b=2\sqrt{2}$, $c=1$ $\Rightarrow$
foci $(\pm1,0)$, vertices $(\pm3,0)$

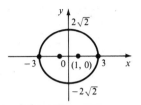

**37.** $6(y^2-6y+9) = -(x+1) \Leftrightarrow (y-3)^2 = -\frac{1}{6}(x+1)$

A parabola with vertex $(-1,3)$, opening to
the left, $p = -1/24$ $\Rightarrow$ focus $(-25/24,3)$ and
directrix $x = -23/24$

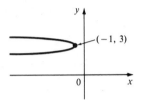

**39.** The parabola opens upward with vertex $(0,4)$ and $p = 2$, so its
equation is $(x-0)^2 = 4\cdot2(y-4) \Leftrightarrow x^2 = 8(y-4)$.

**41.** The hyperbola has center $(0,0)$ and foci on the x-axis. $c=3$ and
$b/a = 1/2$ [from the asymptotes] $\Rightarrow 9 = c^2 = a^2+b^2 = (2b)^2+b^2 = 5b^2$ $\Rightarrow$
$b = 3/\sqrt{5}$ $\Rightarrow$ $a = 6/\sqrt{5}$ $\Rightarrow$ equation $\dfrac{x^2}{36/5} - \dfrac{y^2}{9/5} = 1 \Leftrightarrow 5x^2 - 20y^2 = 36$

**43.** $\cot 2\theta = \dfrac{A-C}{B} = 0 \Rightarrow \theta=\pi/4 \Rightarrow x = X\cos\theta - Y\sin\theta = (X-Y)/\sqrt{2}$ and

$y = X\sin\theta + Y\cos\theta = (X+Y)/\sqrt{2}$. Substitute into equation and
simplify to get $3X^2-Y^2 = 1$. This is a hyperbola with center $(0,0)$,
asymptotes $Y = \pm\sqrt{3}X$, $a=1/\sqrt{3}$, $b=1$, $c=2/\sqrt{3}$ $\Rightarrow$ vertices $(\pm1/\sqrt{3},0)$ and
foci $(\pm2/\sqrt{3},0)$.

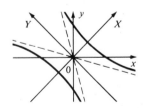

**45.** Directrix $x=4$ $\Rightarrow$ $d=4$, so $e=1/3$ $\Rightarrow$ $r = \dfrac{ed}{1 + e\cos\theta} = \dfrac{4}{3+\cos\theta}$

47. The slopes of the line segments $F_1P$ and $F_2P$ are $y_1/(x_1+c)$ and $y_1/(x_1-c)$, where P is $(x_1,y_1)$. Differentiating implicitly, $2x/a^2 - 2yy'/b^2 = 0 \Rightarrow y' = b^2x/a^2y \Rightarrow$ slope of tangent at P is $b^2x_1/a^2y_1$ so by Formula 1.16

$$\tan \alpha = \frac{(b^2x_1/a^2y_1) - y_1/(x_1+c)}{1 + (b^2x_1y_1)/a^2y_1(x_1+c)} = \frac{b^2x_1(x_1+c) - a^2y_1^2}{a^2y_1(x_1+c) + b^2x_1y_1}$$

$$= \frac{b^2(cx_1+a^2)}{cy_1(cx_1 + a^2)} \quad [\text{using } x_1^2/a^2 - y_1^2/b^2 = 1 \text{ and } a^2+b^2=c^2] \quad = b^2/cy_1$$

$$\tan \beta = \frac{-(b^2x_1/a^2y_1) + y_1/(x_1-c)}{1 + (b^2x_1y_1)/a^2y_1(x_1-c)} = \frac{-b^2x_1(x_1-c) + a^2y_1^2}{a^2y_1(x_1-c) + b^2x_1y_1}$$

$$= \frac{b^2(cx_1-a^2)}{cy_1(cx_1-a^2)} = b^2/cy_1 \text{ and so } \alpha = \beta.$$

CHAPTER TEN

Exercises 10.1

1. $a_n = \frac{n}{2n+1}$ $\left\{\frac{1}{3}, \frac{2}{5}, \frac{3}{7}, \frac{4}{9}, \frac{5}{11}, \cdots\right\}$

3. $a_n = \frac{(-1)^{n-1}n}{2^n}$ $\left\{\frac{1}{2}, -\frac{1}{2}, \frac{3}{8}, -\frac{1}{4}, \frac{5}{32}, \cdots\right\}$

5. $a_n = \frac{1 \cdot 3 \cdot 5 \cdots (2n-1)}{n!}$ $\left\{1, \frac{3}{2}, \frac{5}{2}, \frac{35}{8}, \frac{63}{8}, \cdots\right\}$

7. $a_n = \sin\frac{n\pi}{2}$ $\{1, 0, -1, 0, 1, \ldots\}$

9. $a_1 = 1$, $a_{n+1} = \frac{1}{1+a_n}$ $\left\{1, \frac{1}{2}, \frac{2}{3}, \frac{3}{5}, \frac{5}{8}, \cdots\right\}$

11. $a_n = \frac{1}{2^n}$ 13. $a_n = 3n-2$

15. $a_n = (-1)^n n!$ 17. $a_n = (-1)^{n+1}\left[\frac{n+1}{2n+1}\right]$

19. $\lim\limits_{n\to\infty} \frac{1}{4n^2} = \frac{1}{4} \lim\limits_{n\to\infty} \frac{1}{n^2} = \frac{1}{4} \cdot 0 = 0$. Convergent.

21. $\lim\limits_{n\to\infty} \frac{n^2-1}{n^2+1} = \lim\limits_{n\to\infty} \frac{1 - 1/n^2}{1 + 1/n^2} = 1$. Convergent.

23. $\{a_n\}$ diverges since $\frac{n^2}{n+1} = \frac{n}{1 + 1/n} \to \infty$ as $n\to\infty$.

25. $0 < |a_n| = \frac{n^2}{1+n^3} = \frac{1}{1/n^2 + n} < \frac{1}{n}$ and $\lim\limits_{n\to\infty} \frac{1}{n} = 0$, so by the Squeeze

Theorem, $\lim\limits_{n\to\infty} \frac{n^2}{1+n^3} = 0$, and, by Theorem 10.6, $\lim\limits_{n\to\infty} (-1)^n\left[\frac{n^2}{1+n^3}\right] = 0$.

27. $\lim\limits_{n\to\infty} \frac{1}{5^n} = \lim\limits_{n\to\infty}\left[\frac{1}{5}\right]^n = 0$ by (10.8) with $r = \frac{1}{5}$.

29. $\{a_n\} = \{0, -1, 0, 1, 0, -1, 0, 1, \ldots\}$ This sequence oscillates

among 0, -1, and 1 and so diverges.

31. $a_n = (\pi/3)^n$ so $\{a_n\}$ diverges by 10.8 with $r = \pi/3 > 1$.

33. $\lim\limits_{n\to\infty} \arctan 2n = \pi/2$ since $2n\to\infty$ as $n\to\infty$. Convergent.

35. $0 < \frac{3+(-1)^n}{n^2} \le \frac{4}{n^2}$ and $\lim\limits_{n\to\infty} \frac{4}{n^2} = 0$, so $\left\{\frac{3+(-1)^n}{n^2}\right\}$ converges to 0 by the

Squeeze Theorem.

37. $\lim\limits_{x\to\infty} \dfrac{\ln(x^2)}{x} = \lim\limits_{x\to\infty} \dfrac{2\ln x}{x} \overset{H}{=} \lim\limits_{x\to\infty} \dfrac{2/x}{1} = 0$, so by Theorem 10.2, $\left\{\dfrac{\ln(n^2)}{n}\right\}$ converges to 0.

39. $\sqrt{n+2} - \sqrt{n} = (\sqrt{n+2} - \sqrt{n})\dfrac{\sqrt{n+2} + \sqrt{n}}{\sqrt{n+2} + \sqrt{n}} = \dfrac{2}{\sqrt{n+2} + \sqrt{n}} < \dfrac{2}{2\sqrt{n}} = \dfrac{1}{\sqrt{n}} \to 0$ as

$n\to\infty$. So by the Squeeze Theorem $\{\sqrt{n+2} - \sqrt{n}\}$ converges to 0.

41. $\lim\limits_{x\to\infty} \dfrac{x}{2^x} \overset{H}{=} \lim\limits_{x\to\infty} \dfrac{1}{(\ln 2)2^x} = 0$, so by Theorem 10.2 $\{a_n\}$ converges to 0.

43. Let $y = x^{-1/x}$. Then $\ln y = -(\ln x)/x$ and $\lim\limits_{x\to\infty} (\ln y) \overset{H}{=} \lim\limits_{x\to\infty} -(1/x)/1$

$= 0$, so $\lim\limits_{x\to\infty} y = e^0 = 1$, and so $\{a_n\}$ converges to 1.

45. $0 \le \dfrac{\cos^2 n}{2^n} \le \dfrac{1}{2^n}$ [since $0 \le \cos^2 n \le 1$], so since $\lim\limits_{n\to\infty} \dfrac{1}{2^n} = 0$, $\{a_n\}$

converges to 0 by the Squeeze Theorem.

47. The series converges, since $a_n = \dfrac{1+2+3+\cdots+n}{n^2} = \dfrac{n(n+1)/2}{n^2}$ [Theorem

5.3] $= \dfrac{n+1}{2n} = \dfrac{1 + 1/n}{2} \to \dfrac{1}{2}$ as $n\to\infty$.

49. $a_n = \dfrac{1}{2}\cdot\dfrac{2}{2}\cdot\dfrac{3}{2}\cdots\dfrac{(n-1)}{2}\cdot\dfrac{n}{2} \ge \dfrac{1}{2}\cdot\dfrac{n}{2} = \dfrac{n}{4} \to \infty$ as $n\to\infty$, so $\{a_n\}$ diverges.

51. $0 < a_n = \dfrac{n^3}{n!} = \dfrac{n}{n}\cdot\dfrac{n}{(n-1)}\cdot\dfrac{n}{(n-2)}\cdot\dfrac{1}{(n-3)}\cdots\dfrac{1}{3}\cdot\dfrac{1}{2}\cdot\dfrac{1}{1} \le \dfrac{n^2}{(n-1)(n-2)(n-3)}$ [for

$n\ge 4$] $= \dfrac{1/n}{(1-1/n)(1-2/n)(1-3/n)} \to 0$ as $n\to\infty$, so by the Squeeze Theorem,

$\{a_n\}$ converges to 0.

53. $0 < a_n = \dfrac{1\cdot 3\cdot 5\cdots(2n-1)}{(2n)^n} = \dfrac{1}{2n}\cdot\dfrac{3}{2n}\cdot\dfrac{5}{2n}\cdots\dfrac{2n-1}{2n} \le \dfrac{1}{2n}\cdot(1)\cdot(1)\cdots(1)$

$= \dfrac{1}{2n} \to 0$ as $n\to\infty$, so by the Squeeze Theorem $\{a_n\}$ converges to 0.

55. If $|r|\ge 1$, then $\{r^n\}$ diverges (10.8), so $\{nr^n\}$ diverges also since

$|nr^n| = n|r^n| \ge |r^n|$. If $|r|<1$ then $\lim\limits_{x\to\infty} xr^x = \lim\limits_{x\to\infty} \dfrac{x}{r^{-x}}$

$\overset{H}{=} \lim\limits_{x\to\infty} \dfrac{1}{(-\ln r)r^{-x}} = \lim\limits_{x\to\infty} \dfrac{r^x}{-\ln r} = 0$, so $\lim\limits_{n\to\infty} nr^n = 0$, and hence $\{nr^n\}$

converges whenever $|r|<1$.

57. $3(n+1)+5 > 3n+5$ so $\dfrac{1}{3(n+1)+5} < \dfrac{1}{3n+5} \iff a_{n+1} < a_n$ so $\{a_n\}$ is

decreasing.

59. $\left\{\frac{n-2}{n+2}\right\}$ is increasing since $a_n < a_{n+1} \Leftrightarrow \frac{n-2}{n+2} < \frac{(n+1)-2}{(n+1)+2} \Leftrightarrow$

$(n-2)(n+3) < (n+2)(n-1) \Leftrightarrow n^2+n-6 < n^2+n-2 \Leftrightarrow -6 < -2$, which is of

course true.

61. $a_1=0 > a_2=-1 < a_3=0$, so the sequence is not monotonic.

63. $\left\{\frac{n}{n^2+n-1}\right\}$ is decreasing since $a_{n+1} < a_n \Leftrightarrow \frac{n+1}{(n+1)^2+(n+1)-1} < \frac{n}{n^2+n-1}$

$\Leftrightarrow (n+1)[n^2+n-1] < n(n^2+3n+1) \Leftrightarrow n^3+2n^2-1 < n^3+3n^2+n \Leftrightarrow$

$0 < n^2+n+1 = (n+\frac{1}{2})^2+\frac{3}{4}$, which is obviously true.

65. $a_1 = 2^{1/2}$, $a_2 = 2^{3/4}$, $a_3 = 2^{7/8}, \ldots,$ $a_n = 2^{(2^n-1)/2^n} = 2^{(1-1/2^n)}$

$\lim\limits_{n\to\infty} a_n = \lim\limits_{n\to\infty} 2^{(1-1/2^n)} = 2^1 = 2$

67. (a) We show by induction that $\{a_n\}$ is increasing and bounded above

by 3. Let $P(n)$ be the proposition that $a_{n+1} \geq a_n$ and $a_n \leq 3$. Clearly

$P(1)$ is true. Assume $P(n)$ is true. Then $a_{n+1} \geq a_n \Rightarrow \frac{1}{a_{n+1}} \leq \frac{1}{a_n} \Rightarrow$

$-\frac{1}{a_{n+1}} \leq -\frac{1}{a_n} \Rightarrow a_{n+2} = 3 - \frac{1}{a_{n+1}} \geq 3 - \frac{1}{a_n} = a_{n+1} \Leftrightarrow P(n+1)$. This

proves that $\{a_n\}$ is increasing and bounded above by 3, so

$1 = a_1 \leq a_n \leq 3$, i.e. $\{a_n\}$ is bounded, and hence convergent by

Theorem 10.11.

(b) If $L = \lim\limits_{n\to\infty} a_n$, then $L$ must satisfy $L = 3 - 1/L$, so $L^2-3L+1 = 0$

and the quadratic formula gives $L = \frac{3\pm\sqrt{5}}{2}$. But $L>1$, so $L = \frac{3+\sqrt{5}}{2}$.

69. (a) Let $a_n$ be the number of rabbit pairs in the nth month. Clearly

$a_1=1=a_2$. In the nth month, each pair that is 2 or more months old

(i.e. $a_{n-2}$ pairs) will have a pair of children to add to the $a_{n-1}$

pairs already present. Thus $a_n = a_{n-1}+a_{n-2}$, so that $\{a_n\} = \{f_n\}$,

the Fibonacci sequence.

(b) $a_{n-1} = f_n/f_{n-1} = (f_{n-1}+f_{n-2})/f_{n-1} = 1 + f_{n-2}/f_{n-1} = 1 + 1/a_{n-2}$

If $L = \lim\limits_{n\to\infty} a_n$, then $L$ must satisfy $L = 1 + 1/L$ or $L^2-L-1 = 0$, so

$L = \frac{1+\sqrt{5}}{2}$ (since L must be non-negative).

71. If $\lim\limits_{n\to\infty} |a_n| = 0$ then $\lim\limits_{n\to\infty} -|a_n| = 0$, and since $-|a_n| \leq a_n \leq |a_n|$, we

have that $\lim\limits_{n\to\infty} a_n = 0$ by the Squeeze Theorem.

73. $\frac{1}{n+1} < x < \frac{1}{n} \Rightarrow 1+\frac{1}{n+1} < 1+x < 1+\frac{1}{n}$ and $n+1 > 1/x > n$, so

$\left[1+\frac{1}{n+1}\right]^n < (1+x)^{1/x} < \left[1+\frac{1}{n}\right]^{n+1}$. By problem 72(f), the limits of

the first and last expression exist and are equal to each other

(since $\lim\limits_{n\to\infty}\left[1+\frac{1}{n}\right] = 1$), so since $x \to 0^+$ as $n \to \infty$, $\lim\limits_{x\to0^+}(1+x)^{1/x}$

exists by the Squeeze Theorem.

## Exercises 10.2

1. $\sum\limits_{n=1}^{\infty} 4\left[\frac{2}{5}\right]^{n-1}$ converges to $\frac{4}{1-2/5} = \frac{20}{3}$

3. $\sum\limits_{n=1}^{\infty} \frac{2}{3}\left[-\frac{1}{3}\right]^{n-1}$ converges to $\frac{2/3}{1-(-1/3)} = \frac{1}{2}$

5. $\sum\limits_{n=1}^{\infty} \frac{1}{36}\left[\frac{6}{5}\right]^{n-1}$ diverges since $r = \frac{6}{5} > 1$

7. $a = 2$, $r = 3/4 < 1$, so series converges to $\frac{2}{1-3/4} = 8$

9. $a = 5e/3$, $r = e/3 < 1$, so series converges to $\frac{5e/3}{1-e/3} = \frac{5e}{3-e}$

11. $a = 1$, $r = 5/8 < 1$, so series converges to $\frac{1}{1-5/8} = \frac{8}{3}$

13. $a = 64/3$, $r = 8/3 > 1$, so series diverges

15. $a = \frac{(-2)^4}{5^{-1}} = 80$, $|r| = \left|\frac{-2}{5}\right| < 1$, series converges to $\frac{80}{1-(-2/5)} = \frac{400}{7}$

17. $\lim\limits_{n\to\infty} \frac{n}{n+1} = 1 \neq 0$ so series diverges by the Test for Divergence

19. This series diverges, since if it converged, so would (by Theorem

10.19(a)) $2 \cdot \sum\limits_{n=1}^{\infty} \frac{1}{2n} = \sum\limits_{n=1}^{\infty} \frac{1}{n}$, which we know diverges (Example 7).

21. Converges. $s_n = \sum\limits_{i=1}^{n} \dfrac{1}{(3i-2)(3i+1)} = \sum\limits_{i=1}^{n} \left[\dfrac{1/3}{3i-2} - \dfrac{1/3}{3i+1}\right]$ $\begin{Bmatrix} \text{partial} \\ \text{fractions} \end{Bmatrix}$

$= \left[\dfrac{1}{3}\cdot 1 - \dfrac{1}{3}\cdot\dfrac{1}{4}\right] + \left[\dfrac{1}{3}\cdot\dfrac{1}{4} - \dfrac{1}{3}\cdot\dfrac{1}{7}\right] + \left[\dfrac{1}{3}\cdot\dfrac{1}{7} - \dfrac{1}{3}\cdot\dfrac{1}{10}\right] + \ldots + \left[\dfrac{1}{3}\cdot\dfrac{1}{3n-2} - \dfrac{1}{3}\cdot\dfrac{1}{3n+1}\right]$

$= \dfrac{1}{3} - \dfrac{1}{3(3n+1)}$ [telescoping series]

$\lim\limits_{n\to\infty} s_n = \dfrac{1}{3} \;\Rightarrow\; \sum\limits_{n=1}^{\infty} \dfrac{1}{(3n-2)(3n+1)} = \dfrac{1}{3}$

23. Converges by Theorem 10.19. $\sum\limits_{n=1}^{\infty} (2(0.1)^n + (0.2)^n) =$

$2\sum\limits_{n=1}^{\infty} (0.1)^n + \sum\limits_{n=1}^{\infty} (0.2)^n = 2\left[\dfrac{0.1}{1-0.1}\right] + \dfrac{0.2}{1-0.2} = \dfrac{2}{9} + \dfrac{1}{4} = \dfrac{17}{36}$

25. Diverges by the Test for Divergence. $\lim\limits_{n\to\infty} n/\sqrt{1+n^2} = \lim\limits_{n\to\infty} 1/\sqrt{1 + 1/n^2}$

$= 1 \neq 0.$

27. Converges. $s_n = \sum\limits_{i=1}^{n} \dfrac{1}{i(i+2)} = \sum\limits_{i=1}^{n} \left[\dfrac{1/2}{i} - \dfrac{1/2}{i+2}\right]$ $\begin{Bmatrix} \text{partial} \\ \text{fractions} \end{Bmatrix}$

$= \left[\dfrac{1}{2} - \dfrac{1}{6}\right] + \left[\dfrac{1}{4} - \dfrac{1}{8}\right] + \left[\dfrac{1}{6} - \dfrac{1}{10}\right] + \ldots + \left[\dfrac{1}{2n-2} - \dfrac{1}{2n+2}\right] + \left[\dfrac{1}{2n} - \dfrac{1}{2n+4}\right]$

$= \dfrac{1}{2} + \dfrac{1}{4} - \dfrac{1}{2n+2} - \dfrac{1}{2n+4}$ [telescoping series]

$\sum\limits_{n=1}^{\infty} \dfrac{1}{n(n+2)} = \lim\limits_{n\to\infty} \left[\dfrac{1}{2} + \dfrac{1}{4} - \dfrac{1}{2n+2} - \dfrac{1}{2n+4}\right] = \dfrac{3}{4}$

29. Converges. $\sum\limits_{n=1}^{\infty} \dfrac{3^n+2^n}{6^n} = \sum\limits_{n=1}^{\infty} \left[\dfrac{1}{2}\right]^n + \left[\dfrac{1}{3}\right]^n = \dfrac{1/2}{1 - 1/2} + \dfrac{1/3}{1 - 1/3} = \dfrac{3}{2}$

31. Converges. $s_n = (\sin 1 - \sin 1/2) + (\sin 1/2 - \sin 1/3) + \ldots$

$+ (\sin(1/n) - \sin(1/(n+1))) = \sin 1 - \sin(1/(n+1))$

$\sum\limits_{n=1}^{\infty} (\sin(1/n) - \sin(1/(n+1))) = \lim\limits_{n\to\infty} s_n = \sin 1 - \sin 0 = \sin 1$

33. Diverges since $\lim\limits_{n\to\infty} \arctan n = \dfrac{\pi}{2} \neq 0.$

35. $0.\overline{5} = .5 + .05 + .005 + \cdots = \dfrac{.5}{1-.1} = \dfrac{5}{9}$

37. $0.\overline{307} = .307 + .000307 + .000000307 + \cdots = \dfrac{.307}{1-.001} = \dfrac{307}{999}$

39. $0.123\overline{456} = \dfrac{123}{1000} + \dfrac{.000456}{1-.001} = \dfrac{123}{1000} + \dfrac{456}{999000} = \dfrac{123333}{999000} = \dfrac{41111}{333000}$

41. $\displaystyle\sum_{n=0}^{\infty} (x-3)^n$ is a geometric series with $r = x-3$, so converges whenever

   $|x-3| < 1 \Rightarrow -1 < x-3 < 1 \Longleftrightarrow 2 < x < 4$. The sum is $\dfrac{1}{1-(x-3)} = \dfrac{1}{4-x}$.

43. $\displaystyle\sum_{n=2}^{\infty} \left[\dfrac{x}{5}\right]^n$ is a geometric series with $r = \dfrac{x}{5}$, so converges whenever

   $\left|\dfrac{x}{5}\right| < 1 \Longleftrightarrow -5 < x < 5$. The sum is $\dfrac{(x/5)^2}{1 - x/5} = \dfrac{x^2}{25 - 5x}$.

45. $\displaystyle\sum_{n=0}^{\infty} (2 \sin x)^n$ is geometric so converges whenever $|2 \sin x| < 1 \Longleftrightarrow$

   $-\dfrac{1}{2} < \sin x < \dfrac{1}{2} \Longleftrightarrow n\pi-\dfrac{\pi}{6} < x < n\pi+\dfrac{\pi}{6}$, where the sum is $\dfrac{1}{1 - 2 \sin x}$.

47. Total distance $= 4 + 2 + 2 + 1 + 1 + \dfrac{1}{2} + \dfrac{1}{2} + \dfrac{1}{4} + \dfrac{1}{4} + \ldots$

   $= 4 + 4 + 2 + 1 + \dfrac{1}{2} + \dfrac{1}{4} + \ldots = 10 + \displaystyle\sum_{n=0}^{\infty} \left[\dfrac{1}{2}\right]^n = 10 + \dfrac{1}{1 - 1/2} = 12\text{m}$

49. The series $1 - 1 + 1 - 1 + 1 - 1 + \ldots$ diverges (geometric series with $r = -1$) so we cannot say $0 = 1 - 1 + 1 - 1 + 1 - 1 + \cdots$.

51. $\displaystyle\sum_{n=1}^{\infty} ca_n = \lim_{n\to\infty} \sum_{i=1}^{n} ca_i = \lim_{n\to\infty} c \sum_{i=1}^{n} a_i = c \lim_{n\to\infty} \sum_{i=1}^{n} a_i = c \sum_{n=1}^{\infty} a_n$ which

   exists by hypothesis.

53. Suppose on the contrary that $\displaystyle\sum (a_n+b_n)$ converges. Then by Theorem 10.19(c), so would $\displaystyle\sum [(a_n+b_n) -a_n] = \sum b_n$, a contradiction.

300

**Exercises 10.3**

1.  $\displaystyle\sum_{n=1}^{\infty} 2/\sqrt[3]{n} = 2 \sum_{n=1}^{\infty} \frac{1}{n^{1/3}}$  p-series, $p = \frac{1}{3} < 1$, diverges

3.  $\displaystyle\sum_{n=5}^{\infty} \frac{1}{n^{1.0001}}$  p-series, $p = 1.0001 > 1$, converges

5.  $\displaystyle\sum_{n=5}^{\infty} \frac{1}{(n-4)^2} = \sum_{n=1}^{\infty} \frac{1}{n^2}$  p-series, $p = 2 > 1$, converges

7.  Since $\dfrac{1}{\sqrt{x} + 1}$ is continuous, positive, and decreasing on $[0,\infty)$ we

    can apply the Integral Test. $\displaystyle\int_{1}^{\infty} \frac{1}{\sqrt{x} + 1}\, dx = \lim_{t\to\infty} [2\sqrt{x} - 2\ln(\sqrt{x}+1)]_{1}^{t}$

    [using the substitution $u = \sqrt{x}+1$, so $x = (u-1)^2$ and $dx = 2(u-1)du$]

    $= \displaystyle\lim_{t\to\infty} [(2\sqrt{t} - 2\ln(\sqrt{t}+1)) - (2 - 2\ln 2)]$.   Now $2\sqrt{t} - 2\ln(\sqrt{t}+1)$

    $= 2 \ln\left[\dfrac{e^{\sqrt{t}}}{\sqrt{t}+1}\right]$ and so $\displaystyle\lim_{t\to\infty} (2\sqrt{t} - 2\ln(\sqrt{t}+1)) = \infty$ (using L'Hospital's

    Rule) so both the integral and the original series diverge.

9.  $f(x) = xe^{-x^2}$ is continuous and positive on $[1,\infty)$, and since

    $f'(x) = e^{-x^2}(1-2x^2) < 0$ for $x>1$, $f$ is decreasing as well.   We

    can use the Integral Test. $\displaystyle\int_{1}^{\infty} xe^{-x^2}\, dx = \lim_{t\to\infty} [-\tfrac{1}{2} e^{-x^2}]_{1}^{t} = 0 - \left[-\dfrac{e^{-1}}{2}\right]$

    $= \dfrac{1}{2e}$ so the series converges.

11.  $f(x) = \dfrac{x}{x^2+1}$ is continuous and positive on $[1,\infty)$, and since

    $f'(x) = \dfrac{1-x^2}{(x^2+1)^2} < 0$ for $x>1$, $f$ is also decreasing.   Using the

    Integral Test, $\displaystyle\int_{1}^{\infty} \frac{x}{x^2+1}\, dx = \lim_{t\to\infty} \frac{\ln(x^2+1)}{2}\Big]_{1}^{t} = \infty$, so the series

    diverges.

13.  $f(x) = \dfrac{1}{x \ln x}$ is continuous and positive on $[2,\infty)$, and also

    decreasing since $f'(x) = -\dfrac{1 + \ln x}{x^2(\ln x)^2} < 0$ for $x > 2$, so we can use the Integral Test. $\displaystyle\int_2^\infty \frac{dx}{x \ln x} = \lim_{t\to\infty} [\ln(\ln x)]_2^t$

$= \lim_{t\to\infty} [\ln(\ln t) - \ln(\ln 2)] = \infty$, so the series diverges.

15. $f(x) = \dfrac{\arctan x}{1 + x^2}$ is continuous and positive on $[1,\infty)$.

$f'(x) = \dfrac{1 - 2x \arctan x}{(1 + x^2)^2} < 0$ for $x > 1$, since $2x \arctan x \geq \dfrac{\pi}{2} > 1$ for

$x \geq 1$. So $f$ is decreasing and we can use the Integral Test.

$\displaystyle\int_1^\infty \frac{\arctan x}{1 + x^2} \, dx = \lim_{t\to\infty} [\tfrac{1}{2} \arctan^2 x]_1^t = \frac{(\pi/2)^2}{2} - \frac{(\pi/4)^2}{2} = \frac{3\pi^2}{32}$, so the

series converges.

17. $f(x) = \dfrac{\ln x}{x^2}$ is continuous and positive for $x \geq 2$, and

$f'(x) = \dfrac{1 - 2 \ln x}{x^3} < 0$ for $x \geq 2$ so $f$ is decreasing. $\displaystyle\int_2^\infty \frac{\ln x}{x^2} \, dx$

$= \lim_{t\to\infty} \left[-\dfrac{\ln x}{x} - \dfrac{1}{x}\right]_1^t$ [using integration by parts] $= 1$ [by

L'Hospital's Rule]. Thus $\displaystyle\sum_{n=1}^\infty \frac{\ln n}{n^2} = \sum_{n=2}^\infty \frac{\ln n}{n^2}$ converges by the

Integral Test.

19. $f(x) = \dfrac{\sin(1/x)}{x^2}$ is continuous and positive for $x \geq 1$, and

$f'(x) = -\dfrac{\cos(1/x) + 2x \sin(1/x)}{x^4} < 0$ if $x \geq 1$ (because then

$0 < \dfrac{1}{x} \leq 1 < \dfrac{\pi}{2}$ so that both $\cos(1/x)$ and $\sin(1/x)$ are positive) and

so $f$ is decreasing. Using the Integral Test, $\displaystyle\int_1^\infty \frac{\sin(1/x)}{x^2} \, dx$

$= \lim_{t\to\infty} \cos(1/x) \Big]_1^t = 1 - \cos 1$, and so the series converges.

21. $f(x) = \dfrac{1}{x^2 + 2x + 2}$ is continuous and positive on $[1,\infty)$, and

$f'(x) = -\dfrac{2x+2}{(x^2 + 2x + 2)^2} < 0$ if $x \geq 1$, so $f$ is decreasing and we can

use the Integral Test. $\displaystyle\int_1^\infty \frac{1}{x^2 + 2x + 2} \, dx = \int_1^\infty \frac{1}{(x+1)^2 + 1} \, dx$

$$= \lim_{t \to \infty} \arctan(x+1) \Big]_1^t = \frac{\pi}{2} - \arctan 2, \text{ so the series converges also.}$$

23. We have already shown that when p=1 the series diverges (in Exercise 13 above), so assume p≠1. $f(x) = \dfrac{1}{x(\ln x)^p}$ is continuous and positive on $[2,\infty)$, and $f'(x) = -\dfrac{p + \ln x}{x^2(\ln x)^{p+1}} < 0$ if $x > e^{-p}$, so that f is eventually decreasing and we can use the Integral Test.

$$\int_2^\infty \frac{1}{x(\ln x)^p}\, dx = \lim_{t \to \infty} \frac{(\ln x)^{1-p}}{1-p} \Big]_2^t \quad [\text{for } p \neq 1]$$

$$= \lim_{t \to \infty}\left[\frac{(\ln t)^{1-p}}{1-p}\right] - \frac{(\ln 2)^{1-p}}{1-p}. \quad \text{This limit exists whenever } 1-p < 0$$

⟺ p > 1, so the series converges for p > 1.

25. Clearly the series cannot converge if $p \geq -\frac{1}{2}$, because then $\lim_{n \to \infty} n(1+n^2)^p \neq 0$. Also, if p=-1 the series diverges (see Exercise 11 above). So assume $p < -\frac{1}{2}$, p≠-1. Then $f(x) = x(1+x^2)^p$ is continuous, positive, and eventually decreasing on $[1,\infty)$, and we can use the Integral Test. $\displaystyle\int_1^\infty x(1+x^2)^p dx = \lim_{t \to \infty} \frac{1}{2} \cdot \frac{(1+x^2)^{p+1}}{p+1} \Big]_1^t$

$$= \lim_{t \to \infty} \frac{1}{2} \cdot \frac{(1+x^2)^{p+1}}{p+1} - \frac{2^p}{p+1}. \quad \text{This limit exists and is finite} \Longleftrightarrow$$

p+1 < 0 ⟺ p < -1, so the series converges whenever p < -1.

27. Since this is a p-series with p=x, $\varsigma(x)$ is defined when x > 1.

29. (a) The sum of the areas of the n rectangles in the graph to the right is $1 + \frac{1}{2} + \frac{1}{3} + \dots + \frac{1}{n}$. Now $\displaystyle\int_1^{n+1} \frac{1}{x}\, dx$ is

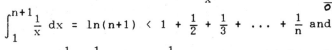

less than this sum because the rectangles extend above the curve $y = \frac{1}{x}$, so

$$\int_1^{n+1} \frac{1}{x}\, dx = \ln(n+1) < 1 + \frac{1}{2} + \frac{1}{3} + \dots + \frac{1}{n} \text{ and}$$

$0 < 1 + \frac{1}{2} + \frac{1}{3} + \dots + \frac{1}{n} - \ln n = t_n$ (since $\ln n < \ln(n+1)$).

(b) The area under $f(x) = \frac{1}{x}$ between x=n and

x=n+1 is $\int_n^{n+1} \frac{1}{x}\,dx = \ln(n+1) - \ln n$, and

this is clearly greater than the area of
the inscribed rectangle in the figure to
the right (which is $\frac{1}{n+1}$), so $t_n - t_{n+1} = (\ln(n+1) - \ln n) - \frac{1}{n+1} > 0$,

and so $t_n > t_{n+1}$, so $\{t_n\}$ is a decreasing sequence.

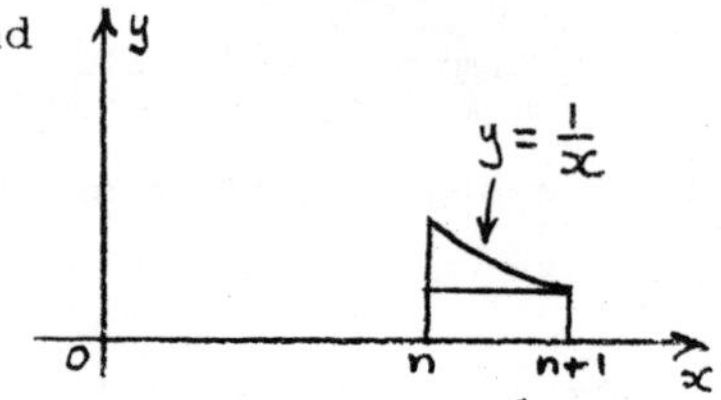

(c) We have shown that $\{t_n\}$ is decreasing and that $t_n > 0$ for all

n. Thus $0 < t_n \leq t_1 = 1$, so $\{t_n\}$ is a bounded monotonic sequence,

and hence converges by Theorem 10.11.

Exercises 10.4

1. $\frac{1}{n^3+n^2} < \frac{1}{n^3}$ since $n^3+n^2 > n^3$ for all n, and since $\sum_{n=1}^{\infty} \frac{1}{n^3}$ is a

convergent p-series (p=3>1), $\sum_{n=1}^{\infty} \frac{1}{n^3+n^2}$ converges also by 10.24(a).

3. $\frac{3}{n2^n} \leq \frac{3}{2^n}$. $\sum_{n=1}^{\infty} \frac{3}{2^n}$ is a geometric series with $|r| = \frac{1}{2} < 1$, and hence

converges, so $\sum_{n=1}^{\infty} \frac{3}{n2^n}$ converges also by the Comparison Test.

5. $\frac{1+5^n}{4^n} > \frac{5^n}{4^n} = \left[\frac{5}{4}\right]^n$. $\sum_{n=0}^{\infty} \left[\frac{5}{4}\right]^n$ is a divergent geometric series

$(|r| = \frac{5}{4} > 1)$ so $\sum_{n=0}^{\infty} \frac{1+5^n}{4^n}$ diverges by the Comparison Test.

7. $\dfrac{3}{n(n+3)} < \dfrac{3}{n^2}$. $\displaystyle\sum_{n=1}^{\infty} \dfrac{3}{n^2} = 3\sum_{n=1}^{\infty} \dfrac{1}{n^2}$ is a convergent p-series (p=2>1) so

$\displaystyle\sum_{n=1}^{\infty} \dfrac{3}{n(n+3)}$ converges by the Comparison Test.

9. $\dfrac{1}{\sqrt{n^3+1}} < \dfrac{1}{\sqrt{n^3}} = \dfrac{1}{n^{3/2}}$. $\displaystyle\sum_{n=1}^{\infty} \dfrac{1}{n^{3/2}}$ is a convergent p-series (p=$\frac{3}{2}$>1) so

$\displaystyle\sum_{n=1}^{\infty} \dfrac{1}{\sqrt{n^3+1}}$ converges by the Comparison Test.

11. $\dfrac{\sqrt{n}}{n-1} > \dfrac{\sqrt{n}}{n} = \dfrac{1}{n^{1/2}}$. $\displaystyle\sum_{n=2}^{\infty} \dfrac{1}{n^{1/2}}$ is a divergent p-series (p=$\frac{1}{2}$<1) so

$\displaystyle\sum_{n=2}^{\infty} \dfrac{\sqrt{n}}{n-1}$ diverges by the Comparison Test.

13. $n^3+1 > n^3 \Rightarrow \dfrac{1}{n^3+1} < \dfrac{1}{n^3} \Rightarrow \dfrac{n}{n^3+1} < \dfrac{n}{n^3} \Rightarrow \dfrac{n-1}{n^3+1} < \dfrac{n}{n^3} = \dfrac{1}{n^2}$. $\displaystyle\sum_{n=1}^{\infty} \dfrac{1}{n^2}$ is a

convergent p-series (p=2>1) so $\displaystyle\sum_{n=1}^{\infty} \dfrac{n-1}{n^3+1}$ converges by Comparison.

15. $\dfrac{3 + \cos n}{3^n} \le \dfrac{4}{3^n}$ since $\cos n \le 1$. $\displaystyle\sum_{n=1}^{\infty} \dfrac{4}{3^n}$ is a geometric series with

$|r| = \dfrac{1}{3} < 1$ so it converges, and so $\displaystyle\sum_{n=1}^{\infty} \dfrac{3 + \cos n}{3^n}$ converges by the

Comparison Test.

17. $(n+1)(2n+1) > n \cdot 2n = 2n^2$ so $\dfrac{4}{(n+1)(2n+1)} < \dfrac{4}{2n^2} = \dfrac{2}{n^2}$. $\displaystyle\sum_{n=1}^{\infty} \dfrac{2}{n^2}$

$= 2\displaystyle\sum_{n=1}^{\infty} \dfrac{1}{n^2}$ is a convergent p-series (p=2>1), so $\displaystyle\sum_{n=1}^{\infty} \dfrac{4}{(n+1)(2n+1)}$

converges by the Comparison Test.

19. $\dfrac{n}{\sqrt{n^5+4}} < \dfrac{n}{\sqrt{n^5}} = \dfrac{1}{n^{3/2}}$. $\displaystyle\sum_{n=1}^{\infty} \dfrac{1}{n^{3/2}}$ is a convergent p-series (p=$\frac{3}{2}$>1) so

$\displaystyle\sum_{n=1}^{\infty} \dfrac{n}{\sqrt{n^5+4}}$ converges by the Comparison Test.

21. $\dfrac{2^n}{1+3^n} < \dfrac{2^n}{3^n} = \left[\dfrac{2}{3}\right]^n$. $\displaystyle\sum_{n=1}^{\infty} \left[\dfrac{2}{3}\right]^n$ is a convergent geometric series

$(|r| = \dfrac{2}{3} < 1)$, so $\displaystyle\sum_{n=1}^{\infty} \dfrac{2^n}{1+3^n}$ converges by the Comparison Test.

23. Let $a_n = \dfrac{1}{1+\sqrt{n}}$ and $b_n = \dfrac{1}{\sqrt{n}}$. Then $\lim\limits_{n\to\infty} \dfrac{a_n}{b_n} = \lim\limits_{n\to\infty} \dfrac{\sqrt{n}}{1+\sqrt{n}} = 1 > 0$. Since

$\displaystyle\sum_{n=1}^{\infty} \dfrac{1}{\sqrt{n}}$ is a divergent p-series $(p=\dfrac{1}{2}<1)$, $\displaystyle\sum_{n=1}^{\infty} \dfrac{1}{1+\sqrt{n}}$ also diverges by the

Limit Comparison Test.

25. Let $a_n = \dfrac{n^2+1}{n^4+1}$ and $b_n = \dfrac{1}{n^2}$. Then $\lim\limits_{n\to\infty} \dfrac{a_n}{b_n} = \lim\limits_{n\to\infty} \dfrac{n^4+n^2}{n^4+1} = 1$. Since

$\displaystyle\sum_{n=1}^{\infty} \dfrac{1}{n^2}$ is a convergent p-series $(p=2>1)$, so is $\displaystyle\sum_{n=1}^{\infty} \dfrac{n^2+1}{n^4+1}$ by the

Limit Comparison Test.

27. Let $a_n = \dfrac{n^2-n+2}{\sqrt[4]{n^{10}+n^5+3}}$ and $b_n = \dfrac{1}{\sqrt{n}}$. Then $\lim\limits_{n\to\infty} \dfrac{a_n}{b_n} = \lim\limits_{n\to\infty} \dfrac{n^{5/2}-n^{3/2}+2n^{1/2}}{\sqrt[4]{n^{10}+n^5+3}}$

$= \lim\limits_{n\to\infty} \dfrac{1-n^{-1}+2n^{-2}}{\sqrt[4]{1+n^{-5}+3n^{-10}}} = 1$. Since $\displaystyle\sum_{n=1}^{\infty} \dfrac{1}{\sqrt{n}}$ is a divergent p-series

$(p=\dfrac{1}{2}<1)$, $\displaystyle\sum_{n=1}^{\infty} \dfrac{n^2-n+2}{\sqrt[4]{n^{10}+n^5+3}}$ diverges by the Limit Comparison Test.

29. Let $a_n = \dfrac{n+1}{n2^n}$ and $b_n = \dfrac{1}{2^n}$. Then $\lim\limits_{n\to\infty} \dfrac{a_n}{b_n} = \lim\limits_{n\to\infty} \dfrac{n+1}{n} = 1$. Since $\displaystyle\sum_{n=1}^{\infty} \dfrac{1}{2^n}$

is a convergent geometric series $(|r| = \dfrac{1}{2} < 1)$, $\displaystyle\sum_{n=1}^{\infty} \dfrac{n+1}{n2^n}$ converges

by the Limit Comparison Test.

31. Let $a_n = \dfrac{\ln n}{n^3}$ and $b_n = \dfrac{1}{n^2}$. Then $\lim\limits_{n\to\infty} \dfrac{a_n}{b_n} = \lim\limits_{n\to\infty} \dfrac{\ln n}{n} = \lim\limits_{n\to\infty} \dfrac{1/n}{1} = 0$.

So since $\displaystyle\sum_{n=1}^{\infty} \dfrac{1}{n^2}$ converges (p-series, p=2), so does $\displaystyle\sum_{n=1}^{\infty} \dfrac{\ln n}{n^3}$ by part

(b) of the Limit Comparison Test.

33. Clearly $n! = n(n-1)(n-2)\cdots(3)(2) \geq 2\cdot2\cdot2\cdots2\cdot2 = 2^{n-1}$, so

$\dfrac{1}{n!} \leq \dfrac{1}{2^{n-1}}$. $\displaystyle\sum_{n=1}^{\infty} \dfrac{1}{2^{n-1}}$ is a convergent geometric series $(|r| = \dfrac{1}{2} < 1)$

so $\displaystyle\sum_{n=1}^{\infty} \frac{1}{n!}$ converges by the Comparison Test.

35. $\dfrac{n!}{n^2} = \dfrac{n(n-1)(n-2)\cdots 1}{n^2} = \dfrac{(n-1)(n-2)\cdots 1}{n} \geq \dfrac{1}{n}$ and $\displaystyle\sum_{n=1}^{\infty} \frac{1}{n}$ diverges

(harmonic series), so $\displaystyle\sum_{n=1}^{\infty} \frac{n!}{n^2}$ diverges by the Comparison Test.

[OR: $\displaystyle\lim_{n\to\infty} \frac{n!}{n^2} = \lim_{n\to\infty} \frac{n(n-1)!}{n\cdot n} = \lim_{n\to\infty} \frac{(n-1)!}{n} = \lim_{n\to\infty} \left[1 - \frac{1}{n}\right](n-2)! = \infty$, so

the series $\displaystyle\sum_{n=1}^{\infty} \frac{n!}{n^2}$ diverges by the Test for Divergence (10.18).]

37. Let $a_n = \sin\left[\dfrac{1}{n}\right]$ and $b_n = \dfrac{1}{n}$. Then $\displaystyle\lim_{n\to\infty} \frac{a_n}{b_n} = \lim_{n\to\infty} \frac{\sin(1/n)}{1/n} = \lim_{\theta\to 0} \frac{\sin\theta}{\theta}$

$= 1$, so since $\displaystyle\sum_{n=1}^{\infty} b_n$ is the harmonic series (which diverges),

$\displaystyle\sum_{n=1}^{\infty} \sin\left[\frac{1}{n}\right]$ diverges as well by the Limit Comparison Test.

39. Since $\dfrac{d_n}{10^n} \leq \dfrac{9}{10^n}$ for each n, and since $\displaystyle\sum_{n=1}^{\infty} \frac{9}{10^n}$ is a convergent

geometric series ($|r| = \dfrac{1}{10} < 1$), $0.d_1 d_2 d_3 \cdots = \displaystyle\sum_{n=1}^{\infty} \frac{d_n}{10^n}$ will always

converge by the Comparison Test.

41. Since $\sum a_n$ converges, $\displaystyle\lim_{n\to\infty} a_n = 0$, so there exists N such that

$|a_n - 0| < 1$ for all $n > N \Rightarrow 0 \leq a_n < 1$ for all $n > N \Rightarrow 0 \leq a_n^2 \leq a_n$.
Since $\sum a_n$ converges, so will $\sum a_n^2$ by the Comparison Test.

43. We wish to prove that if $\displaystyle\lim_{n\to\infty} \frac{a_n}{b_n} = \infty$ and $\sum b_n$ diverges, then so does

$\sum a_n$. So suppose on the contrary that $\sum a_n$ converges. Since

$\displaystyle\lim_{n\to\infty} \frac{a_n}{b_n} = \infty$, we have that $\displaystyle\lim_{n\to\infty} \frac{b_n}{a_n} = 0$, so by part (b) of the Limit

Comparison Test (proved in Exercise 42), if $\sum a_n$ converges, so must

$\sum b_n$. But this contradicts our hypothesis, so $\sum a_n$ must diverge.

Exercises 10.5

1. $\displaystyle\sum_{n=1}^{\infty} (-1)^{n-1} \frac{3}{n+4}$ $\quad a_n = \frac{3}{n+4} > 0$ and $a_{n+1} < a_n$ for all n; $\displaystyle\lim_{n\to\infty} a_n = 0$ so

the series converges by the Alternating Series Test.

3. $\displaystyle\sum_{n=1}^{\infty} (-1)^n \frac{n}{n+1}$ $\quad \displaystyle\lim_{n\to\infty} \frac{n}{n+1} = 1$ so $\displaystyle\lim_{n\to\infty} (-1)^n \frac{n}{n+1}$ does not exist and the

series diverges by the Test for Divergence.

5. $\displaystyle\sum_{n=1}^{\infty} (-1)^{n-1} \frac{1}{n^2}$ $\quad a_n = \frac{1}{n^2} > 0$ and $a_{n+1} < a_n$ for all n, and $\displaystyle\lim_{n\to\infty} 1/n^2 =$

0, so the series converges by the Alternating Series Test.

7. $\displaystyle\sum_{n=1}^{\infty} (-1)^{n+1} \frac{n}{5n+1}$ $\quad \displaystyle\lim_{n\to\infty} \frac{n}{5n+1} = \frac{1}{5}$ so $\displaystyle\lim_{n\to\infty} (-1)^{n+1} \frac{n}{5n+1}$ does not exist

and the series diverges by the Test for Divergence.

9. $\displaystyle\sum_{n=1}^{\infty} (-1)^n \frac{n}{n^2+1}$ $\quad a_n = \frac{n}{n^2+1} > 0$ for all n. $a_{n+1} < a_n \iff$

$\dfrac{n+1}{(n+1)^2+1} < \dfrac{n}{n^2+1} \iff (n+1)(n^2+1) < [(n+1)^2+1]n \iff$

$n^3+n^2+n+1 < n^3+2n^2+2n \iff 0 < n^2+n-1$, which is true for all $n \geq 1$,

so we can apply the Alternating Series Test. $\displaystyle\lim_{n\to\infty} \frac{n}{n^2+1}$

$= \displaystyle\lim_{n\to\infty} \frac{1/n}{1 + 1/n^2} = 0$, so the series converges.

11. $\displaystyle\sum_{n=1}^{\infty} (-1)^{n-1} \frac{\sqrt{n}}{n+4}$ $\quad a_n = \frac{\sqrt{n}}{n+4} > 0$ for all n. Let $f(x) = \frac{\sqrt{x}}{x+4}$. Then

$f'(x) = \dfrac{4-x}{2\sqrt{x}(x+4)^2} < 0$ if $x > 4$, so $\{a_n\}$ is decreasing after n=4.

$\displaystyle\lim_{n\to\infty} \frac{\sqrt{n}}{n+4} = \lim_{n\to\infty} \frac{1}{\sqrt{n} + 4/\sqrt{n}} = 0$, and the series converges by the

Alternating Series Test.

13. $\displaystyle\sum_{n=2}^{\infty} (-1)^n \frac{n}{\ln n}$ $\quad \displaystyle\lim_{n\to\infty} \frac{n}{\ln n} = \lim_{n\to\infty} \frac{1}{1/n} = \infty$ so the series diverges.

15. $\sum_{n=1}^{\infty} (-1)^{n+1} \frac{n+10}{n(n+1)}$    $a_n = \frac{n+10}{n(n+1)} > 0$ for all n. Let $f(x) = \frac{x+10}{x(x+1)}$.

Then $f'(x) = -\frac{x^2+20x+10}{(x^2+x)^2} < 0$ for $x \geq 1$, so $\{a_n\}$ is decreasing.

$\lim_{n\to\infty} \frac{n+10}{n(n+1)} = \lim_{n\to\infty} \frac{1/n + 10/n^2}{1 + 1/n} = 0$, so the series converges by the

Alternating Series Test.

17. $\sum_{n=1}^{\infty} \frac{\cos n\pi}{n^{3/4}} = \sum_{n=1}^{\infty} \frac{(-1)^n}{n^{3/4}}$    $a_n = \frac{1}{n^{3/4}}$ is decreasing and positive, and

$\lim_{n\to\infty} \frac{1}{n^{3/4}} = 0$ so the series converges by the Alternating Series Test.

19. $\sum_{n=1}^{\infty} (-1)^n \sin\left[\frac{\pi}{n}\right]$    $a_n = \sin\left[\frac{\pi}{n}\right] > 0$ for $n \geq 2$ and $\sin\left[\frac{\pi}{n}\right] \geq \sin\left[\frac{\pi}{n+1}\right]$,

and $\lim_{n\to\infty} \sin\left[\frac{\pi}{n}\right] = \sin 0 = 0$, so the series converges by the

Alternating Series Test.

21. $\sum_{n=1}^{\infty} (-1)^n \frac{1}{\sqrt[n]{n}}$    Let $L = \lim_{n\to\infty} \frac{1}{\sqrt[n]{n}}$ (if it exists). Then $\ln L =$

$\lim_{n\to\infty} \ln(n^{-1/n}) = \lim_{n\to\infty} -\frac{\ln n}{n} = \lim_{n\to\infty} -\frac{1/n}{1} = 0$, so $L = e^0 = 1$, so

$\lim_{n\to\infty} (-1)^n \frac{1}{\sqrt[n]{n}}$ does not exist and the series diverges by the Test for

Divergence.

23. Let $\sum b_n$ be the series for which $b_n = 0$ if n is odd and $b_n = \frac{1}{n^2}$ if n

is even. Then $\sum b_n = \sum \frac{1}{(2n)^2}$ clearly converges (by comparison with

the p-series for p=2). So suppose that $\sum (-1)^{n-1} a_n$ converges.

Then by Theorem 10.19(b) so does $\sum [(-1)^{n-1} a_n + b_n] =$

$1 + \frac{1}{3} + \frac{1}{5} + \ldots = \sum \frac{1}{2n-1}$. But this diverges (comparison with

harmonic series), a contradiction. So $\sum (-1)^{n-1} a_n$ must diverge.

The Alternating Series Test does not apply since $\{a_n\}$ is not

decreasing.

25. Clearly $a_n = \frac{1}{n+p}$ is decreasing and eventually positive and

$\lim\limits_{n\to\infty} a_n = 0$ for any p. So the series will converge (by the

Alternating Series Test) for any p for which all the $a_n$'s are

defined — i.e. $n+p \neq 0$ for $n \geq 1$, or p is not a negative integer.

27. If $a_n = \frac{1}{n^2}$, then $a_{11} = \frac{1}{121} < 0.01$, so by Theorem 10.28,

$$\sum_{n=1}^{\infty} \frac{1}{n^2} \approx \sum_{n=1}^{10} \frac{1}{n^2} \approx 0.82.$$

29. $\sum\limits_{n=0}^{\infty} (-1)^n \frac{2^n}{n!}$   Since $\frac{2}{n} < \frac{2}{3}$ for $n \geq 4$, $0 < \frac{2^n}{n!} < \frac{2 \cdot 2 \cdot 2}{1 \cdot 2 \cdot 3} \cdot \left[\frac{2}{3}\right]^{n-3} \to 0$ as $n \to \infty$,

so by the Squeeze Theorem, $\lim\limits_{n\to\infty} \frac{2^n}{n!} = 0$, and hence $\sum\limits_{n=0}^{\infty} (-1)^n \frac{2^n}{n!}$ is a

convergent alternating series.   $\frac{2^8}{8!} = \frac{256}{40320} < 0.01$, so $\sum\limits_{n=0}^{\infty} (-1)^n \frac{2^n}{n!}$

$\approx \sum\limits_{n=0}^{7} (-1)^n \frac{2^n}{n!} \approx 0.13.$

31. $\sum\limits_{n=1}^{\infty} \frac{(-1)^{n-1}}{(2n-1)!}$   $a_5 = \frac{1}{(2 \cdot 5 - 1)!} = \frac{1}{362880} < 0.00001$, so $\sum\limits_{n=1}^{\infty} \frac{(-1)^{n-1}}{(2n-1)!}$

$\approx \sum\limits_{n=1}^{4} \frac{(-1)^{n-1}}{(2n-1)!} \approx 0.8415.$

33. $\sum\limits_{n=0}^{\infty} \frac{(-1)^n}{2^n n!}$   $a_6 = \frac{1}{2^6 6!} = \frac{1}{46080} < 0.000022$, so $\sum\limits_{n=0}^{\infty} \frac{(-1)^n}{2^n n!} \approx \sum\limits_{n=0}^{5} \frac{(-1)^n}{2^n n!}$

$\approx 0.6065.$

35. (a) We will prove this by induction.  Let P(n) be the proposition

that $s_{2n} = h_{2n} - h_n$.  P(1) is true by an easy calculation.  So

suppose that P(n) is true.  We will show that P(n+1) must be true

as a consequence.  $h_{2n+2} - h_{n+1} = (h_{2n} + \frac{1}{2n+1} + \frac{1}{2n+2}) - (h_n + \frac{1}{n+1})$

$= (h_{2n} - h_n) + \frac{1}{2n+1} - \frac{1}{2n+2} = s_{2n} + \frac{1}{2n+1} - \frac{1}{2n+2} = s_{2n+2}$, which is

P(n+1), and proves that $s_{2n} = h_{2n} - h_n$ for all n.

(b) We know $h_{2n} - \ln(2n) \to \gamma$ and $h_n - \ln(n) \to \gamma$ as $n \to \infty$. So

$$s_{2n} = h_{2n} - h_n = (h_{2n} - \ln(2n)) - (h_n - \ln(n)) + (\ln(2n) - \ln n),$$

and $\lim_{n \to \infty} s_{2n} = \gamma - \gamma + \lim_{n \to \infty} (\ln(2n) - \ln n) =$

$\lim_{n \to \infty} (\ln 2 + \ln n - \ln n) = \ln 2.$

Section 10.6

Exercises 10.6

1.  $\sum_{n=1}^{\infty} \frac{1}{n\sqrt{n}} = \sum_{n=1}^{\infty} \frac{1}{n^{3/2}}$ is a convergent p-series ($p=\frac{3}{2}>1$), so the given

series is absolutely convergent.

3.  $\lim_{n \to \infty} \left| \frac{a_{n+1}}{a_n} \right| = \lim_{n \to \infty} \left| \frac{(-3)^{n+1}/(n+1)^3}{(-3)^n/n^3} \right| = 3 \lim_{n \to \infty} \left[ \frac{n}{n+1} \right]^3 = 3 > 1$, so the

series diverges by the Ratio Test.

5.  $\sum_{n=1}^{\infty} \frac{1}{2n+1}$ diverges (use the Integral Test or the Limit Comparison

Test with $b_n = \frac{1}{n}$), but since $\lim_{n \to \infty} \frac{1}{2n+1} = 0$, $\sum_{n=1}^{\infty} \frac{(-1)^{n+1}}{2n+1}$ converges by

the Alternating Series Test, and so is conditionally convergent.

7.  $\lim_{n \to \infty} \left| \frac{a_{n+1}}{a_n} \right| = \lim_{n \to \infty} \left[ \frac{1/(2n+1)!}{1/(2n-1)!} \right] = \lim_{n \to \infty} \frac{1}{(2n+1)2n} = 0$, so by the Ratio

Test the series is absolutely convergent.

9.  $\sum_{n=1}^{\infty} \frac{n}{n^2+4}$ diverges (use the Limit Comparison Test with $b_n = \frac{1}{n}$). But

since $0 \le \frac{n+1}{(n+1)^2+4} < \frac{n}{n^2+4} \Longleftrightarrow n^3+n^2+4n+4 < n^3+2n^2+5n \Longleftrightarrow 0 < n^2+n-4$

(which is true for $n \ge 2$), and since $\lim_{n \to \infty} \frac{n}{n^2+4} = 0$, $\sum_{n=1}^{\infty} (-1)^n \frac{n}{n^2+4}$

converges by the Alternating Series Test, and so converges

conditionally.

11. $\lim\limits_{n\to\infty} \dfrac{2n}{3n-4} = \dfrac{2}{3}$ so $\sum\limits_{n=1}^{\infty} (-1)^n \dfrac{2n}{3n-4}$ diverges by the Test for Divergence.

13. $\left|\dfrac{\sin 2n}{n^2}\right| \leq \dfrac{1}{n^2}$ and $\sum\limits_{n=1}^{\infty} \dfrac{1}{n^2}$ converges (p-series, p=2>1), so $\sum\limits_{n=1}^{\infty} \dfrac{\sin 2n}{n^2}$

converges absolutely (by the Comparison Test).

15. $\lim\limits_{n\to\infty} \left|\dfrac{a_{n+1}}{a_n}\right| = \lim\limits_{n\to\infty} \left|\dfrac{2^{n+1}/(n+1)3^{n+2}}{2^n/n3^{n+1}}\right| = \dfrac{2}{3} \lim\limits_{n\to\infty} \dfrac{n}{n+1} = \dfrac{2}{3} < 1$ so the series

converges absolutely by the Ratio Test.

17. $\lim\limits_{n\to\infty} \left|\dfrac{a_{n+1}}{a_n}\right| = \lim\limits_{n\to\infty} \dfrac{(n+2)5^{n+1}/(n+1)3^{2(n+1)}}{(n+1)5^n/n3^{2n}} = \lim\limits_{n\to\infty} \dfrac{5n(n+2)}{9(n+1)^2} = \dfrac{5}{9} < 1$ so

the series converges absolutely by the Ratio Test.

19. $\sum\limits_{n=2}^{\infty} \dfrac{1}{\ln n}$ diverges (since $\dfrac{1}{\ln n} > \dfrac{1}{n}$ and $\sum\limits_{n=1}^{\infty} \dfrac{1}{n}$ diverges), but $\sum\limits_{n=2}^{\infty} \dfrac{(-1)^n}{\ln n}$

converges by the Alternating Series Test (since $\lim\limits_{n\to\infty} \dfrac{1}{\ln n} = 0$), so

the series converges conditionally.

21. $\lim\limits_{n\to\infty} \left|\dfrac{a_{n+1}}{a_n}\right| = \lim\limits_{n\to\infty} \dfrac{(n+1)!/10^{n+1}}{n!/10^n} = \lim\limits_{n\to\infty} \dfrac{n+1}{10} = \infty$ so the series diverges

by the Ratio Test.

23. $\dfrac{|\cos(n\pi/3)|}{n!} \leq \dfrac{1}{n!}$ and $\sum\limits_{n=1}^{\infty} \dfrac{1}{n!}$ converges (Exercise 33, Section 10.4),

so the given series converges absolutely by the Comparison Test.

25. $\lim\limits_{n\to\infty} \left|\dfrac{a_{n+1}}{a_n}\right| = \lim\limits_{n\to\infty} \dfrac{(n+1)^{n+1}/5^{2n+5}}{n^n/5^{2n+3}} = \lim\limits_{n\to\infty} \dfrac{1}{25}\left[\dfrac{n+1}{n}\right]^n (n+1) = \infty$, so the

series diverges by the Ratio Test.

27. $\lim\limits_{n\to\infty} \sqrt[n]{|a_n|} = \lim\limits_{n\to\infty} \left|\dfrac{1-3n}{3+4n}\right| = \dfrac{3}{4} < 1$, so the series converges absolutely

by the Root Test.

29. $\lim\limits_{n\to\infty} \left|\dfrac{a_{n+1}}{a_n}\right| = \lim\limits_{n\to\infty} \dfrac{(n+1)!/(1\cdot3\cdot5\cdots(2n+1))}{n!/(1\cdot3\cdot5\cdots(2n-1))} = \lim\limits_{n\to\infty} \dfrac{n+1}{2n+1} = \dfrac{1}{2} < 1$, so the

series converges absolutely by the Ratio Test.

31. $\sum\limits_{n=1}^{\infty} \dfrac{2\cdot4\cdot6\cdots(2n)}{n!} = \sum\limits_{n=1}^{\infty} \dfrac{2^n n!}{n!} = \sum\limits_{n=1}^{\infty} 2^n$ which diverges since $\lim\limits_{n\to\infty} 2^n = \infty$.

33. $\lim\limits_{n\to\infty} \left|\dfrac{a_{n+1}}{a_n}\right| = \lim\limits_{n\to\infty} \dfrac{(n+3)!/(n+1)!10^{n+1}}{(n+2)!/n!10^n} = \dfrac{1}{10} \lim\limits_{n\to\infty} \dfrac{n+3}{n+1} = \dfrac{1}{10} < 1$ so the

series converges absolutely by the Ratio Test.

35. $\dfrac{|\sin 3n|\, n^2}{(1.1)^n} \le \dfrac{1 \cdot n^2}{(1.1)^n}$ for all n, so we test the series $\displaystyle\sum_{n=1}^{\infty} \dfrac{n^2}{(1.1)^n}$

using the Ratio Test. $\displaystyle\lim_{n\to\infty} \left|\dfrac{a_{n+1}}{a_n}\right| = \dfrac{(n+1)^2/(1.1)^{n+1}}{n^2/(1.1)^n} =$

$\dfrac{1}{1.1} \displaystyle\lim_{n\to\infty} \left[\dfrac{n+1}{n}\right]^2 = \dfrac{1}{1.1} < 1$ so $\displaystyle\sum_{n=1}^{\infty} \dfrac{n^2}{(1.1)^n}$ converges absolutely, and by

the Comparison Test, so does $\displaystyle\sum_{n=1}^{\infty} \dfrac{(\sin 3n)\, n^2}{(1.1)^n}$.

37. (a) $\displaystyle\lim_{n\to\infty} \left|\dfrac{a_{n+1}}{a_n}\right| = \displaystyle\lim_{n\to\infty} \dfrac{|x|^{n+1}/(n+1)!}{|x|^n/n!} = |x| \displaystyle\lim_{n\to\infty} \dfrac{1}{n+1} = 0$, so by the

Ratio Test the series converges for any x.

(b) Since the series of part (a) always converges, we must have

$\displaystyle\lim_{n\to\infty} x^n/n! = 0$ by Theorem 10.17.

39. (a) Following the hint, we get that $|a_n| < r^n$ for $n \ge N$, and so

since the geometric series $\displaystyle\sum_{n=1}^{\infty} r^n$ converges $(0 < r < 1)$, the series

$\displaystyle\sum_{n=N}^{\infty} |a_n|$ will converge as well by the Comparison Test, and hence so

does $\displaystyle\sum_{n=1}^{\infty} |a_n|$, so $\displaystyle\sum_{n=1}^{\infty} a_n$ is absolutely convergent.

(b) If $\displaystyle\lim_{n\to\infty} \sqrt[n]{|a_n|} = L > 1$, then there is an integer N such that

$\sqrt[n]{|a_n|} > 1$ for all $n \ge N$, and so $|a_n| > 1$ for $n \ge N$. Thus $\displaystyle\lim_{n\to\infty} a_n \ne 0$, so

$\displaystyle\sum_{n=1}^{\infty} a_n$ diverges by the Test for Divergence.

41. Let $\displaystyle\sum b_n$ be the rearranged series constructed in the hint. (This

series can be constructed by virtue of the result of Exercise

40(b)). This series will have partial sums $s_n$ that oscillate in

value back and forth across r. Since $\displaystyle\lim_{n\to\infty} a_n = 0$ (by Theorem

10.17), and since the size of the oscillations $|s_n - r|$ is always

less than $|a_n|$ because of the way $\displaystyle\sum b_n$ was constructed, we have

that $\displaystyle\sum b_n = \displaystyle\lim_{n\to\infty} s_n = r$.         313

Exercises 10.7

1. Use the Limit Comparison Test, with $a_n = \dfrac{\sqrt{n}}{n^2+1}$ and $b_n = \dfrac{1}{n^{3/2}}$.

$\lim\limits_{n\to\infty} \dfrac{a_n}{b_n} = \lim\limits_{n\to\infty} \dfrac{\sqrt{n}/(n^2+1)}{1/n^{3/2}} = \lim\limits_{n\to\infty} \dfrac{n^2}{n^2+1} = 1$ and $\displaystyle\sum_{n=1}^{\infty} b_n$ is a convergent

p-series $(p=\frac{3}{2}>1)$, so $\displaystyle\sum_{n=1}^{\infty} a_n = \sum_{n=1}^{\infty} \dfrac{\sqrt{n}}{n^2+1}$ converges as well.

3. $\displaystyle\sum_{n=1}^{\infty} \dfrac{4^n}{3^{2n-1}} = 3\sum_{n=1}^{\infty} \left[\dfrac{4}{9}\right]^n$ which is a convergent geometric series

$\left(|r| = \dfrac{4}{9} < 1\right)$.

5. Converges by the Alternating Series Test, since $a_n = \dfrac{1}{(\ln\ n)^2}$ is

decreasing ($\ln\ x$ is an increasing function) and $\lim\limits_{n\to\infty} a_n = 0$.

7. $\displaystyle\sum_{k=1}^{\infty} \dfrac{1}{k^{1.7}}$ is a convergent p-series ($p = 1.7 > 1$).

9. $\lim\limits_{n\to\infty} \left|\dfrac{a_{n+1}}{a_n}\right| = \lim\limits_{n\to\infty} \dfrac{(n+1)/e^{n+1}}{n/e^n} = \dfrac{1}{e} \lim\limits_{n\to\infty} \dfrac{n+1}{n} = \dfrac{1}{e} < 1$, so the series

converges by the Ratio Test.

11. Use the Limit Comparison Test with $a_n = \dfrac{n^3+1}{n^4-1}$ and $b_n = \dfrac{1}{n}$. $\lim\limits_{n\to\infty} \dfrac{a_n}{b_n}$

$= \lim\limits_{n\to\infty} \dfrac{n^4+n}{n^4-1} = \lim\limits_{n\to\infty} \dfrac{1 + 1/n^3}{1 - 1/n^4} = 1$, and since $\displaystyle\sum_{n=2}^{\infty} b_n$ diverges (harmonic

series) so does $\displaystyle\sum_{n=2}^{\infty} \dfrac{n^3+1}{n^4-1}$.

13. Let $f(x) = \dfrac{2}{x(\ln\ x)^3}$. $f(x)$ is clearly positive and decreasing for

$x \geq 2$, so we apply the Integral Test. $\displaystyle\int_2^{\infty} \dfrac{2}{x(\ln\ x)^3}\ dx =$

$\lim\limits_{t\to\infty} \dfrac{-1}{(\ln\ x)^2}\Big]_2^t = 0 - \dfrac{-1}{(\ln\ 2)^2}$ which is finite. So $\displaystyle\sum_{n=2}^{\infty} \dfrac{2}{n(\ln\ n)^3}$

converges.

15. $\lim\limits_{n\to\infty} \left| \dfrac{a_{n+1}}{a_n} \right| = \lim\limits_{n\to\infty} \dfrac{3^{n+1}(n+1)^2/(n+1)!}{3^n n^2/n!} = 3 \lim\limits_{n\to\infty} \dfrac{n+1}{n^2} = 0$, so the series

converges by the Ratio Test.

17. $\dfrac{3^n}{5^n + n} \leq \dfrac{3^n}{5^n} = \left[\dfrac{3}{5}\right]^n$ Since $\sum\limits_{n=1}^{\infty} \left[\dfrac{3}{5}\right]^n$ is a convergent geometric series

$(|r| = \dfrac{3}{5} < 1)$, $\sum\limits_{n=1}^{\infty} \dfrac{3^n}{5^n + n}$ will converge by the Comparison Test.

19. $\lim\limits_{n\to\infty} \left| \dfrac{a_{n+1}}{a_n} \right| = \lim\limits_{n\to\infty} \dfrac{(n+1)!/(2\cdot5\cdot8\cdots(3n+5))}{n!/(2\cdot5\cdot8\cdots(3n+2))} = \lim\limits_{n\to\infty} \dfrac{n+1}{3n+5} = \dfrac{1}{3} < 1$, so the

series converges by the Ratio Test.

21. Use the Limit Comparison Test with $a_i = \dfrac{1}{\sqrt{i(i+1)}}$ and $b_i = \dfrac{1}{i}$. $\lim\limits_{i\to\infty} \dfrac{a_i}{b_i}$

$= \lim\limits_{i\to\infty} \dfrac{i}{\sqrt{i(i+1)}} = \lim\limits_{i\to\infty} \dfrac{1}{\sqrt{1 + 1/i}} = 1$. Since $\sum\limits_{i=1}^{\infty} b_i$ diverges (harmonic

series) so does $\sum\limits_{i=1}^{\infty} \dfrac{1}{\sqrt{i(i+1)}}$.

23. $\lim\limits_{n\to\infty} 2^{1/n} = 2^0 = 1$, so $\lim\limits_{n\to\infty} (-1)^n 2^{1/n}$ does not exist and the series

diverges by the Test for Divergence.

25. Let $f(x) = \dfrac{\ln x}{\sqrt{x}}$. Then $f'(x) = \dfrac{2 - \ln x}{2x^{3/2}} < 0$ when $\ln x > 2$ or $x > e^2$

so $\dfrac{\ln n}{\sqrt{n}}$ is decreasing for $n > e^2$. By l'Hospital's Rule, $\lim\limits_{n\to\infty} \dfrac{\ln n}{\sqrt{n}}$

$= \lim\limits_{n\to\infty} \dfrac{1/n}{1/2\sqrt{n}} = \lim\limits_{n\to\infty} \dfrac{2}{\sqrt{n}} = 0$, so the series converges by the

Alternating Series Test.

27. The series diverges since it is a geometric series with $r = -\pi$ and

$|r| = \pi > 1$. [Or use the Test for Divergence.]

29. $\sum\limits_{n=1}^{\infty} \dfrac{(-2)^{2n}}{n^n} = \sum\limits_{n=1}^{\infty} \left[\dfrac{4}{n}\right]^n$. $\lim\limits_{n\to\infty} \sqrt[n]{|a_n|} = \lim\limits_{n\to\infty} \dfrac{4}{n} = 0$, so the series

converges by the Root Test.

31. Since $\dfrac{k \ln k}{(k+1)^3} < \dfrac{k \ln k}{k^3} = \dfrac{\ln k}{k^2}$, and since $\sum\limits_{n=1}^{\infty} \dfrac{\ln n}{n^2}$ converges (Exercise

17, Section 10.3), the given series converges by the Comparison

Test.

33. $\lim\limits_{n\to\infty} \left|\dfrac{a_{n+1}}{a_n}\right| = \lim\limits_{n\to\infty} \dfrac{2^{n+1}/(2n+3)!}{2^n/(2n+1)!} = 2\lim\limits_{n\to\infty}\dfrac{1}{(2n+3)(2n+2)} = 0$, so the series converges by the Ratio test.

35. $0 < \dfrac{\tan^{-1}n}{n^{3/2}} < \dfrac{\pi/2}{n^{3/2}}$. $\displaystyle\sum_{n=1}^{\infty}\dfrac{\pi/2}{n^{3/2}} = \dfrac{\pi}{2}\sum_{n=1}^{\infty}\dfrac{1}{n^{3/2}}$ which is a convergent

p-series $(p=\dfrac{3}{2}<1)$, so $\displaystyle\sum_{n=1}^{\infty}\dfrac{\tan^{-1}n}{n^{3/2}}$ converges by the Comparison Test.

37. Since $\dfrac{3}{\pi} < 1$, $\lim\limits_{n\to\infty}\dfrac{1}{1+(3/\pi)^n} = \dfrac{1}{1+0} = 1 \neq 0$, so the series diverges by the Test for Divergence.

39. $\lim\limits_{n\to\infty}\sqrt[n]{|a_n|} = \lim\limits_{n\to\infty}(2^{1/n}-1) = 1-1 = 0$, so the series converges by the Root Test.

Section 10.8

Exercises 10.8

NOTE: "R" stands for "radius of convergence" and "I" stands for "interval of convergence" in this section.

1. If $u_n = \dfrac{x^n}{n+2}$, then $\lim\limits_{n\to\infty}\left|\dfrac{u_{n+1}}{u_n}\right| = \lim\limits_{n\to\infty}\left|\dfrac{x^{n+1}}{n+3}\cdot\dfrac{n+2}{x^n}\right| = |x|\lim\limits_{n\to\infty}\dfrac{n+2}{n+3} = |x|$

$< 1$ for convergence (by the Ratio Test). So R = 1. When x=1, the series is $\displaystyle\sum_{n=0}^{\infty}\dfrac{1}{n+2}$ which diverges (by the Integral Test or Comparison

Test), and when x = -1, it is $\displaystyle\sum_{n=0}^{\infty}\dfrac{(-1)^n}{n+2}$ which converges (by the

Alternating Series Test), so I = [-1,1).

3. If $u_n = nx^n$, then $\lim\limits_{n\to\infty}\left|\dfrac{u_{n+1}}{u_n}\right| = \lim\limits_{n\to\infty}\left|\dfrac{(n+1)x^{n+1}}{nx^n}\right| = |x|\lim\limits_{n\to\infty}\dfrac{n+1}{n} = |x|$

$< 1$ for convergence (by the Ratio Test). So R = 1. When x = 1 or

-1, $\lim\limits_{n\to\infty} nx^n$ does not exist, so $\displaystyle\sum_{n=0}^{\infty}nx^n$ diverges for these values.

So I = (-1,1).

5.  If $u_n = \dfrac{x^n}{n!}$, then $\lim\limits_{n\to\infty} \left|\dfrac{u_{n+1}}{u_n}\right| = \lim\limits_{n\to\infty} \left|\dfrac{x^{n+1}/(n+1)!}{x^n/n!}\right| = |x| \lim\limits_{n\to\infty} \dfrac{1}{n+1} = 0$

$< 1$ for all $x$. So, by the Ratio Test, $R = \infty$, and $I = (-\infty, \infty)$.

7.  If $u_n = \dfrac{(-1)^n x^n}{n\, 2^n}$, then $\lim\limits_{n\to\infty} \left|\dfrac{u_{n+1}}{u_n}\right| = \lim\limits_{n\to\infty} \left|\dfrac{x^{n+1}/((n+1)2^{n+1})}{x^n/(n2^n)}\right|$

$= \left|\dfrac{x}{2}\right| \lim\limits_{n\to\infty} \dfrac{n}{n+1} = \left|\dfrac{x}{2}\right| < 1$ for convergence, so $|x| < 2$ and $R = 2$.

When $x=2$, $\displaystyle\sum_{n=1}^{\infty} \dfrac{(-1)^n x^n}{n\, 2^n} = \sum_{n=1}^{\infty} \dfrac{(-1)^n}{n}$ which converges by the Alternating

Series Test. When $x = -2$, $\displaystyle\sum_{n=1}^{\infty} \dfrac{(-1)^n x^n}{n\, 2^n} = \sum_{n=1}^{\infty} \dfrac{1}{n}$ which diverges

(harmonic series), so $I = (-2, 2]$.

9.  If $u_n = \dfrac{3^n x^n}{(n+1)^2}$, then $\lim\limits_{n\to\infty} \left|\dfrac{u_{n+1}}{u_n}\right| = \lim\limits_{n\to\infty} \left|\dfrac{3^{n+1} x^{n+1}}{(n+2)^2} \cdot \dfrac{(n+1)^2}{3^n x^n}\right|$

$= 3|x| \lim\limits_{n\to\infty} \left[\dfrac{n+1}{n+2}\right]^2 = 3|x| < 1$ for convergence, so $|x| < \dfrac{1}{3}$ and $R = \dfrac{1}{3}$.

When $x = \dfrac{1}{3}$, $\displaystyle\sum_{n=0}^{\infty} \dfrac{3^n x^n}{(n+1)^2} = \sum_{n=0}^{\infty} \dfrac{1}{(n+1)^2} = \sum_{n=1}^{\infty} \dfrac{1}{n^2}$ which is a convergent

p-series ($p=2>1$). When $x = -\dfrac{1}{3}$, $\displaystyle\sum_{n=0}^{\infty} \dfrac{3^n x^n}{(n+1)^2} = \sum_{n=0}^{\infty} \dfrac{(-1)^n}{(n+1)^2}$ which

converges by the Alternating Series Test, so $I = [-1/3, 1/3]$.

11. If $u_n = \dfrac{x^n}{\ln n}$, then $\lim\limits_{n\to\infty} \left|\dfrac{u_{n+1}}{u_n}\right| = \lim\limits_{n\to\infty} \left|\dfrac{x^{n+1}}{\ln(n+1)} \cdot \dfrac{\ln n}{x^n}\right|$

$= |x| \lim\limits_{n\to\infty} \dfrac{\ln n}{\ln(n+1)} = |x|$ (using L'Hospital's Rule), so $R = 1$. When

$x = 1$, $\displaystyle\sum_{n=2}^{\infty} \dfrac{x^n}{\ln n} = \sum_{n=2}^{\infty} \dfrac{1}{\ln n}$ which diverges because $\dfrac{1}{\ln n} > \dfrac{1}{n}$ and

$\displaystyle\sum_{n=2}^{\infty} \dfrac{1}{n}$ is the divergent harmonic series. When $x = -1$, $\displaystyle\sum_{n=2}^{\infty} \dfrac{x^n}{\ln n}$

$= \displaystyle\sum_{n=2}^{\infty} \dfrac{(-1)^n}{\ln n}$ which converges by the Alternating Series Test. So

$I = [-1, 1)$.

13. If $u_n = \dfrac{(-1)^n (x-1)^n}{\sqrt{n}}$, then $\lim\limits_{n\to\infty} \left| \dfrac{u_{n+1}}{u_n} \right| = \lim\limits_{n\to\infty} \left| \dfrac{(x-1)^{n+1}}{\sqrt{n+1}} \cdot \dfrac{\sqrt{n}}{(x-1)^n} \right|$

$= |x-1| \lim\limits_{n\to\infty} \sqrt{\dfrac{n}{n+1}} = |x-1| < 1$ for convergence, or $0 < x < 2$, and

$R = 1$. When $x=0$, $\sum\limits_{n=1}^{\infty} \dfrac{(-1)^n (x-1)^n}{\sqrt{n}} = \sum\limits_{n=1}^{\infty} \dfrac{1}{\sqrt{n}}$ which is a divergent

p-series $(p=\tfrac{1}{2}<1)$. When $x=2$, $\sum\limits_{n=1}^{\infty} \dfrac{(-1)^n}{\sqrt{n}}$ which converges by the

Alternating Series Test. So $I = (0,2]$.

15. If $u_n = \dfrac{(x-2)^n}{n^n}$, then $\lim\limits_{n\to\infty} \sqrt[n]{|u_n|} = \lim\limits_{n\to\infty} \dfrac{x-2}{n} = 0$, so the series

converges for all $x$ (by the Root Test). $R = \infty$ and $I = (-\infty, \infty)$.

17. If $u_n = \dfrac{2^n (x-3)^n}{n+3}$, then $\lim\limits_{n\to\infty} \left| \dfrac{u_{n+1}}{u_n} \right| = \lim\limits_{n\to\infty} \left| \dfrac{2^{n+1} (x-3)^{n+1}}{n+4} \cdot \dfrac{n+3}{2^n (x-3)^n} \right|$

$= 2|x-3| \lim\limits_{n\to\infty} \dfrac{n+3}{n+4} = 2|x-3| < 1$ for convergence, or $|x-3| < \dfrac{1}{2} \Longleftrightarrow$

$\dfrac{5}{2} < x < \dfrac{7}{2}$, and $R = \dfrac{1}{2}$. When $x = \dfrac{5}{2}$, $\sum\limits_{n=0}^{\infty} \dfrac{2^n (x-3)^n}{n+3} = \sum\limits_{n=0}^{\infty} \dfrac{(-1)^n}{n+3}$ which

converges by the Alternating Series Test. When $x = \dfrac{7}{2}$, $\sum\limits_{n=0}^{\infty} \dfrac{2^n (x-3)^n}{n+3}$

$= \sum\limits_{n=0}^{\infty} \dfrac{1}{n+3} = \sum\limits_{n=3}^{\infty} \dfrac{1}{n}$, the harmonic series, which diverges. So

$I = [5/2, 7/2)$.

19. If $u_n = \dfrac{n(x+10)^n}{(n^2+1) 4^n}$, then $\lim\limits_{n\to\infty} \left| \dfrac{u_{n+1}}{u_n} \right| = \lim\limits_{n\to\infty} \left| \dfrac{(n+1)(x+10)^{n+1}}{((n+1)^2+1) 4^{n+1}} \cdot \dfrac{(n^2+1) 4^n}{n(x+10)^n} \right|$

$= \dfrac{|x+10|}{4} \lim\limits_{n\to\infty} \dfrac{n^3+n^2+n+1}{n^3+2n^2+2n} = \dfrac{|x+10|}{4} < 1$ for convergence, so $|x+10| < 4$,

$-14 < x < -6$, and $R = 4$. When $x = -14$, $\sum\limits_{n=0}^{\infty} \dfrac{n(x+10)^n}{(n^2+1) 4^n} = \sum\limits_{n=0}^{\infty} \dfrac{(-1)^n n}{(n^2+1)}$

which converges by the Alternating Series Test. When $x = -6$,

$\sum\limits_{n=0}^{\infty} \dfrac{n(x+10)^n}{(n^2+1) 4^n} = \sum\limits_{n=0}^{\infty} \dfrac{n}{n^2+1}$ which diverges (by the Integral Test or the

Limit Comparison Test with $b_n = 1/n$). So $I = [-14, -6)$.

21. If $u_n = \left[\frac{n}{2}\right]^n (x+6)^n$, then $\lim\limits_{n\to\infty} \sqrt[n]{|u_n|} = \lim\limits_{n\to\infty} \frac{n(x+6)}{2} = \infty$ unless $x = -6$,

in which case the limit is 0. So by the Root Test, the series

converges only for $x = -6$. $R = 0$ and $I = \{-6\}$.

23. If $u_n = \frac{(2x-1)^n}{n^3}$, then $\lim\limits_{n\to\infty} \left|\frac{u_{n+1}}{u_n}\right| = |2x-1| \lim\limits_{n\to\infty} \left[\frac{n}{n+1}\right]^3 = |2x-1| < 1$

for convergence, so $|x-1/2| < 1/2 \iff 0 < x < 1$, and $R = 1/2$. The

series $\sum\limits_{n=1}^{\infty} \frac{(2x-1)^n}{n^3}$ converges both for $x=0$ and $x=1$ (in the first case

because of the Alternating Series Test and in the last case because

we get a p-series with p=3>1). So $I = [0,1]$.

25. If $u_n = \frac{n}{\sqrt{n+1}}(x-e)^n$, then $\lim\limits_{n\to\infty} \left|\frac{u_{n+1}}{u_n}\right| = \lim\limits_{n\to\infty} \left|\frac{(n+1)(x-e)^{n+1}/\sqrt{n+2}}{n(x-e)^n/\sqrt{n+1}}\right|$

$= |x-e| \lim\limits_{n\to\infty} \left[\frac{n^3+3n^2+3n+1}{n^3+2n^2}\right]^{1/2} = |x-e| < 1$ for convergence, so

$e-1 < x < e+1$ and $R = 1$. When $x = e\pm1$, $\sum\limits_{n=0}^{\infty} \frac{n}{\sqrt{n+1}}(x-e)^n$ will diverge

by the Test for Divergence since $\lim\limits_{n\to\infty} \frac{n}{\sqrt{n+1}} = \infty$. So $I = (e-1,e+1)$.

27. If $u_n = \frac{n!\ x^n}{(2n)!}$, then $\lim\limits_{n\to\infty} \left|\frac{u_{n+1}}{u_n}\right| = \lim\limits_{n\to\infty} \left|\frac{(n+1)!\ x^{n+1}}{(2n+2)!} \cdot \frac{(2n)!}{n!\ x^n}\right|$

$= \lim\limits_{n\to\infty} \frac{n+1}{(2n+1)(2n+2)} |x| = 0 < 1$ for all $x$, so $R = \infty$ and $I = (-\infty,\infty)$.

29. If $u_n = \frac{(-1)^n x^{2n+1}}{n!(n+1)!2^{2n+1}}$, then $\lim\limits_{n\to\infty} \left|\frac{u_{n+1}}{u_n}\right| = \left[\frac{x}{2}\right]^2 \lim\limits_{n\to\infty} \frac{1}{(n+1)(n+2)} = 0$ for

all $x$. So $J_1(x)$ converges for all $x$; the domain is $(-\infty,\infty)$.

31. We use the Root Test on the series $\sum a_n x^n$. $\lim\limits_{n\to\infty} \sqrt[n]{|a_n x^n|}$

$= |x| \lim\limits_{n\to\infty} \sqrt[n]{|a_n|} = a|x| < 1$ for convergence, or $|x| < \frac{1}{a}$, so $R = \frac{1}{a}$.

Exercises 10.9

1. $f(x) = \cos x$ $\qquad\qquad$ $f(0) = 1$

$f'(x) = -\sin x$ $\qquad\qquad$ $f'(0) = 0$

$f''(x) = -\cos x$ $\qquad\qquad$ $f''(0) = -1$

$f^{(3)}(x) = \sin x$ $\qquad\qquad$ $f^{(3)}(0) = 0$

$f^{(4)}(x) = \cos x$ $\qquad\qquad$ $f^{(4)}(0) = 1$

$\cdots$ $\qquad\qquad\qquad$ $\cdots$

$\cos x = f(0) + f'(0)x + \dfrac{f''(0)}{2!}x^2 + \dfrac{f^{(3)}(0)}{3!}x^3 + \dfrac{f^{(4)}(0)}{4!}x^4 + \cdots$

$\qquad = 1 - \dfrac{x^2}{2!} + \dfrac{x^4}{4!} - \cdots = \displaystyle\sum_{n=0}^{\infty} \dfrac{(-1)^n x^{2n}}{(2n)!}$

If $u_n = \dfrac{(-1)^n x^{2n}}{(2n)!}$, then $\displaystyle\lim_{n\to\infty}\left|\dfrac{u_{n+1}}{u_n}\right| = x^2 \lim_{n\to\infty} \dfrac{1}{(2n+2)(2n+1)} = 0$ for all

$x$. So $R = \infty$.

3. $f(x) = \sin x$ $\qquad\qquad$ $f(\pi/4) = \sqrt{2}/2$

$f'(x) = \cos x$ $\qquad\qquad$ $f'(\pi/4) = \sqrt{2}/2$

$f''(x) = -\sin x$ $\qquad\qquad$ $f''(\pi/4) = -\sqrt{2}/2$

$f^{(3)}(x) = -\cos x$ $\qquad\qquad$ $f^{(3)}(\pi/4) = -\sqrt{2}/2$

$f^{(4)}(x) = \sin x$ $\qquad\qquad$ $f^{(4)}(\pi/4) = \sqrt{2}/2$

$\cdots$ $\qquad\qquad\qquad$ $\cdots$

$\sin x = f(\tfrac{\pi}{4}) + f'(\tfrac{\pi}{4})(x-\tfrac{\pi}{4}) + \dfrac{f''(\pi/4)}{2!}(x-\tfrac{\pi}{4})^2 + \dfrac{f^{(3)}(\pi/4)}{3!}(x-\tfrac{\pi}{4})^3 +$

$\qquad\qquad \dfrac{f^{(4)}(\pi/4)}{4!}(x-\tfrac{\pi}{4})^4 + \cdots$

$\qquad = \dfrac{\sqrt{2}}{2}\left[1 + (x-\tfrac{\pi}{4}) - \dfrac{1}{2!}(x-\tfrac{\pi}{4})^2 - \dfrac{1}{3!}(x-\tfrac{\pi}{4})^3 + \dfrac{1}{4!}(x-\tfrac{\pi}{4})^4 + \cdots\right]$

$\qquad = \dfrac{\sqrt{2}}{2}\displaystyle\sum_{n=0}^{\infty}\dfrac{(-1)^{n(n-1)/2}(x-\pi/4)^n}{n!}$

If $u_n = \dfrac{(-1)^{n(n-1)/2}(x-\pi/4)^n}{n!}$, then $\displaystyle\lim_{n\to\infty}\left|\dfrac{u_{n+1}}{u_n}\right| = \lim_{n\to\infty}\dfrac{|x-\pi/4|}{n+1} = 0$

for all $x$, so $R = \infty$.

5. $f(x) = (1+x)^{-2}$ $\qquad\qquad$ $f(0) = 1$

$f'(x) = -2(1+x)^{-3}$ $\qquad\qquad$ $f'(0) = -2$

$f''(x) = 2\cdot 3(1+x)^{-4}$ $\qquad\qquad$ $f''(0) = 2\cdot 3$

$$f^{(3)}(x) = -2 \cdot 3 \cdot 4(1+x)^{-5} \qquad f^{(3)}(0) = -2 \cdot 3 \cdot 4$$
$$f^{(4)}(x) = 2 \cdot 3 \cdot 4 \cdot 5(1+x)^{-6} \qquad f^{(4)}(0) = 2 \cdot 3 \cdot 4 \cdot 5$$

$\cdots$ $\qquad\qquad\qquad\qquad \cdots$

So $f^{(n)}(0) = (-1)^n(n+1)!$, and

$$\frac{1}{(1+x)^2} = \sum_{n=0}^{\infty} \frac{(-1)^n(n+1)!}{n!} x^n = \sum_{n=0}^{\infty} (-1)^n(n+1)x^n.$$

If $u_n = (-1)^n(n+1)x^n$, then $\lim_{n\to\infty} \left| \frac{u_{n+1}}{u_n} \right| = |x|$ so $R = 1$.

7. $f(x) = x^{-1}$ $\qquad\qquad\qquad\qquad f(1) = 1$

   $f'(x) = -x^{-2}$ $\qquad\qquad\qquad\quad f'(1) = -1$

   $f''(x) = 2x^{-3}$ $\qquad\qquad\qquad\quad f''(1) = 2$

   $f^{(3)}(x) = -3 \cdot 2x^{-4}$ $\qquad\qquad\quad f^{(3)}(1) = -3 \cdot 2$

   $f^{(4)}(x) = 4 \cdot 3 \cdot 2x^{-5}$ $\qquad\qquad f^{(4)}(1) = 4 \cdot 3 \cdot 2$

   $\cdots$ $\qquad\qquad\qquad\qquad\quad \cdots$

   So $f^{(n)}(1) = (-1)^n n!$, and $\dfrac{1}{x} = \displaystyle\sum_{n=0}^{\infty} \dfrac{(-1)^n n!}{n!} (x-1)^n = \sum_{n=0}^{\infty} (-1)^n(x-1)^n$.

   If $u_n = (-1)^n(x-1)^n$ then $\lim_{n\to\infty} \left| \frac{u_{n+1}}{u_n} \right| = |x-1| < 1$ for convergence, so

   $0 < x < 2$ and $R = 1$.

9. Clearly $f^{(n)}(x) = e^x$, so $f^{(n)}(3) = e^3$ and $e^x = \displaystyle\sum_{n=0}^{\infty} \dfrac{e^3}{n!}(x-3)^n$. If $u_n$

   $= \dfrac{e^3}{n!}(x-3)^n$ then $\lim_{n\to\infty} \left| \frac{u_{n+1}}{u_n} \right| = \lim_{n\to\infty} \dfrac{|x-3|}{n+1} = 0$ for all $x$, so $R = \infty$.

11. $f(x) = \sinh x$ $\qquad\qquad\qquad\quad f(0) = 0$

    $f'(x) = \cosh x$ $\qquad\qquad\qquad\quad f'(0) = 1$

    $f''(x) = \sinh x$ $\qquad\qquad\qquad\quad f''(0) = 0$

    $f^{(3)}(x) = \cosh x$ $\qquad\qquad\qquad f^{(3)}(0) = 1$

    $f^{(4)}(x) = \sinh x$ $\qquad\qquad\qquad f^{(4)}(0) = 0$

    $\cdots$ $\qquad\qquad\qquad\qquad\quad \cdots$

    So $f^{(n)}(0) = \begin{cases} 0 & \text{if } n \text{ is even} \\ 1 & \text{if } n \text{ is odd} \end{cases}$, and $\sinh x = \displaystyle\sum_{n=0}^{\infty} \dfrac{x^{2n+1}}{(2n+1)!}$. If $u_n =$

    $\dfrac{x^{2n+1}}{(2n+1)!}$ then $\lim_{n\to\infty} \left| \frac{u_{n+1}}{u_n} \right| = x^2 \lim_{n\to\infty} \dfrac{1}{(2n+3)(2n+2)} = 0$ for all $x$, so $R = \infty$.

13. $f(x) = \dfrac{1}{1+x} = \dfrac{1}{1-(-x)} = \displaystyle\sum_{n=0}^{\infty} (-1)^n x^n$ with $|-x| < 1 \Longleftrightarrow |x| < 1$ so R=1.

15. $f(x) = \dfrac{1}{(1+x)^2} = -\dfrac{d}{dx}\left[\dfrac{1}{1+x}\right] = -\dfrac{d}{dx}\left[\displaystyle\sum_{n=0}^{\infty} (-1)^n x^n\right]$ [from Exercise 13]

$= \displaystyle\sum_{n=1}^{\infty} (-1)^{n+1} n x^{n-1} = \sum_{n=0}^{\infty} (-1)^n (n+1) x^n$ with R = 1.

17. $f(x) = \dfrac{1}{1 + 4x^2} = \displaystyle\sum_{n=0}^{\infty} (-1)^n (4x^2)^n$ [substituting $4x^2$ for x in the

series from Exercise 13 above] $= \displaystyle\sum_{n=0}^{\infty} (-1)^n 4^n x^{2n}$, with $|4x^2| < 1$ so

$x^2 < \dfrac{1}{4}$, $|x| < \dfrac{1}{2}$ and R = $\dfrac{1}{2}$.

19. $f(x) = \dfrac{1}{4 + x^2} = \dfrac{1}{4}\left[\dfrac{1}{1 + x^2/4}\right] = \dfrac{1}{4}\displaystyle\sum_{n=0}^{\infty} (-1)^n \left[\dfrac{x^2}{4}\right]^n$ [using Exercise 13]

$= \displaystyle\sum_{n=0}^{\infty} \dfrac{(-1)^n x^{2n}}{4^{n+1}}$, with $\left|\dfrac{x^2}{4}\right| < 1 \Longleftrightarrow x^2 < 4 \Longleftrightarrow |x| < 2$, so R = 2.

21. $f(x) = \dfrac{1}{1 - x^2} = \displaystyle\sum_{n=0}^{\infty} (x^2)^n = \sum_{n=0}^{\infty} x^{2n}$, with $|x^2| < 1 \Longleftrightarrow |x| < 1$ so R=1.

23. $f(x) = \ln(1+x) - \ln(1-x) = \displaystyle\int \dfrac{dx}{1+x} + \int \dfrac{dx}{1-x} = \int \left[\sum_{n=0}^{\infty} (-1)^n x^n + \sum_{n=0}^{\infty} x^n\right] dx$

$= \displaystyle\int \sum_{n=0}^{\infty} 2x^{2n}\, dx = \sum_{n=0}^{\infty} \dfrac{2x^{2n+1}}{2n+1}$, with R = 1.

25. $e^{3x} = \displaystyle\sum_{n=0}^{\infty} \dfrac{(3x)^n}{n!} = \sum_{n=0}^{\infty} \dfrac{3^n x^n}{n!}$, with R = $\infty$.

27. $x^2 \cos x = x^2 \displaystyle\sum_{n=0}^{\infty} \dfrac{(-1)^n x^{2n}}{(2n)!} = \sum_{n=0}^{\infty} \dfrac{(-1)^n x^{2n+2}}{(2n)!}$, with R = $\infty$.

29. $x \sin\left[\dfrac{x}{2}\right] = x \displaystyle\sum_{n=0}^{\infty} \dfrac{(-1)^n (x/2)^{2n+1}}{(2n+1)!} = \sum_{n=0}^{\infty} \dfrac{(-1)^n\, x^{2n+2}}{(2n+1)!\, 2^{2n+1}}$, with R = $\infty$.

31. $\sin^2 x = \dfrac{1}{2}[1 - \cos 2x] = \dfrac{1}{2}\left[1 - \displaystyle\sum_{n=0}^{\infty} \dfrac{(-1)^n (2x)^{2n}}{(2n)!}\right]$

$= \dfrac{1}{2}\left[1 - 1 - \displaystyle\sum_{n=1}^{\infty} \dfrac{(-1)^n (2x)^{2n}}{(2n)!}\right] = \sum_{n=1}^{\infty} \dfrac{(-1)^{n+1} 2^{2n-1} x^{2n}}{(2n)!}$, with R = $\infty$.

33. $\dfrac{\sin x}{x} = \dfrac{1}{x} \displaystyle\sum_{n=0}^{\infty} \dfrac{(-1)^n x^{2n+1}}{(2n+1)!} = \displaystyle\sum_{n=0}^{\infty} \dfrac{(-1)^n x^{2n}}{(2n+1)!}$ and this series also gives

the required value at $x=0$, so $R = \infty$.

35. $f(x) = (1+x)^{1/2}$ $\qquad\qquad f(0) = 1$

$f'(x) = \dfrac{1}{2}(1+x)^{-1/2}$ $\qquad\quad f'(0) = \dfrac{1}{2}$

$f''(x) = -\dfrac{1}{4}(1+x)^{-3/2}$ $\qquad\; f''(0) = -\dfrac{1}{4}$

$f^{(3)}(x) = \dfrac{3}{8}(1+x)^{-5/2}$ $\qquad\; f^{(3)}(0) = \dfrac{3}{8}$

$f^{(4)}(x) = -\dfrac{15}{16}(1+x)^{-7/2}$ $\qquad f^{(4)}(0) = -\dfrac{15}{16}$

$\cdots$ $\qquad\qquad\qquad\qquad\qquad \cdots$

So $f^{(n)}(0) = \dfrac{(-1)^{n-1} 1 \cdot 3 \cdot 5 \cdots (2n-3)}{2^n}$ for $n \ge 2$, and

$\sqrt{1+x} = 1 + \dfrac{x}{2} + \displaystyle\sum_{n=2}^{\infty} \dfrac{(-1)^{n-1} 1 \cdot 3 \cdot 5 \cdots (2n-3)}{2^n \, n!} x^n$. If $u_n =$

$\dfrac{(-1)^{n+1} 1 \cdot 3 \cdot 5 \cdots (2n-3)}{2^n \, n!} x^n$ then $\displaystyle\lim_{n\to\infty} \left| \dfrac{u_{n+1}}{u_n} \right| = \dfrac{|x|}{2} \lim_{n\to\infty} \dfrac{2n-1}{n+1} = |x| < 1$

for convergence, so $R = 1$.

37. $f(x) = (1-x)^{-1/3}$ $\qquad\qquad f(0) = 1$

$f'(x) = \dfrac{1}{3}(1-x)^{-4/3}$ $\qquad\quad f'(0) = \dfrac{1}{3}$

$f''(x) = \dfrac{4}{9}(1-x)^{-7/3}$ $\qquad\quad f''(0) = \dfrac{4}{9}$

$f^{(3)}(x) = \dfrac{28}{27}(1-x)^{-10/3}$ $\qquad f^{(3)}(0) = \dfrac{28}{27}$

$f^{(4)}(x) = \dfrac{280}{81}(1-x)^{-13/3}$ $\qquad f^{(4)}(0) = \dfrac{280}{81}$

$\cdots$ $\qquad\qquad\qquad\qquad\qquad \cdots$

So $f^{(n)}(0) = \dfrac{1 \cdot 4 \cdot 7 \cdot 10 \cdot 13 \cdots (3n-2)}{3^n}$ for $n \ge 1$, and

$f(x) = 1 + \displaystyle\sum_{n=0}^{\infty} \dfrac{1 \cdot 4 \cdot 7 \cdot 10 \cdot 13 \cdots (3n-2)}{3^n \, n!} x^n$. If $u_n =$

$\dfrac{1 \cdot 4 \cdot 7 \cdot 10 \cdot 13 \cdots (3n-2)}{3^n \, n!} x^n$ then $\displaystyle\lim_{n\to\infty} \left| \dfrac{u_{n+1}}{u_n} \right| = \dfrac{|x|}{3} \lim_{n\to\infty} \dfrac{3n+1}{n+1} = |x| < 1$

for convergence, so $R = 1$.

**39.**  $f(x) = (1+x)^{-3} = -\frac{1}{2}\frac{d}{dx}\left[\frac{1}{(1+x)^2}\right] = -\frac{1}{2}\frac{d}{dx}\left[\sum_{n=0}^{\infty}(-1)^n(n+1)x^n\right]$ [from

Exercise 15 above] $= -\frac{1}{2}\sum_{n=1}^{\infty}(-1)^n n(n+1)x^{n-1} = \sum_{n=0}^{\infty}\frac{(-1)^n(n+1)(n+2)x^n}{2}$,

with R = 1 (since that is the R in Exercise 15).

**41.**  $\ln(5+x) = \ln(5(1 + x/5)) = \ln(5) + \ln(1 + x/5)$

$= \ln(5) + \frac{1}{5}\int\frac{dx}{1 + x/5} = \ln(5) + \frac{1}{5}\int\sum_{n=0}^{\infty}(-1)^n\left[\frac{x}{5}\right]^n dx$

$= \ln(5) + \sum_{n=0}^{\infty}\frac{(-1)^n x^{n+1}}{(n+1)5^{n+1}} = \ln(5) + \sum_{n=1}^{\infty}\frac{(-1)^{n-1}x^n}{n5^n}$, with R = 5.

**43.**  $\ln(1+x) = \int\frac{dx}{1+x} = \int\sum_{n=0}^{\infty}(-1)^n x^n dx = \sum_{n=1}^{\infty}\frac{(-1)^{n-1}x^n}{n}$ with R = 1, so

$\ln(1.1) = \sum_{n=1}^{\infty}\frac{(-1)^{n-1}(0.1)^n}{n}$.  This is an alternating series with $a_5$

$= \frac{(0.1)^5}{5} = 0.000002$, so to 5 decimals, $\ln(1.1) \approx \sum_{n=1}^{4}\frac{(-1)^{n-1}(0.1)^n}{n}$

$\approx 0.09531$.

**45.**  $\int\sin(x^2)dx = \int\sum_{n=0}^{\infty}(-1)^n\frac{(x^2)^{2n+1}}{(2n+1)!}dx = \int\sum_{n=0}^{\infty}\frac{(-1)^n x^{4n+2}}{(2n+1)!}dx$

$= C + \sum_{n=0}^{\infty}\frac{(-1)^n x^{4n+3}}{(4n+3)(2n+1)!}$

**47.**  $\int\frac{dx}{1+x^4} = \int\sum_{n=0}^{\infty}(-1)^n x^{4n}dx = C + \sum_{n=0}^{\infty}\frac{(-1)^n x^{4n+1}}{4n+1}$

**49.**  Using the series we obtained in Exercise 35, we get

$\sqrt{x^3+1} = 1 + \frac{x^3}{2} + \sum_{n=2}^{\infty}\frac{(-1)^{n-1}1\cdot3\cdot5\cdots(2n-3)}{2^n n!}x^{3n}$ so

$\int\sqrt{x^3+1}\ dx = \int\left[1 + \frac{x^3}{2} + \sum_{n=2}^{\infty}\frac{(-1)^{n-1}1\cdot3\cdot5\cdots(2n-3)}{2^n n!}x^{3n}\right]dx$

$= C + x + \frac{x^4}{8} + \sum_{n=2}^{\infty}\frac{(-1)^{n-1}1\cdot3\cdot5\cdots(2n-3)}{2^n n!\ (3n+1)}x^{3n+1}$

51. Using our series from Exercise 45, we get $\int_0^1 \sin(x^2)dx =$

$$\sum_{n=0}^{\infty} \frac{(-1)^n x^{4n+3}}{(4n+3)(2n+1)!}\Big]_0^1 = \sum_{n=0}^{\infty} \frac{(-1)^n}{(4n+3)(2n+1)!} \text{ and } |a_3| = \frac{1}{75600} < 0.0005 \text{ so}$$

we use $\displaystyle\sum_{n=0}^{2} \frac{(-1)^n}{(4n+3)(2n+1)!} = \frac{1}{3} - \frac{1}{42} + \frac{1}{1320} \approx 0.310.$

53. $\displaystyle\int_0^{0.5} \frac{dx}{1+x^6} = \int_0^{0.5} \sum_{n=0}^{\infty} (-1)^n x^{6n} dx = \sum_{n=0}^{\infty} \frac{(-1)^n x^{6n+1}}{6n+1}\Big]_0^{1/2} = \sum_{n=0}^{\infty} \frac{(-1)^n}{(6n+1)2^{6n+1}}$

and $a_2 = \dfrac{1}{106496} < 0.00001$ so we use $\displaystyle\sum_{n=0}^{1} \frac{(-1)^n}{(6n+1)2^{6n+1}} = \frac{1}{2} - \frac{1}{896}$

$\approx 0.4989.$

55. $\displaystyle\int_0^{0.5} x^2 e^{-x^2} dx = \int_0^{0.5} \sum_{n=0}^{\infty} \frac{(-1)^n x^{2n+2}}{n!} dx = \sum_{n=0}^{\infty} \frac{(-1)^n x^{2n+3}}{n!(2n+3)}\Big]_0^{1/2}$

$= \displaystyle\sum_{n=0}^{\infty} \frac{(-1)^n}{n!(2n+3)2^{2n+3}}$ and since $a_2 = \dfrac{1}{1792} < 0.001$ we use

$\displaystyle\sum_{n=0}^{1} \frac{(-1)^n}{n!(2n+3)2^{2n+3}} = \frac{1}{24} - \frac{1}{160} \approx 0.0354.$

57. $J_0(x) = \displaystyle\sum_{n=0}^{\infty} \frac{(-1)^n x^{2n}}{2^{2n}(n!)^2}$, $J_0'(x) = \displaystyle\sum_{n=1}^{\infty} \frac{(-1)^n 2n x^{2n-1}}{2^{2n}(n!)^2}$, and

$J_0''(x) = \displaystyle\sum_{n=1}^{\infty} \frac{(-1)^n 2n(2n-1)x^{2n-2}}{2^{2n}(n!)^2}$, so $x^2 J_0''(x) + x J_0'(x) + x^2 J_0(x)$

$= \displaystyle\sum_{n=1}^{\infty} \frac{(-1)^n 2n(2n-1)x^{2n}}{2^{2n}(n!)^2} + \sum_{n=1}^{\infty} \frac{(-1)^n 2n x^{2n}}{2^{2n}(n!)^2} + \sum_{n=0}^{\infty} \frac{(-1)^n x^{2n+2}}{2^{2n}(n!)^2}$

$= \displaystyle\sum_{n=1}^{\infty} \frac{(-1)^n 2n(2n-1)x^{2n}}{2^{2n}(n!)^2} + \sum_{n=1}^{\infty} \frac{(-1)^n 2n x^{2n}}{2^{2n}(n!)^2} + \sum_{n=1}^{\infty} \frac{(-1)^{n-1} x^{2n}}{2^{2n-2}((n-1)!)^2}$

$= \displaystyle\sum_{n=1}^{\infty} (-1)^n \left[\frac{2n(2n-1) + 2n - 2^2 n^2}{2^{2n}(n!)^2}\right] x^{2n} = \sum_{n=1}^{\infty} (-1)^n \left[\frac{4n^2 - 2n + 2n - 4n^2}{2^{2n}(n!)^2}\right] x^{2n}$

$= 0.$

59. If $u_n = \dfrac{x^n}{n^2}$, then $\displaystyle\lim_{n\to\infty} \left|\dfrac{u_{n+1}}{u_n}\right| = |x| \lim_{n\to\infty}\left[\dfrac{n}{n+1}\right]^2 = |x| < 1$ for

convergence so $R = 1$. When $x = \pm 1$, $\displaystyle\sum_{n=1}^{\infty}\left|\dfrac{x^n}{n^2}\right| = \sum_{n=1}^{\infty}\dfrac{1}{n^2}$ which is a

convergent p-series ($p=2>1$), so the interval of convergence for f

is $[-1,1]$. By Theorem 10.39, the radii of convergence of f' and f"

are both 1, so we need only check endpoints.

$f'(x) = \displaystyle\sum_{n=1}^{\infty}\dfrac{nx^{n-1}}{n^2} = \sum_{n=0}^{\infty}\dfrac{x^n}{n+1}$, and this series diverges for $x=1$

(harmonic series) and converges for $x = -1$ (Alternating Series

Test), so the interval of convergence is $[-1,1)$.

$f"(x) = \displaystyle\sum_{n=1}^{\infty}\dfrac{nx^{n-1}}{n+1}$ diverges at both 1 and -1 (Test for Divergence)

since $\displaystyle\lim_{n\to\infty}\dfrac{n}{n+1} = 1 \neq 0$, so its interval of convergence is $(-1,1)$.

61. $f(x) = \begin{cases} e^{-1/x^2} & \text{if } x\neq 0 \\ 0 & \text{if } x=0 \end{cases}$, so $f'(0) = \displaystyle\lim_{x\to 0}\dfrac{f(x)-f(0)}{x-0} = \lim_{x\to 0}\dfrac{e^{-1/x^2}}{x}$

$= \displaystyle\lim_{x\to 0}\dfrac{1/x}{e^{1/x^2}} = \lim_{x\to 0}\dfrac{x}{2e^{1/x^2}}$ [using l'Hospital's Rule and simplifying]

$= 0$. Similarly, we can use the definition of the derivative and

l'Hospital's Rule to show that $f"(0) = 0$, $f^{(3)}(0) = 0,\ldots$

$f^{(n)}(0) = 0$, so that the Maclaurin series for f consists entirely

of zero terms. But since $f(x) \neq 0$ except for $x=0$, we see that f

cannot equal its Maclaurin series except at $x=0$.

## Exercises 10.10

1. $(1+x)^{1/2} = \sum_{n=0}^{\infty} \begin{bmatrix} 1/2 \\ n \end{bmatrix} x^n = 1 + \left[\frac{1}{2}\right]x + \frac{\left[\frac{1}{2}\right]\left[-\frac{1}{2}\right]}{2!}x^2 + \frac{\left[\frac{1}{2}\right]\left[-\frac{1}{2}\right]\left[-\frac{3}{2}\right]}{3!}x^3 + \ldots$

$= 1 + \frac{x}{2} - \frac{x^2}{2^2 \cdot 2!} + \frac{1 \cdot 3}{2^3 \cdot 3!}x^3 - \frac{1 \cdot 3 \cdot 5}{2^4 \cdot 4!}x^4 + \ldots$

$= 1 + \frac{x}{2} + \sum_{n=2}^{\infty} \frac{(-1)^{n-1} 1 \cdot 3 \cdot 5 \cdots (2n-3)\ x^n}{2^n \cdot n!}$    $R = 1$

3. $[1 + (2x)]^{-4} = 1 + (-4)(2x) + \frac{(-4)(-5)}{2!}(2x)^2 + \frac{(-4)(-5)(-6)}{3!}(2x)^3 + \ldots$

$= 1 + \sum_{n=1}^{\infty} \frac{(-1)^n 2^n 4 \cdot 5 \cdot 6 \cdots (n+3)}{n!} x^n = \sum_{n=0}^{\infty} (-1)^n \frac{2^n (n+1)(n+2)(n+3)}{6} x^n$

$|2x| < 1 \iff |x| < \frac{1}{2}$   so $R = \frac{1}{2}$

5. $(1 + (-x))^{-1/2} = \sum_{n=0}^{\infty} \begin{bmatrix} -1/2 \\ n \end{bmatrix} (-n)^n = 1 + \left[-\frac{1}{2}\right](-x) + \frac{\left[-\frac{1}{2}\right]\left[-\frac{3}{2}\right]}{2!}(-x)^2 + \ldots$

$= 1 + \frac{x}{2} + \frac{1 \cdot 3}{2^2 2!} \cdot x^2 + \frac{1 \cdot 3 \cdot 5}{2^3 3!} \cdot x^3 + \frac{1 \cdot 3 \cdot 5 \cdot 7}{2^4 4!} \cdot x^4 + \ldots$

$= 1 + \sum_{n=1}^{\infty} \frac{1 \cdot 3 \cdot 5 \cdots (2n-1)}{2^n \cdot n!} x^n$   so   $\frac{x}{\sqrt{1-x}} = x + \sum_{n=1}^{\infty} \frac{1 \cdot 3 \cdot 5 \cdots (2n-1)}{2^n \cdot n!} x^{n+1}$

with $R = 1$

7. $(8+x)^{-1/3} = \frac{1}{2}\left[1 + \frac{x}{8}\right]^{-1/3} = \frac{1}{2}\left[1 + \left[-\frac{1}{3}\right]\left[\frac{x}{8}\right] + \frac{\left[-\frac{1}{3}\right]\left[-\frac{4}{3}\right]}{2!}\left[\frac{x}{8}\right]^2 + \ldots\right]$

$= \frac{1}{2}\left[1 + \sum_{n=1}^{\infty} \frac{(-1)^n 1 \cdot 4 \cdot 7 \cdots (3n-2)}{3^n\ n!\ 8^n} x^n\right]$   and   $\left|\frac{x}{8}\right| < 1 \iff |x| < 8$, so $R = 8$

9. $(1-x^4)^{1/4} = 1 + \left[\frac{1}{4}\right](-x^4) + \frac{\left[\frac{1}{4}\right]\left[-\frac{3}{4}\right]}{2!}(-x^4)^2 + \frac{\left[\frac{1}{4}\right]\left[-\frac{3}{4}\right]\left[-\frac{7}{4}\right]}{3!}(-x^4)^3 + \ldots$

$= 1 - \frac{x^4}{4} - \sum_{n=2}^{\infty} \frac{3 \cdot 7 \cdot 11 \cdots (4n-5)}{4^n\ n!} x^{4n}$   with $R = 1$

11. $(1-x)^{-5} = 1 + (-5)(-x) + \frac{(-5)(-6)}{2!}(-x)^2 + \frac{(-5)(-6)(-7)}{3!}(-x)^3 + \ldots$

$= 1 + \sum_{n=1}^{\infty} \frac{5 \cdot 6 \cdot 7 \cdots (n+4)}{n!} x^n = \sum_{n=0}^{\infty} \frac{(n+4)!}{4!\ n!} x^n$   →

$$\frac{x^5}{(1-x)^5} = \sum_{n=0}^{\infty} \frac{(n+4)!}{4!n!} x^{n+5} \left[\text{or} \sum_{n=0}^{\infty} \frac{(n+1)(n+2)(n+3)(n+4)}{24} x^{n+5}\right] \text{ with } R = 1$$

13. (a) $(1-x^2)^{-1/2} = 1 + \left[-\frac{1}{2}\right](-x^2) + \frac{\left[-\frac{1}{2}\right]\left[-\frac{3}{2}\right]}{2!}(-x^2)^2 + \frac{\left[-\frac{1}{2}\right]\left[-\frac{3}{2}\right]\left[-\frac{5}{2}\right]}{3!}(-x^2)^3 + \cdots$

$$= 1 + \sum_{n=1}^{\infty} \frac{1 \cdot 3 \cdot 5 \cdots (2n-1)}{2^n n!} x^{2n}$$

(b) $\sin^{-1}x = \int \frac{1}{\sqrt{1-x^2}} dx = C + x + \sum_{n=1}^{\infty} \frac{1 \cdot 3 \cdot 5 \cdots (2n-1)}{(2n+1) \, 2^n n!} x^{2n+1}$

$$= x + \sum_{n=1}^{\infty} \frac{1 \cdot 3 \cdot 5 \cdots (2n-1)}{(2n+1) \, 2^n n!} x^{2n+1} \text{ since } 0 = \sin^{-1}0 = C.$$

15. (a) $(1+x)^{-1/2} = 1 + \left[-\frac{1}{2}\right]x + \frac{\left[-\frac{1}{2}\right]\left[-\frac{3}{2}\right]}{2!}x^2 + \frac{\left[-\frac{1}{2}\right]\left[-\frac{3}{2}\right]\left[-\frac{5}{2}\right]}{3!}x^3 + \cdots$

$$= 1 + \sum_{n=1}^{\infty} \frac{(-1)^n 1 \cdot 3 \cdot 5 \cdots (2n-1)}{2^n n!} x^n$$

(b) Take $x = 0.1$ in the above series. $\frac{1 \cdot 3 \cdot 5 \cdot 7}{2^4 4!}(0.1)^4 < 0.00003$, so

$$\frac{1}{\sqrt{1.1}} \approx 1 - \frac{0.1}{2} + \frac{1 \cdot 3}{2^2 \cdot 2!}(0.1)^2 - \frac{1 \cdot 3 \cdot 5}{2^3 3!}(0.1)^3 \approx 0.953.$$

17. (a) $(1-x)^{-2} = 1 + (-2)(-x) + \frac{(-2)(-3)}{2!}(-x)^2 + \cdots = \sum_{n=0}^{\infty}(n+1)x^n$, so

$$\frac{x}{(1-x)^2} = \sum_{n=0}^{\infty}(n+1)x^{n+1} = \sum_{n=1}^{\infty} nx^n.$$

(b) With $x = \frac{1}{2}$ in part (a), we have $\sum_{n=1}^{\infty}\frac{n}{2^n} = \frac{1/2}{(1-1/2)^2} = 2$.

19. (a) $(1+x^2)^{1/2} = 1 + \left[\frac{1}{2}\right](x^2) + \frac{\left[\frac{1}{2}\right]\left[-\frac{1}{2}\right]}{2!}(x^2)^2 + \frac{\left[\frac{1}{2}\right]\left[-\frac{1}{2}\right]\left[-\frac{3}{2}\right]}{3!}(x^2)^3 + \cdots$

$$= 1 + \frac{x^2}{2} + \sum_{n=2}^{\infty}\frac{(-1)^{n-1} 1 \cdot 3 \cdot 5 \cdots (2n-3)}{2^n n!} x^{2n}$$

(b) The coefficient of $x^{10}$ in the above Maclaurin series will be $\frac{f^{(10)}(0)}{10!}$, so $f^{(10)}(0) = 10!\left[\frac{1 \cdot 3 \cdot 5 \cdot 7}{2^5 5!}\right] = 99,225.$

21. (a) $g'(x) = \sum_{n=1}^{\infty} \begin{bmatrix} k \\ n \end{bmatrix} n x^{n-1}$. $(1+x)g'(x) = (1+x) \sum_{n=1}^{\infty} \begin{bmatrix} k \\ n \end{bmatrix} n x^{n-1}$

$$= \sum_{n=1}^{\infty} \begin{bmatrix} k \\ n \end{bmatrix} n x^{n-1} + \sum_{n=1}^{\infty} \begin{bmatrix} k \\ n \end{bmatrix} n x^{n} = \sum_{n=0}^{\infty} \begin{bmatrix} k \\ n+1 \end{bmatrix} (n+1) x^{n} + \sum_{n=0}^{\infty} \begin{bmatrix} k \\ n \end{bmatrix} n x^{n}$$

$$= \sum_{n=0}^{\infty} \left[ (n+1) \frac{k(k-1)(k-2)\cdots(k-n)}{(n+1)!} \right] x^{n} + \sum_{n=0}^{\infty} \left[ (n) \frac{k(k-1)(k-2)\cdots(k-n+1)}{n!} \right] x^{n}$$

$$= \sum_{n=0}^{\infty} \frac{(n+1)k(k-1)(k-2)\cdots(k-n+1)}{(n+1)!} [(k-n)+n] x^{n}$$

$$= \sum_{n=0}^{\infty} \frac{k^{2}(k-1)(k-2)\cdots(k-n+1)}{n!} x^{n} = k \sum_{n=0}^{\infty} \begin{bmatrix} k \\ n \end{bmatrix} x^{n} = k g(x)$$

So $g'(x) = \dfrac{kg(x)}{1+x}$.

(b) $h'(x) = -k(1+x)^{-k-1} g(x) + (1+x)^{-k} g'(x)$

$= -k(1+x)^{-k-1} g(x) + (1+x)^{-k} \left[ \dfrac{kg(x)}{1+x} \right]$

$= -k(1+x)^{-k-1} g(x) + k(1+x)^{-k-1} g(x) = 0$

(c) From part (b) we see that h must be constant for $x \in (-1,1)$, so $h(x) = h(0) = 1$ for $x \in (-1,1)$. Thus $h(x) = 1 = (1+x)^{-k} g(x) \iff g(x) = (1+x)^{k}$ for $x \in (-1,1)$.

Exercises 10.11

1. $f(x) = 1+2x+3x^{2}+4x^{3}$      $f(-1) = -2$

$f'(x) = 2+6x+12x^{2}$      $f'(-1) = 8$

$f''(x) = 6+24x$      $f''(-1) = -18$

$f^{(3)}(x) = 24$      $f^{(3)}(-1) = 24$

$f^{(4)}(x) = 0$      $f^{(4)}(-1) = 0$

$T_{4}(x) = \sum_{n=0}^{4} \frac{f^{(n)}(-1)}{n!} (x+1)^{n} = -2 + 8(x+1) - 9(x+1)^{2} + 4(x+1)^{3}$

3.  $f(x) = \sin x$ $\qquad$ $f(\frac{\pi}{6}) = \frac{1}{2}$

$f'(x) = \cos x$ $\qquad$ $f'(\frac{\pi}{6}) = \frac{\sqrt{3}}{2}$

$f''(x) = -\sin x$ $\qquad$ $f''(\frac{\pi}{6}) = -\frac{1}{2}$

$f^{(3)}(x) = -\cos x$ $\qquad$ $f^{(3)}(\frac{\pi}{6}) = -\frac{\sqrt{3}}{2}$

$$T_3(x) = \sum_{n=0}^{3} \frac{f^{(n)}(\pi/6)}{n!}(x-\frac{\pi}{6})^n = \frac{1}{2} + \frac{\sqrt{3}}{2}(x-\frac{\pi}{6}) - \frac{1}{4}(x-\frac{\pi}{6})^2 - \frac{\sqrt{3}}{12}(x-\frac{\pi}{6})^3$$

5.  $f(x) = \tan x$ $\qquad\qquad$ $f(0) = 0$

$f'(x) = \sec^2 x$ $\qquad\qquad$ $f'(0) = 1$

$f''(x) = 2\sec^2 x \tan x$ $\qquad\qquad$ $f''(0) = 0$

$f^{(3)}(x) = 4\sec^2 x \tan^2 x + 2\sec^4 x$ $\quad$ $f^{(3)}(0) = 2$

$f^{(4)}(x) = 8\sec^2 x \tan^3 x + 16\sec^4 x \tan x$ $\quad$ $f^{(4)}(0) = 0$

$$T_4(x) = \sum_{n=0}^{4} \frac{f^{(n)}(0)}{n!} x^n = x + \frac{2}{3!}x^3 = x + \frac{x^3}{3}$$

7.  $f(x) = e^x \sin x$ $\qquad\qquad$ $f(0) = 0$

$f'(x) = e^x(\sin x + \cos x)$ $\qquad$ $f'(0) = 1$

$f''(x) = 2e^x \cos x$ $\qquad\qquad$ $f''(0) = 2$

$f^{(3)}(x) = 2e^x(\cos x - \sin x)$ $\quad$ $f^{(3)}(0) = 2$

$$T_3(x) = \sum_{n=0}^{3} \frac{f^{(n)}(0)}{n!} x^n = x + x^2 + \frac{x^3}{3}$$

9.  $f(x) = x^{1/2}$ $\qquad\qquad$ $f(9) = 3$

$f'(x) = \frac{1}{2}x^{-1/2}$ $\qquad\qquad$ $f'(9) = \frac{1}{6}$

$f''(x) = -\frac{1}{4}x^{-3/2}$ $\qquad\qquad$ $f''(9) = -\frac{1}{108}$

$f^{(3)}(x) = \frac{3}{8}x^{-5/2}$ $\qquad\qquad$ $f^{(3)}(9) = \frac{1}{648}$

$$T_3(x) = \sum_{n=0}^{3} \frac{f^{(n)}(9)}{n!}(x-9)^n = 3 + \frac{1}{6}(x-9) - \frac{1}{216}(x-9)^2 + \frac{1}{3888}(x-9)^3$$

11. $f(x) = \ln \sin x$ $\qquad f(\frac{\pi}{2}) = 0$

$f'(x) = \cot x$ $\qquad f'(\frac{\pi}{2}) = 0$

$f''(x) = -\csc^2 x$ $\qquad f''(\frac{\pi}{2}) = -1$

$f^{(3)}(x) = 2\csc^2 x \cot x$ $\qquad f^{(3)}(\frac{\pi}{2}) = 0$

$$T_3(x) = \sum_{n=0}^{3} \frac{f^{(n)}(\pi/2)}{n!}(x-\pi/2)^n = -\frac{1}{2}\left[x-\frac{\pi}{2}\right]^2$$

13. $f(x) = \cos x$ $\qquad f(0) = 1$

$f'(x) = -\sin x$ $\qquad f'(0) = 0$

$f''(x) = -\cos x$ $\qquad f''(0) = -1$

$f^{(3)}(x) = \sin x$ $\qquad f^{(3)}(0) = 0$

$f^{(4)}(x) = \cos x$ $\qquad f^{(4)}(0) = 1$

$T_1(x) = 1 \qquad T_2(x) = 1 - \frac{x^2}{2}$

$T_3(x) = 1 - \frac{x^2}{2}$

$T_4(x) = 1 - \frac{x^2}{2} + \frac{x^4}{24}$

15. $f(x) = (1+x)^{1/2}$ $\qquad f(0) = 1$

$f'(x) = \frac{1}{2}(1+x)^{-1/2}$ $\qquad f'(0) = \frac{1}{2}$

$f''(x) = -\frac{1}{4}(1+x)^{-3/2}$

(a) $(1+x)^{1/2} = 1 + \frac{1}{2}x + R_1(x)$ where $R_1(x) = \frac{f''(z)}{2!}x^2 = -\frac{1}{8(1+z)^{3/2}}x^2$

and z lies between 0 and x. (b) $0 \le x \le 0.1 \Rightarrow 0 \le x^2 \le 0.01$, and $0 < z < 0.1 \Rightarrow 1 < 1+z < 1.1$ so $|R_1(x)| < \frac{0.01}{8 \cdot 1} = 0.00125$.

17. $f(x) = \sin x$ $\qquad f(\frac{\pi}{4}) = \frac{\sqrt{2}}{2}$

$f'(x) = \cos x$ $\qquad f'(\frac{\pi}{4}) = \frac{\sqrt{2}}{2}$

$f''(x) = -\sin x$ $\qquad f''(\frac{\pi}{4}) = -\frac{\sqrt{2}}{2}$

$f^{(3)}(x) = -\cos x$ $\qquad f^{(3)}(\frac{\pi}{4}) = -\frac{\sqrt{2}}{2}$

$f^{(4)}(x) = \sin x$ $\qquad f^{(4)}(\frac{\pi}{4}) = \frac{\sqrt{2}}{2}$

$f^{(5)}(x) = \cos x$ $\qquad f^{(5)}(\frac{\pi}{4}) = \frac{\sqrt{2}}{2}$

$f^{(6)}(x) = -\sin x$

(a) $\sin x = \frac{\sqrt{2}}{2} + \frac{\sqrt{2}}{2}(x-\frac{\pi}{4}) - \frac{\sqrt{2}}{4}(x-\frac{\pi}{4})^2 - \frac{\sqrt{2}}{12}(x-\frac{\pi}{4})^3 + \frac{\sqrt{2}}{48}(x-\frac{\pi}{4})^4 +$

$\frac{\sqrt{2}}{240}(x-\frac{\pi}{4})^5 + R_5(x)$ where $R_5(x) = \frac{f^{(6)}(z)}{6!}(x-\frac{\pi}{4})^6 = \frac{-\sin z}{720}(x-\frac{\pi}{4})^6$ and

$z$ lies between $\frac{\pi}{4}$ and $x$. (b) Since $0 \leq x \leq \frac{\pi}{2}$, $-\frac{\pi}{4} \leq x-\frac{\pi}{4} \leq \frac{\pi}{4}$

$\Rightarrow 0 \leq (x-\frac{\pi}{4})^6 \leq (\frac{\pi}{4})^6$, and since $0 < z < \frac{\pi}{2}$, $0 < \sin z < 1$, so

$|R_5(x)| < \frac{(\pi/4)^6}{720} \approx 0.00033$.

19.    $f(x) = (1+2x)^{-4}$            $f(0) = 1$

       $f'(x) = -8(1+2x)^{-5}$         $f'(0) = -8$

       $f''(x) = 80(1+2x)^{-6}$       $f''(0) = 80$

       $f^{(3)}(x) = -960(1+2x)^{-7}$     $f^{(3)}(0) = -960$

       $f^{(4)}(x) = 13440(1+2x)^{-8}$

       (a) $(1+2x)^{-4} = 1-8x+40x^2-160x^3 + R_3(x)$ where $R_3(x) = \frac{f^{(4)}(z)}{4!}x^4$

$= \frac{13440(1+2z)^{-8}}{4!}x^4 = \frac{560 x^4}{(1+2z)^8}$. (b) $|x| \leq 0.1 \Rightarrow 0 \leq x^4 \leq 0.0001$ and

$|z| < 0.1 \Rightarrow 0.8 < 1+2z < 1.2$, so $|R_3(x)| < \frac{560 \cdot (0.0001)}{(0.8)^3} < 0.34$.

21.    $f(x) = \tan x$                 $f(0) = 0$

       $f'(x) = \sec^2 x$             $f'(0) = 1$

       $f''(x) = 2\sec^2 x \tan x$       $f''(0) = 0$

       $f^{(3)}(x) = 4\sec^2 x \tan^2 x + 2\sec^4 x$    $f^{(3)}(0) = 2$

       $f^{(4)}(x) = 8\sec^2 x \tan^3 x + 16\sec^4 x \tan x$

       (a) $\tan x = x + \frac{x^3}{3} + R_3(x)$ where $R_3(x) = \frac{f^{(4)}(z)}{4!} x^4 =$

$\frac{8\sec^2 z \tan^3 z + 16\sec^4 z \tan z}{4!} x^4 = \frac{\sec^2 z \tan^3 z + 2\sec^4 z \tan z}{3} x^4$ where

$z$ lies between $0$ and $x$. (b) $0 \leq x^4 \leq [\frac{\pi}{6}]^4$ and $0 < z < \frac{\pi}{6} \Rightarrow$

$\sec^2 z < \frac{4}{3}$ and $\tan z < \frac{\sqrt{3}}{3}$ so

$|R_3(x)| < \frac{(4/3)(1/3\sqrt{3}) + 2(16/9)(1/\sqrt{3})}{3} [\frac{\pi}{6}]^4 = \frac{4\sqrt{3}}{9} [\frac{\pi}{6}]^4 < .06$

23. $f(x) = e^{x^2}$  $\qquad$ $f(0) = 1$

$f'(x) = e^{x^2}(2x)$  $\qquad$ $f'(0) = 0$

$f''(x) = e^{x^2}(2+4x^2)$  $\qquad$ $f''(0) = 2$

$f^{(3)}(x) = e^{x^2}(12x+8x^3)$  $\qquad$ $f^{(3)}(0) = 0$

$f^{(4)}(x) = e^{x^2}(12+48x^2+16x^4)$

(a) $e^{x^2} = 1 + x^2 + R_3(x)$ where $R_3(x) = \dfrac{f^{(4)}(z)}{4!} x^4 =$

$\dfrac{e^{z^2}(3+12z^2+4z^4)}{6} x^4$ and z is between 0 and x.   (b) $0 \le x \le 0.1 \Rightarrow$

$|R_3(x)| < \dfrac{e^{0.01}(3+0.12+0.0004)}{6}(0.0001) < 0.00006$

25. $f(x) = x^{3/4}$  $\qquad$ $f(16) = 8$

$f'(x) = \dfrac{3}{4}x^{-1/4}$  $\qquad$ $f'(16) = \dfrac{3}{8}$

$f''(x) = \dfrac{-3}{16}x^{-5/4}$  $\qquad$ $f''(16) = \dfrac{-3}{512}$

$f^{(3)}(x) = \dfrac{15}{64}x^{-9/4}$  $\qquad$ $f^{(3)}(16) = \dfrac{15}{32768}$

$f^{(4)}(x) = -\dfrac{135}{256}x^{-13/4}$

(a) $x^{3/4} = 8 + \dfrac{3}{8}(x-16) - \dfrac{3}{1024}(x-16)^2 + \dfrac{5}{65536}(x-16)^3 + R_3(x)$ where

$R_3(x) = \dfrac{f^{(4)}(z)}{4!}(x-16)^4 = -\dfrac{135(x-16)^4}{256 \cdot 4! z^{13/4}}$ and z lies between 16 and x.

(b) $|x-16| \le 1$ and $z > 15 \Rightarrow |R_3(x)| < \dfrac{135}{256 \cdot 24 \cdot 15^{13/4}} < 0.0000034.$

27. $e^x = 1 + x + \dfrac{x^2}{2!} + \ldots + \dfrac{x^n}{n!} + R_n(x)$ where $R_n(x) = \dfrac{e^z}{(n+1)!}x^{n+1}$ and z

lies between 0 and x.   So taking $x = 0.1$, we have $z < 0.1 \Rightarrow$

$e^z < 3^{0.1} < 2 \Rightarrow |R_3(0.1)| < \dfrac{2}{4!}(0.0001) < 0.00001$ and

$e^{0.1} \approx 1 + 0.1 + \dfrac{(0.1)^2}{2} + \dfrac{(0.1)^3}{6} \approx 1.10517.$

29. $f(x) = (1+x)^{1/5} = 1 + \dfrac{1}{5}x + \dfrac{\left[\frac{1}{5}\right]\left[-\frac{4}{5}\right]}{2!} x^2 + R_2(x)$

$= 1 + \dfrac{1}{5}x - \dfrac{2}{25}x^2 + R_2(x)$ where $|R_2(x)| = \dfrac{1 \cdot 4 \cdot 9}{5^3 \cdot 3!}|1+z|^{-2-4/5}|x|^3$ and z

lies between 0 and x, so $|R_2(0.1)| < \dfrac{1 \cdot 4 \cdot 9}{5^3 \cdot 3!}(0.001) = 0.000048 <$

$0.00005$.   Thus $(1.1)^{1/5} \approx 1 + \dfrac{0.1}{5} - \dfrac{2}{25}(0.1)^2 = 1.0192$ (correct to 4

decimals).

31. If $f(x) = \ln(1+x)$ then $f^{(n)}(x) = (-1)^n(n-1)!(1+x)^{-n}$ so $\ln(1+x)$

$= x - \dfrac{x^2}{2} + \dfrac{x^3}{3} - \dfrac{x^4}{4} + \dfrac{x^5}{5} + R_5(x)$ where $|R_5(x)| = \dfrac{|x|^6}{6|1+z|^6}$ and z is

between 0 and x. So $|R_5(0.4)| < \dfrac{(0.4)^6}{6} \approx 0.00068 < 0.001$ and

$\ln(1.4) \approx 0.4 - \dfrac{(0.4)^2}{2} + \dfrac{(0.4)^3}{3} - \dfrac{(0.4)^4}{4} + \dfrac{(0.4)^5}{5} \approx 0.336$.

33. $\sin x = x - \dfrac{x^3}{3!} + \dfrac{x^5}{5!} + R_5(x)$, where $R_5(x) = \dfrac{-\sin z}{6!} x^6$ and z lies

between 0 and x. So $|R_5(0.5)| < \dfrac{1 \cdot (0.5)^6}{720} \approx 0.00002 < 0.0001$, and

$\sin(0.5) \approx 0.5 - \dfrac{(0.5)^3}{3!} + \dfrac{(0.5)^5}{5!} \approx 0.4794$.

35. $\cos x = 1 - \dfrac{x^2}{2!} + \dfrac{x^4}{4!} + R_4(x)$ where $R_4(x) = \dfrac{\cos z}{5!} x^5$ and z is between

0 and x, so $|R_4(10^\circ)| = |R_4(\dfrac{\pi}{18})| < \dfrac{(\pi/18)^5}{5!} < 0.0000014$ so

$\cos(10^\circ) \approx 1 - \dfrac{(\pi/18)^2}{2!} + \dfrac{(\pi/18)^4}{4!} \approx 0.98481$.

37. Using the information of Exercise 3, above, we see that

$\sin x = \dfrac{1}{2} + \dfrac{\sqrt{3}}{2}(x-\dfrac{\pi}{6}) - \dfrac{1}{4}(x-\dfrac{\pi}{6})^2 - \dfrac{\sqrt{3}}{12}(x-\dfrac{\pi}{6})^3 + R_3(x)$, where $R_3(x) =$

$\dfrac{\sin z}{4!}(x-\dfrac{\pi}{6})^4$ and z lies between $\dfrac{\pi}{6}$ and x. Now $35^\circ$ is $\dfrac{\pi}{6} + \dfrac{\pi}{36}$ in

radian measure, so $|R_3(\dfrac{\pi}{36})| < \dfrac{(\pi/36)^4}{4!} < 0.000003$ and

$\sin 35^\circ \approx \dfrac{1}{2} + \dfrac{\sqrt{3}}{2}(\dfrac{\pi}{36}) - \dfrac{1}{4}(\dfrac{\pi}{36})^2 - \dfrac{\sqrt{3}}{12}(\dfrac{\pi}{36})^3 \approx 0.57358$.

39. $\sin x = x - \dfrac{x^3}{6} + R_4(x)$ where $R_4(x) = \dfrac{\sin z}{5!} x^5$ and z is between 0

and x. So $|R_4(x)| < \dfrac{|x|^5}{120} < 0.01 \Rightarrow |x|^5 < 1.2 \Rightarrow |x| < (1.2)^{1/5}$

$\approx 1.037$. This will certainly be true if $|x| \le 1$.

41. We will use Theorem 10.56. $R_n(x) = \dfrac{f^{(n+1)}(z)}{(n+1)!} x^{n+1}$ with

$f^{(n+1)}(z) = \pm\sin z$ or $\pm\cos z$, and with z between 0 and x. In every

case $|f^{(n+1)}(z)| \le 1$. Thus $|R_n(x)| \le \dfrac{|x|^{n+1}}{(n+1)!} \to 0$ as $n \to \infty$ by (10.57),

so sin x is equal to its Taylor series by Theorem 10.56.

43. $R_n(x) = \frac{f^{(n+1)}(z)}{(n+1)!} x^{n+1}$ where $f^{(n+1)}(z) = \sinh z$ or $\cosh z$. Since $z$ lies between 0 and $x$, $|\sinh z| < |\sinh x|$ and $|\cosh z| < |\cosh x|$, so in the case $f^{(n+1)}(z) = \sinh z$ we have $|R_n(x)| < |\sinh x| \frac{|x|^{n+1}}{(n+1)!}$ $\to 0$ as $n \to \infty$ by (10.57) (and similarly if $f^{(n+1)}(z) = \cosh z$). So by Theorem 10.56, $\sinh x$ is equal to its Taylor series.

45. $y = T_1(x) = f(c) + f'(c)(x-c) \iff y - f(c) = f'(c)(x-c)$ is the equation of the line passing through $(c, f(c))$ with slope $f'(c)$, and this describes the tangent line.

47. (a) $g(x) = f(x) - f(x) - \sum_{i=1}^{n} \frac{f^{(i)}(x)}{i!}(x-x)^i - \frac{K}{(n+1)!}(x-x)^{n+1} = 0$

$g(c) = f(x) - f(c) - \sum_{i=1}^{n} \frac{f^{(i)}(c)}{i!}(x-c)^i - \frac{K}{(n+1)!}(x-c)^{n+1}$

$= f(x) - f(x) = 0$

(b) $g'(t) = -f'(t) - [f''(t)(x-t) + f'(t)(-1)]$

$- \left[ \frac{f^{(3)}(t)}{2!}(x-t)^2 + \frac{f''(t)}{2!} \cdot 2(x-t)(-1) \right] - \cdots$

$- \left[ \frac{f^{(n+1)}(t)}{n!}(x-t)^n + \frac{f^{(n)}(t)}{n!} \cdot n(x-t)^{n-1}(-1) \right] - \frac{K}{(n+1)!}(n+1)(x-t)^n(-1)$

$= -\frac{f^{(n+1)}(t)}{n!}(x-t)^n + \frac{K}{n!}(x-t)^n$

(c) By Rolle's Theorem, there must exist $z$ between $x$ and $c$ such that $g'(z) = 0 \Rightarrow -\frac{f^{(n+1)}(z)}{n!}(x-z)^n + \frac{K}{n!}(x-z)^n = 0 \Rightarrow f^{(n+1)}(z) = K$.

49. $q!(e-s_q) = q! \left[ \frac{p}{q} - 1 - \frac{1}{1!} - \frac{1}{2!} - \cdots - \frac{1}{q!} \right] = p(q-1)! - q! - q! - \frac{q!}{2!}$

$- \cdots - 1$, which is clearly an integer, and $q!(e-s_q) = q! \left[ \frac{e^z}{(q+1)!} \right]$

$= \frac{e^z}{q+1}$. We have $0 < \frac{e^z}{q+1} < \frac{e}{q+1} < \frac{e}{3} < 1$ since $0 < z < 1$ and $q > 2$, and so $0 < q!(e-s_q) < 1$, which is a contradiction since we have already shown $q!(e-s_q)$ must be an integer. So $e$ cannot be rational.

Review Exercises for Chapter 10

1. $\lim\limits_{n\to\infty} \dfrac{n}{2n+5} = \lim\limits_{n\to\infty} \dfrac{1}{2 + 5/n} = \dfrac{1}{2}$ and the sequence is convergent.

3. $\{2n+5\}$ is divergent since $2n+5 \to \infty$ as $n \to \infty$.

5. $\{\sin n\}$ is divergent since $\lim\limits_{n\to\infty} \sin n$ does not exist.

7. $\left\{ \left[ 1 + \dfrac{3}{n} \right]^{4n} \right\}$ is convergent. Let $y = \left[ 1 + \dfrac{3}{x} \right]^{4x}$. Then $\lim\limits_{x\to\infty} \ln y$

$= \lim\limits_{x\to\infty} 4x \ln(1 + 3/x) = \lim\limits_{x\to\infty} \dfrac{\ln(1 + 3/x)}{1/4x} = \lim\limits_{x\to\infty} \dfrac{\dfrac{1}{1 + 3/x} \cdot (-3/x^2)}{-1/4x^2}$

$= \lim\limits_{x\to\infty} \dfrac{12}{1 + 3/x} = 12$, so $\lim\limits_{x\to\infty} y = \lim\limits_{n\to\infty} \left[ 1 + \dfrac{3}{n} \right]^{4n} = e^{12}$.

9. Use the Limit Comparison Test with $a_n = \dfrac{n^2}{n^3+1}$ and $b_n = \dfrac{1}{n}$.

$\lim\limits_{n\to\infty} \dfrac{a_n}{b_n} = \lim\limits_{n\to\infty} \dfrac{n^2/(n^3+1)}{1/n} = \lim\limits_{n\to\infty} \dfrac{n^3}{n^3+1} = \lim\limits_{n\to\infty} \dfrac{1}{1 + 1/n^3} = 1$. Since $\sum\limits_{n=1}^{\infty} \dfrac{1}{n}$

(the harmonic series) diverges, $\sum\limits_{n=1}^{\infty} \dfrac{n^2}{n^3+1}$ diverges also.

11. An alternating series with $a_n = \dfrac{1}{n^{1/4}}$, $a_n > 0$ for all $n$, and

$a_n > a_{n+1}$. $\lim\limits_{n\to\infty} a_n = \dfrac{1}{n^{1/4}} = 0$ so the series converges by the

Alternating Series Test.

13. $\lim\limits_{n\to\infty} \sqrt[n]{|a_n|} = \lim\limits_{n\to\infty} \dfrac{n}{3n+1} = \dfrac{1}{3} < 1$, so series converges by the Root Test.

15. $\dfrac{|\sin n|}{1 + n^2} \leq \dfrac{1}{1 + n^2} < \dfrac{1}{n^2}$ and since $\sum\limits_{n=1}^{\infty} \dfrac{1}{n^2}$ converges (p-series with

$p=2>1$), so does $\sum\limits_{n=1}^{\infty} \dfrac{|\sin n|}{1 + n^2}$ by the Comparison Test.

17. $\lim\limits_{n\to\infty} \left| \dfrac{a_{n+1}}{a_n} \right| = \lim\limits_{n\to\infty} \dfrac{1 \cdot 3 \cdot 5 \cdots (2n-1)(2n+1)}{5^{n+1}(n+1)!} \cdot \dfrac{5^n n!}{1 \cdot 3 \cdot 5 \cdots (2n-1)} = \lim\limits_{n\to\infty} \dfrac{2n+1}{5(n+1)}$

$= \dfrac{2}{5} < 1$, so the series converges by the Ratio Test.

19. $\lim\limits_{n\to\infty} \left| \dfrac{a_{n+1}}{a_n} \right| = \lim\limits_{n\to\infty} \dfrac{4^{n+1}}{(n+1)3^{n+1}} \cdot \dfrac{n3^n}{4^n} = \dfrac{4}{3} \lim\limits_{n\to\infty} \dfrac{n}{n+1} = \dfrac{4}{3} > 1$ so the series diverges by the Ratio Test.

21. Convergent geometric series. $\displaystyle\sum_{n=1}^{\infty} \dfrac{2^{2n+1}}{5^n} = 2 \sum_{n=1}^{\infty} \dfrac{4^n}{5^n} = 2 \left[ \dfrac{4/5}{1 - 4/5} \right] = 8$

23. $1.2 + 0.0\overline{345} = \dfrac{12}{10} + \dfrac{345/10000}{1 - 1/1000} = \dfrac{12}{10} + \dfrac{345}{9990} = \dfrac{4111}{3330}$

25. $\displaystyle\sum_{n=1}^{\infty} \dfrac{(-1)^{n+1}}{n^5} = 1 - \dfrac{1}{32} + \dfrac{1}{243} - \dfrac{1}{1024} + \dfrac{1}{3125} - \dfrac{1}{7776} + \dfrac{1}{16807} - \dfrac{1}{32768} + \cdots$

Since $\dfrac{1}{32768} < 0.000031$, $\displaystyle\sum_{n=1}^{\infty} \dfrac{(-1)^{n+1}}{n^5} \approx \sum_{n=1}^{7} \dfrac{(-1)^{n+1}}{n^5} \approx 0.9721$.

27. Use the Limit Comparison Test. $\lim\limits_{n\to\infty} \left| \dfrac{\left[\frac{n+1}{n}\right] a_n}{a_n} \right| = \lim\limits_{n\to\infty} \dfrac{n+1}{n} = \lim\limits_{n\to\infty} \left[ 1 + \dfrac{1}{n} \right]$

$= 1 > 0$. Since $\sum |a_n|$ is convergent, so is $\sum \left| \left[\dfrac{n+1}{n}\right] a_n \right|$ by the Limit Comparison Test.

29. $\lim\limits_{n\to\infty} \left| \dfrac{u_{n+1}}{u_n} \right| = \lim\limits_{n\to\infty} \left| \dfrac{x^{n+1}}{3^{n+1}(n+1)^3} \cdot \dfrac{3^n n^3}{x^n} \right| = \dfrac{|x|}{3} \lim\limits_{n\to\infty} \left[ \dfrac{n}{n+1} \right]^3 = \dfrac{|x|}{3} < 1$ for convergence (Ratio Test) $\Rightarrow |x| < 3$ and the radius of convergence is 3. When $x = \pm 3$, $\displaystyle\sum_{n=1}^{\infty} |u_n| = \sum_{n=1}^{\infty} \dfrac{1}{n^3}$ which is a convergent p-series $(p=3>1)$, so the interval of convergence is $[-3,3]$.

31. $\lim\limits_{n\to\infty} \left| \dfrac{u_{n+1}}{u_n} \right| = \lim\limits_{n\to\infty} \left| \dfrac{2^{n+1}(x-3)^{n+1}}{\sqrt{n+4}} \cdot \dfrac{\sqrt{n+3}}{2^n(x-3)^n} \right| = 2|x-3| \lim\limits_{n\to\infty} \sqrt{\dfrac{n+3}{n+4}}$

$= 2|x-3| < 1 \iff |x-3| < 1/2$ so the radius of convergence is 1/2. For $x = \dfrac{7}{2}$ the series becomes $\displaystyle\sum_{n=0}^{\infty} \dfrac{1}{\sqrt{n+3}} = \sum_{n=3}^{\infty} \dfrac{1}{n^{1/2}}$ which diverges $(p=\frac{1}{2}<1)$, but for $x = \dfrac{5}{2}$ we get $\displaystyle\sum_{n=0}^{\infty} \dfrac{(-1)^n}{\sqrt{n+3}}$ which is a convergent alternating series, so the interval of convergence is $[5/2, 7/2)$.

33. $f(x) = \sin x$ $\qquad\qquad\qquad f\left(\dfrac{\pi}{6}\right) = \dfrac{1}{2}$

$f'(x) = \cos x$ $\qquad\qquad\qquad f'\left(\dfrac{\pi}{6}\right) = \dfrac{\sqrt{3}}{2}$

$$f''(x) = -\sin x \qquad\qquad f''(\tfrac{\pi}{6}) = -\tfrac{1}{2}$$

$$f^{(3)}(x) = -\cos x \qquad\qquad f^{(3)}(\tfrac{\pi}{6}) = -\tfrac{\sqrt{3}}{2}$$

$$f^{(4)}(x) = \sin x \qquad\qquad f^{(4)}(\tfrac{\pi}{6}) = \tfrac{1}{2}$$

$$\cdots \qquad\qquad\qquad \cdots$$

$$f^{(2n)}(\tfrac{\pi}{6}) = (-1)^n \cdot \tfrac{1}{2} \text{ and } f^{(2n+1)}(\tfrac{\pi}{6}) = (-1)^n \cdot \tfrac{\sqrt{3}}{2}$$

$$\sin x = \sum_{n=0}^{\infty} \frac{f^{(n)}(\pi/6)}{n!}(x-\tfrac{\pi}{6})^n = \sum_{n=0}^{\infty} \frac{(-1)^n}{2(2n)!}(x-\tfrac{\pi}{6})^{2n} + \sum_{n=0}^{\infty} \frac{(-1)^n\sqrt{3}}{2(2n+1)!}(x-\tfrac{\pi}{6})^{2n+1}$$

35. $\dfrac{1}{1+x} = \dfrac{1}{1-(-x)} = \displaystyle\sum_{n=0}^{\infty}(-1)^n x^n$ for $|x|<1 \Rightarrow \dfrac{x^2}{1+x} = \displaystyle\sum_{n=0}^{\infty}(-1)^n x^{n+2}$ with $R = 1$

37. $\dfrac{1}{1-x} = \displaystyle\sum_{n=0}^{\infty} x^n$ for $|x|<1 \Rightarrow \ln(1-x) = -\displaystyle\int\dfrac{dx}{1-x} = -\displaystyle\int\sum_{n=0}^{\infty} x^n\,dx = C - \sum_{n=0}^{\infty}\dfrac{x^{n+1}}{n+1}$

$\ln(1-0) = C - 0 \Rightarrow C=0 \Rightarrow \ln(1-x) = -\displaystyle\sum_{n=0}^{\infty}\dfrac{x^{n+1}}{n+1} = \sum_{n=1}^{\infty}\dfrac{-x^n}{n}$ with $R = 1$

39. $\sin x = \displaystyle\sum_{n=0}^{\infty}\dfrac{(-1)^n x^{2n+1}}{(2n+1)!} \Rightarrow \sin(x^4) = \sum_{n=0}^{\infty}\dfrac{(-1)^n(x^4)^{2n+1}}{(2n+1)!} = \sum_{n=0}^{\infty}\dfrac{(-1)^n x^{8n+4}}{(2n+1)!}$

for all x, so radius of convergence is $\infty$.

41. $(16-x)^{-1/4} = \dfrac{1}{2}\left[1 - \dfrac{x}{16}\right]^{-1/4} = \dfrac{1}{2}\left[1 + \left[-\dfrac{1}{4}\right]\left[-\dfrac{x}{16}\right] + \dfrac{\left[-\frac{1}{4}\right]\left[-\frac{5}{4}\right]}{2!}\left[-\dfrac{x}{16}\right]^2 + \cdots\right]$

$= \displaystyle\sum_{n=0}^{\infty}\dfrac{1\cdot5\cdot9\cdots(4n-3)}{2\cdot4^n\cdot n!\cdot16^n}x^n = \sum_{n=0}^{\infty}\dfrac{1\cdot5\cdot9\cdots(4n-3)}{2^{6n+1}n!}x^n$ for $\left|-\dfrac{x}{16}\right| < 1 \Rightarrow R=16$

43. $e^x = \displaystyle\sum_{n=0}^{\infty}\dfrac{x^n}{n!}$ so $\dfrac{e^x}{x} = \dfrac{1}{x} + \sum_{n=1}^{\infty}\dfrac{x^{n-1}}{n!}$ and $\displaystyle\int\dfrac{e^x}{x}dx = C + \ln|x| + \sum_{n=1}^{\infty}\dfrac{x^n}{n\cdot n!}$

45. $e^{-1/4} = \displaystyle\sum_{k=0}^{n}\dfrac{(-1/4)^k}{k!} + R_n(-1/4)$ where $R_n(-1/4) = \dfrac{e^z}{(n+1)!}\left[-\dfrac{1}{4}\right]^{n+1}$ and

$-\dfrac{1}{4} < z < 0 \Rightarrow |R_n(-1/4)| < \dfrac{e^0}{(n+1)!}\left[\dfrac{1}{4}\right]^{n+1} = \dfrac{1}{(n+1)!}\left[\dfrac{1}{4}\right]^{n+1} \Rightarrow$

$|R_4(-1/4)| < \dfrac{1}{5!\cdot4^5} < 0.0000082 < 0.0001$ so

$e^{-1/4} \approx 1 - \dfrac{1}{4} + \dfrac{1}{32} - \dfrac{1}{384} + \dfrac{1}{6144} \approx 0.7788$

47. $f(x) = x^{1/2} \qquad\qquad\qquad f(1) = 1$

$$f'(x) = \frac{1}{2}x^{-1/2} \qquad\qquad f'(1) = \frac{1}{2}$$

$$f''(x) = -\frac{1}{4}x^{-3/2} \qquad\qquad f''(1) = -\frac{1}{4}$$

$$f^{(3)}(x) = \frac{3}{8}x^{-5/2} \qquad\qquad f^{(3)}(1) = \frac{3}{8}$$

$$f^{(4)}(x) = -\frac{15}{16}x^{-7/2}$$

$$\sqrt{x} = 1 + \frac{1}{2}(x-1) - \frac{1}{8}(x-1)^2 + \frac{1}{16}(x-1)^3 + R_3(x) \text{ where } R_3(x)$$

$$= \frac{f^{(4)}(z)}{4!}(x-1)^4 = -\frac{5(x-1)^4}{128z^{7/2}} \text{ with } z \text{ between } x \text{ and } 1. \quad \text{If}$$

$0.9 \le x \le 1.1$ then $0 \le |x-1| \le 0.1$ and $z^{7/2} > (0.9)^{7/2}$ so

$$|R_3(x)| < \frac{5(0.1)^4}{128(0.9)^{7/2}} < 0.000006.$$

**49.** $e^x = \displaystyle\sum_{n=0}^{\infty} \frac{x^n}{n!} \Rightarrow e^{x^2} = \sum_{n=0}^{\infty} \frac{(x^2)^n}{n!} = \sum_{n=0}^{\infty} \frac{x^{2n}}{n!} = \sum_{k=0}^{\infty} \frac{f^{(k)}(0)}{k!}x^k \Rightarrow \frac{f^{(2n)}(0)}{(2n)!} = \frac{1}{n!}$

$\Rightarrow f^{(2n)}(0) = \dfrac{(2n)!}{n!}$